The Ecology of Australia's Wet Tropics

Proceedings of a symposium held at the University of Queensland, Brisbane August 25-27, 1986

Edited by

R. L. Kitching
Department of Ecosystem Management
University of New England

1988

Proceedings of the Ecological Society of Australia — Volume 15

ISBN 0 909436 06 1
National Library Canberra

PRINTED AND PUBLISHED IN AUSTRALIA
FOR THE ECOLOGICAL SOCIETY OF AUSTRALIA BY
SURREY BEATTY & SONS PTY LTD
43 Rickard Road, Chipping Norton, NSW 2170 Australia

Contents

Freshwater ecology

Coastal and marine ecology

Plant biology

Disturbance and recovery

Conservation and management

Foreword

In August 1986 the Ecological Society of Australia held its annual scientific meeting in Brisbane in conjunction with the Australian Systematic Botany Society. The theme symposium of that meeting was entitled "The Ecology of Australia's Wet Tropics". This volume presents the edited proceedings of that symposium.

For those who attended the meeting a few words of explanation are in order. First, four plenary lectures were delivered at the Symposium: by Dr Barlow (CSIRO) on the origin and evolution of the tropical flora, by Associate Professor Archer (University of New South Wales) on the origin and evolution of the tropical fauna, by Professor Specht (University of Queensland) on tropical terrestrial ecosystems in Australia, and by Professor Choat (James Cook University) on tropical marine ecosystems. Sadly neither Professor Archer nor Professor Choat felt able to submit manuscripts for publication: hence, the somewhat phytophilic appearance of the Plenary lectures as here published. In addition publication plans for the associated symposium of the Australian Systematic Botany Society on "The Evolution and Systematics of Australia's tropical Flora" came to nought. Accordingly the Ecological Society agreed to incorporate certain of the contributions to that botanical symposium in this volume. As editor I have attempted to integrate these within the loosely logical ordering of contributions in this book. On occasion, however, this was stretching the logic even more than somewhat and I beg forbearance on the part of critics in that regard.

Lastly I would like to thank Mr Ivor Beatty Snr of Surrey Beatty Publishers for patience, promptness and efficiency on his part in bringing this work, finally, to a conclusion. This and earlier Symposium Volumes of the Ecological Society of Australia have had somewhat mixed publishing histories. The Society has now arrived at an ongoing agreement with Surrey Beatty concerning the publication of our Symposium volumes and the delays and uncertainties concerning the how and, in particular, the when of publication of our Proceedings are now behind us. I regret that this volume had to be the proving case and, accordingly, appeared much later than expected. I trust the high quality, variety and current relevance of the papers here published on our important and magnificent tropical ecosystems will more than make up for this delay.

R. L. Kitching
Editor

Armidale, October 1988

Proc. Ecol. Soc. Aust. 15:1-17

The origins of the flora of Australia's wet tropics

B. A. BARLOW[1] and B. P. M. HYLAND[2]

[1]*Australian National Herbarium, Division of Plant Industry, C.S.I.R.O., G.P.O. Box 1600, Canberra, A.C.T. 2601*

[2]*Australian National Herbarium, Division of Plant Industry, C.S.I.R.O., P.O. Box 780, Atherton, Qld. 4883*

Abstract

The tropical rainforests of northern Australia are the complex products of a sequence of paleogeographic and paleoclimatic events. These include: (1) the early differentiation of warm and cool, and of moist and dry elements in a mesic pan-Gondwanan flora, (2) the survival since the early Tertiary period of many primitive components of this flora in durable refugia in north-eastern and eastern Australia, (3) the early Tertiary differentiation of open forest and heathland derivatives in response to edaphic evolution, (4) interdigitation in New Guinea and Australia of the autochthonous rainforest flora with a differentiated Malesian rainforest flora, following a long period of isolation which was weakened by geophysical and climatic events in late Tertiary time, and (5) local sifting of communities, involving Australia and New Guinea, in response to increasing amplitude of climatic oscillations in recent times. The result is an intricate pattern of humid forest communities which, within a mantle of structural similarity, show significant differences in composition, age, endemism, relictuality and richness.

Introduction

The revolution in studies and hypotheses on the origin of the Australian flora appears to be well and truly over, at least for the time being. Most of what has been written in the last five years shows agreement as to the general floristic and ecological history of the region, although differences in detail are to be expected. It would be sufficient, therefore, to begin a review of the origins of tropical Australasian flora with the words "in the beginning there was Gondwanaland . . . ".

The concept of continental displacement has been firmly with us for most of this century, and in fact dates back to 1858 (Smith 1982). Whilst its relevance to Australian biogeography was always very obvious, it mostly loomed in the background as a monster which we sought to destroy. Perhaps rightly, we pointed to the circular reasoning involved in accepting it. Continental displacement was thought to raise more biogeographical problems than it solved. And even if it did occur, it must have happened far too early to have had any bearing on angiosperm dispersal.

Thus for a century after Hooker's (1860) floristic analysis the history of the Australian flora was explained in terms of the elements he recognized, and of successive colonizations from without. Spatial relationships between the continents remained constant, and angiosperm invasions occurred over land bridges fortuitously made and broken. The history of the invasion theory has been summarized elsewhere (Barlow 1981), and can thus be reviewed very briefly. The first invading angiosperm stock, from the north-west, diversified rapidly under conditions of low selection to produce a pan-Australian flora of distinctive scleromorphic and xeromorphic character, with high generic endemism; this became the Paleoaustralian element of the flora.

Later invasions from the north and the south introduced, respectively, the Indomalayan and Subantarctic elements. Owing to competition from the already established Paleoaustralian element, these later immigrant stocks were confined to habitats to which they were already preadapted. Thus they diversified less, and retained closer floristic resemblances to plants of the regions from which they migrated. The Indomalayan element was constrained as a flora of tropical mesic and monsoonal habitats in the north and east, while the Subantarctic element persisted only as a flora of cool temperate and alpine habitats in the east and south. Thus the origins of the flora were explained in macroecological terms.

Now, of course, the question of the origin of the tropical Australian flora is almost the question of the origin of the angiosperms. A large mass of evidence supports the conclusion that before its fragmentation, Gondwanaland supported a widespread and diversified angiosperm flora. In the prevailing equable climatic conditions, this flora took the form predominantly or entirely of mesic forests. There may have been differentiation into cool and warm forest types (with correlated taxonomic differentiation. Thus the so-called Subantarctic and Indomalayan elements of the Australian flora are not relatively late invasive stocks in the region; they are the surviving, least modified representatives of the original Gondwanan flora. The so-called Paleoaustralian element is not the much-modified original colonizing stock but an endemic derivative of the same flora. It is more satisfactorily termed the autochthonous subelement of the Gondwanan element of the flora (cf. Nelson 1981). It developed in response to the onset of climatic events subsequent to the separation and northward drift of Australia from Antarctica, and particularly to cycles of aridity which increased in frequency and amplitude from the middle of the Tertiary period (Bowler 1982). It also developed in response to edaphic evolution, which may have been the catalyst for the widespread evolution of scleromorphy (Specht 1972; Beadle 1966; Johnson & Briggs 1975).

At the present time, then, most of the Australian continental area previously occupied by a warm to cool mesic Gondwanan forest flora is now occupied by its derivative — a scleromorphic and xeromorphic flora of open forests, woodlands, arid shrublands, steppes and heathlands. The less-modified mesic forest flora now persists in restricted areas mostly near the eastern margin of the continent and in widely distributed but localized northern monsoon scrubs. In this general sense the entire wet tropical flora could be described as relictual, and as representing in part or in whole the ancestral Gondwanan flora.

History of Land Masses and Landforms

At the close of the Cretaceous period, Antarctica, Australia, New Zealand and the components of the present New Guinea were closely assembled at a higher latitude, and had a warm, humid climate. Although the northward separation of Australia from a stationary Antarctica commenced about 125 million years ago (Ma), movement remained slow (4.5 mm y^{-1}) until 55 Ma, when the rate increased to 60-70 mm y^{-1} (Crook & Taylor 1985; Wellman & McDougall 1974). A narrow land connection between Australia and Antarctica may have persisted until about 40 Ma (Kemp 1978). The Tasman Sea started to open about 80 Ma, with New Zealand and the Lord Howe Rise separating from Australia leaving a clean faulted edge on the Australian side (Ollier 1986). The Coral Sea also started to open at about this time, with the separation of Papua New Guinea from north-eastern Australia. The southern part of western Papua is thus a continuation of the Queensland part of the Australian plate. The Papuan Ultramafic Belt is possibly an obducted slab thrust over the Australian plate in a collision event, and the part of New Guinea north of the Sepik-Ramu line is another exotic terrane attached following collision (Ollier 1986). The exotic terranes may have originally been located further south than the rest of New Guinea, perhaps near New Caledonia or even New Zealand.

The uplands of eastern Australia were previously considered to be of recent elevation (Kosciusko Uplift), but are now thought to have been formed much earlier. Whilst the exact cause is still uncertain, there is some evidence of uplift throughout Tertiary time (Ollier 1986), and that it may be the result of warping associated with the rifting which preceded the opening of the Tasman Sea (Ollier 1982). Dissected plateaus, narrow valleys, edaphic mosaics, localized high rainfall and fire proofing, all associated with the survival of closed forests, have presumably been a continuous feature of the eastern Australian Tertiary-Quaternary environment.

In contrast, most of the New Guinean uplands appear to be very young, and much of New

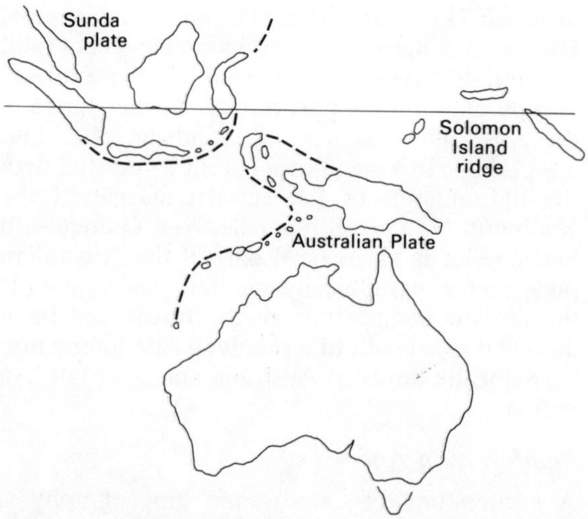

FIG. 1. Possible configurations of the Australian and Sundaland plates 20 Ma, prior to collision (modified from Audley-Charles *et al.* 1981).

Guinea was low-lying even at the end of the Tertiary period. In addition to the emplacement of exotic terranes there has been vertical uplift along the axis of the island at a rate of *c.* 2 mm y^{-1} (Chappell 1974), and remains of Plio-Pleistocene erosion surfaces can be found as plateaus even on the high points of the Owen Stanley Range (Pain & Ollier 1984). Thus whilst New Guinea is clearly part of the Australian plate with respect to floristic history, the extensive ever-wet uplands of the present day did not exist in this form through most of the period of angiosperm history of the region. Nevertheless it must be presumed that the smaller, less elevated Tertiary New Guinea which formed the northern edge of the Australian plate has, like the region immediately to the south, continuously provided conditions favourable for wet closed forest communities.

When the northward-moving Australian plate came into collision with the westward-moving Sunda plate in the Miocene (Fig. 1), the northern (west New Guinean) margin of the Australian plate was involved first (Powell *et al.* 1981; Audley-Charles *et al.* 1981). As the Australian plate continued northwards the Sunda arcs were contorted into a great fold which probably increased the total extent of land surfaces in these arcs. With the continued westwards rotation of the Sunda arcs their distance from Australia has subsequently increased again.

This collision is clearly of major significance in understanding the origins and relationships of the flora of tropical Australia. Although the geological histories of the north-western margins of the

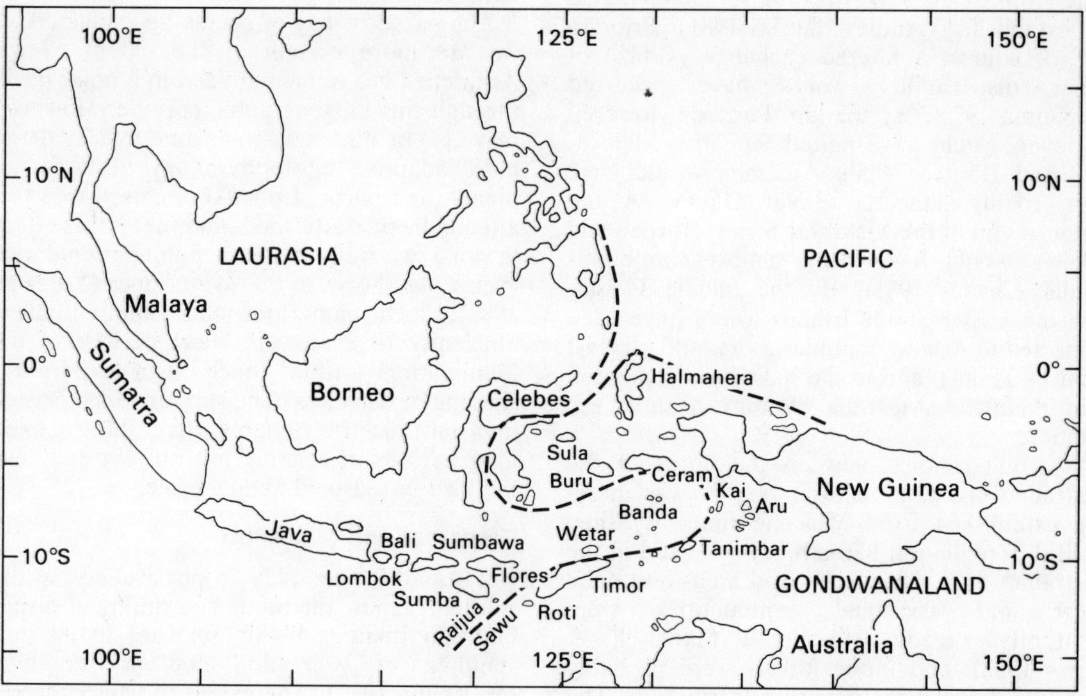

FIG. 2. Relationships of Australian and Sundaland plate areas at the present day (modified from Audley-Charles 1981).

Australian plate, of the Sunda Arc, and of the Greater Sunda region, during these collision events are not yet fully understood, Audley-Charles (1981) has re-examined the concept of Wallace's Line in terms of the geophysical data. He has further developed an hypothesis presented earlier (Audley-Charles *et al.* 1972), which now receives broad general support. The data suggest that a significant portion of eastern Malesia is Gondwanan, although in some areas the rocks of the Gondwana margin have been overthrust by Laurasian elements, with associated enlargement of land surfaces. The north-west margin of Gondwanaland is now represented by the Outer Banda Arc, which extends as a great deformation from Sawu (near Timor) to Buru, thence the south-eastern and eastern arms of Celebes, the other Moluccan islands including Halmahera, and finally the Vogelkop of New Guinea (Fig. 2). The Makassar Strait between Borneo and Celebes, long recognized as the most robust part of the biogeographic boundary between east and west Malesia, has existed as a seaway for a long period, and is confirmed as the strongest part of Wallace's line.

Although the earliest collision has been dated as mid-Miocene, it was probably not before latest Miocene or early Pliocene that a land migration route from Celebes to Australia via New Guinea was established (Audley-Charles 1981). Prior to this, of course, a filtered exchange of taxa of higher dispersibility would have occurred (McKenna 1972). By the late Pliocene, lowered sea levels would have united Sumatra, Malaya, Java and Borneo, while Australia would have been broadly connected to New Guinea. At the southern end of the Makassar Strait, Borneo and Celebes would have been almost continuous (Audley-Charles 1981). By the middle Pleistocene the Lesser Sunda Islands would have been connected to Asia by continuous dry land (at least as far as Timor), and would have been separated from mainland Australia by only a short sea distance.

A strong contact between the floras of the Australian and Indomalayan regions was therefore established from Miocene time. By then much differentiation had already occurred in the Australian flora. Seasonally arid areas and open forest and woodland communities were apparently already widespread (Truswell & Harris 1982), and mesic forests were probably already somewhat restricted in distribution. The scenario which emerges for the closed forests in Australia at the end of the Tertiary period is one in which (1) some differentiation of the original Gondwanan mesic forest into warm and cool (and wet and dry) types has occurred, (2) these forests occupy only a small part of their former range in the Australian fragment of Gondwanaland, and are perhaps to a significant extent associated with the old uplands of the eastern margin of the continent, (3) a possibly smaller New Guinea with lower relief is an integral part of the Australian plate both geographically and floristically, and (4) the floristic composition of the forests has been modified as a result of a relatively late interaction between the floras of Australia and the Malesian region.

Rainforests in Antarctica?

A cornerstone of Gondwanan biogeography is clearly the presumption that humid forests existed continuously in the Antarctic part of Gondwanaland, at least from late Cretaceous to mid-Tertiary times. There is fossil evidence of *Nothofagus*, southern gymnosperms, Proteaceae and Myrtaceae to at least 77°S latitude (Kemp & Barrett 1975; Thomson & Burn 1977). On the other hand there is the question of how such vegetation and diversity could exist, even under warm humid conditions, in a region where day length fluctuations would include continuous darkness for much of each year.

The paradox may of course be explained when we are more certain of the extent to which Antarctica has remained fixed in a polar position through this biogeographic episode. However, it may also be necessary to assume that by then the basic adaptive photophysiology of understorey plants (and parts of plants) in closed forests had already been established, and that it was effective in long as well as short illumination/darkness cycles. As shown in this symposium (Yates *et al.* 1988), such plants respond rapidly and very efficiently to even very short periods of direct illumination within much longer periods of shading or darkness. The species which existed at high latitudes presumably were able to tolerate long periods of shading not only in daily cycles but also in seasonal annual cycles.

Dispersal and Migration

Because of the complex geophysical history of the southern lands, the mode and timing of dispersal between them is highly relevant to an understanding of vegetation and floristic history (McKenna 1972). The extent to which species of closed forests are dispersible over long distances is still strongly debated. It has been argued as negligible (Steenis 1979), but given cautious support by Smith (1978, 1986). The need for a suitable habitat at the site of arrival is axiomatic, and has been emphasized for tropical forests by Webb & Tracey (1981). The latter have also argued the importance of post-germination factors such as availability of pollen vectors and the implications of compatibility systems for successful establishment. Thus the predictable conclusion was reached that plant migration is most likely to occur over short distances and to involve dispersal of many seeds in close proximity.

There is abundant evidence for shifts in areas of plant communities by expansion or contractions at the margins. It has been documented for tropical rainforest in north Queensland, for example, by Unwin *et al.* (1986), and for temperate rainforest in Tasmania by Ellis (1986). If all rainforest dispersal and migration could be explained by such shifts there would be few difficulties in explaining the present regional pattern.

In discussing the limits to dispersal capacity of rainforest taxa, Webb *et al.* (1986) have drawn attention to the large size of many diaspores, and to their frequently limited viability. Bird- and bat-dispersed species predominate in all rainforest types, and only in wind prone sites such as coastal systems and mountain tops is a significant proportion of wind dispersed taxa found (Webb & Tracey 1981). A detailed account of the importance of cassowaries as dispersal vectors in north Queensland rainforests has been given by Stocker and Irvine (1983).

Migratory fruit-eating birds are nevertheless associated with tropical forests. A notable migrant in northern Australia is the Torres Strait pigeon, and there is good evidence that it transports seeds of rainforest trees (Crome 1975). Being a race of the imperial pigeon, the distribution of which crosses Wallace's Line, this bird may be an example of the likely importance of larger migratory birds in distance dispersal of rainforest taxa.

Webb & Tracey (1981) and Webb *et al.* (1986) have argued strongly in support of the view of Takhtajan (1969) that obligate association between rainforest components requires that they migrate "in unison". They have suggested, therefore, that this is the major factor limiting distance dispersal of rainforest taxa. However, this constraint may be far from universal in its application. Species of secondary forest quickly became established on Krakatau (Steenis 1972). Rainforest taxa readily establish under a eucalypt canopy (Unwin *et al.* 1985; Ellis 1986). Numerous tropical forest species are grown in open plantations for timber, fruit and other forest products. Whilst some host/parasite and other interplant specificities certainly exist, and may favour migration in unison, it appears likely that most rainforest species can establish in atypical situations. Whilst dispersal over very long distances may be constrained by short seed viability and by the limited distances which birds actually carry seeds, migrations over shorter and intermediate distances seem very likely within the time frames available. The practical dispersal ceiling of 100-200 km argued by Steenis (1972) therefore appears realistic, and rare somewhat longer dispersal events not unreasonable. The landform and climatic history of the Malesian/Papuasian region has apparently been such that dispersal of this scope has made possible the floristic exchange which has occurred (see below).

Australian Rainforest Communities and their Relationships

Another prerequisite for detailed discussion on the origins of the Australian rainforest flora is an understanding of its component floristic groups. The most detailed analysis yet undertaken is that of Webb & Tracey (1981), later extended (Webb *et al.* 1984), and summarized in Webb *et al.* (1986). Numerical analysis of 1 316 species in 406 genera from 561 sites identified three major "ecofloristic" regions, these together comprising 7-9 ecofloristic provinces (Fig. 3). The three ecofloristic regions, generated by analysis at the species level, are (i) temperate and subtropical humid rainforests of south-eastern Australia, (ii) warm humid rainforests of north-eastern Australia, and (iii) "dry" monsoon forests of northern and eastern Australia. These three regions overlap substantially, but a core area can be identified for each. Outside the core areas each floristic entity tends to be represented by outliers, often well within the area of another entity. The outlying occurrences have mostly been regarded as relictual, and the pattern has been interpreted as being the result of climatic oscillations and sifting of the ecofloristic entities. The core areas correlate in a general way with the optimal growth regimes in Australia calculated by Nix (1982).

Such is the difference in composition of these ecofloristic entities that Webb and Tracey have argued that they are not simply different segregates of a single rainforest flora, derived by climatic filtering and attenuation. Rather, they were considered to reflect ancient floristic

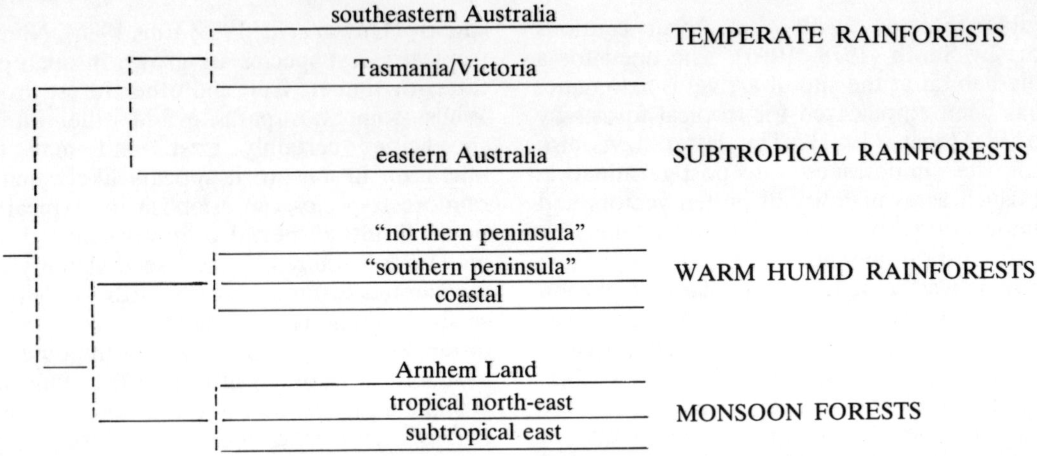

FIG. 3. Schematic representation of ecofloristic regions and provinces as proposed by Webb & Tracey (1981).

elements differentiated very early from Gondwanan stocks, and with different histories in the Australasian region.

Within the temperate and subtropical region three ecofloristic provinces have been recognized. Two of these are the temperate rainforests of southernmost and south-east Australia respectively. The third comprises the subtropical rainforests of eastern Australia between latitudes of approximately 22° and 30°S. Webb & Tracey (1981) and Webb *et al.* (1986) have strongly argued a separate identity of this latter rainforest province, and in postulating biogeographical relationships have treated it for convenience as a fourth rainforest component of the Australian vegetation (Fig. 3).

Of these four, the two ancient southern humid subelements are recognized as the direct derivates, in Australia, of the ancestral Gondwanan flora. Together they comprise the first major ecofloristic region described above. One subelement is therefore the cool temperate and montane mossy (*Nothofagus*) forest, and the other is the warmer subtropical vine forest, characterized by *Elaeocarpus, Ceratopetalum, Doryphora* and *Cinnamomum*. Perhaps the latter is the extant relict of the inappropriately named "*Cinnamomum* flora" of the early Tertiary (cf. Christophel 1981). Webb & Tracey have argued that these two subelements may represent the original differentiation of the Gondwanan flora into cool and warm forest formations. The "antarctic elements" in the rainforest flora (Thorne 1986a) are mostly represented in these subelements.

The two northern ecofloristic entities are apparently much more modified by later interaction with the Sundaland flora. The almost entirely relictual monsoon forest flora, with its low endemism, absence of primitive families, and strong Sundaland links, is discussed below. Webb & Tracey (1981) recognized three ecofloristic provinces within this region, based on floristic composition; there are differences between the provinces in the level of floristic affinity with the hot humid rainforests.

The other northern ecofloristic region thus comprises the true rainforests of Australia's wet tropics – the hot, humid vine forests of north Queensland. Here Webb & Tracey (1981) again recognized three ecofloristic provinces. This division effectively distinguishes the rainforests of the Cooktown-Atherton Tableland block from, first, the specialized forests of sand-dune systems on eastern Cape York Peninsula, and second, the seasonal forests which occur mainly in areas of lower relief in northern Cape York Peninsula. A further analysis of these floristic provinces is presented below.

Primitive angiosperms in Australian rainforests

The tropical and subtropical rainforests of Australia include a remarkable array of primitive angiosperms. In terms of the biogeographical history of mesic forests in Australia, all of these primitive taxa could be argued to be relictual at the present day. This attribute of the Australian closed forests has featured prominently in recent reviews and analyses, especially those concerned with the conservation of the remaining rainforest communities in Australia.

Of the 14 families of primitive angiosperms recognized by Takhtajan (1969), eight occur in north-eastern Australia (Webb & Tracey

1981). These include two narrowly endemic families (Austrobaileyaceae, Idiospermaceae), and another which is shared only with New Guinea (Eupomatiaceae). Although there are greater concentrations of primitive genera and species in South-East Asia, the concentration of primitive families in the Australasian region is the greatest in the world (Webb & Tracey 1981).

There is no evidence that these primitive families evolved *in situ* in north-eastern Australia (including New Guinea). The fossil record suggests that angiosperms first spread to the Australian portion of Gondwanaland perhaps 5 million years after their inception elsewhere in Gondwanaland (Dettman 1981). Owing to the incompleteness of fossil records, this time lag, of course, may eventually prove to be much shorter. The centre of origin of the angiosperms is usually suggested as being western Gondwanaland (Raven & Axelrod 1974; Brenner 1976). It is likely, therefore, that the primitive angiosperms still extant all achieved wide distributions in Gondwanaland in Cretaceous times. The north-eastern Australian refugium is thus of great antiquity.

It therefore appears that the eastern Australian region has continuously provided habitats favourable for survival of these ancient stocks. The extant primitive angiosperms now survive in equable, seasonal uplands (Takhtajan 1969). Uplands *per se* may be important as refugia because of the secure habitats which may persist within the mosaic of sites which they generate. The Great Divide and Great Escarpment of eastern Australia have apparently existed at least since the beginning of the Tertiary period (Ollier 1986). The other old uplands in Australia have all been overcome by major climatic cold and arid cycles, and the present day uplands of New Guinea and especially New Zealand are too young to have become major refugia. Thus the north-eastern and eastern uplands of the Australian plate (including the perhaps limited uplands of the ancestral New Guinea) may be the only large part of the entire Australasian/Pacific region where Gondwanan warm mesic forests have persisted continuously since Cretaceous times. In other regions where this has presumably happened (e.g., Fiji, New Caledonia), the concentration of surviving primitive angiosperms may be lower simply because the refugia themselves are smaller.

Monsoon forest flora

Monsoon forests occur in tropical areas in which there is a pronounced dry season. They often lack some of the attributes usually associated with tropical rainforests (Webb & Tracey 1981). The trees may be relatively short, and become increasingly deciduous with increase in length or intensity of the dry season. The canopy is more open, plank buttresses are lost and growth forms such as bottle trees and spinescent shrubs are more conspicuous.

The ecological and floristic affinities of monsoon forests are nevertheless with the rainforests (Webb & Tracey 1981). Webb and Tracey have concluded from their floristic analyses that the monsoon forests of the warm seasonally dry region were derived by differentiation from the rainforest flora of the tropical humid region.

The monsoon forest flora is distinguished by low species diversity and low endemism. There are no primitive angiosperms, and conifers are rare. The genera are typically widespread, usually to Malesia, and often to Africa (Whitmore 1981a; Steenis 1979). The monsoon forest may thus represent a dry paleotropical floristic element which may have had an extensive and relatively continuous distribution in the past. In Australia at the present time it may be restricted to virtually relictual occurrences in local areas, on a large scale as a consequence of Recent arid maxima, but on a finer scale because of its fire sensitivity. These occurrences extend from Arnhem Land eastwards to Cape York Peninsula and thence southwards to the Upper Hunter region of New South Wales. Webb and Tracey (1981) have identified a core area in central/southern Queensland.

In the Malesian region tropical rainforest occurs in two great blocks (Whitmore 1981a); one in Sundaland, centred on Sumatra, Malaya and Borneo, and the other in Papuasia, centred on New Guinea. They are separated by a great corridor of seasonally dry climates covering much of the Philippines, Celebes, Moluccas, Lesser Sunda and Banda Arc islands, part of Java and part of southern New Guinea (Fig. 4). In this corridor monsoon forest forms a mosaic with rainforest, the latter being confined to gallery and other wetter sites. The monsoon forest species nevertheless have disjunct distributions, and Whitmore (1981a) has suggested that extensive dispersal between present day populations is unlikely.

The Australian occurrence of monsoon forests is clearly an extension of this pattern, and it appears that northern Australia and some seasonally drier parts of eastern Australia are part of the same corridor. It thus appears probable that

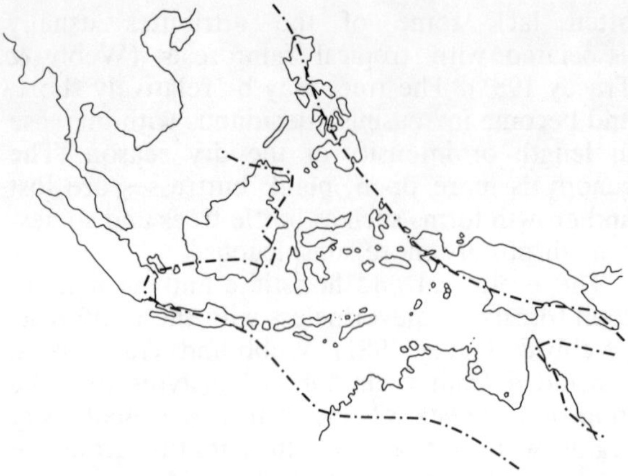

FIG. 4. Corridor of seasonal and strongly seasonal habitats in the Malesian-Australian region (modified from Whitmore 1981a).

this entire corridor shared a relatively more continuous monsoon forest vegetation in the recent geological past.

As summarized above, the maximum extent of continuous land in the Australo-Papuasian/Malesian region probably existed in the late Pliocene and Pleistocene. By this time progressive increase in seasonality of the climate would have resulted in expansion of monsoon forests at the expense of closed mesic forests. Thus at a time when the potential for migration over continuous land in the region was maximized, the predominant vegetation in the corridor may have been the monsoon forest type. The low endemism and wide distribution of the major genera of this formation is consistent with such a history.

Webb & Tracey (1981) and Webb *et al.* (1986) have interpreted the floristic homogeneity of paleotropical monsoon forests as evidence of "ancient floral links perhaps extending to Gondwanan time, . . . preserved . . . as the result of similar processes of climatic sifting". Herbert (1960, 1967) similarly argued the existence of an ancient paleotropical monsoon forest flora. By virtue of its derivation from closed mesic forest, the monsoon forest flora is almost certainly of Gondwanan derivation. Given the post-Gondwanan opportunities for floristic exchange between Africa, the rafting India, Sundaland and Papuasia, however, an early Tertiary origin in the differentiating paleotropics seems equally as consistent with the floristic homogeneity which exists as does a Cretaceous origin in an intact Gondwanaland.

The Intrusive Element in the Tropical Flora

The Miocene collision of the Australian and Sunda plates is now geophysical dogma. However, the nature and the consequences of the interaction between their floras remain a topic of uncertainty and of different interpretation. Some authors have seen the event as a contact between megafloras; the concept of contact and interdigitation of Laurasian and Gondwanan floras has been notably developed by Whitmore (1981a). Others have been content to describe an intermingling of Malesian and Australian stocks (e.g., Specht 1981b; Lange 1982). At the other extreme it has been argued that the entire tropical flora of the Asian-Australian region is Gondwanan, and

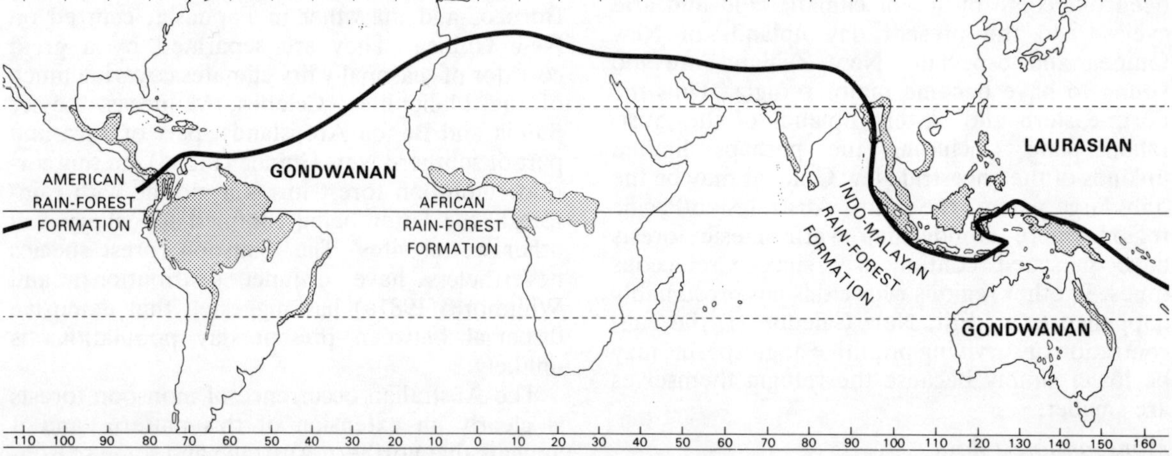

FIG. 5. Distribution of tropical rainforests on Laurasian and Gondwanan land surfaces. Map modified from Richards (1952).

the collision event has simply re-established contact between derivatives of the one basic stock (Webb & Tracey 1981; Webb *et al.* 1986).

To some extent these different viewpoints may simply be an exercise in semantics, and there may be little merit in applying "Gondwanan" and "Laurasian" labels too positively (Thorne 1986b). On the one hand there is clear evidence of taxonomic difference between the Asian and Australian tropical floras, and in this sense they may be conveniently described as Laurasian and Gondwanan. On the other hand there are obvious taxonomic links between the tropical floras of Africa, Asia and Australia. It is probable that *originally* all of these tropical floras were derived from an ancestral Gondwanan flora, and in this sense Webb *et al.* are nomenclaturally correct in their interpretation. The world distribution of tropical rainforest, as illustrated by Richards (1952) shows that most of its present area is still on Gondwanan land surfaces (Fig. 5). The largest area of apparently Laurasian land surface covered by tropical forest is in fact the South-East Asian/Malesian area. However the close proximity of Malesia to northward-drifting India about 40 Ma (Fig. 6), at a time of equable climate when mesic forests were presumably still very extensive, provides an indication of a possible source of the Malesian tropical flora, and of its floristic links with Africa. The importance of India in rafting a Gondwanan floristic element to Laurasia has been emphasized by Specht (1981b). Irrespective of terminology, however, a situation appears to have existed where two floras, separated by a long period of time and differentiation, were brought into contact as a result of the collision event.

The integrity of the Malesian flora itself is uncertain. The extent of land surfaces in the Malesian area was greatly increased as a consequence of the Miocene collision (Audley-Charles *et al.* 1981). Thus the colonization of these new areas occurred in conjunction with floristic interchange in the region. Whitmore (1981a) has described the formation of both Malesia and its flora as dating from the Miocene collision.

The concept and significance of Wallace's Line has recently been reviewed in terms of the collision scenario (Whitmore 1981a; Audley-Charles 1981). In some groups, such as the palms (Dransfield 1981), Sundaland and Papuasian stocks can be identified, with limited post-Miocene exchange. Several other similar examples were presented by Whitmore (1981b).

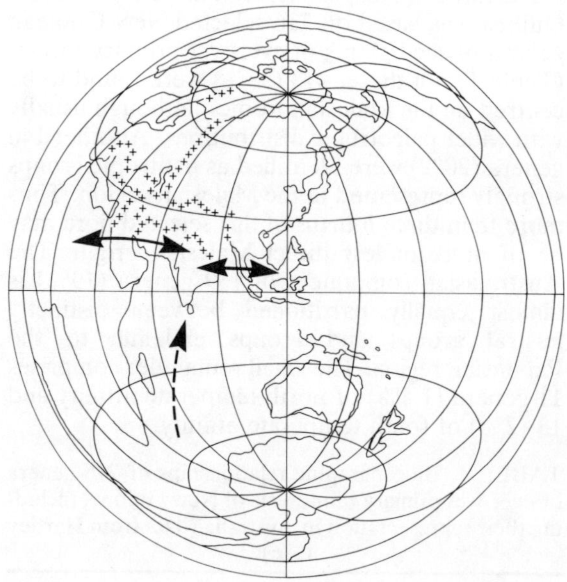

FIG. 6. Postulated configuration of paleotropic land surfaces 40 Ma (modified from Audley-Charles *et al.* 1981). The solid arrows indicate the close proximity of India to the Sundaland region and to northern Africa, and the potential for floristic exchange. The broken arrow indicates the displacement path of India as a Gondwanaland fragment.

In such cases, Wallace's Line appears to define a relatively robust contact zone.

The modern wet forest flora of New Guinea, however, provides evidence that Wallace's Line does not represent a boundary across which migration of such plants has been severely restricted. According to Whitmore (1981a), "Laurasian floristic elements are strongly predominant [in New Guinea], reflecting a massive invasion of this northern fringe of Gondwanaland from the west." From his analysis of the New Guinean rainforest flora, Hartley (1986) concluded that "New Guinea has received more of its rainforest genera from areas to the north and west than to the south. Although New Guinea must have initially been occupied by Australian plants, it appears that it was subsequently more or less over-run by South-East Asian plants". A possible explanation was encapsulated by Hartley in the suggestion that at the time of the collision between the Australian and Sunda plates, warm humid tropical conditions were extensive at the northern margin of the Australian plate. The original Australian (Gondwanan) flora, being temperate and subtropical in origin, may have been less competitive in this situation than the Malesian flora to the west.

Hartley's (1986) floristic analysis for New Guinea was based on 716 selected New Guinean genera of seed plants of ever-wet primary forests (Table 1). Of these, 413 (58%) were found to be centred on the Malesian tropics, although usually with wider paleotropic distributions. Another 140 genera (20%) were identified as pantropic groups strongly represented in the Malesian region. Thus more than three fourths of this selected flora may be of more or less direct Malesian origin. The Australasian component of 139 genera (19%) is almost equally partitioned between distinctly austral groups and groups endemic to the Papuasian region. The small remainder comprises 11 genera (1.5%) of north temperate affinity and 13 (2%) of south temperate affinity.

TABLE 1. Biogeographic relationships of 716 genera of ever-wet primary rainforests of New Guinea, including their representation in Australia (data from Hartley 1986).

Geographic distribution group	Numbers of Genera in New Guinea		also in Australia	
Pantropic	140		115	
African and/or Indomalayan	413		222	
African-Indomalayan		89		66
Madagascan-Indomalayan		27		22
South-East Asian		211		106
Western Malesian		86		28
Australasian	139		66	
North temperate	11		4	
South temperate	13		13	
TOTALS	716		420	

Both Whitmore (1981a) and Hartley (1986) have drawn attention to the residual ecological differentiation of the major elements in the New Guinean rainforest flora. Whilst the pantropical/Malesian element is seemingly ubiquitous in the lowland ever-wet forests, the Australian element predominantly occurs in lower montane and montane forests, mostly above 900 m elevation. However, the differentiation is not absolute. Whilst many tree genera of New Guinean rainforests above 1 500 m are of southern affinities (Schodde & Calaby 1972), there are genera of wider ecological amplitude which extend into this zone, and which are often of Malesian affinity (Hartley 1986). Interestingly, this paleoaustral rainforest province appears to be just as clearly evident at the present day in New Guinea as it is in north Queensland, indicating the status of the Tertiary New Guinea as part of the Australian refugial area.

Hartley's analysis can be extended to provide an indication of the extent of further intrusion of Malesian rainforest stocks into the Australian region. The numbers of genera in each of five distribution groups of the New Guinean primary ever-wet rainforest flora which also extend to Australia are set out in Table 1, with subdivision into subgroups for the largest component. It can be seen that very high proportions of pantropic, African-Indomalayan and Madagascan-Indomalayan genera in the New Guinean rainforest flora also extend to Australia (79% of the aggregate). Interestingly, the proportions of more local Malesian groups which reach Australia are lower: 50% for South-East Asian groups and 33% for western Malesian groups. The latter figures presumably reflect the more restricted distribution of young, endemic, truly Malesian genera, and perhaps also indicate the significance of an endemic Malesian floristic element in the biogeographic history of the region.

These data thus indicate the existence of a strong Malesian influence in the rainforest flora of tropical Australia, and support the analysis derived directly from the northern Australian floristic region (see below). This Malesian source is itself largely a secondary one, being predominantly a part of a wider, Afro-Asian or pantropical stock. As already mentioned, there has been some confusion caused by the application of both the terms Gondwanan and Laurasian to this stock. As far as the Australian flora is concerned, however, the interpretation seems clear. There is a tropical intrusive element in the rainforest flora of northern Australia. While it may ultimately be Gondwanan in origin, it has reached Australia/New Guinea from the north-west (Laurasia) in relatively recent times, and taxonomically and adaptively is distinct from the autochthonous rainforest with which it is now interdigitated. It appears, therefore, that the major area of interaction of the intrusive and autochthonous rainforest floras is well to the east of Wallace's Line, in New Guinea and northern Australia.

Webb *et al.* (1986) have placed a different interpretation on the floristic differences between Australasian rainforest provinces, and have argued that the concept of an Indomalesian source stock is not valid. They have accepted only a very limited exchange across Wallace's Line and have argued that the floristic differences are the consequence of differential extinction, possibly due to aridity in the later Tertiary. *Ilex* was cited as an example of a genus now occurring in northern Australia and Malesia but known from the pollen record to have been widespread in southern Australia in the Cretaceous (Martin 1981, 1982). This interpretation thus appears to be in conflict with those of Whitmore (1981a), Thorne (1986b) and Hartley (1986). Its acceptance depends essentially on judgment as to the relative importance of extinction, on the one hand, and refugia, dispersal and migration on the other. Unfortunately our present knowledge of any of these processes provides little basis for sound judgment. Given the likely long duration of rainforest habitats in eastern Australia, however, the broadly harmonious distribution patterns may favour migration rather than extinction as an operative force. The youth of the landforms at the northern margin of the Australian plate also perhaps favours migration from the west (and the south) as a source of its flora. Overall, the floristic data tend to support the view of Thorne (1986b) that current views on Gondwanan biogeography have downgraded the impact of an Indomalayan intrusive flora too much, and this impact should be recognized as a major contributor to the rainforest flora of the Australasian wet tropics.

Quaternary Changes – the Final Sifting

In terms of general structure, composition and regional distribution, the character of Australian wet tropical forests had been effectively determined by the end of the Tertiary period. The climatic oscillations of the Quaternary period have functioned mainly to regulate the extent and geographical relationships of the vegetation types of the region. Thus the distribution patterns of rainforest now evident are at the most only a few thousand years old.

Over the last 700 000 years nine glacial/interglacial intervals have been identified (Singh 1982). Temperatures equal to or higher than the present only occurred about 10% of the time. With the cooling there came aridity; new land areas, including land connections to New Guinea exposed by falling sea levels, probably would seldom have supported humid forests.

The studies of Kershaw (1978, 1981) suggest a sequence of vegetation changes on the Atherton Tableland, north Queensland. In the period 120 000-80 000 years ago the area was apparently occupied by rainforest. Only in the Holocene, from 7 000 years ago, has tropical rainforest returned to the area. In the intervening period rainforest angiosperms first became reduced, to be replaced by a gymnosperm-dominated rainforest with *Araucaria* predominant. The proportion of sclerophylls also increased. From the pollen record between 76 000 and 63 000 years ago, Webb & Tracey (1981) interpreted the formation as low mesophyll vine forest similar to that now present as cool subtropical rainforest of south-east Queensland. Following a return to warm subtropical rainforest with increased angiosperm composition 63 000-50 000 years ago, the formation returned to the *Araucaria* dominated cooler forest until 38 000 years ago. From then until 10 000 years ago sclerophyll woodland, with *Eucalyptus* and *Casuarina*, appears to have predominated. Since then the present-day forests have returned to occupy the area.

Even allowing for the difficulties of identifying and interpreting pollen spectra, Kershaw's study provides a good illustration of likely changes of vegetation through time in a relatively small area of tropical Australia. However, it can not show clearly whether the changes involved major shifts in vegetation or just local movements in a vegetation mosaic. Given the mosaic of vegetation types which occupies the area today, the latter seems more likely (Barlow 1981).

The integrity of the various forest provinces has not been maintained absolutely through these vicissitudes; the shifting mosaics have left some waifs as evidence of their movement. Webb & Tracey (1981) have listed more than 20 species which are widely distributed in a "core area" and have a limited "relict" occurrence remote from it, or in a different floristic province. They are associated with all three of the major floristic regions defined by Webb and Tracey. Thus one consequence of the climatic cycles of the Quaternary has been some blurring of the floristic distinctions between the tropical forest provinces of the region.

Much greater uncertainty surrounds the extent to which floristic exchange may have occurred between New Guinean and Australian rainforests in the Quaternary. Whilst the two lands have been broadly connected above sea level through much of Quaternary time, the low-lying Torresian land surface may have been mostly unsuitable for humid forest colonization. Sea levels are now near their highest during Quaternary time, although they may have reached 3-7 m higher (Whitmore 1981a). The upper altitudinal limits of forest are presently high, temperatures are near the Quaternary maximum, and precipitation is close to its highest. If conditions promoting rainforest communities were seldom more favourable than at present, the Torres Strait region may have functioned through all or most of this time as an ecological barrier to floristic exchange between rainforest blocks (Walker 1972).

However, during fluctuations in climate and vegetation distribution through the late Quaternary, a continuous or almost continuous rainforest belt from New Guinea to Queensland may have existed briefly and sporadically. In their modelling of possible past distribution of structural vegetation types in the region, based on climatic criteria, Nix & Kalma (1972) described a scenario 8 000 years ago in which closed forests may have extended along a narrow Torresian isthmus almost to the point of continuity. The zonations Nix and Kalma generated were by nature broad, and there is no doubt that rainforest persisted at least locally in some areas where their analysis indicated open forest as the major vegetation type. This, and the likely occurrence of humid forests at least in gallery situations in otherwise unfavourable terrain, suggest that in the recent past there may have been pockets or even blocks of rainforest close enough in space effectively to have generated contact between the humid forest floras of New Guinea and Australia. As Webb & Tracey (1981) have pointed out, the long successional cycles and consequent long generation cycles of climax tree species in such forests might mean that a brief pluvial period may not facilitate much differentiation or exchange. However, if such a brief event was possible 8 000 years ago, then more protracted contact may have occurred around 100 000 years ago when tropical forests had more continuous tenure. Webb & Tracey (1981) have in fact assumed that northern wet tropical forests in Australia have in the past had "continuity with and replenishment from" the forests of New Guinea.

The northern Peninsula flora

As described above, the analysis by Webb & Tracey (1981) indicates that three floristic provinces may be recognized in the tropical humid forest region of Australia (Fig. 4). The larger two of these occur in northern Cape York Peninsula and the Cooktown-Atherton Tableland areas respectively, and are separated by a corridor in the Princess Charlotte Bay area. Work by one of us (B.P.M.H.) also indicates that there are significant floristic differences between the two regions, and that they are relevant to an understanding of the floristic history of the region.

The data presented below have been summarized from a file which has been assembled (by B.P.M.H.) of the species which occur in tropical mesic or closed forests in Australia. It includes data on geographic distribution and growth habit. Species lists have been extracted for the humid forests in two areas: one comprising that part of Cape York Peninsula north of Princess Charlotte Bay (termed "northern" below and in the tables), and the other comprising the part of north Queensland between Princess Charlotte Bay and Townsville (termed "southern"). In order to concentrate on the significant structural components of the tropical flora, and to make the analysis reasonably manageable, the comparisons have been confined to dicotyledonous angiosperms. Within these, data have been extracted separately in two groups: (1) trees and shrubs, and (2) all vines, together with perennial epiphytes or epiphytic parasites, such as ant plants and mistletoes.

The total numbers of genera and species in the above categories recorded for the two areas are summarized in Table 2. The southern area is floristically much richer than the northern, but is of course larger and has more diversity.

TABLE 2. Numbers of dicotyledonous humid forest genera and species in north Queensland, by region. NORTHERN = Cape York Peninsula north of Princess Charlotte Bay; SOUTHERN = north Queensland between Princess Charlotte Bay and Townsville.

	NORTHERN		SOUTHERN	
	Genera	Species	Genera	Species
Trees and shrubs	251	492	354	888
Vines, etc.	118	189	132	259
Total	369	679*	486	1 145*

*Totals are adjusted to allow for two species which occur both as vines and trees/shrubs.

The numbers and proportions of genera and species which occur in only one of the two areas are summarized in Table 3. At the species level the floristic compositions are clearly different, and are proportionately similar for trees/shrubs and for vines, etc. Only about 40% of the total species are common to both areas; of the 60% of species which occur in only one of the areas, more are in the southern area than the northern. As would be expected, the difference is not so great at the generic level, but nevertheless about 40% of the genera occur in only one of the

TABLE 3. Distribution of dicotyledonous mesic forest genera and species in north Queensland by region and growth habit. Northern and southern areas are as defined in Table 2. Figures in bold are percentages.

	Genera			Species/Infraspecies		
	Northern	Both areas	Southern	Northern	Both areas	Southern
Trees and shrubs	27	224	130	123	369	519
	7	**59**	**34**	**12**	**36**	**51**
Vines etc.	21	97	35	60	129	130
	14	**63**	**23**	**19**	**40**	**41**
Total	48	321	165	183	496*	649
	9	**60**	**31**	**14**	**37**	**49**

*Total is adjusted to allow for two species which occur both as vines and trees/shrubs.

two areas. Among the genera with restricted distributions in the region, there is much greater imbalance in tree/shrubs (27:130) than in vines, etc. (21:35).

Presentation of a complete list of species and generic distributions by family is hardly justified in a review such as this. Table 4 therefore presents data for the largest families which occur in the region (having 20 or more species or eight or more tree/shrub genera or seven or more genera of vines, etc.). The data are generally consistent, and as expected tend to conform with the trend identified above. Among species of trees/shrubs, fewer than 50% are common to both northern and southern areas in all larger families except Moraceae, where endemism appears to be unusually low, and Meliaceae and Rubiaceae, where the figure is close to 50%. Among species of

TABLE 4. Distribution of dicotyledonous humid forest genera and species in selected families in north Queensland. Family acronyms are as established by Weber (1982). Northern and southern areas are as defined in Table 1.

	Trees and Shrubs						Vines, etc.					
FAM	Genera			Species/infraspp			Genera			Species/infraspp		
	Nth	Both	Sth	Nth	Both	Sth	Nth	Both	Sth	Nth	Both	Sth
ANN	2	5	5	6	3	11	4	2	1	6	1	2
APO	1	7	1	3	9	8	1	5		4	6	9
ASC								6	1	8	6	7
CEL	1	4	7	1	5	14	1	2	1	1	4	2
CNV							1	4	2	2	4	5
CUN		2	6		2	12						
ELC		2	2		6	26						
EUP	5	19	7	15	35	23	1	4		2	2	8
LAU		7		10	23	55		1		1	1	1
LEG	5	13	5	9	19	17	1	13	4	5	13	12
LOR							2	5		5	14	9
MEL		9	1	3	15	9						
MNS							2	6	2	5	9	6
MOR		3		3	23	8		3			3	
MRT	1	15	11	23	36	66			1			1
PRT		4	20	1	3	42						
RHM		3	1		6	1	2	4	1	3	3	3
RUB		13	3	2	16	12	2	4	1	6	6	5
RUT	2	13	3	7	19	24		1		1		1
SAP	3	15	5	12	20	45						
SPT		7	2	3	9	16						
STR	1	6	1	4	6	9						
VIT								4		2	13	12

vines, etc., the result is similar, with only Moraceae showing low endemism. When generic occurrences are compared the results are more variable, but some families show strong demarcation between the two areas. Notable are Annonaceae and Celastraceae, in which fewer than 50% of genera are common to the two areas, and Cunoniaceae and Proteaceae, in which there are very few genera present in the northern area and none at all restricted to it.

These comparisons suggest that biogeographic histories for the two areas have been different. Levels of generic endemism and the wider distributions of the genera also tend to support this assumption. Of the 27 genera of trees/shrubs present in the northern but not the southern area, only one (*Welchiodendron*, a monotypic genus segregated from *Tristania*) is endemic; the other 26 all occur in New Guinea or more widely in the Malesian region. All 21 genera of vines, etc. confined to the northern area have similar wider distributions.

The 165 genera which occur in the southern but not the northern area show contrasting biogeographic patterns. Of the 130 genera of trees/shrubs, 49 are endemic in Australia, many of them being distributed widely in eastern Australia. A further 14 genera have apparently Gondwanan distributions including New Caledonia and sometimes Fiji, New Zealand or South America, but do not extend to Malesia. Thus a total of 63 (about half) of these tree/shrub genera reach their northern limits in the Atherton Tableland-Cooktown area, and southern relationships are evident for most of them. Most of the primitive angiosperms discussed above are in this category. Among the genera of vines, etc., the pattern is not so strongly evident, with only four of the relevant 35 genera being endemic to the region, and with none showing strong southern affinities.

Almost all of the remaining 67 genera of trees/shrubs and 31 of the 35 genera of vines, etc. are further distributed in New Guinea and/or the Malesian region. Thus their absences from the northern area simply represent disjunctions in their distributions. The northern area emerges as a floristic province characterized both by low endemism and a depauperate flora. This is consistent with the idea that the northern region has undergone more pronounced floristic perturbations in the Recent than the areas immediately to the north and south, which are more stable and more refugial in character.

It is suggested, therefore, that during periods of maximum cooling and aridity the mesic forests may have disappeared almost completely from the lands of low relief in the Torresian region. Only a few gallery communities may have persisted, perhaps along with very localized refugial forests in places like the McIlwraith Range. Such an event may have occurred most recently 20 000-17 000 years ago (Nix & Kalma 1972). The present forests thus represent the limited results of recolonization, from both New Guinean and Athertonian refugia, which probably reached a maximum about 8 000 years ago. The data presented above suggest that as a group, the tree/shrub genera of the southern area have been subjected to the strongest filtering effects during the recolonization processes.

Studies in two species groups lend support to the view that Recent colonization of the northern Cape York Peninsula has involved New Guinea as a source area. In a cluster analysis of *Syzygium* and allied genera Whiffin & Hyland (1986) identified geographic/floristic units which generally agreed with the classification of rainforests in Australia by Webb & Tracey (1981). These units included a "Torres Strait" element linked closely with a "Cape York" element and more distantly with a "north-east Queensland" element, but also showing a strong relationship with New Guinean taxa. Similar results were obtained in *Cryptocarya* (Whiffin & Hyland 1986); a similar pattern of geographic/floristic units was generated, and again a "Torres Strait" element was identified, having taxonomic links with New Guinea.

Examples from the family Loranthaceae illustrate a distribution pattern which is becoming more common as botanical exploration in Cape York Peninsula continues. Common and widespread mistletoes of the lowland rainforests in New Guinea which have been discovered in the northern Peninsula area include *Amyema friesianum, A. seemenianum, Dactyliophora novaeguineae* and *Decaisnina hollrungii* (Barlow 1984). Only the latter species has also been found in the southern area, although it is much more strongly established in the northern. Perhaps more surprising is the occurrence of the monotypic *Cecarria obtusifolia* in the northern area; this species has a number of disjunct occurrences in Papuasia and the Philippines, and is thought to be relictual.

Open Forests and Heaths

In discussions on origins of the tropical flora the extensive occurrence of wet open forest and heath communities tends to be overlooked. In the Australian-Papuasian region these communities are clearly Gondwanan in derivation, and are dominated by myrtaceous genera, including *Melaleuca, Tristania, Xanthostemon* and *Eucalyptus* (sens. lat.), with genera of Rutaceae and Proteaceae also prominent.

Most evidence indicates that these communities are themselves of tropical origin. The phylogenetic studies of the eucalypts (Pryor & Johnson 1981; Johnson & Briggs 1984) suggest that as a group they have originated at the margins of mesic closed forests. The distributions of their constituent subgroups, including *Arillastrum* in New Caledonia and others in tropical Australia, suggest a subtropical or tropical origin. A similar pattern is evident in *Melaleuca*, which replaces the eucalypts as the dominant tree in seasonally inundated tropical lowlands. The *M. leucadendra* complex is close to the ancestral state of the genus, and is primarily tropical in distribution. Relationships in *Melaleuca* are discussed elsewhere in these proceedings (Barlow 1988).

Predominance of Myrtaceae, Proteaceae and Rutaceae in scleromorph heaths, and the occurrence of these heaths in tropical habitats adjacent to rainforest, support the suggestion that scleromorph communities are differentiated from tropical forests along gradients of decreasing soil fertility (Beadle 1981; Johnson & Briggs 1981; Specht 1981a; Barlow 1981). The origins of the open forest and scleromorph communities are usually thought to date from the early to middle Tertiary, and they would have differentiated at times when warm to cool mesic forests were widespread in Australia. The communities in tropical northern Australia may therefore in a sense be relictual. The much greater development of such communities in temperate seasonal and arid situations in Australia reflects the massive later radiation which has occurred as these climatic conditions have overtaken the southern part of the continent.

Whilst open forest and heath communities may not have differentiated *in situ* in northern Australia, they are probably tropical/subtropical in origin and represent an early phase in the evolution of the autochthonous subelement of the Gondwanan flora in Australia. Families such as Rutaceae and Proteaceae are keys, then, to the identity of an autochthonous element in rainforests of Australia and New Guinea (Hartley 1986; Johnson & Briggs 1981), and also to the origins of an autochthonous scleromorphic flora.

Conclusion

The tropical rainforests of northern Australia are the complex products of a sequence of paleogeographic and paleoclimatic events. These include: (1) the early differentation of warm and cool and of moist and dry elements from a mesic Gondwanan flora; (2) the survival of many primitive components of this flora in durable refugia in northeastern and eastern Australia; (3) the differentiation of open forest and heathland derivatives in response to edaphic evolution; (4) interdigitation in New Guinea and Australia of the autochthonous rainforest flora with a Malesian rainforest flora after a long period of isolation, and (5) local sifting of communities, involving Australia and New Guinea, in response to increasing amplitude of climatic oscillations in recent times. The result is an intricate pattern of humid forest communities which, within a mantle of structural similarity, show significant differences in composition, age, endemism, relictuality and richness.

References

Audley-Charles M. G. (1981) Geological history of the region of Wallace's Line. In: *Wallace's Line and Plate Tectonics* (ed. T. C. Whitmore) pp. 24-35. Clarendon, Oxford.

Audley-Charles M. G., Carter D. J. & Milsom J. S. (1972) Tectonic development of eastern Indonesia in relation to Gondwanaland dispersal. *Nature Phys. Sci.* **239**, 35-39.

Audley-Charles M. G., Hurley A. M. & Smith A. G. (1981) Continental movements in the Mesozoic and Cenozoic. In: *Wallace's Line and Plate Tectonics* (ed. T. C. Whitmore) pp. 9-23. Claredon, Oxford.

Barlow B. A. (1981) The Australian Flora: its origin and evolution. In: *Flora of Australia* (ed. A. S. George) Vol. 1, pp. 25-75. Aust. Govt. Publ. Service, Canberra.

Barlow B. A. (1984) Loranthaceae. In: *Flora of Australia* (ed. A. S. George) Vol. 22, pp. 68-131. Aust. Govt. Publ. Service, Canberra.

Barlow B. A. (1988) Patterns of differentiation in tropical species of *Melaleuca. Proc. ecol. Soc. Aust.* **15**, 239-47.

Beadle N. C. W. (1966) Soil phosphate and its role in molding segments of the Australian flora and vegetation, with special reference to xeromorphy and sclerophylly. *Ecology* **47**, 992-1007.

Beadle N. C. W. (1981) Origins of the Australian angiosperm flora. In: *Ecological Biogeography of Australia* (ed. A. Keast) pp. 407-26. W. Junk, The Hague.

Bowler J. M. (1982) Age, origin and landform expression of aridity in Australia. In: *Evolution of the Flora and Fauna of Arid Australia* (ed. W. R. Barker & P. J. M. Greenslade) pp. 35-45. Peacock, Adelaide.

Brenner G. J. (1976) Middle Cretaceous floral provinces and early migrations of angiosperms. In: *Origin and Early Evolution of Angiosperms* (ed. C. B. Beck). Columbia University Press, New York.

Chappell J. (1974) Geology of coral terraces, Huon Peninsula, New Guinea: a study of Quaternary tectonic movements and sea-level changes. *Bull. Geol. Soc. Amer.* **85**, 555-70.

Christophel D. C. (1981) Tertiary megafossil floras of Australia as indicators of floristic association and palaeoclimate. In: *Ecological Biogeography of Australia* (ed A. Keast), pp. 377-90. W. Junk, The Hague.

Crome F. H. J. (1975) The ecology of fruit pigeons in tropical northern Queensland. *Aust. Wildl. Res.* **2**, 155-85.

Crook K. A. W. & Taylor G. R. (1985) Early opening history of the Southern Ocean between Australia and Antarctica. In: *Otway 85: Resources of the Otway Basin: Summary papers and excursion guide. Geol. Soc. Aust., S. Aust. & Vic. Divs.*, No. 1.

Dettman M. E. (1981) The Cretaceous flora. In *Ecological Biogeography of Australia* (ed. A. Keast) pp. 355-75. W. Junk, The Hague.

Dransfield J. (1981) Palms and Wallace's Line. In: *Wallace's Line and Plate Tectonics* (ed. T. C. Whitmore) pp. 43-56. Clarendon, Oxford.

Ellis R. C. (1986) The relationship between subalpine grassland and forest communities in Tasmania. In: *Flora and Fauna of Alpine Australasia: Ages and Origins* (ed. B. A. Barlow) pp. 243-48. C.S.I.R.O., Melbourne.

Hartley T. G. (1986) Floristic relationships of the rainforest flora of New Guinea. *Telopea* **2**, 619-30.

Herbert D. A. (1960) Tropical and subtropical rainforest in Australia. *Aust. J. Sci.* **22**, 283-90.

Herbert D. A. (1967) Ecological segregation and Australian phytogeographic elements. *Proc. Roy. Soc. Queensland* **78**, 101-11.

Hooker J. D. (1960) Introductory Essay. *Botany of the Antarctic Voyage of H. M. Discovery Ships* "Erebus" *and* "Terror" *in the years 1839-1843, III. Flora Tasmaniae.* Reeve, London.

Johnson L. A. S. & Briggs B. G. (1975) On the Proteaceae — the evolution and classification of a southern family. *Bot. J. Linn. Soc.* **70**, 83-112.

Johnson L. A. S. & Briggs B. G. (1981) Three old southern families — Myrtaceae, Proteaceae and Restionaceae. In: *Ecological Biogeography of Australia* (ed. A. Keast) pp. 427-69. W. Junk, The Hague.

Johnson L. A. S. & Briggs B. G. (1984) Myrtales and Myrtaceae — a phylogenetic analysis. *Ann. Miss. Bot. Gard.* **71**, 700-56.

Kemp E. M. (1978) Tertiary climatic evolution and vegetation history in the south-east Indian Ocean region. *Palaeogeogr. Palaeoclimatol. Palaeoecol.* **24**, 169-208.

Kemp E. M. & Barrett P. J. (1975) Antarctic glaciation and early Tertiary vegetation. *Nature (Lond.)* **258**, 507-08.

Kershaw A. P. (1978) A record of the last interglacial-glacial cycle from north-eastern Queensland, Australia. *Nature (Lond.)* **272**, 159-61.

Kershaw A. P. (1981) Quaternary vegetation and environments. In: *Ecological Biogeography of Australia* (ed. A. Keast) pp. 81-101. W. Junk, The Hague.

Lange R. T. (1982) Australian Tertiary vegetation. In: *A History of Australasian Vegetation* (ed. J. M. B. Smith) pp. 44-89. McGraw Hill, Sydney.

McKenna M. C. (1972) Possible biological consequences of plate tectonics. *Bio-Science* **22**, 519-25.

Martin H. A. (1981) The Tertiary flora. In: *Ecological Biogeography of Australia* (ed. A. Keast) pp. 391-406. W. Junk, The Hague.

Martin H. A. (1982) Changing Cenozoic barriers and the Australian palaeobotanical record. *Ann. Miss. Bot. Gard.* **69**, 625-67.

Nelson E. C. (1981) Phytogeography of southern Australia. In: *Ecological Biogeography of Australia* (ed. A. Keast) pp. 733-59. W. Junk, The Hague.

Nix H. A. (1982) Environmental determinants of biogeography and evolution in Terra Australis. In: *Evolution of the Flora and Fauna of Arid Australia* (eds W. R. Barker & P. J. M. Greenslade) pp. 47-66. Peacock, Adelaide.

Nix H. A. & Kalma J. D. (1972) Climate as a dominant control in the biogeography of northern Australia and New Guinea. In: *Bridge and Barrier: the natural and cultural history of Torres Strait* (ed. D. Walker) pp. 61-91. Australian National University, Canberra.

Ollier C. D. (1982) The Great Escarpment of eastern Australia: tectonic and geomorphic significance. *J. Geol. Soc. Aust.* **39**, 431-35.

Ollier C. D. (1986) The origin of alpine landforms in Australasia. In: *Flora and Fauna of Alpine Australasia: Ages and Origins* (ed. B. A. Barlow) pp. 3-25. C.S.I.R.O., Melbourne.

Pain C. F. & Ollier C. D. (1984) Drainage patterns and tectonics around Milne Bay, Papua New Guinea. *Rev. geomorph. Dyn.* **32**, 113-20.

Powell C. McA., Johnson B. D. & Veevers J. J. (1981) The early Cretaceous break-up of eastern Gondwanaland, the separation of Australia and India, and their interaction with south-east Asia. In: *Ecological Biogeography of Australia* (ed. A. Keast) pp. 15-29. W. Junk, The Hague.

Pryor L. D. & Johnson L. A. S. (1981) *Eucalyptus,* the universal Australian. In: *Ecological Biogeography of Australia* (ed. A. Keast) pp. 499-536. W. Junk, The Hague.

Raven P. H. & Axelrod D. I. (1974) Angiosperm biogeography and past continental movements. *Ann. Miss. Bot. Gard.* **61,** 539-673.

Richards P. W. (1952) *The Tropical Rain Forest.* Cambridge University Press, Cambridge.

Schodde R. & Calaby J. H. (1972) The biogeography of the Australo-Papuan bird and mammal faunas in relation to Torres Strait. In: *Bridge and Barrier: the natural and cultural history of Torres Strait* (ed. D. Walker) pp. 257-300. Australian National University, Canberra.

Singh G. (1982) Environmental upheaval. In: *A history of Australasian Vegetation* (ed. J. M. B. Smith) pp. 90-108. McGraw Hill, Sydney.

Smith J. M. B. (1978) Dispersal and establishment of plants. In: *Biology and Quaternary Environments* (eds D. Walker & J. C. Guppy) pp. 207-22. Australian Academy of Science, Canberra.

Smith J. M. B. (1982) An introduction to the history of Australasian vegetation. In: *A History of Australasian Vegetation* (ed. J. M. B. Smith) pp. 1-31. McGraw Hill, Sydney.

Smith J. M. B. (1986) Origins of Australasian tropicalpine and alpine floras. In: *Flora and Fauna of Alpine Australasia: Ages and Origins* (ed. B. A. Barlow), pp. 109-28. C.S.I.R.O., Melbourne.

Specht R. L. (1972) *The Vegetation of South Australia.* Govt. Printer, Adelaide.

Specht R. L. (1981a) Ecophysical principles determining the biogeography of major vegetation formations in Australia. In: *Ecological Biogeography of Australia* (ed. A. Keast) pp. 299-333. W. Junk, The Hague.

Specht R. L. (1981b) Evolution of the Australian flora: some generalizations. In: *Ecological Biogeography of Australia* (ed. A. Keast) pp. 783-805. W. Junk, The Hague.

Steenis C. G. G. J. van (1972) *The Mountain Flora of Java.* Brill, Leiden.

Steenis C. G. G. J. van (1979) Plant geography of east Malesia. *Bot. J. Linn. Soc.* **79,** 97-178.

Stocker G. C. & Irvine A. K. (1983) Seed dispersal by Cassowaries *(Casuarius casuarius)* in north Queensland's rainforests. *Biotropica* **15,** 170-76.

Takhtajan A. (1969) *Flowering Plants: Origin and Dispersal.* Oliver & Boyd, Edinburgh.

Thomson M. R. A. & Burn R. W. (1977) Angiosperm fossils from latitude 70°S. *Nature (Lond.)* **269,** 139-41.

Thorne R. F. (1986a) Antarctic elements in Australasian rainforests. *Telopea* **2,** 611-17.

Thorne R. F. (1986b) Summary statement. *Telopea* **2,** 697-704.

Truswell E. M. & Harris W. K. (1982) The Cainozoic palaeobotanical record in arid Australia: fossil evidence for the origin of an arid-adapted flora. In: *Evolution of the Flora and Fauna of Arid Australia* (eds W. R. Barker & P. J. M. Greenslade) pp. 67-76. Peacock, Adelaide.

Unwin G. L., Stocker G. C. & Sanderson K. D. (1985) Fire and the forest ecotone in the Herberton highland, north Queensland. *Proc. ecol. Soc. Aust.* **13,** 215-24.

Walker D. (1972) Bridge and Barrier. In: *Bridge and Barrier: the natural and cultural history of Torres Strait* (ed. D. Walker) pp. 399-405. Australian National University, Canberra.

Webb L. J. & Tracey J. G. (1981) Australian rainforests: pattern and change. In: *Ecological Biogeography of Australia* (ed. A. Keast) pp. 605-94. W. Junk, The Hague.

Webb L. J., Tracey J. G. & Williams W. T. (1984) A floristic framework of Australian rainforests. *Aust. J. Ecol.* **9,** 169-98.

Webb L. J., Tracey J. G. & Jessup L. W. (1986) Recent evidence for autochthony of Australian tropical and subtropical rainforest floristic elements. *Telopea* **2,** 575-89.

Weber W. A. (1982) Mnemonic three-letter acronyms for the families of vascular plants: a device for more effective herbarium curation. *Taxon* **31,** 74-88.

Wellman P. & McDougall I. (1974) Cainozoic igneous activity in eastern Australia. *Tectonophysics* **23,** 49-65.

Whiffin T. & Hyland B. P. M. (1986) Taxonomic and biogeographic evidence on the relationships of Australian rainforest plants. *Telopea* **2,** 591-610.

Whitmore T. C. (1981a) Palaeoclimate and vegetation history. In: *Wallace's Line and Plate Tectonics* (ed. T. C. Whitmore) pp. 36-42. Clarendon, Oxford.

Whitmore T. C. (1981b) Wallace's Line and some other plants. In: *Wallace's Line and Plate Tectonics* (ed. T. C. Whitmore) pp. 70-80. Clarendon, Oxford.

Yates D. J., Unwin G. L. & Doley D. (1988) Rainforest environment and physiology. *Proc. ecol. Soc. Aust.* **15,** 31-37.

Proc. Ecol. Soc. Aust. 15:19-30

Origin and evolution of terrestrial plant communities in the wet-dry tropics of Australia

R. L. SPECHT

Botany Department, University of Queensland, St Lucia, Qld. 4067

Abstract

The generic and species composition of tropical Australian plant communities indicates that many taxa have remained unaltered since the breakup of Gondwanaland.

The seasonally wet tropical-subtropical climate of the Tertiary produced lateritic earths and podzols, krasnozems and skeletal sandstone and granitic soils, on which rainforests, montane Nothofagus *forests,* Heidewald *and derived plant communities developed. Tropical mangrove vegetation occurred around the coastline.*

Periods of aridity began in the mid-Miocene and have continued to the Present, causing disjunctions in plant communities and evolution of canopy foliage characteristics able to survive water stress. The species richness of overstorey and understorey strata developed in relation to the annual growth rate of the community, controlled by the annual water balance and temperature.

Quaternary basalt flows and alluvial deposits enabled the savanna elements to expand. Infertile sands blown inland from exposed sea beds during glacial periods led to expansion of the sclerophyll flora.

The Gondwanan Heritage

It was fortunate that Dr. Ferdinand von Mueller published his reports (Mueller 1856, 1857, 1858), on the botanical observations made on A. C. Gregory's North Australian expedition in 1855-56, before Sir Joseph Hooker wrote his essay on the flora of Australia (Hooker 1860). Hooker was thus able to distinguish six major floristic groups in tropical Australia:

1. tropical rainforests
2. brigalow scrubs
3. open grass downs, nearly devoid of trees.
4. desert vegetation
5. sandstone tablelands
6. seacoast vegetation

From his deep knowledge of the flora of India, Hooker recognized that the tropical rainforests contained "numerous Indian genera of umbrageous trees". The sandstone tablelands of the Kimberleys, Western Australia and Arnhem Land, Northern Territory, and the coastal vegetation all showed the same strong floristic affinities with India.

"The sandstone tableland presents an arid, cheerless landscape, described by Dr. Mueller in terms that apply perfectly well to the sandstone tablelands of the peninsula of India, and indeed many of the characteristic genera are common to both. These consist of *Terminalia, Melia, Cochlospermum, Sterculia, Buchanania, Zizyphus, Nauclea, Bauhinia, Indigofera, Erythrina, Gardenia, Strychnos, Santalum,* a profusion of Andropogoneous Grasses, and other shrubs and herbs, all of which the Indian botanist recognizes at once as the prominent feature of the sandstone ranges of western Bengal and central India" (Hooker 1860 p.xl).

Regarding the flora of tropical northern Australia as a whole, Hooker (1860, p. xlix) pointed out that "the Indian botanist will recognize the double element, one consisting of a littoral and the other of an inland flora, the former prevalent over the shores of both Indian peninsulas, the Malay and Philippine islands . . . ; the other an inland Indian Flora characteristic of the Carnatic, the sandstone tablelands of the western Peninsula of India, and which reappears in the Upper Birma *[sic]* valley, where the climate of this becomes dry for a considerable portion of the year". "The rarity of Cryptogams*, which are almost wholly absent in western and central tropical Australia, and in the islands of the Gulf of Carpentaria" was also noted (Hooker 1860, p. xxxix).

The vegetations of monsoonal northern Australia and India, with a wet summer alternating with a very dry winter, show close relationships in structure and floristics. Hooker (1860) listed 53 families of vascular plants, including 88 genera, common to both India and monsoonal Australia. These floras, which derived from the original Gondwanan vegetation, maintained many of their attributes over the 125 million years which have passed since India separated from Australia in the early Cretaceous (Powell *et al.* 1981).

In spite of this long geological separation, the Gondwanan flora of monsoonal Australia, still contains 26% of the species (not just the genera, as noted by Hooker (1860)) in common with the Indian Peninsula (Specht 1958b). Some plant formations include even higher percentages, *viz.* monsoon rainforests (35%), coastal wetlands (62%), coastal dunes (26%), wetlands (32%), eucalypt open-forest (17%), sandstone complex (14%).

The distribution of many of these species extends to tropical Africa (Specht 1958b) — monsoon rainforests (15%), coastal wetlands (31%), coastal dunes (16%), wetlands (16%), eucalypt open-forest (9%), sandstone complex (7%). Of the genera of vascular plants listed for Australian tropical rainforests (Specht 1981a), 108 (of 51 families) are also found in southern Africa (Palgrave 1977). Thirty-seven of the tropical Australian grass genera must be regarded as Gondwanan, evolved before the break-up of this southern continent (Clifford & Simon 1981); even species within some Gondwanan grass genera (e.g., *Themeda triandra, Heteropogon contortus*) may have remained relatively unaltered over this long period of time.

Without our current understanding of plate tectonics and continental drift, it was natural that the biogeographers of the nineteenth century considered that a large proportion of the Australian flora was derived from Indo-Malaya. Hooker (1860) listed nearly 500 species under 273 genera from tropical Australia which were either identical, or closely allied to Indian species. Only 11 genera were considered by him to have evolved in Australia and to have reached India or southeast Asia — *Baeckea, Casuarina, Dacrydium, Haloragis, Lagenifera, Leptospermum, Leucopogon, Metrosideros, Philydrum, Stylidium, Tristania.*

India began to separate from a combined Antarctica-Australasian plate in the early Cretaceous, 125 million years ago (Powell *et al.* 1981). By the mid-Cretaceous, the Indian Ocean was over 1 000 km wide, forming an effective barrier to the dispersal of most land-based flora and fauna. By the Paleocene/Eocene, 55 million years ago, India had crossed the equator and collided with Eurasia (Powell *et al.* 1981).

During the Paleocene/Eocene the Australasian tectonic plate (including New Guinea) separated from Antarctica and began its drift northward at an average rate of 66 mm per year (Wellman & McDougall 1974). During the mid-Miocene (15 million years ago), the western end of New Guinea (on the Australasian plate) came in contact with the southern margin of Indonesia on the south-east Asian continental block, called Sundaland (Powell *et al.* 1981). Intermingling of the floras of Sundaland and Australasia could then occur.

As Hooker (1860) indicated, little of the uniquely Australian flora migrated into Sundaland. How much of the Australian Gondwanan flora migrated northwards and, conversely, how much of the south-east Asian flora migrated southwards? An analysis of the biogeography of the ericaceous genus *Rhododendron*, with seed readily dispersed by wind and birds, may provide some answers to this question (Sleumer 1966-67; Specht & Womersley 1979; Specht 1981d). Fig. 1. assumes that the ancestral species of *Rhododendron* were present on Gondwanaland before its break-up. The *elepidote* section of *Rhododendron* on the Indian tectonic plate spread northward and eastward from India. When the Australasian plate made contact with Sundaland 15 million years ago the *lepidote* section of the genus expanded from the northern part of Australasia (New Guinea/North Queensland) into Sundaland (as far west as the Malay Peninsula and the extreme south of Indo-China). Minimal intermingling of the Indian-Eurasian sections and the Australasian-Sundaland section of the genus *Rhododendron* resulted. Studies of Webb *et al.* (1986) on the biogeography of Australian rainforests add support to the Gondwanan origin of the flora.

*The paucity of lichens in open plant communities of tropical Australia has been noted by Specht (1958a), Specht *et al.* (1977) and Rogers & Stevens (1981).

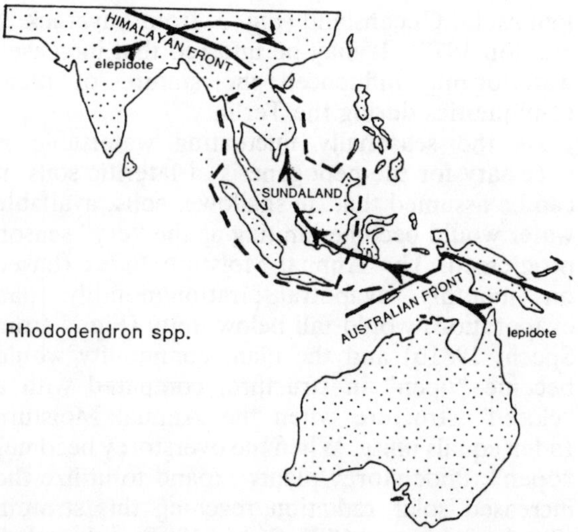

FIG. 1. Dispersal of the *elepidote* and *lepidote* sections of the genus *Rhododendron* from the Indian and Australasian tectonic plates of the ancient southern continent of Gondwanaland.

The origins of many of the genera of vascular plants, found today in tropical Australia, must pre-date the split-up of the ancient southern continent of Gondwanaland. So far, the fossil material from Cretaceous deposits in Australia (Dettmann 1981) provide little concrete evidence of this ancient flora; micro- and mega-fossils of the Early Tertiary indicate that the Gondwanan flora was, by then, well developed (Christophel 1981; Martin 1981).

The range of taxa recorded within any one fossil deposit can provide an idea of the floristic associations which may have existed around that sedimentary basin. Continuity of floristic associations (at the family/generic level) throughout the Tertiary into the Quaternary can thus be postulated from fossil records. The presence of tropical rainforests, tropical and temperate *Nothofagus* forests, *Eucalyptus* and *Acacia* communities, sclerophyll (heathy) and savanna communities, mangrove and salt-marsh (and allied chenopod shrubland) communities can be inferred. As present-day floristic associations, at this level of resolution, provide little evidence of the range of structural variability (with many dominant *Eucalyptus* spp. ranging in size from over 30 m trees to 2 m shrubs – in communities ranging from tall open-forests to heathlands), only broad generalizations regarding vegetation structure can be assumed.

The Tertiary Heritage

Laterites

Probably the most impressive fossil records from the Tertiary are the extensive remnants of deep lateritic earths in northern Australia and lateritic podzolic soils in southern Australia. These Tertiary soils, in various stages of erosion, influence floristic associations today, as they apparently did throughout the Tertiary.

It appears that the lateritic soils formed under a subtropical-tropical climate which induced a fluctuating watertable (with associated anaerobic and aerobic conditions) within the soil profile: precipitation of iron, aluminium and silicon oxides results in this zone, while the clay mineral, kaolinite, developed in the deep subsoil. The process of lateritization began late in the Cretaceous, continued through much of the Tertiary (Stephens 1971). Extensive areas of lateritic soils developed across much of Australia, India and Africa, also in Brazil and the Guianas of South America (Prescott and Pendleton 1952). Under the mesothermal climate, the high, but markedly seasonal, rainfall, induced deep leaching of nutrients within the soil and acidic conditions (pH about 6.0). Soils low in phosphorus, nitrogen, potassium and molybdenum (and other plant nutrients) resulted, though lateritic earths developed on argillaceous material, appear to be somewhat more fertile than lateritic podzols, developed on siliceous parent material (Fig. 2). Water was available for growth of the plant community throughout the year, but growth would have been reduced or inhibited during the period of anaerobic waterlogging. Such an environment appears similar to that reported for the *Heidewald* (heath rainforest) in Borneo (Winkler 1914; Richards 1936; Brünig 1968; Specht & Womersley 1979). Similar vegetation may be found from Cape York Peninsula, Queensland (Lavarack & Stanton 1977) to the South China Sea (Steenis 1957). Few characteristic "heathland" species are present, though sub-sclerophyllous, leptophyllous species of *Gymnostoma* (syn. *Casuarina), Dacrydium, Agathis* and *Baeckea* become dominant on shallow sandy soils with fluctuating watertables; a wide variety of Gondwanan tree genera co-exist with these genera (Brünig 1970; Specht & Womersley 1979).

As the watertable approaches the soil surface, the Heidewald becomes progressively stunted — a dense "padang" heathland occupies the wettest habitats in Sarawak (see Specht & Womersley 1979) and in the Jardine River of Cape York

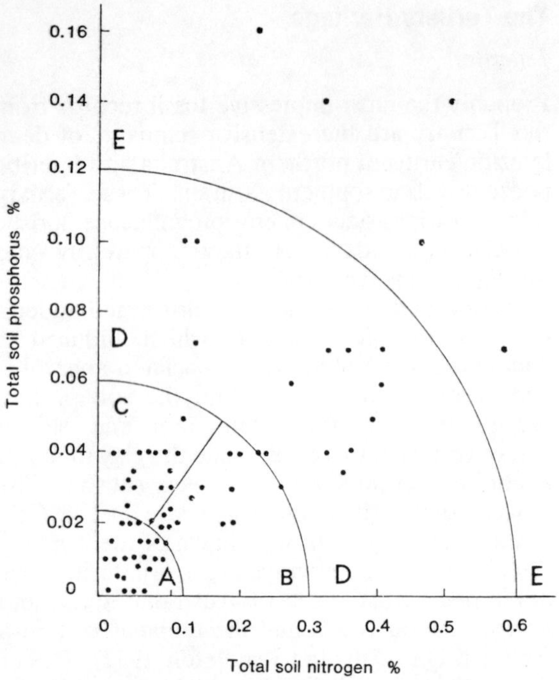

FIG. 2. Edaphic range of Australian plant communities with respect to Total Nitrogen (%) and Total Phosphorus (%) in surface soil. *Zone A:* Sclerophyll (heathy) communities, *Acacia* shrublands and *Triodia* hummock grasslands. *Zone B:* Savanna communities, including grassy open forests/woodlands on lateritic earths. *Zone C:* Chenopod and *Stipa-Aristida* communities. *Zone D:* Grasslands on cracking clays; wet sclerophyll forests — rainforests on krasnozems.

Peninsula, Queensland (Fig. 3, from Lavarack & Stanton 1977). It may be inferred that increased waterlogging influenced the stature of plant communities during the Tertiary.

As the seasonally fluctuating watertable is necessary for the pedogenesis of lateritic soils, it can be assumed that, in shallower soils, available water would become limiting as the "dry" season progressed. The Annual Moisture Index (based on monthly evapotranspiration/monthly pan evaporation) would fall below unity (Fig. 4 from Specht 1981b) and the plant community would become "open" in structure, compared with a "closed" structure when the Annual Moisture Index equals unity. When the overstorey becomes "open", understorey plants expand to utilize the increased solar radiation reaching this stratum (Specht & Morgan 1981; Specht 1983):- sclerophyll (heathy) species on sandy lateritic podzols; savanna species on the marginally more fertile lateritic earths (Fig. 2).

Sandstone tablelands

Ancient siliceous sedimentary rocks, pre-dating the Tertiary laterites, are extensive in tropical northern Australia (especially in Arnhem Land and the Kimberleys, Lawn Hill and the Carnavon Ranges) and are found as far south as Sydney (Hawkesbury Sandstone), the Grampians in Victoria, Adelaide (Stonyfell Quartzites) and Albany-Esperance (Mt Barren Quartzite).

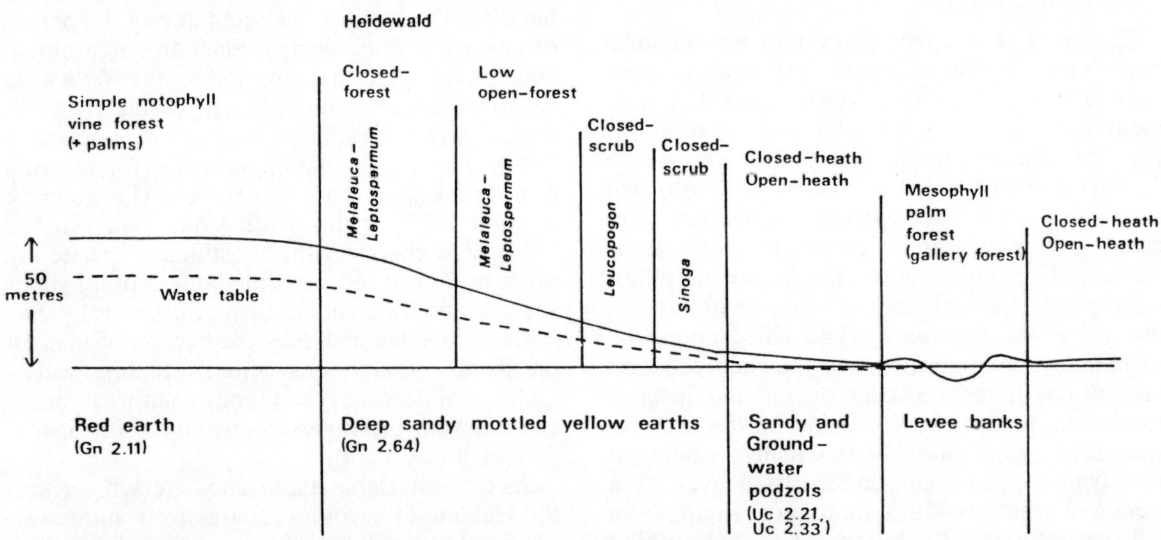

FIG. 3. Diagrammatic cross-section of the sequence of sclerophyllous communities observed on the Jardine River Catchment, Cape York Peninsula, Queensland (after Lavarack & Stanton 1977).

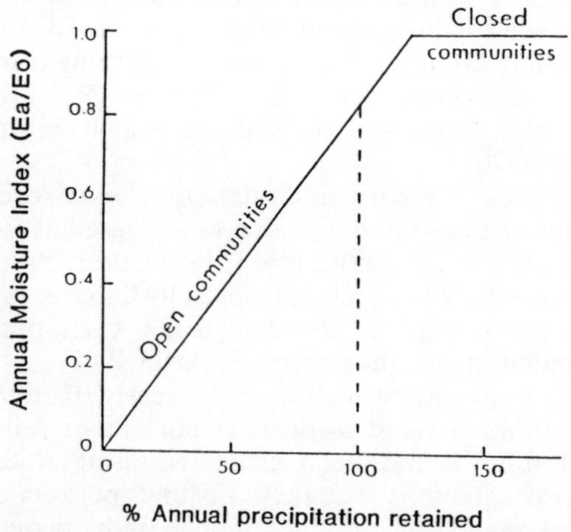

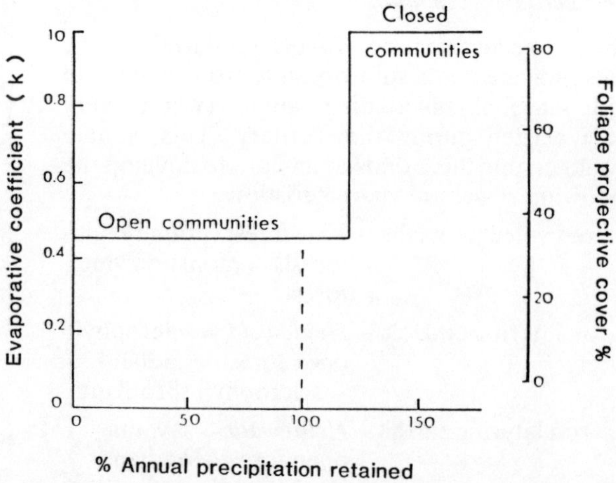

FIG. 4. (a) Mean annual Moisture Index and (b) the evaporative coefficient (k) and Foliage Projective Cover of the overstorey, plotted against the percentage of annual precipitation retained in the soil within the rooting zone. (Specht 1981b).

Erosion has kept pace with the slow weathering of these hard siliceous rocks, forming craggy hills deeply cleft along the joint lines in which dissecting streams flow. Most of the sandstone landscape is virtually bare rock with soils developing only on the plateaux, in rock clefts and in valleys. Vegetation develops in every crevice in which soil accumulates. It is only on the less-eroded plateaux that Tertiary laterites have developed.

By inference from present day conditions, it is assumed that extensive run-off and erosion has occurred from the harsh sandstone landscape; the Annual Moisture Index of many microhabitats approached zero (Fig. 4) enabling very open plant communities of sclerophyllous Gondwanan genera to develop on these extremely infertile soils (Fig. 2). In extremely xeric sites, the sclerophyllous grass *Triodia* (an endemic genus) probably colonized the sclerophyllous shrublands.

Granitic outcrops

Granite batholiths, crystallized in Paleozoic rocks, have been exposed by uplift and erosion in several places on the Australian continent (Stevens 1985). The fragmented granitic rocks of the Bellenden Ker Range in north Queensland have formed a centre, throughout the Tertiary, for the survival and evolution of the rainforest vegetation and the adjacent eucalypt forests. Further south, massive granitic outcrops from Stanthorpe, Queensland to Armidale, New South Wales, the Mount Kosciusko and Mount Buffalo districts of New South Wales and Victoria, Wilson's Promontory, Victoria and scattered outcrops across the southern part of the continent, became centres for the survival and speciation of the low-nutrient sclerophyll (heathy) vegetation.

Krasnozems (red volcanic loams)

As the Australian tectonic plate drifted northward during the Tertiary, the eastern part of the continent passed over a "hot-spot" — at periodic intervals, volcanic activity resulted (dated by Wellman & McDougall 1974), *viz.*

Hillsborough Region, Qld	32 million years
Springsure Region, Qld	27 million years
Border Range Region, Qld-NSW	23 million years
Comboyne Region, NSW	16 million years
Orange-Dubbo Region, NSW	12 million years
Macedon Region, Vic.	6 million years

The volcanic material which covered the ancient rocks was in general, base-rich in nutrients (Fig. 2). The long period of weathering dating from the volcanic activity to the present-day produced deep, friable, red loamy soils, with degeneration of the clay minerals to kaolinite (complexing P and Mo in unavailable form).

The deep, friable loams trapped rainfall, ensuring that available water was optimal for all months of the year (the Annual Moisture Index was unity, Fig. 4). A closed community of Gondwanan tree and vine genera here flourished, sharply delimited from the vegetation on infertile sandstone/lateritic soils (Fig. 3).

The rainforest flora now found on these volcanic rocks must have been derived from the Paleocene-Eocene flora.

The Tertiary vegetation

The soil pedogenic process which produced lateritic soils indicate that a subtropical to tropical climate with seasonally-fluctuating rainfall (wet to very wet) existed during the Tertiary. This climate would enable the Gondwanan flora to develop the following structural characteristics:

on red volcanic earths – Vine forest (primary rainforest) → monsoon vine forest.

on lateritic podzols – *Heidewald* → sclerophyll open forest/woodland → sclerophyll shrubland

on lateritic earths – *Heidewald* → savanna open forest/woodland/ shrubland (including *Triodia*)

In the mid-Tertiary, *Eucalyptus* spp. were evolving in the transitional area around the vine forest and the Heidewald (Pryor & Johnson 1981).

Microfossil evidence showed that cooler montane habitats all over Australia supported a closed-forest dominated by the tropical section *brassii* of the genus *Nothofagus*, now confined to New Guinea and New Caledonia (Hekel 1972; Martin 1981).

In subalpine climates, sclerophyll plants (*Rhododendron, Vaccinium,* etc.), epiphytic in the top-most branches of the primary rainforest at lower altitudes, became ground plants at higher altitudes to form sclerophyll shrublands (Specht & Womersley 1979; Specht 1981d).

Tropical-subtropical mangrove vegetation was well developed around the coastline as far south as Perth, Western Australia (Churchill 1973; Specht 1981c).

Aridity (mid-Miocene to Present)

Increasingly frequent periods of aridity began in the mid-Miocene and have continued to the Present. Plant communities characteristic of the early Tertiary landscape adjusted to the drier conditions (lower Annual Moisture Index and evaporative coefficient) by evolving more open-structured foliage canopies with xerophytic attributes (Fig. 5 from Specht & Specht in press a). Foliage projective cover (F.P.C.) of the overstorey decreased (Fig. 5a), the evergreen leaves became smaller in area* (Fig. 5b) and thicker (Leaf specific area, Fig. 5c). The number of tropical overstorey species (even *Eucalyptus* spp.) with large, seasonally-deciduous foliage increased with increasing aridity, the community becoming completely deciduous on rocky outcrops (Perry & Lazarides 1964) and on soils of heavy texture (Story 1970).

The fossil record indicates that overstorey species of *Eucalyptus* and *Acacia* increased as the landscape became more arid (Martin 1981; Pryor & Johnson 1981). Grasses (especially C4 grasses), herbaceous genera and arid zone chenopods expanded in the understorey (Martin 1981).

The more mesic section of the early Tertiary vegetation survived in disjunct pockets of relict vegetation. It has been estimated using water balance equations that these disjunct pockets of similar vegetation were continuous under a rainfall regime 25-30% higher than the present — the disjunct pockets of *Eucalyptus goniocalyx* in southeastern Australia (Specht *et al.* 1961) and the isolated remnants of monsoon vine forests in northern Australia (Specht *et al.* 1977) were formerly continuous and could become continuous again in the future.

The balance between microfossils of rainforest and eucalypt open forest/woodland communities deposited in the sediments of Lynches Crater, north Queensland has changed over the last 100 000 years B.P. (Kershaw 1981). There is an indication of increasing aridity until almost 10 000 B.P., followed by an increase in the rainforest proportion to about 75% of the vegetation recorded at the base of the deposit (100 000 years B.P.). A further 25% increase would restore the original landscape dominated by rainforest.

Late Tertiary-Quaternary Landscape Rejuvenation

Throughout much of the Tertiary, the landscape became increasingly infertile under the subtropical-tropical climate which produced lateritic soils and krasnozems. Late in the Tertiary, erosion associated with epeirogenic uplift and basaltic flows tended to "rejuvenate" some of the landscape. More fertile soils developed on the basaltic and alluvial materials, often forming cracking, clay soils (grey and brown soils of heavy texture). Some of the deciduous/semi-deciduous overstorey trees were able to establish on these soils, but, generally, tap-rooted eucalypt seedlings failed to survive — extensive areas of fibrous rooted grasslands of *Astrebla* and *Dichanthium* spp. resulted.

*The leaves decrease in area from the tropics to the subtropics to temperate regions (Specht & Specht in press a).

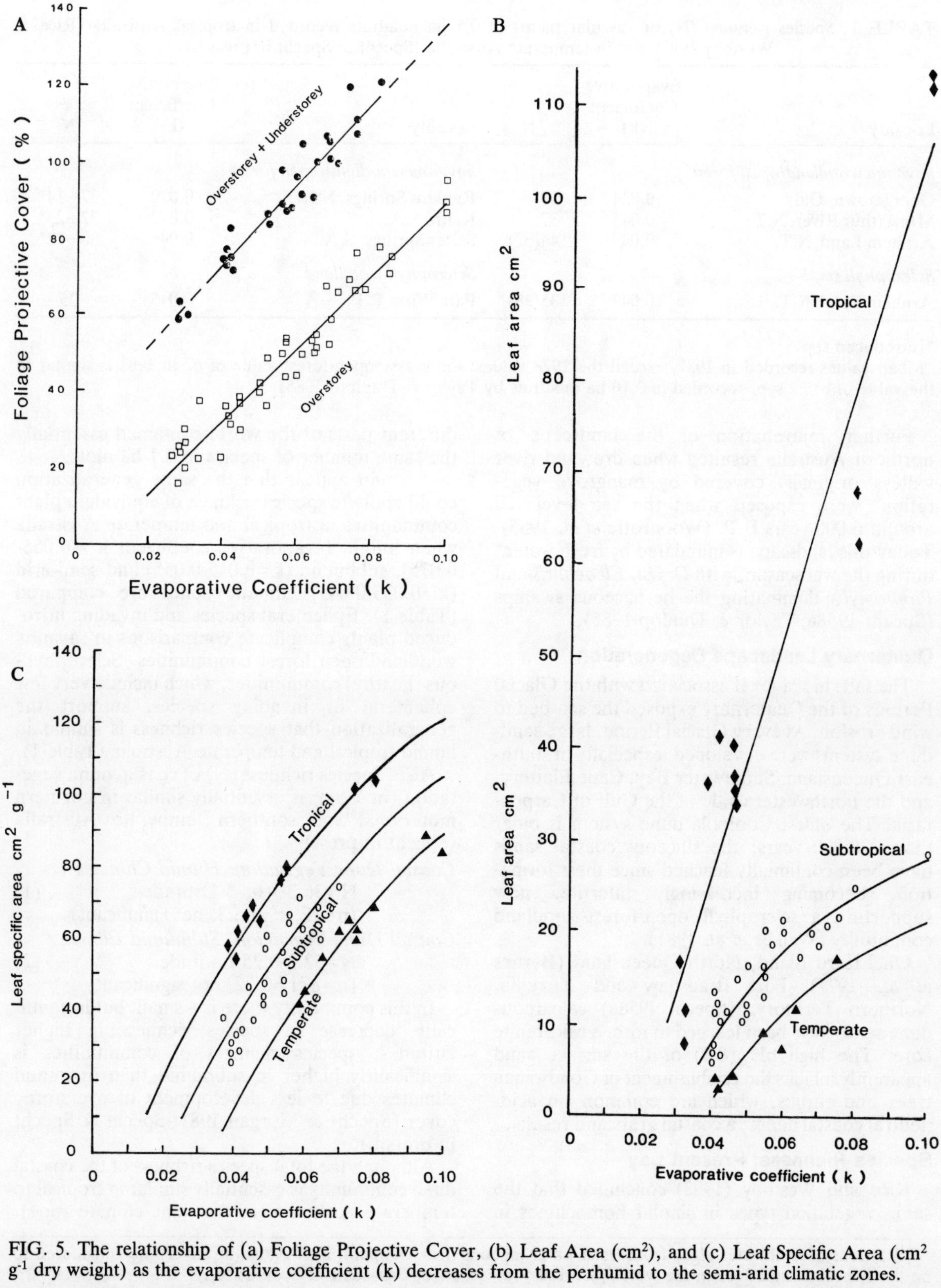

FIG. 5. The relationship of (a) Foliage Projective Cover, (b) Leaf Area (cm^2), and (c) Leaf Specific Area (cm^2 g^{-1} dry weight) as the evaporative coefficient (k) decreases from the perhumid to the semi-arid climatic zones.

TABLE 1. Species richness (N) of vascular plants in 0.1 ha quadrats recorded in tropical Australia (Rice & Westoby 1983) and in temperate Australia (Specht & Specht in press b).

Locality	Evaporative Coefficient (k)	N	Locality	Evaporative Coefficient (k)	N
Savanna woodland/open forest			*Savanna woodland/open forest*		
Georgetown, Qld	0.034	36	Rankins Springs, NSW	0.039	37+14*
MacArthur River, N.T.	0.045	37	Keith, S.A.	0.042	32+23*
Arnhem Land, N.T.	0.047	44(62)⁺	Salter Springs, S.A.	0.048	30+12*
Sclerophyll scrub			*Sclerophyll woodland*		
Arnhem Land, N.T.	0.047	35(38)⁺	Para Wirra R.P., S.A.	0.045	35

*Introduced spp.
⁺Mean values recorded in 1981, exceed the 1978 values; the grassy open-forest value of 62 in 1981 is similar to the value of 61±2 spp. recorded in 0.16 ha quadrats by Taylor & Dunlop (1985).

Further rejuvenation of the landscape of northern Australia resulted when drowned river valleys originally covered by mangrove vegetation, were exposed when the sea level fell around 6 000 years B.P. (Woodroffe *et al.* 1985). Today this landscape is inundated by fresh water* during the wet season, with *Oryza, Eleocharis* and *Fimbristylis* dominating the herbaceous swamps (Specht 1958a; Taylor & Dunlop 1985).

Quaternary Landscape Degeneration

The falls in sea-level associated with the Glacial Periods of the Quaternary exposed the sea-bed to wind erosion. At every Glacial Period, large sand-dune systems were developed, especially in south-east Queensland, Shoalwater Bay, Cape Flattery, and the northwestern side of the Gulf of Carpentaria. The oldest Cooloola dune system is older than 120 000 years; the siliceous coastal sands have been continually leached since their formation, becoming increasingly infertile, now supporting a sclerophyll open-forest/woodland community (Walker *et al.* 1981).

On Lizard Island, North Queensland (Byrnes *et al.* 1977), Port Bradshaw and Yirrkala, Northern Territory, (Specht 1958a) calcareous dune sands have been leached to form a calcarenite core. The high pH (8-9) of the surface sand apparently inhibits the establishment of Gondwanan trees and shrubs, which are common on acid-neutral coastal dunes; a coastal grassland results.

Species Richness: Present Day

Rice and Westoby (1983) concluded that the same vegetation types in similar homoclimes in different parts of the world contained essentially the same number of species in 0.1 ha plots.

It would appear that the same generalization could apply to species richness of equivalent plant communities in tropical and temperate Australia when humid (evaporative coefficient k =0.055-0.075), subhumid (k=0.045-0.055) and semi-arid (k=0.035-0.045) climatic zones are compared (Table 1). Ephemeral species and invading introduced plants complicate comparisons in savanna woodland/open forest communities. Sclerophyllous (heathy) communities, which include very few ephemeral or invading species, support the generalization that species richness is similar in humid tropical and temperate Australia (Table 1).

Also, species richness (N) of coastal dune vegetation (in 1 km^2 is essentially similar in northern monsoonal and southern temperate Australia (Specht in press).

Coastal Dune Vegetation: Humid Climates

$$N=36.3 - 0.25 \text{ Latitude} \quad (1)$$
$$(n=12,\ r^2=0.23, \text{ not significant})$$

Coastal Dune Vegetation: Subhumid Climates

$$N=52.3-0.25 \text{ Latitude} \quad (2)$$
$$(n=8,\ r=0.32, \text{ not significant})$$

In this community there is a small, but insignificant, decrease in species richness in higher latitudes; species richness of communities is significantly higher in subhumid than in humid climates due to less development of overstorey cover (Specht & Morgan 1981; Specht & Specht in press b).

Although the total species richness of the coastal dune community is essentially similar in tropical to temperate Australia (in the same climate zone),

*Collections of freshwater algae from fresh water lagoons near Oenpelli, East Alligator River, are considered to be the richest desmid flora in the world (Scott & Prescott 1958; Croasdale & Scott 1976).

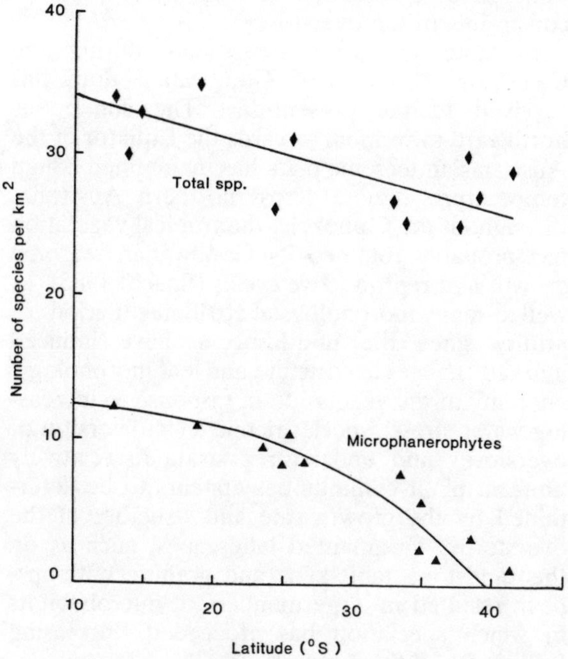

FIG. 6. The number of species per km^2 of tall shrubs/trees, compared with total spp., in coastal dune vegetation, from tropical to temperate climatic regions of eastern Australia (Specht in press).

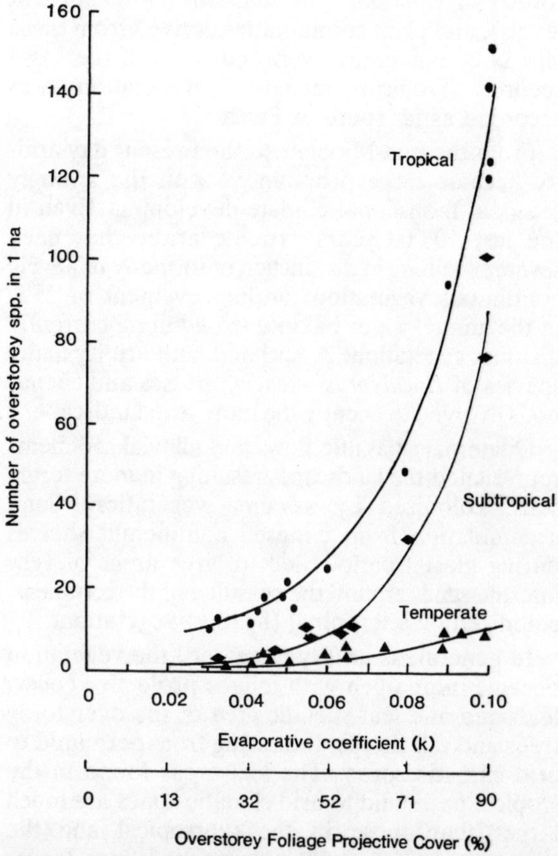

FIG. 7. Species richness (per hectare) of trees/tall shrubs in the overstorey of climax communities in tropical, subtropical and temperate Australia (Specht & Specht in press c).

the number of species of tall shrubs/trees is highest in the tropics (Fig. 6), suppressing species richness in the understorey (Specht in press). The development of species richness of the overstorey (trees and tall shrubs) in tropical, subtropical and temperate Australia is demonstrated in Fig. 7 (Specht & Specht in press c). The exponential increase observed in the perhumid zone (k =0.075-0.100) indicates total species richness (of overstorey and understorey) of plant communities in the climatic zone would differ markedly from tropical to temperate regions, a difference not obvious in less humid climatic zones (see Fig. 6), where species richness (N) of the plant community may be related to the evaporative coefficient (k) and Annual Moisture Index (AMI) by the following equation (Specht 1984):

$$N=(19.4/k-212)\ AMI \quad (3)$$

Species richness of the overstorey appears positively related to the annual rate of development of the foliage canopy of the overstorey regenerating after disturbance (Specht & Specht in press c). Increased annual growth rate of the overstorey accelerates the rate of loss of understorey species (Specht & Specht in press c).

Conclusions

The flora of tropical Australia (Specht 1981d, Webb *et al.* 1984) has evolved from ancient Gondwanan taxa which has remained largely unaltered, at the generic level, since the Cretaceous.

The seasonally wet subtropical-tropical climate of much of the Tertiary produced oscillating water-tables in the soils; seasonally-alternating reducing and oxidizing conditions at depth in the soil resulted in lateritic earths and lateritic podzols widespread across the continent. Extreme erosion on ancient sandstone/quartzitic rocks induced a multitude of microhabitats in which disjunction of widespread species occurred and evolution of the flora proceeded in isolation. Volcanic activity, when the continent passed over a "hot spot", enabled deep red loam soils (krasnozems) to develop on the volcanic outflows. A range of plant communities from vine forests (rainforests), montane *(Nothofagus)* rainforests, *Heidewald* (on soils with low nutrient levels), and plant communities derived from these climax communities, were indicated in the fossil record. Tropical mangrove vegetation was recorded as far south as Perth.

From the mid-Miocene to the present day aridity became more pronounced with the strongly seasonal monsoonal climate developing. Even in the last 10 000 years extreme aridity has been severe, resulting in disjunction of formerly relatively continuous vegetation; an improvement of 25% in the annual water balance would unite currently disjunct vegetation. Associated with aridity many species of *Eucalyptus, Acacia,* grasses and chenopods evolved to occupy the more arid landscape.

Quaternary basaltic flows and alluvial sediments rejuvenated the landscape, resulting in more fertile soils, colonized by savanna vegetation. Sand accumulation from exposed continental shelves during glacial periods led to large areas of very infertile sands around the coastline of the continent, colonized by sclerophyll (heathy) vegetation.

In general, as aridity increased, the vegetation became more open with foliage projective cover, leaf area and leaf specific area of the overstorey trees and tall shrubs decreasing from perhumid to arid climatic zones. The leaf areas found in the tropical perhumid to arid climatic zones are much larger than those in the subtropical and the temperate regions of the continent. Large leaves (characteristic of the tropics — Fig. 5) often become deciduous during the dry monsoonal season, especially in the subhumid to semi-arid zones. Evolution of the plant community and its component species appears to be determined by the resistance to evaporation of available soil water imposed by the overstorey canopy (in particular, foliage projective cover and leaf resistances).

The high temperatures experienced throughout the year in the tropics appears to increase cellular metabolism, reducing the annual growth rates of the tropical communities compared with plant communities at higher latitudes. As annual growth rates decrease, species richness of the overstorey (trees and tall shrubs) increase. The increased shading of the overstorey causes species richness of the understorey to decline. Thus the tropical region favours the survival and evolution of a wealth of tall woody species at the expense of understorey species.

It is only in fragmented areas (the sandstone tablelands) with many microhabitats and on Recent alluvial and basaltic soils in which woody overstorey growth is inhibited, that a wealth of understorey species can develop, away from the constraints of the overstorey.

In spite of many vicissitudes during the Cainozoic, the ancient Gondwanan flora has survived to the present-day. The continuous northward movement towards the Equator of the Australasian tectonic plate has maintained a high temperature climate across northern Australia, throughout the Cainozoic; the tropical vegetation has probably retained its Gondwanan seasonal growth and reproductive cycles (Specht 1986), as well as many morphological attributes. Periods of aridity since the mid-Miocene have induced marked changes in structure and leaf morphology/anatomy in the vegetation in response to increasing water stress. Species richness (α-diversity) of overstorey and understorey strata in relatively uniform plant communities appears to be determined by the growth rate and structure of the overstorey. Fragmented landscapes, such as on the sandstone tablelands and granitic outcrops, have resulted in large numbers of microhabitats in which speciation has proceeded, increasing β-diversity of the region.

References

Brünig E. F. (1968) Some observations on the status of heath forests on Sarawak and Brunei. *Proc. Symp. Rec. Adv. trop. Ecol.* **1968,** 451-57.

Brünig E. F. (1970) Stand structure, physiognomy and environmental factors in some lowland forests in Sarawak. *Trop. Ecol.* **11,** 26-43.

Burbidge N. T. (1960) The phytogeography of the Australian region. *Aust. J. Bot.* **8,** 75-209.

Byrnes N., Everist S. L., Reynolds S. T., Specht A., & Specht R. L. (1977) Vegetation of Lizard Island, North Queensland. *Proc. R. Soc. Qld* **88,** 1-15.

Christophel D. C. (1981) Tertiary megafossil floras of Australia as indicators of floristic associations and palaeoclimate. In: *Ecological Biogeography of Australia* (ed. A. Keast) pp. 377-90. Junk, The Hague.

Churchill D. M. (1973) The ecological significance of tropical mangroves in the Early Tertiary floras of Southern Australia. *Geol. Soc. Aust., Spec. Publ.* **4,** 79-86.

Clifford H. T. & Simon B. K. (1981) The biogeography of Australian grasses. In: *Ecological Biogeography of Australia* (ed. A. Keast) pp. 537-54. Junk, The Hague.

Croasdale H. & Scott A. M. (1976) New or otherwise interesting desmids from Northern Australia. *Nova Hedwigia* **27,** 501-95.

Dettmann M. E. (1981) The Cretaceous flora. In: *Ecological Biogeography of Australia* (ed. A. Keast) pp. 355-75. Junk, The Hague.

Hekel H. (1972) Pollen and spore assemblages from Queensland Tertiary sediment. *Qld Geol. Surv. Publ.* **355.**

Hooker J. D. (1860) On the flora of Australia, being part of an introductory essay to *Flora Tasmaniae,* The botany of the Antarctic voyage of H. M. discovery ships *Erebus* and *Terror,* in the years 1839-1843, part III, 1: XXVII-CXXVIII.

Kershaw A. P. (1981) Quarternary vegetation and environments, In: *Ecological Biogeography of Australia* (ed. A. Keast) pp. 81-101. Junk, The Hague.

Lavarak P. S. & Stanton J. P. (1977) Vegetation of the Jardine River Catchment and adjacent coastal areas. *Proc. R. Soc. Qld* **88,** 39-48.

Martin H. A. (1981) The Tertiary flora. In: *Ecological Biogeography of Australia* (ed. A. Keast) pp. 391-406. Junk, The Hague.

Mueller F. von (1856) North Australian botany, observations on, by Dr Frederick Mueller, botanist to the North-West Australian Government Expedition, under the command of Mr Surveyor Gregory, *Hooker's J. Bot.* **8,** 321-31.

Mueller F. von (1857) Notes made during the recent expedition across the northern portion of Australia, under the command of Mr Surveyor Gregory. *Hooker's J. Bot.* **9,** 165-73.

Mueller F. von (1858) Botanical report on the North Australian expedition, under the command of A. C. Gregory, Esq. *J. Proc. Linn. Soc. (Bot.)* **2,** 137-63.

Palgrave K. C. (1977) *Trees of Southern Africa* (ed. E. J. Moll). C. Struik Publishers, Cape Town.

Pedley L. & Isbell R. F. (1971) Plant communities of Cape York Peninsula. *Proc. R. Soc. Qld* **82,** 51-74.

Perry R. A. & Lazarides M. (1964) Vegetation of the Leichhardt-Gilbert Area. *C.S.I.R.O. Land Res. Ser.* **11,** 152-91.

Powell C. M., Johnson B. D. & Veevers J. J. (1981) The Early Cretaceous break-up of Eastern Gondwanaland, the separation of Australia and India, and their interaction with Southeast Asia. In: *Ecological Biogeography of Australia* (ed. A. Keast) pp. 15-29. Junk, The Hague.

Prescott J. A. & Pendleton R. L. (1952) Laterite and lateric soils. *Comm. Bur. Soil Sci., Tech. Commun.* **47.**

Pryor L. D. & Johnson L. A. S. (1981) *Eucalyptus,* the universal Australian. In: *Ecological Biogeography of Australia* (ed. A. Keast) pp. 499-536. Junk, The Hague.

Rice B. & Westoby M. (1983) Plant species richness at the 0.1 hectare scale in Australian vegetation compared to other continents. *Vegetatio* **52,** 129-40.

Richards P. W. (1936) Ecological observations on the rainforest of Mount Dulit, Sarawak. Part I. *J. Ecol.* **24,** 1-37.

Rogers R. W. & Stevens G. N. (1981) Lichens. In: *Ecological Biogeography of Australia* (ed. A. Keast) pp. 591-603. Junk, The Hague.

Schwarz O. (1928) Analytische Studie uber die Beziehungen der Phanerogamenflora von Arnhemsland (Nordaustralien). *Fedde Rep.* **51,** 59-113.

Scott A. M. & Prescott G. W. (1958) Some freshwater algae from Arnhem Land in the Northern Territory of Australia. In: *Records of the American-Australian Scientific Expedition to Arnhem Land.* Vol. 3. *Botany and Plant Ecology* (eds R. L. Specht & C. P. Mountford) pp. 9-136. Melbourne University Press, Melbourne.

Sleumer H. Q. (1966-67) Ericaceae. *Flora Malesiana Ser. I.* **6,** 469-914.

Specht R. L. (1958a) The climate, geology, soils and plant ecology of the northern portion of Arnhem Land. In: *Records of the American-Australian Scientific Expedition to Arnhem Land.* Vol. 3. *Botany and Plant Ecology* (eds R. L. Specht & C. P. Mountford) pp. 333-414. Melbourne University Press, Melbourne.

Specht R. L. (1958b) The geographical relationship of the flora of Arnhem Land. In: *Records of the American-Australian Scientific Expedition to Arnhem Land.* Vol. 3. *Botany and Plant Ecology* (eds R. L. Specht & C. P. Mountford) pp. 415-78. Melbourne University Press, Melbourne.

Specht R. L. (1972) Water use by perennial evergreen plant communitites in Australia and Papua New Guinea. *Aust. J. Bot.* **20,** 273-99.

Specht R. L. (1981a) Major vegetation formations in Australia. In: *Ecological Biogeography of Australia* (ed. A. Keast) pp. 163-297. Junk, The Hague.

Specht R. L. (1981b) Ecophysiological principles determining the biogeography of major vegetation formations in Australia. In: *Ecological Biogeography of Australia* (ed. A. Keast) pp. 299-333. Junk, The Hague.

Specht R. L. (1981c) Biogeography of halophytic angiosperms (salt-marsh, mangrove and sea-grasses.) In: *Ecological Biogeography of Australia* (ed. A. Keast) pp. 575-89. Junk, The Hague.

Specht R. L. (1981d) Evolution of the Australian flora: some generalizations. In: *Ecological Biogeography of Australia* (ed. A. Keast) pp. 785-805. Junk, The Hague.

Specht R. L. (1983) Foliage projective covers of overstorey and understorey strata of mature vegetation in Australia. *Aust. J. Ecol.* **8,** 433-39.

Specht R. L. (1984) Species-richness of plant communities in the mediterranean climate of south eastern Australia. In: *MEDECOS IV. Proceedings of the 4th International Conference on Mediterranean Ecosystems* (ed. B. Dell) pp. 143-44. Botany Department, University of Western Australia, Nedlands.

Specht R. L. (1986) Functioning of tropical plant communities. 6. Phenology. In: *Tropical Plant Communities. Their Resilience, Functioning and Management in Northern Australia* (eds H. T. Clifford & R. L. Specht) pp. 78-90. Botany Department, University of Queensland, St Lucia.

Specht R. L. (in press) (1) Coastal dune vegetation of Australia — an overview. (2) Ecosystem dynamics in coastal dunes of eastern Australia. In: *Ecosystems of the World.* Vol. 2. *Dry Coastal Ecosystems* (ed. E. van der Maarel) Elsevier, Amsterdam.

Specht R. L. (ed) (in press) Major plant communitites in Australia: An objective assessment. *Aust. J. Bot.*

Specht R. L. & Morgan D. G. (1981) The balance between the foliage projective covers of overstorey and understorey strata in Australian vegetation. *Aust. J. Ecol.* **6**, 193-202.

Specht R. L. & Specht A. (in press a) Canopy structure of *Eucalyptus*-dominated communities in Australia along climatic gradients. *Oecologia Plantarum.*

Specht R. L. & Specht A. (in press b) Species richness of sclerophyll (heathy) plant communities in Australia — The influence of overstorey cover. *Aust. J. Bot.*

Specht R. L. & Specht A. (in press c) Species richness of overstorey strata in Australian plant communities — The influence of overstorey growth rates. *Aust. J. Bot.*

Specht R. L. & Womersley J. S. (1979) Heathlands and related shrublands of Malesia (with particular reference to Borneo and New Guinea). In: *Ecosystems of the World.* Vol. 9A. *Heathlands and Related Shrublands. Descriptive Studies* (ed R. L. Specht) pp. 321-38. Elsevier, Amsterdam.

Specht R. L., Brownell P. F. & Hewitt P. N. (1961) The plant ecology of the Mount Lofty Ranges, South Australia. 2. The distribution of *Eucalyptus elaeophora. Trans. R. Soc. S. Aust.* **85**, 115-76.

Specht R. L., Salt R. B. & Reynolds S. T. (1977) Vegetation in the vicinity of Weipa, North Queensland. *Proc. R. Soc. Qd* **88**, 17-38.

Steenis C. G. G. J. van (1957) Outline of vegetation types in Indonesia and some adjacent regions. *Proc. Eighth Pac. Sci. Congr. (1953)* **4**, 61-97.

Stephens C. G. (1971) Laterite and silcrete in Australia. *Geoderma* **5**, 5-52.

Stevens N. C. (1985) Landscape and geology. *Proc. R. Soc. Qd* **96**, 1-17.

Story R. (1970) Vegetation of the Mitchell-Normanby Area. *C.S.I.R.O. Land Res. Ser.* **26**, 75-88.

Taylor J. A. & Dunlop C. R. (1985) Plant communities of the wet-dry tropics of Australia: the Alligator Rivers region, Northern Territory. *Proc. ecol. Soc. Aust.* **13**, 83-127.

Tracey J. G. (1982) *The vegetation of the Humid Tropical Region of North Queensland.* C.S.I.R.O., Melbourne.

Walker J., Thompson C. H., Fergus I. F. & Tunstall B. R. (1981) Plant succession and soil development in coastal sand dunes of subtropical eastern Australia. In: *Forest Succession: Concepts and Application* (eds D. C. West, H. H. Shugart & D. B. Botkin) pp. 107-31. Springer Verlag, New York.

Webb L. J. (1959) A physiognomic classification of Australian rainforests. *J. Ecol.* **47**, 551-70.

Webb L. J., Tracey J. G. & Williams W. T. (1984) A floristic framework of Australian rainforests. *Aust. J. Ecol.* **9**, 169-98.

Webb L. J., Tracey J. G. & Jessup L. W. (1986) Recent evidence for autochthony of Australian tropical and subtropical rainforest floristic elements. *Telopea* **2**, 575-89.

Wellman P. & McDougall I. (1974) Cainozoic igneous activity in eastern Australia. *Tectonophysics* **23**, 49-65.

Winkler H. (1914) Die Pflanzendecke Sudost-Borneos. *Bot. Jahrb.* **50**, (Festband), 188-208.

Woodroffe C. D., Chappell J. M. A., Thom B. G. & Wallensky E. (1985) Stratigraphy of the South Alligator tidal rivers and plains, Northern Territory. In: *Coasts and Tidal Wetlands of the Australian Monsoon Region.* (eds K. N. Bardsley, J. D. S. Davie & C. D. Woodroffe) pp. 17-30. Australian National University North Australian Research Unit. Mangrove Monograph No. 1, Darwin.

Proc. Ecol. Soc. Aust. 15:31-37

Rainforest environment and physiology

D. J. YATES[1], G. L. UNWIN[2] and D. DOLEY[1]

[1]Botany Department, University of Queensland, St Lucia, Qld. 4067

[2]CSIRO Division of Forest Research, P.O. Box 780, Atherton, Qld. 4883

Abstract

The physiological processes of plants in tropical rainforests may be influenced by conditions quite different from those experienced by humans at ground level. Also, the survival of plants may be determined more by extreme than by median environmental conditions. Within the forest, there are large variations between the upper canopy and forest floor in windspeed, radiation duration and intensity, atmospheric relative humidity. Leaf temperature and water balance all affect the rates of carbon gain and water loss by leaves. Rainforest species have been shown to tolerate extremes of heat and cold, and leaf water potentials more commonly associated with arid environments. This confirms the conclusion that the rainforests of Australia are not identical with the humid equatorial forests, and their physiological attributes require further detailed study.

Introduction

Other papers in this volume refer to the importance of changed climatic conditions in influencing plant distribution throughout geological time. Ashton (unpubl.) has emphasized the influence of animal, particularly herbivore, distribution on plant distribution. He has also referred to the desirability of combining systematics with ecology to increase the understanding of patterns of plant distribution. Many of the changes in plant distribution have been mediated by physiological phenomena at the single plant level and for this reason the combination of physiological considerations with ecological studies is also desirable. Ashton (unpubl.) has highlighted the value in considering the physiological ecology of closely related sympatric species in ecological studies.

Despite several reviews of the ecophysiology of tropical rainforest species (Tomlinson & Zimmermann 1978; Medina 1983; Medina *et al.* 1984), there is very little material published dealing with the physiology of Australia's humid tropical rainforests. For this reason, we intend to look at several facets of the rainforest, its macroclimate and microclimate and their implications for the physiologies of plants.

Plants in tropical rainforests experience an environment determined by the macroclimate just as do plants in other plant communities. There are peculiarities of moist tropical and subtropical forests and their climates which need to be considered when thinking about physiological performance and other aspects of rainforest ecology.

The popular "steaming jungle" image of tropical forests diverts attention from some of the more important questions relating to their environment and effects on plants. It portrays conditions which, being warm and moist, may be humanly uncomfortable, but are generally conducive to rapid and luxuriant plant growth. These conditions do occur to a greater or lesser extent in tropical forests, however, it is conditions which are less conducive to photosynthesis and growth which must be important in determining whether a plant exists in a particular environment.

There are several characteristics of the rainforest environment that deserve consideration here. Some of these are a reflection of human limitations and require mentioning because of some of the less well-informed views of the rainforests, but others are of considerable ecophysiological importance.

There is a large spatial separation between the upper canopy assimilation processes which determine present biomass and the lower canopy where processes occur which determine the structure and composition of the future canopy. There is a similar separation of the principal sites of canopy water supply and loss.

Much casual observation occurs from the lower two metres of the rainforest canopy which is representative of neither the microclimate nor of the flora. These observations must influence one's appreciation of the ecosystem.

Human perception of the passage of time and rates of change etc., may bear no relation to their real importance in the forest.

The abilities of the human eye to respond rapidly to a wide range of lighting conditions literally blinds observers to the absolute variation of radiation levels within rainforests. Human eyes function equally well at 100% and 1% of full sunlight and the sense of colour does not even approximate coarsely any of the spectral responses of plants.

The "steaming jungle" notion brings with it the impression of ample water supply and limited atmospheric demand for water.

Tropical conditions are not always warm or hot with mild nights, especially where increased altitude or distance from the sea become important, or where, for topographic reasons, local cold air drainage occurs.

Seasonality occurs in humid tropical rainforests just as it does at higher latitudes although contrasts between seasons are not as apparent in the wet tropics.

The survival and geographic distribution of plants is usually influenced by extreme conditions or climate variability rather than by the median conditions.

Several aspects of the environment that may be critical for plant functioning will now be considered.

Windspeed

Within a rainforest canopy, particularly toward the lower levels, air movement is markedly reduced to the point that many workers have described the microclimate as being still. In the upper canopy, windspeed will be as high as in any other vegetation type. Olsen & Lamb (1988) and Unwin, Applegate, Stocker & Nicholson (1988) have discussed the effects of cyclonic winds on rainforests. Apart from emphasising that microclimatic conditions within a damaged canopy are very different from those within the undisturbed canopy, such wind is out of the scope of the present considerations.

Because of the uneven nature of the canopy surface, under less strong wind conditions, turbulence at the surface will be important in influencing the rates of mass and energy exchange from the upper canopy at least. Boundary layer conductances for leaves nearer the upper canopy are greater under natural conditions than they are lower in the canopy. It is likely however, that these differences are of relatively minor importance in comparison with differences in radiation regime throughout the depth of the canopy.

A possible and unusual effect of wind in rainforest relates to the disturbance of the upper canopy and its modification of the radiation regime in the understorey. In forests where lighting is inadequate for seedling growth under still conditions, wind may cause sufficient disturbance to provide the increased light necessary for adequate rates of photosynthesis in seedlings. There is some suggestion that this may be the case in some Malaysian forests. (D. Lamb, pers. comm).

Radiation

Spatial variation

The highest radiation loads occur at the top of the canopy and because of the latitudinal range of Australian rainforests, their absolute values are also high under clear sky conditions (Gentilli 1971) as compared with areas closer to the equator (Pinker *et al.* 1980).

At the floor of the closed forest, average radiation levels are as low as 0.5-2% of those above. Consequently, leaves of a particular species may experience average maximum radiation levels differing by a factor of ×100-×200. Many seedlings at the forest floor experience the low levels of radiation for days or weeks at a time and are exposed only very briefly or perhaps seasonally to high levels of radiation. The occurrence of sunflecks would appear critical to the survival of many plants at the forest floor. Gaps in the canopy above provide markedly different radiation environments at ground level from those prevailing beneath the intact canopy.

Bjorkman & Ludlow (1972) examined aspects of the light climate below a rainforest on the Lamington Plateau (south-east Queensland). Some of their data are presented in Fig. 1. On a relatively clear day in March, the upper canopy received 58.9 mole quanta $m^{-2}.day^{-1}$ with a roughly sinusiodal daily distribution (Fig. 1a)

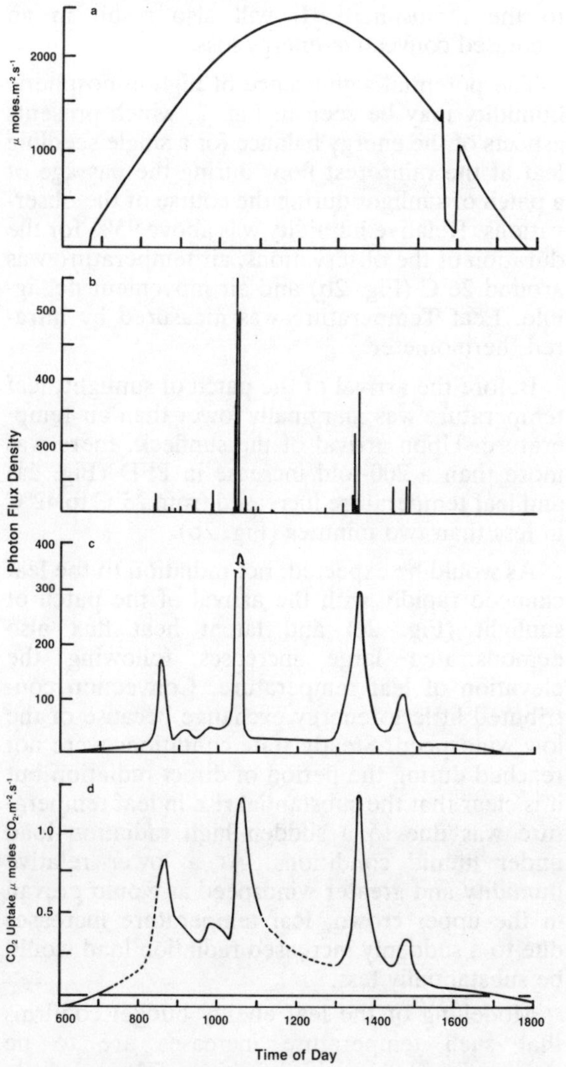

FIG. 1. Photon Flux Density (PFD) for a rainforest community and photosynthesis of *Alocasia macrorrhiza* on the Lamington Plateau (taken from Bjorkman & Ludlow 1972 and Bjorkman *et al.* 1972). (a) Time course of PFD above the forest canopy. (b) Time course of sunflecks to PFD at the forest floor. (c) Contribution of PFD at the forest floor. (d) Time course of carbon dioxide uptake for a leaf of *Alocasia macrorrhiza.*

whilst the floor of the forest received 0.242 mole quanta $m^{-2}.day^{-1}$ (i.e., 0.4% of that at the upper canopy, Fig. 1b). Of the radiation reaching the forest floor, 62% occurred in sunflecks (Fig. 1c) which occurred during a few minutes of the day. Turton (1988) reports similar figures for rainforests growing on the Atherton Tableland, north Queensland and Bazzaz & Pickett (1980) report similar extreme values from several authors.

Bjorkman *et al.* (1972) also studied the carbon dioxide uptake by two species from the forest floor on the Lamington Plateau, *Alocasia macrorrhiza* and *Cordyline rubra.* A striking result was the very low rates of net photosynthesis for these species and the miniscule daily integral of carbon dioxide uptake. The rate of carbon dioxide uptake (Fig. 1d) was closely related to the Photon Flux Denisty (PFD) experienced by the leaf and the importance of the occurrence of sunflecks is clear. Over the period of a day, approximately 40% of the carbon dioxide uptake occurred during brief sunflecks. On a clear day, net photosynthetic efficiency was approximately 12% and on a cloudy day, 16%. These are very high values and similar to those determined under ideal *in vitro* conditions. One of the reasons for the high net efficiency was that light compensation points were extremely low.

Temporal variation

Sunflecks are receiving attention in the literature because so much of the photosynthesis occurring toward the floor of the forest occurs as a result of sunflecks. Sunflecks last from fractions of a second to several minutes. There may be a 100-fold change in the radiation reaching a leaf partway down the canopy over a period of a few seconds to several minutes. Figure 2a indicates the timecourse of a long-lived patch of sunlight experienced by a leaf at the forest floor which consisted of a number of relatively shortlived sunflecks. Its significance to the leaf will be discussed below.

Atmospheric Humidity

Whilst atmospheric moisture content is commonly expressed in terms of relative humidity, from a biological point of view other measures of atmospheric moisture are more relevant. Physiologists generally consider transpiration in terms of the gradient of water vapour concentration from leaves to the atmosphere. This is determined not only by atmospheric conditions but also by leaf temperature because leaf temperature determines the concentration of water vapour in the intercellular air spaces of the leaf. Under conditions of high humidity and high radiation loads, there is the probability of leaf temperatures higher than ambient. The prime reason for this is that transpirational energy loss must be relatively low when leaf temperature is close to air temperature. With increased radiation loads, storage of sensible energy will occur in the leaf and its temperature will increase. Under relatively constant radiation conditions, this will cause a steeper water vapour concentration gradient from the leaf to the atmosphere. It will also result in an increased convective energy loss.

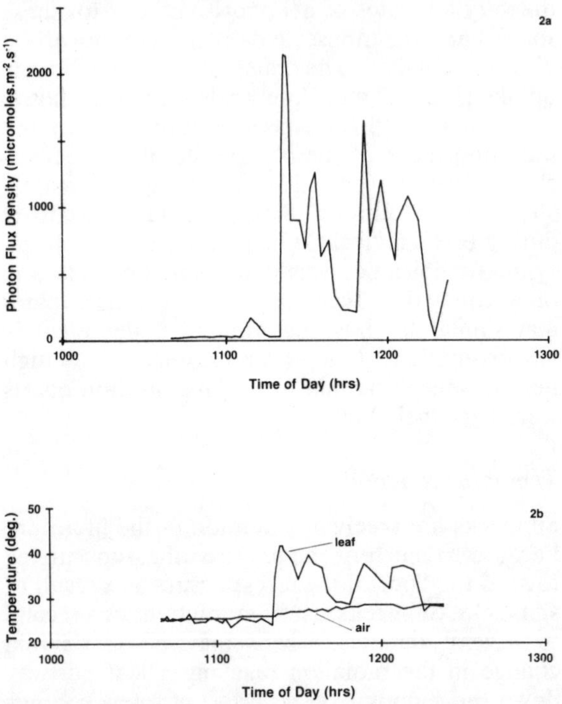

FIG. 2. The influence of the arrival of a sunfleck on components of the energy balance of a seedling leaf at the forest floor (see text for details). (a) Photon flux density incident on the leaf. (b) Leaf temperature and air temperature. (c) Net radiation to the leaf.

The potential significance of high atmospheric humidity may be seen in Fig. 2, which presents aspects of the energy balance for a single seedling leaf at the rainforest floor during the passage of a patch of sunlight during the course of the observations. Relative humidity was above 95% for the duration of the observations, air temperature was around 26°C (Fig. 2b) and air movement negligible. Leaf Temperature was measured by infrared thermometer.

Before the arrival of the patch of sunlight, leaf temperature was marginally lower than air temperature. Upon arrival of the sunfleck, there was more than a 200-fold increase in PFD (Fig. 2a) and leaf temperature increased from 25°C to 42°C in less than two minutes (Fig. 2b).

As would be expected, net radiation to the leaf changed rapidly with the arrival of the patch of sunlight (Fig. 2c) and latent heat flux also demonstrated large increases following the elevation of leaf temperature. Convection contributed little to energy exchange because of the low windspeed. Steady state conditions were not reached during the period of direct radiation but it is clear that the substantial rise in leaf temperature was due to a sudden high radiation load under humid conditions. At a lower relative humidity and greater windspeed as would prevail in the upper crown, leaf temperature increases due to a suddenly increased radiation load would be substantially less.

Modelling of the leaf energy budget confirms that such temperature increases are to be expected under humid conditions. There is little published information dealing with these questions in humid rainforest environments. However, under humid conditions, leaves exposed to sunflecks or patches of full sunlight may well spend up to two hours at temperatures well above those associated with temperature optima for photosynthesis, and experienced during most of the remainder of the day. Considering that the greatest portion of daily carbon gains also occurs during sunflecks, the physiological significance of leaf temperature during sunflecks deserves further examination.

Moisture Availability

Despite the high radiation levels at the upper surface of the rainforest canopy, rates of carbon dioxide uptake may not be high. Photosynthesis under these conditions is likely to be limited by one of several factors. In recent work on the Atherton Tableland with *Argyrodendron peralatum* toward the end of the dry season, we have found that with adequate soil moisture, carbon dioxide exchange throughout the canopy is primarily dependent on PFD. Within four days after heavy rain, but under atmospheric conditions apparently ideal for photosynthesis, maximum photosynthesis rates of upper canopy leaves fell rapidly to zero and remained so until subsequent storms. This limitation of photosynthesis was due to stomatal closure resulting from reduced water availability in the upper crown. Lower in the canopy, some limited carbon gains were made under these conditions. It is apparent that at least toward the end of the dry season, branches of the lower crown make an important contribution to the total carbon balance of *A. peralatum,* despite the relatively low levels of radiation they receive.

As suggested earlier, the folklore surrounding rainforests would lead one to believe that there is always, or usually, an ample water supply and limited atmospheric demand for water within rainforests. This further implies, and has been supposed in the past, that rainforest trees do not limit their loss of water. Over several years (1975-1982) at Brookfield on the outskirts of Brisbane, observations have been made on a range of trees and herbs in a low (15 m) notophyll vine forest. The rainfall distribution is distinctly seasonal, the site slopes to the south at about 20° and the soils are about 1.5 m deep and stony. Over the period of observations, there was not a close relationship between minimum leaf water potential of trees and herbs and precipitation for periods of one to three months before the observations. However, minimum leaf water potential in most species was approximately related to the total rainfall in the period May-August (Fig. 3a), that is the period from about the end of the normal wet season to the middle of the dry season. In *Araucaria cunninghamii,* minimum shoot water potential was insensitive to environmental conditions, and in this regard it was different from all other species studied.

The daily range of shoot water potential in *Araucaria cunninghamii* was smaller than that in almost all other species examined (Fig. 3b) and particularly so under conditions of environmental water deficit. This indicates very effective control of water loss under all conditions. In *Flindersia*

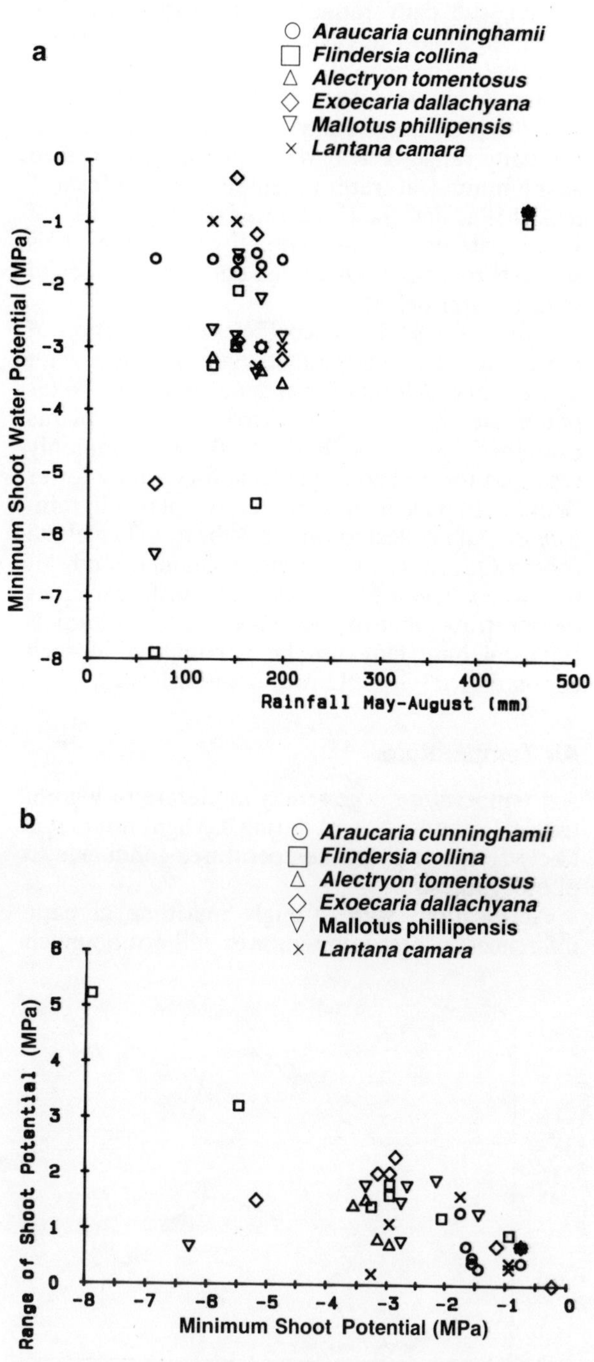

FIG. 3. Aspects of the water relations of six species growing in a notophyll vine forest at Brookfield (south-east Queensland) during the years 1975-1982. (a) Minimum daily shoot water potential (MPa) in relation to total rainfall for the period May-August. (b) Daily range of shoot water potential (MPa) as a function of minimum shoot water potential (MPa).

collina, the daily range of leaf water potential increased almost linearly as minimum leaf water potential decreased indicating nocturnal recharge, but relatively ineffective control over water loss during the day. In most other species, the daily range of leaf water potential increased as minimum leaf water potential declined from -1 to -3 MPa, and then decreased as minimum leaf water potential declined to -5 or -6 MPa. This suggests incomplete recharge under conditions of severe water deficit.

From 1975-1982, the coefficient of variation of rainfall for the period July-August was 61% with figures varying from 68 mm to 451 mm. The water potentials reached by many of the plants examined were as low as those commonly reported for arid zone plants and certainly lower than might have been expected for plants in rainforest. Data collected on the Atherton Tableland (north Queensland) in a more detailed study of the water relations of selected rainforest trees demonstrates that the low shoot water potentials reported here may not be uncommon in such communities (G. L. Unwin unpublished).

Air Temperature

Air temperature is generally moderate to high in tropical rainforests and during daylight hours it is likely to result in leaf temperatures conducive to photosynthesis.

At locations with strongly maritime climatic influences, winter temperatures will not be much lowered, but at higher altitudes or greater distances from the sea, where climate is modified by continental influences, winter air temperatures may fall to values well below those optimal for photosynthesis. Under such conditions, the importance of higher radiation loads is likely to be seen through their effect on leaf energy balance and temperatures that are elevated to a range favourable for rapid photosynthesis.

Meteorological data have been collected since July 1984 on a tower within rainforest near Yungaburra on the Atherton Tableland. Over a period of two years, maximum air temperature at a height of 40 m (corresponding approximately with the height of the bulk of the crown) was 37.5°C in February 1985 at 1300 hrs, and the minimum of less than 2°C was recorded at 700 hrs in July 1984 during a cold winter (Fig. 4). At a height of 10 m from the ground, maximum air temperature was 34°C in December 1984 and the minimum (in July 1984) less than 1°C. Leaves on many of the canopy trees are retained for more than one year and would thus experience annual temperature ranges of up to 35°C (upper canopy) and 33°C (lower canopy). Whilst the minimum temperatures reported here do not occur every year, their frequency of perhaps one in ten years, and their recurrence over a week or more at a time, means that plant tissues must be tolerant of cold, if not of freezing temperatures as well as tolerant of relatively high temperatures.

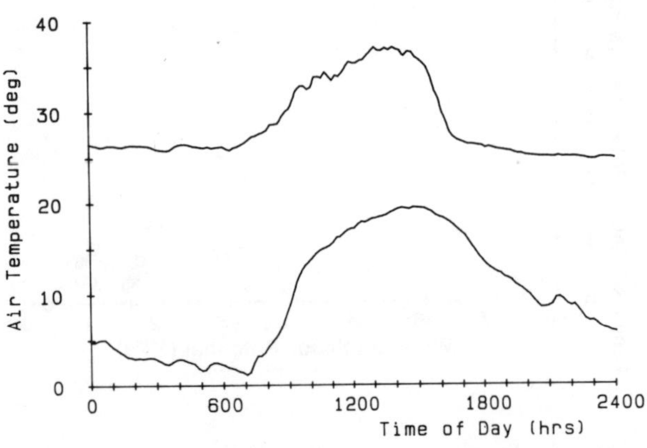

FIG. 4. The time course of air temperature measured 40 m above the ground in rainforest on the Atherton Tableland, north Queensland. The two curves represent the maximum (February 1985) and the minimum (July 1984) air temperatures reached over a period of one year.

Conclusions

We have demonstrated that at some times and places, environmental conditions within rainforests are not as conducive to growth and development as they are often imagined to be. Ashton (unpubl.) has emphasized the need for combining not only taxonomic but also physiological studies with ecological studies in examining the floristic history of a region. Barlow (1988) has placed emphasis on atmospheric humidity and temperature in determining the distribution of rainforest flora and has referred to the importance of remnants and refugia of ancient communities in determining present floristic compositions. Under conditions of environmental change, the abilities of plant physiological processes to cope with adverse conditions will determine which species survive the changes. Knowledge of these responses not only helps us understand the natural occurrence of plants, it should also help us to appreciate the impact of human activities on tropical forests and, if we are serious in attempting to do so, in aiding effort to conserve and manage this dwindling resource.

References

Barlow B. A. & Hyland B. P. M. (1988) The origins of the flora of Australia's wet tropics. *Proc. ecol. Soc. Aust.* **15**, 1-17.

Bazzaz F. A. & Pickett S. T. A. (1980) Physiological ecology of tropical succession: a comparative review. *Ann. Rev. Ecol. Syst.* **11**, 287-310.

Bjorkman O. & Ludlow M. M. (1972) Characterization of the light climate on the floor of a Queensland rainforest. *Carnegie Institution of Washington Year Book* **71**, 85-94.

Bjorkman O., Ludlow M. M. & Morrow P. A. (1972) Photosynthetic performance of two rainforest species in their native habitat and analysis of their gas exchange. *Carnegie Institution of Washington Year Book* **71**, 94-102.

Gentilli J. (ed.) (1971) *Climates of Australia and New Zealand: World survey of Climatology.* Vol 13. Elsevier Publishing Co., Amsterdam.

Medina E. (1983) Adaptions of tropical trees to moisture stress. In: *Ecosystems of the World Vol 14a: Tropical Forest ecosystems: Structure and Function.* (ed. F. B. Golley) pp. 225-37. Elsevier Publishing Co., Amsterdam.

Medina E., Mooney H. A. & Vasquez Yanez C. (eds.) (1984) *Physiological Ecology of Plants of the Wet Tropics.* Dr W. Junk. The Hague.

Olsen M. & Lamb D. (1988) Recovery of subtropical rainforest following storm damage. *Proc. ecol. Soc. Aust.* **15**, 295-99.

Pinker R. T., Thompson O. E. & Eck T. F. (1980). The energy balance of a tropical evergreen forest. *J. appl. Meteorol.* **19**, 1341-49.

Tomlinson P. B. & Zimmermann M. H. (eds.) (1978) *Tropical Trees as Living Systems.* Cambridge Univ. Press, Cambridge.

Turton S. M. (1988) Solar radiation regimes in a north Queensland rainforest. *Proc. ecol. Soc. Aust.* **15**, 101-05.

Unwin G. L., Applegate G., Stocker G. C. & Nicholson D. I. (1988) Initial effects of tropical cyclone "Winifred" on forests of north Queensland. *Proc. ecol. Soc. Aust.* **156**, 281-94.

Proc. Ecol. Soc. Aust. 15:39-47

Tree species diversity in rainforests – establishment and maintenance

G. C. STOCKER

CSIRO Division of Forest Research, Atherton, Qld. 4883

Abstract

Favourable climatic and pyric conditions have allowed development of closed forests in which a characteristic disturbance regime is created by the more or less random deaths of individual trees. Latitude and gap width in relation to canopy height are important variables controlling the light regime on the forest floor beneath such canopy gaps. To exploit variability in this light environment, many rainforest tree species have evolved with attributes which favour their occurrence within a particular segment of the light availability continuum. When canopy gaps of the size produced by the death of a single tree, or small group of trees, are considered, the high solar elevations above tropical rainforests create a wider continuum than occurs in temperate rainforests.

Evolution of a range of tree species, each with a characteristic shade tolerance, appears to have had a number of consequences tending to enhance morphological and physiological diversity. Features commonly related to shade tolerance include crown architecture, seed size, germination mode, growth rate and interval between seed crops. Although variability in soil characteristics may explain some local differences in tree species diversity, it does not seem to account for major latitudinal trends.

As well as influencing light regimes on the forest floor, latitude also has an important effect on the degree to which seasonal factors (mainly unfavourable temperatures and soil water regimes) limit plant establishment and growth. At any particular site the spatial and temporal patterns of disturbance, plus weather conditions during seedling establishment, add to the number of habitat niches potentially available. At low latitudes, regional diversity may be further enhanced via altitudinally dependent thermal regimes.

Introduction

Tropical rainforest diversity[1] is a major feature which is immediately obvious to the visitor from temperate regions and has attracted long-standing interest. An extensive literature has accumulated. Most early explanations were based on the presumed great age of the tropical rainforest and the freedom for species to speciate in the constantly favourable environment (e.g., Richards 1952). Recent studies, such as those reviewed by Leigh (1982) and Whitmore (1983a), have not arrived at any real concensus and further progress appears dependent on a better understanding of the mechanisms controlling regeneration of rainforest species.

The apparent importance of canopy gaps in rainforest regenerative processes has often been examined (Whitmore 1974, 1975, 1978, 1982; Hartshorn 1978, 1980; Hallé *et al.* 1978; Denslow 1980; Rollet 1983). Canopy gaps have been observed to increase light levels on the forest floor, thereby improving the growth and survival

[1]This paper follows past authors concerned with this topic (e.g. Connell 1978; Leigh 1982) by equating "diversity" with "species richness".

of regenerating tree species. Although it has long been realized that tree species may be classified according to their shade tolerance (Spurr 1952), the possible contribution to floristic diversity of regeneration partitioning the light environment on gap floors, does not appear to have been examined in detail. This and several other ways in which the habitat niche[1] can be divided to promote diversity among rainforest tree species, are discussed below.

For the purpose of the following discussion the diversity scales adopted are:

(i) "gap diversity" — that within an area of 10-1 000 m^2 (i.e., the normal area available beneath a canopy gap for the establishment and growth of tree seedlings);

(ii) "site diversity" — that within an area containing limited edaphic and topographic variability (e.g., the 0.5 ha plots which provide the data for much of the discussion below); and,

(iii) "regional diversity" — large relatively discrete rainforest blocks containing a wide range of environments (e.g., the sector of the east coast of Australia between Cooktown and Townsville).

Forest Floor Light Environments

The amount of PAR (photosynthetically active radiation) available to tree seedlings on a rainforest floor can vary by several orders of magnitude. For example beneath the multilayered overlapping crowns of primary species in upland rainforest near Atherton, PAR was sometimes found to be less than 1% of that in the centre of large tree-fall gaps (Table 1). Although many shade tolerant plants can persist, sometimes for many years, with this level of illumination, they grow very slowly, if at all (e.g. *Darlingia darlingiana* — Table 1). In the rainforests of this region only a few highly specialized plants (e.g., *Harpullia frutescens*) appear able to complete their life cycles in heavily shaded situations.

The relative importance of various possible disturbance regimes is indicated by mortality patterns from a series of 19 long-term plots (West *et al.* 1988) provide some information on location, establishment and maintenance of these plots) established in unlogged rainforest by Stocker, Unwin and associates of CSIRO's Division of Forest Research, at Atherton. These patterns (Fig. 1) suggest that canopy gaps are primarily formed by the occasional death of individual upper canopy and emergent trees. In most of this region's rainforests, trees would usually have to be larger than 30-40 cm dbh (diameter at breast height) before their crowns would reach upper canopy levels. Overall mortality records for the plots (Table 2) reveal an annual death rate of about 1% of the trees in any size class (or 7.0 stems ha^{-1} $year^{-1}$).

Patterns of recruitment to the 10-20 cm dbh class (Fig. 1) and the overall level of recruitment (7.6 stems ha^{-1} $year^{-1}$) support the notion of more or less continuous opportunities for regeneration. Observations that, in most plots, all stems recruited were of species with intermediate to high shade tolerance, also indicate the importance of small gaps in rainforest dynamics. However, a brief study of the demography of some of the commoner tree species in the plots (West *et al.* 1988) indicated that in a few plots periodic catastrophes appear to have caused pulses of mortality and consequent recruitment in which shade intolerant species have been conspicuous. Studies of the regeneration floristics in a very large gap (1 ha) created by an imposed catastrophe (felling and burning) also showed that while shade intolerant species were the most obvious (they were the fastest growing), there were many species of intermediate to high shade tolerance present (Stocker 1981).

TABLE 1. Mean heights (m) and relative light availability[a] ($\mu E\ m^{-2} \times 10^{-6}$) at age four years for survivors in a field establishment and growth trial.

	Canopy treatment		
	Intact	Small gap	Large gap
Relative light availability	68(88)	896(392)	4 352(771)
Height			
Darlingia darlingiana	0.2(0.1)	0.8(0.5)	0.4(0.4)
Toona australis	f	1.1(0.5)	3.1(2.0)
Acacia aulacocarpa	f	f	6.3(0.7)
Alphitonia petriei	f	f	6.5(0)

(standard deviation). [a] = estimated from integrated measurements of photosynthetically active radiation at each treatment site during the first year. f = not establishing in any replicate (total four) of that treatment.

[1]Niche definitions follow Grubb (1977).

Table 2. Mean annual mortality by size classes for trees 10 cm dbh in sample plots established in north Queensland's rainforests (deaths ha^{-1} annum^{-1}).

	Size class (dbh cm)									
	10-20	20-30	30-40	40-50	50-60	60-70	70-80	80-90	90-100	100+
Absolute	4.21	1.41	0.55	0.43	0.22	0.09	0.05	0.06	0.03	0.03
% size class total	0.85	0.86	0.75	0.95	1.1	0.83	0.71	1.8	1.1	0.83

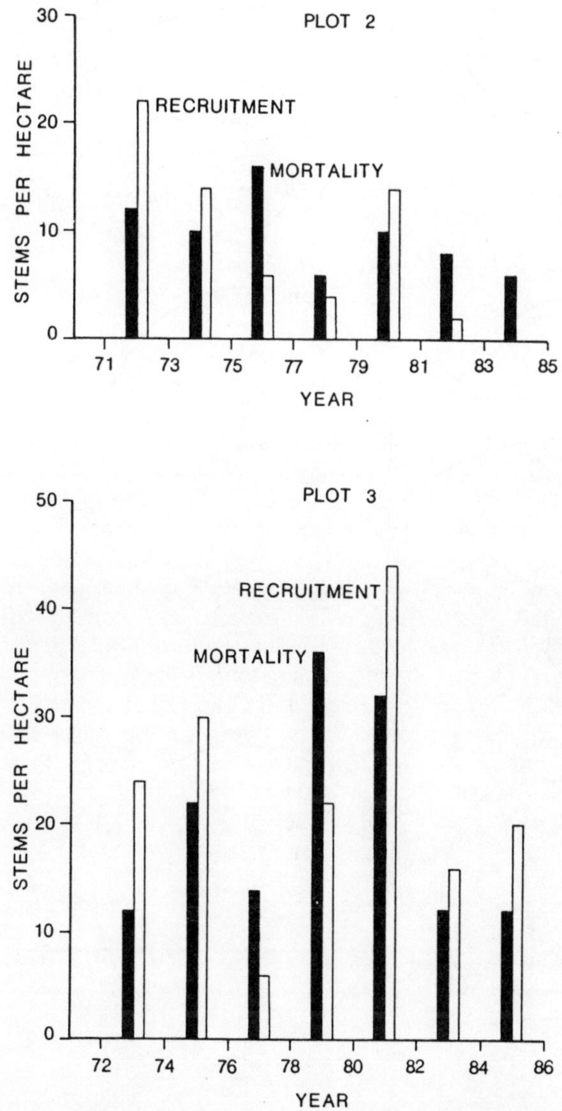

FIG. 1. Examples of patterns of mortality and recruitment for stems > 10 cm dbh within unlogged rainforest plots.

When models of light environments on the forest floor beneath canopy gaps are examined, they show predictable spatial and temporal variability which might be exploited by a number of species each favoured by different segments of the PAR continuum (Stocker and Taylor unpublished). In these studies gaps were assumed to be cylindrical holes through a flat opaque canopy to the forest floor (following Hartshorn 1980). Gap diameter in relation to canopy height and latitude were found to be major factors determining characteristics of the light environment (Figs. 2-4). From the viewpoint of the potential for seedlings to partition the forest floor light resource, an important feature is that at low latitudes, canopy gaps produced by the death of a single tree would usually provide a range of light environments on the forest floor beneath. At higher latitudes, gaps of similar size produce an increasingly restricted range of light environments.

Simultaneous sowings of a range of species at spots beneath undisturbed rainforest canopies and in the centres of freshly created gaps of various size, indicated that rainforest tree seedlings have a range of species-specific survival and growth patterns which appear dependent on gap size (Table 1). There appears to be a continuum of response from the very slow growing shade tolerant understorey specialists which grow slowly in all regimes, to the fast growing shade intolerant weeds which can only survive in large, well illuminated gaps. Glasshouse trials, in which seedlings were grown under a light gradient, indicate that PAR availability can largely explain the growth and survival patterns observed in the field experiments (Duff and Stocker, unpublished).

Temperature and Soil Moisture

Given the short periods of viability (usually only a few weeks) characteristic of the seed of most rainforest tree species (Whitmore 1983b) and their production of seed crops at intervals of three of more years (Baur 1964), soil temperature and moisture conditions which limit the period when successful seedling esablishment can take place, will restrict the potential maximum diversity which can be established on that site. It is also apparent that just as gap characteristics influence the light regime on forest floors beneath gaps, associated solar radiation would create

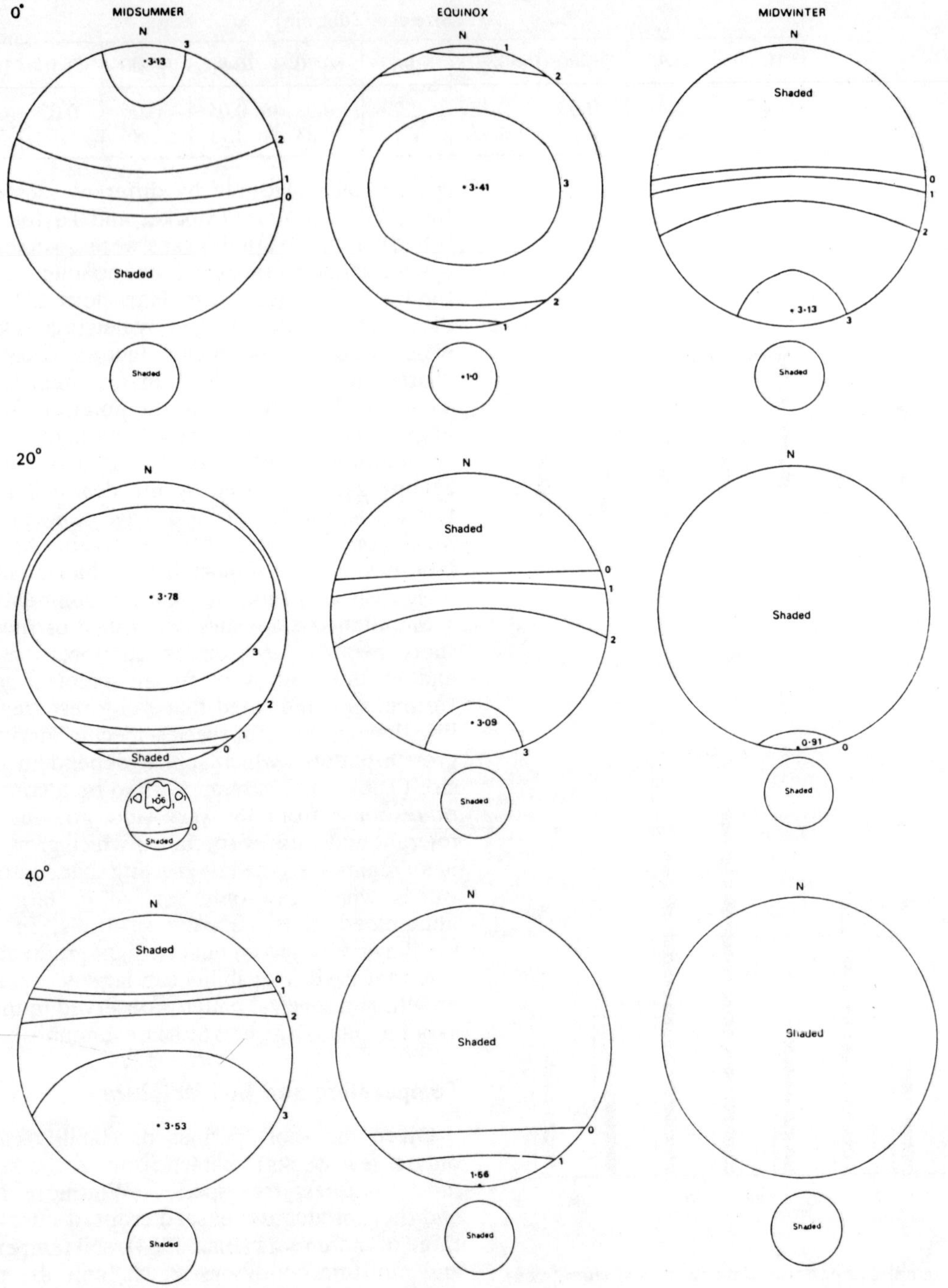

FIG. 2. The spatial sunlight patterns on the floors of gaps at three different south latitudes, for two gaps sizes and three contrasting days of the year are shown. The larger circles represent gaps of height/diameter = 1; the smaller, gaps with height/diameter = 4. Isopleths join points with 0, 1, 2 and 3 hours of direct sunlight on the days indicated (assuming cloud free conditions). Dot points indicate maxima (from G. Stocker & W. Taylor unpublished).

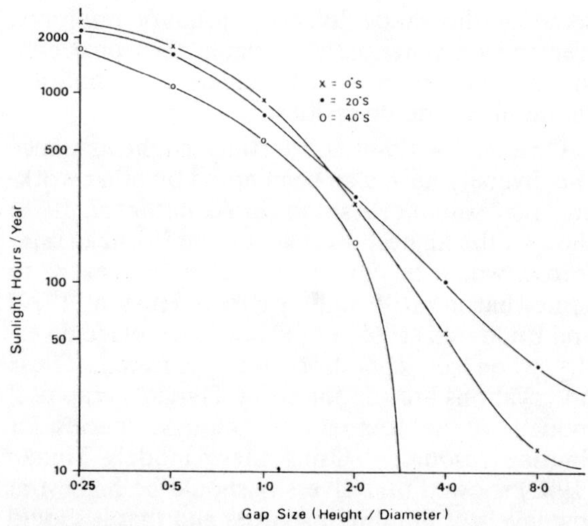

FIG. 3. The maximum potential number of annual sunlight hours received on the floors of idealized gaps of various sizes, are plotted for latitudes 0, 20 and 40 degrees south (from G. Stocker & W. Taylor unpublished).

predictable patterns of soil temperature with some consequent soil drying. Although some partitioning of soil temperature and moisture patterns in gaps and weather conditions during establishment might be possible, the interregional effects on diversity of seasonal variations in temperature and soil moisture appear to be of greater importance.

An example of general trends of seasonal conditions which would influence seedling establishment, may be obtained by examining meterological records (Anon. 1958; McAlpine *et al.* 1975) for localities representative of regional climates in Papua New Guinea and eastern Australia. These records indicate that the first limiting influence is "winter" drought which intrudes rapidly at about 8°S. Although the rainfall is again aseasonal at around 30-35°S, summer drought is apparent at still higher latitudes. The latitudinal influence causing winter temperatures to be unfavourable for plant establishment or growth, follows a more consistent pattern, gradually increasing its effects with increasing latitude.

Nix (1982) pointed out that latitude also has an influence on the available range of altitudinally determined thermal regimes for optimal plant establishment and growth. The effect of these regimes on diversity is essentially regional for, with increasing latitude (and decreasing mean temperatures at sea level), the number of potential regimes decreases. However, it should be noted that in many parts of the tropics, the altitudinal range is insufficient to provide the cooler thermal regimes. Because of opportunities for chance establishment of species outside their normal thermal range, some additional enhancement of gap and site diversity may be expected where a range of thermal response regimes is available in a restricted geographical area.

Soil Physical and Chemical Properties

Even within the area of a gap, disturbance due to tree-falls can create a variety of soil surface characteristics which may be partitioned by seedlings of different species. For example bare soil may be exposed around the base of toppled trees whereas, in another sector of the gap, crown debris may have smashed and smothered advanced seedlings and hindered seedling regeneration, especially for small seeded species.

Partitioning of the soil resource may also occur at regional scales due to variability created by interaction among parent material, geomorphology and climate. An examination of the plot data for this region showed that although there were some general relationships between floristics and soil fertility, it was not possible to isolate the species groups involved (West *et al.* 1988). P. W. West (pers. comm.) also examined the relationships between

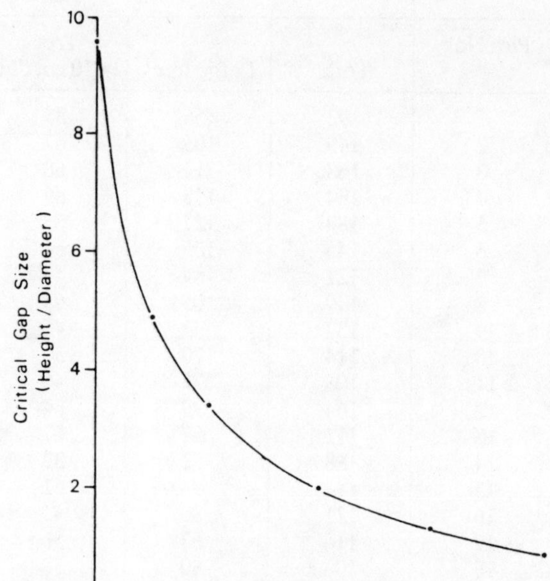

FIG. 4. Plotted for various southern latitudes are the minimum sizes of idealized gaps which would allow direct sunlight to strike the gap floor.

the species richness of plots and environmental factors and showed that site diversity was inversely related to the level of some exchangeable bases (calcium and magnesium). Table 3 illustrates the variability in diversity among these plots.

Discussion

Gap diversity appears to reflect partitioning of gap resources, primarily light, among the available species; each able to achieve optimal growth and survival in different sectors of a gap or in gaps of different size. Variability in the light resource is due to the high density of mature rainforest canopies and the geometric characteristics of sunlight penetration of canopy gaps. Rainforests gaps are usually created by more or less continuous disturbances of characteristic spatial scale. Some additional partitioning on soil surface characteristics which influence seedling establishment and growth, seems possible.

Although site diversity will largely be controlled by factors contributing to gap diversity, species richness at site scales could be greater because of the infrequent availability of the propagules of some of the shade tolerant, primary rainforest tree species potentially available. Seasonal conditions which reduce establishment opportunities, should limit site diversities.

Observations that soil fertility might influence site diversity have also been noted by other workers. For example, Ashton (in Ashton *et al.* 1978) thought the highest diversity on the Bornean rainforests would be found on sites with average to somewhat infertile soils, whereas Huston (1979) and Budowski (1965) noted low assemblage diversity when soil conditions were extreme. These observations are supported by Tilman's graphical models of the competition amongst species for limiting resources. Using these models Tilman (1982) showed that diversity should be highest at low levels of limiting resources and that it should decrease at both high and very low levels.

In plant assemblages where the gap regeneration processes outlined above were prominent, life forms having the ability to grow from the forest floor to the upper canopy, would appear best able to exploit the range of regeneration niches available. Indeed species which grow into large trees

TABLE 3. Numbers of vascular plant species listed according to life form classes on 0.5 ha rainforest plots established in certain rainforests of north Queensland.

Plot No.	Numbers of vascular plant species						
	Total	Large tree[1]	Tree > 10 cm dbh	Small tree[2] and shrub	Liane[3]	Epiphyte[3]	Other[3]
1	92	56	35	10	14	3	9
2	149	105	67	12	14	11	7
3	163	112	68	14	24	5	8
4	184	135	89	14	19	9	7
5	184	127	83	13	20	24	0
6	145	105	64	9	19	5	7
7	121	84	49	16	10	5	6
8	149	105	65	13	17	9	5
9	137	93	46	20	15	2	7
10	114	70	50	8	15	12	9
11	108	75	46	12	15	1	5
12	164	99	64	25	20	5	15
13	111	87	67	15	4	0	5
14	88	52	32	5	12	8	11
15	144	94	61	12	23	4	11
16	121	80	47	18	14	4	5
17	116	83	56	14	11	7	1
18	108	80	58	6	13	2	7
19	153	112	59	12	15	5	11

1. Includes all tree species capable of reaching the upper canopy.
2. Includes tree species which are rarely, if ever, found in the upper canopy. These species seldom grow > 10 cm dbh.
3. Species counts in these categories are based on a preliminary examination. Although many may prove to be underestimates, closer examination is not expected to significantly alter the trends recorded above.

or robust lianes were more numerous than those in other life form groups recorded for the study sites referred to above (Table 3). It might also be expected that low tree species site diversity would be a characteristic of plant assemblages in which most regeneration occurs following an infrequent catastrophe such as a severe fire or where canopy structure allows a relatively uniform light environment on the forest floor. Such comparisons can be made in north-east Australia where open eucalypt forests of low diversity often border species rich rainforests. For example, a 0.5 ha rainforest plot near Atherton contained 61 tree species greater than 10 cm dbh. On a very similar site about 200 m to the east, a 2.25 ha open eucalypt forest plot contained only five tree species greater than 10 cm dbh (Unwin 1983). It might be supposed that such differences could be the result of large differences on the period available for these assemblages to have developed. However, this seems unlikely and the little evidence available (Kershaw 1983) suggests that open eucalypt forest species have been in this immediate area longer than those of the rainforest.

Regional scales allow partitioning on many other environmental axes (e.g., altitude, slope, aspect, rainfall and soils). While West *et al.* (1988) indicated that niche axes relating to rainfall and soil fertility may be the most important of those controlling tree species distribution in the rainforests of north-east Australia, clear patterns were not obvious in the data set derived from the study sites.

The potential for latitude to influence diversity at all scales through the controls it exerts on the characteristics of gap floor light environments and on climatic factors which seasonally limit seedling establishment, is noteworthy. This phenomenon might explain the general decrease in regional rainforest diversity from tropical to temperate eastern Australia (e.g., Stocker & Irvine 1983). Although comparable data are not readily available, general observations indicate that this trend is absent from, or of much lesser magnitude for, assemblages where gap regenerative processes are of little, if any, importance (i.e., the open eucalypt forests and woodlands).

An examination of possible evolutionary trends associated with gap regeneration processes should be considered. During the establishment of the rich floristic diversity characteristic of tropical rainforests, it is envisaged that a range of species evolved, each with a characteristic shade tolerance. Their degree of shade tolerance appears controlled by the extent of various physiological and morphological adaptations allowing seedlings of shade tolerant species to survive, if not grow, in very low light environments. The physiological attributes primarily seem to minimize respiration rates. Although they may increase the possible lifespan of an individual, their influence seems to be at the expense of an ability to grow rapidly when light is not limiting and may be the reason for the very slow growth rates observed among shade tolerant tree species even when these become emergents (West *et al.* 1988).

Physiological attributes associated with shade tolerance may also cause internal resource partitioning, culminating in a relatively long interval between successive seed crops, characteristic of this group of tree species. Since fruiting tends to be gregarious, seed predation, if not actually contributing to the evolution of this feature, certainly reinforces its effects on forest dynamics by increasing the interval between effective seed crops. The generally favourable establishment environment for shade tolerant species and predation pressure, especially on large seeds characteristic of this group, have probably led to evolution of immediate, rather than delayed germination.

Morphological characteristics associated with shade tolerance include seed size and crown architecture. The reserves contained in large seeds seem to enable seedlings to persist for long periods (perhaps as long as ten years for some species) in a low light environment and thus increase the possibility that light conditions will become more favourable before they die. They might also allow expansion of a large initial leaf area to collect what little light is available beneath an undisturbed canopy. Shade tolerant and shade intolerant species have contrasting crown architectures. The multilayered crowns of the former maximize the amount of light intercepted. This reduces illumination of the forest floor beneath intact primary forest canopies to very low levels, such as those recorded in the seedling establishment and growth trial noted above.

While evolution may act periodically to reduce diversity by producing a species which is faster growing than others of equivalent or lesser shade tolerance, Tilman's models (Tilman 1982) indicate that the effect would be short-lived because such "super-species" would soon begin to specialize and then speciate, redressing the balance. Perhaps this is the mechanism responsible for speciation among the dipterocarps in South-east Asian forests and even the dominance change in tropical rainforest floras from gymnosperms during the Cretaceous to the angiosperms of more recent times.

Conclusions

Previous interest in gap resource partitioning to promote rainforest diversity has centred on those edaphic factors which are thought to enable specific rainforest species to occupy that particular portion of the available habitat providing it with the best competitive edge (e.g., Ashton 1969). Ricklefs (1977) suggested that tropical trees may have divided this niche-axis finely. Although Connell (1978) saw variation along niche-axis (such as food, habitat space and time of activity) as providing many individual niches for animal species to occupy, he saw plants as being less specialized because they all have similar basic resource requirements (such as light, water, carbon dioxide and mineral nutrients). These outlooks do not appear to have appreciated the scale of the light-resource niche-axis existing in assemblages with the structural and disturbance characteristics of rainforests. They also fail to appreciate the number of ways each regeneration site can be divided among species having similar "food" niche requirements. For example, species might show divergence in other niche characteristics particularly those of the "phenological niche" which allows propagule availability to play a considerable role in forest dynamics, and the "life-form niche" which relates to productivity and growth form characteristics.

The basis of a better understanding appears to have evolved steadily over the past ten years. Important landmarks were:-

(a) the conclusion by Grubb (1977) that diversity had more to do with the requirements for regeneration than with the tolerances of adults; and,

(b) recognition of the importance of environmental heterogeneity and, in particular, that of canopy gap environments by Whitmore 1974, 1975, 1978; Ricklefs 1977; Hartshorn 1978, 1980; Denslow 1980 and others.

Discussions here and in Whitehead (1982) indicate that diversity may not be generally related to assemblage features such as stability, maturity, productivity, evolutionary time and predation pressure (Hill 1973).

Many of the processes which are postulated above to influence rainforest diversity, have still to be thoroughly examined. Further studies of canopy disturbance patterns, the reproductive biology of rainforest trees and the physiological parameters influencing tree survival and growth should be particularly useful. Mathematical models of interactions among processes may offer some insights into the relative importance of these processes and the role of chance in the establishment and maintenance of rainforest diversity.

Acknowledgements

The author is grateful for comments on an earlier version of this manuscript by Greg Unwin, Dr Paul Kriedemann, Dr Mike Hopkins and Dr Mike Austin. Keith Sanderson prepared the illustrations.

References

Anonymous (1956) *Climatic Averages Australia.* Commonwealth Bureau of Meteorology, Melbourne.

Ashton P. S. (1969) Speciation among tropical forest trees: some deductions in the light of recent evidence. *Biol. J. Linn. Soc.* **1,** 155-96.

Ashton P. S., Hopkins M. J., Webb L. J., Williams W. T. & Palmer J. (1978) Description, functioning and evolution of tropical forest ecosystems. The natural forest: plant biology, regeneration and tree growth. In: *Natural Resources Research XIV. Tropical Forest Ecosystems. A state of knowledge report.* UNESCO, Paris.

Baur G. N. (1964) *The Ecological Basis of Rainforest Management.* Forestry Commission of N.S.W., Sydney.

Budowski G. (1965) Distribution of tropical American rainforest species in the light of successional processes. *Turrialba* **15,** 40-42.

Connell J. H. (1978) Diversity on tropical rainforests and coral reefs. *Science* **199,** 1302-10.

Denslow J. S. (1980) Gap partitioning among tropical rainforest trees. *Suppl. Biotropica* **12,** 47-55.

Grubb P. J. (1977) The maintenance of species-richness in plant communities: The importance of the regeneration niche. *Biol. Rev.* **52,** 107-45.

Hallé F., Olderman R. A. A. & Tomlinson P. B. (1978) *Tropical Trees and Forests — an Architectural Analysis.* Springer-Verlag, Berlin.

Hartshorn G. S. (1978) Tree falls and tropical forest dynamics. In: *Tropical Trees as Living Systems* (eds P. B. Tomlinson & M. H. Zimmermann). Cambridge Univeristy Press, Cambridge.

Hartshorn G. S. (1980) Neotropical forest dynamics. *Suppl. Biotropica* **12,** 23-30.

Hill M. O. (1973) Diversity of evenness: A unifying notion and its consequences. *Ecology* **54,** 427-32.

Huston M. (1979) A general hypothesis of species diversity. *Am. Nat.* **113,** 81-101.

Kershaw A. P. (1983) A Holocene pollen diagram from Lynch's Crater, north-eastern Queensland, Australia. *New Phytol.* **94,** 669-82.

Leigh E. G. (1982) Why are there so many kinds of tropical trees? In: *The Ecology of a Tropical Forest, Seasonal Rhythms and Long-term Changes* (eds E. G. Leigh, A. S. Rand & A. S. Windsor). Smithsonian Institute Press, Washington DC.

McAlpine J. R., Keig G. & Short K. (1975) Climatic tables for Papua New Guinea. *CSIRO Australia, Div. Land Use Res. Tech. Pap.* **37.**

Nix H. A. (1982) Environmental determinants of biogeography and evolution in Terra Australis. In: *Evolution of the Fauna and Flora of Arid Australia* (eds W. R. Barker & P. J. M. Greenslade). Peacock Publications, South Australia.

Rollet B. (1983) La régénération naturelle dans les trouées. Un processus général de la dynamique des forêts tropicales humides. *Revue Bios et Forêts des Tropiques* **201,** 3-34.

Richards P. W. (1952) The *Tropical Rain Forest.* University Press, Cambridge.

Ricklefs R. E. (1977) Environmental heterogeneity and plant species diversity. A hypothesis. *Am. Nat.* **111,** 376-81.

Spurr S. H. (1952) Origins of the concept of forest succession. *Ecology* **33,** 426-27.

Stocker G. C. (1981) The regeneration of a north Queensland rainforest following felling and burning. *Biotropica* **13,** 86-92.

Stocker G. C. & Irvine A. K. (1983) Seed dispersal by Cassowaries *(Casuarius casuarius)* in north Queensland's rainforests. *Biotropica* **15,** 170-76.

Tilman D. (1982) *Resource Competition and Community Structure.* Princeton University Press, Princeton.

Unwin G. L. (1983) *Dynamics of the rainforest/eucalypt forest boundary in the Herberton Highland, north Queensland.* M.Sc. Thesis, James Cook University, Townsville.

West P. W., Stocker G. C. & Unwin G. L. (1988) Environmental relationships and floristic and structural change in some unlogged tropical rainforest plots of north Queensland. *Proc. ecol. Soc. Aust.* **15,** 49-60.

Whitehead D. (1982) Ecological aspects of natural and plantation forests. *For. Abs.* **43,** 615-24.

Whitmore T. C. (1974) Change with time and the role of cyclones in tropical rainforest on Kolombangara, Solomon Islands. *Commonwealth Forestry Insitutue, Paper 46.*

Whitmore T. C. (1975) *Tropical Rain Forests of the Far East.* Clarendon Press, Oxford.

Whitmore T. C. (1978) Gaps in the rainforest canopy. In: *Tropical Trees as Living Systems* (eds P. B. Tomlinson & M. H. Zimmermann). Cambridge University Press, Cambridge.

Whitmore T. C. (1982) On pattern and process in forests. In: *The Plant Community as a Working Mechanism. Special publication 1,* (ed I. E. Newman). Br. ecol. Soc., Blackwell, Oxford.

Whitmore T. C. (1983a) Plant species diversity in tropical rainforests. In: *The Significance of Species Diversity in Tropical Forest Ecosystems. Biology International,* Special issue **6.** (eds G. Maury-Lechon, M. Hadley & T. Younes).

Whitmore T. C. (1983b) Secondary succession from seed in tropical rainforests. *For. Abs.* **44,** 767-79.

Proc. Ecol. Soc. Aust. 15:49-60

Environmental relationships and floristic and structural change in some unlogged tropical rainforest plots of north Queensland

P. W. WEST[1], G. C. STOCKER[2] and G. L. UNWIN[2]

[1]*C.S.I.R.O. Division of Forestry and Forest Products, Stowell Avenue, Hobart, Tas. 7004*

[2]*C.S.I.R.O. Tropical Forest Research Centre, P.O. Box 780, Atherton, Qld. 4883*

Abstract

Much is now known of the ecological relationships of Australian rainforest at a biogeographical scale. The present work considers the problems of examining such relationships at a local scale, such as that required by forest managers. Data are considered for 19 unlogged rainforest plots scattered along the east coast of north Queensland. Classification of the plots indicates there is great species diversity among the plots, consistent with the known floristic diveristy of the region. Species ordination and regression analysis indicate that species occurrence bears some relationship to rainfall and soil fertility. Structural and floristic changes that have occurred in two-year growth intervals in the plots over periods up to 10 years are examined. Changes are found to occur as a result of the small size of the plot relative to the scale of the local floristic unit or to be cyclic or unidirectional. Some of these changes may occur as a result of irregular catastrophes such as cyclones, droughts or fire. The use of data from repeatedly measured rainforest plots in providing information useful to managers is discussed.

Introduction

Successful management of tropical rainforest depends ultimately on how well the mechanisms that determine floristic pattern in forests are understood. Only then can the effects of management practices be predicted. Current understanding of these mechanisms has been deduced from observed relationships among floristic patterns and climatic and edaphic factors (e.g., Webb & Tracey 1981). Many of these relationships have been observed at scales of peripheral interest to forest managers. For example, although floristic patterns may appear deterministic at regional scales, a manager regenerating a logging coupe might find that important features of the developing floristic pattern cannot be predicted because of chance factors (e.g., those relating to seed production and seedling establishment) operating within a much broader deterministic framework. These chance factors might cause either cyclic or unidirectional drifts in local floristics.

Development of useful models of floristic change at a logging coupe scale will require much more information than is available at present on the attributes of the various rainforest species, particularly those attributes relating to their propagule production, dispersal and predation characteristics and relative shade tolerance. In the meanwhile, it may be possible to draw some inferences on the strength of the relationships between environmental parameters and floristics at local and regional scales by examining data from sites for which sufficient structural, growth, recruitment and mortality data are available to indicate the magnitude and direction of any floristic changes which might be occurring. The present work describes some attempts to do this and the problems that are associated with it.

Source of data and methods

The data used in the present work come from 19 plots, each 0.5 ha, established in 1971 at unlogged sites in north Queensland (Fig. 1). Seventeen of the plots are in ecofloristic province B_2 as defined by Webb, Tracey and Williams (1984) and two (plots 10 and 18) are in province B_1. In terms of the structural classification of rainforest (Webb 1981), the plots are all simple or complex, mesophyll or nanophyll vine forests. Reference to Webb (1981) and references quoted there will give a general description of the floristic types and environmental circumstances of the plots in terms of this structural classification.

In establishing the plots, unlogged sites were used in the belief that the processes underlying the dynamics of these forests would be more clearly seen on sites which had not been greatly influenced by man. Plots were often established ahead of a logging face. They were selected as the opportunity arose, not as a result of any formal sampling scheme. Most were seen intuitively, as being broadly representative of the forest types of the region. They cover a broad range of the structural and environmental conditions (e.g., rainfall, soil type, elevation and successional status) found in the region.

Data recorded for each plot included species by life form of all vascular plants, and identity, diameter at breast height (dbh), height and postition of all trees of 10 cm dbh or greater. In total, 432 tree species were in the 19 plots. A variety of environmental variables was also recorded in each plot. Their ranges and mean values over all the plots, together with mnemonics used for later reference, are shown in Table 1. Because the distributions of values about their means were very skewed for some of the variables, values of these variables were transformed for use in all subsequent work as indicated in the table. Plot aspect was transformed to its cosine so aspects north of east or west had positive values in the range 0-1 and those south of east or west had negative values in the range 0-1. This is a simple transformation appropriate to apply to aspect as suggested by Roise & Betters (1981).

The soil variables shown in Table 1 were determined by combining samples, from 0-30 cm depth, from five regularly spaced locations within each plot. Soil pH was determined with a glass-calomel electrode using a 1:5 soil:water suspension. Organic carbon was determined by rapid titration (Walkey & Black 1934) and a correction factor of x1.32 may be applied (Teakle 1950). Total nitrogen was determined by the Honda (1962) modification of the semi-micro Kjeldahl digestion procedure and total phosphorus by the

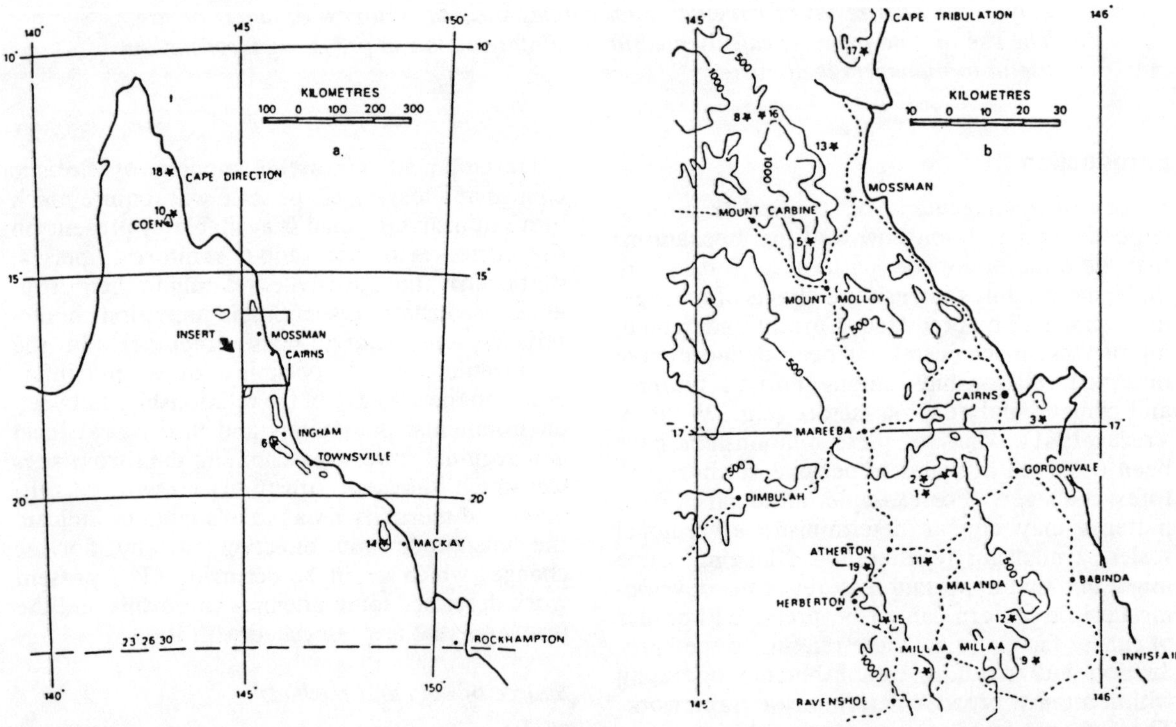

FIG. 1. Location of plots.

colorimetric method of Murphy & Riley (1962) using a perchloric-nitric acid digest (Baseden, Wooley & Moody 1974). Exchangeable basic cations (Ca, Mg, Na and K) were extracted with normal (pH 7) ammonium acetate and determined by atomic absorption spectroscopy. Cation exchange capacity was determined after cation extraction with ammonium acetate and replacement of the ammonium with 0.1M hydrochloric acid and titration (Chapman 1965). No values were determined for sodium in four plots. These missing values were estimated using the method of Beale and Little (1975).

Changes in tree diameters and numbers (i.e., growth, recruitment and mortality) have been recorded on several occasions since establishment of each plot at approximately two year intervals. Data collected over, at most, ten years were considered in the present work.

Results

Classification of plots by species compostition

To begin examining ecological relationships within these plots, an attempt was made to group those with similar species suites by numerical classification. The hierarchial clustering programme SAHN in the CSIRO TAXON suite of programmes, with the nearest neighbour sorting strategy, was used; the method is that of Lance & Williams (1966). The similarity matrix was computed using the method of Gower (1971) with double absences of species ignored.

The resulting dendrogram, to the nine group level, is Fig. 2. In this case the similarity matrix was computed using the numbers of individuals of each species in each plot. The striking features of the dendogram are the very narrow range of similarities over which the separation between groups has occurred and the low values of the similarities (0.4-0.5). When the similarity matrix was computed using species presence-absence data, the result was similar except that the range of similarities over which separation occurred was much lower (0.1-0.3).

The conclusion from these results is that the plots differ so widely in their species suites that no realistic groupings of them exist. This is not surprising, considering the known floristic diveristy of this ecofloristic province (Webb *et al.* 1984) and the necessarily intuitive manner in which the plots were established.

Relationship between species occurrence and environment

An attempt was then made to relate species occurrence in these plots to the environmental variables listed in Table 1. Initially, an attempt was made to reduce the number of environmental variables being considered by grouping together those which contained similar information about the plot sites. Firstly, the correlation matrix of the 17 measures was computed. Secondly, this matrix was subject to a principal components analysis. Four components had eigenvalues in excess of unity and explained 37%, 18%, 14% and 11% of the variation in the data, respectively. Thirdly, these four axes were rotated with the varimax rotation. The correlations of the original 17

TABLE 1. Site factors measured in the 19 plots.

Factor	Mnemonic	Transformation	Minimum-mean-maximum (untransformed)
Altitude above sea level (m)	A	–	15 – 651 – 1 200
Latitude (°S)	L	–	13 – 17 – 21
Distance from coast (km)	D	–	2 – 35 – 69
Average annual rainfall (cm)	R	–	120 – 233 – 400
Slope (radians)	T	cosine	0 – 1.12 – 0.31
Aspect from north (radians)	H	–	0 – 2.6 – 4.7
Soil pH	pH	–	4.0 – 4.8 – 6.5
Soil organic carbon (%)	C	–	1.2 – 3.5 – 7.8
Soil total nitrogen (%)	N	–	0.08 – 0.31 – 0.57
Soil total phosphorus (%)	P	log	0.01 – 0.07 – 0.38
Soil cation exchange capcacity (m.eq. 100 g^{-1})	CEC	log	4 – 15 – 31
Soil Ca (m.eq. 100 g^{-1})	Ca	log	0.1 – 2.9 – 20.3
Soil Mg (m.eq. 100 g^{-1})	Mg	log	0.2 – 1.2 – 3.7
Soil K (m.eq. 100 g^{-1})	K	log	0.1 – 0.3 – 1.0
Soil Na (m.eq. 100 g^{-1})	Na	log	0.05 – 0.11 – 0.22
Soil total bases (m.eq. 100 g^{-1})	B	log	1 – 5 – 25
Soil base saturation (proportion)	S	arcsine$\sqrt{}$	0.07 – 0.27 – 0.94

variates with these rotated axes are shown in Table 2. Variates with a high correlation (greater than about 0.5) with any one axis may be considered to contain similar information about the plots. A more thorough discussion of the method and meaning of this type of analysis is given in chapters 4 and 5 of Cooley & Lohnes (1971) and chapter 4 of Chatfield & Collins (1980).

In Table 2, the environmental variables have been arranged according to the groupings suggested by the above analysis. Variables not included in groups appear to be associated with more than one group, since they have substantial correlation with more than one rotated factor. Group 1 appears to represent geographical factors of altitude and latitude, group 2 largely represents soil organic constituents, group 3 mainly represents soil bases (which probably reflect fertility), and group 4 appears to represent moisture availability as expressed by rainfall. Given this, the 19 plot × 17 environmental variable matrix of observations was then transformed to a 19 × 4 reduced matrix, the columns of which were orthogonal and represented the four groups of environmental variables. The method of making this transformation can be found in Cooley & Lohnes (1971). The use of these transformed variables is described below.

To obtain a set of scores which represent the species composition of the plots and which might be related to the environmental variables, an ordination of the numbers of individuals of each species in the plots was undertaken. The method used was detrended correspondence analysis (Hill & Gauch 1980). In effect, such an analysis produces a score for each plot based on the species present in the plot. Plots unlike in their species suites should have scores which are quite different, while like plots should have similar scores. In fact, several sets of scores (called axes or ordinates) are produced, each set being orthogonal with the others. The object then is to relate each set of scores to one or more of the environmental variables. Because there are several sets of scores, it may be possible to see how different trends in species occurrence across plots are related

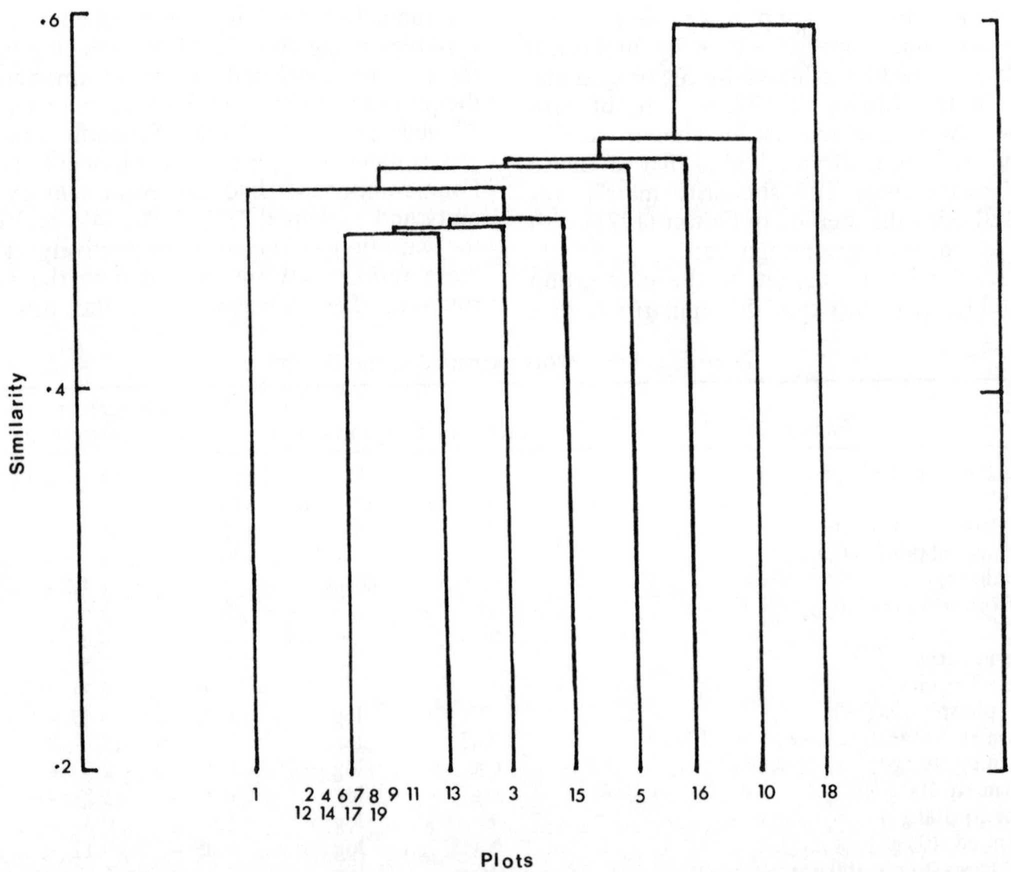

FIG. 2. Dendogram of the classification by frequency of species occurrence of the 19 plots used in this work.

TABLE 2. Correlations of plot environmental variables with rotated factors from varimax rotation. Groupings indicated by the analysis are indicated.

Environmental variable		Factor 1	2	3	4
Group 1	A	0.37	-0.16	0.49	-0.72
	L	-0.15	-0.20	-0.23	-0.80
	D	-0.03	-0.07	0.29	-0.87
	P	-0.56	-0.31	-0.28	-0.43
Group 2	C	0.34	-0.83	0.08	-0.18
	N	-0.13	-0.91	-0.04	-0.29
	CEC	-0.29	-0.87	-0.01	-0.22
	Na	-0.18	-0.90	-0.05	0.01
	Mg	-0.45	-0.71	0.32	0.13
	pH	-0.64	-0.53	0.11	-0.08
	B	-0.72	-0.51	0.41	0.08
Group 3	T	0.77	-0.37	0.15	-0.07
	Ca	-0.81	-0.31	0.33	0.01
	K	-0.70	-0.04	0.16	-0.43
	S	-0.65	0.06	0.62	0.28
Group 4	R	-0.02	-0.00	-0.85	0.18
	H	-0.15	-0.07	0.71	-0.04

to trends in the various environmental variables. For the present work, only the first two sets of scores were used; these ordinates had eigenvalues of 0.89 and 0.71, the other ordinates all having eigenvalues less than 0.5.

To see if relationships existed between the sets of ordination scores and the environmental variables, each of the two sets of ordination scores were related to the transformed environmental variables by linear, least-squares, multiple regression. Because of the orthogonality of the transformed environmental variables, the contribution of each variable to the regression is quite independent of that of each other variable. This is not the case when untransformed variables are used because of the correlations that exist among them. This makes it possible to determine unequivocally which environmental variables are related to the dependent variable in the regression (the ordination scores in this case). The regression analysis showed that the first set of ordination scores was related significantly ($P<0.05$ at least) only to the moisture availability group of environmental variables, whereas the second set was related significantly only to the soil fertility group.

These results suggest that species occurrence in these plots is associated with both rainfall and soil fertility but not with altitude (hence temperature) or soil organic constituents. In many respects, the present work repeats that of Webb *et al.* (1967) who studied a similar set of plots from the same region. They too found that both moisture and soil fertility appeared to affect species occurrence. Attempts to explore further these results to determine exactly which species groups were affected by changes in rainfall and soil fertility were unsuccessful. This reflects the floristic diveristy of the plots as found in the previous section.

Floristic change in plots

Analyses of environmental relationships such as those described above may be of use to a forest manager by providing an ecological framework on which management systems may be used. However, when such results are derived from a single (in the present work the initial) measurement and floristic description of vegetation, it is quite possible that environmental relationships may be obscured simply because of the random point of sampling in real time in relation to the life-cycle of the vegetation. In this section of the present work, some observations are made on the structural and floristic changes with time that have occurred in these plots. The intention of this is to determine how complex these changes are and whether or not better understanding of the environmental relationships that exist in these forests can be obtained by considering such changes.

Even though data recorded in the present work extends only over, at most, 10 years, several plot features may be used to indicate indirectly that long term floristic changes are taking place. The most important of those apparent at an initial examination is the pattern created by the number of stems within defined size classes and the size class distribution of important species. Subsequent remeasurements provide growth, recruitment and mortality data for important species and overall trends in species numbers and structural characteristics. A knowledge of autecology of the important species and of the probable disturbance history enables an assessment of the probable nature of the change, that is, whether cyclic, unidirectional or merely an artefact of plot size in relation to the local floristic unit. Examples of how these features may be used to assess shifts in plot floristics are provided in the following discussion.

Although floristic changes appear to be occurring in most of the plots, those in plots 5, 11 and 13 seem representative of wider trends. The structural features of these three plots are illustrated in Figs. 3-5 and characteristics of their "important" species are given in Tables 3-5. Important species are defined as those which contribute individually 3% or more of plot basal area or of the total number of stems equal to or in excess of 10 cm dbh in the plot.

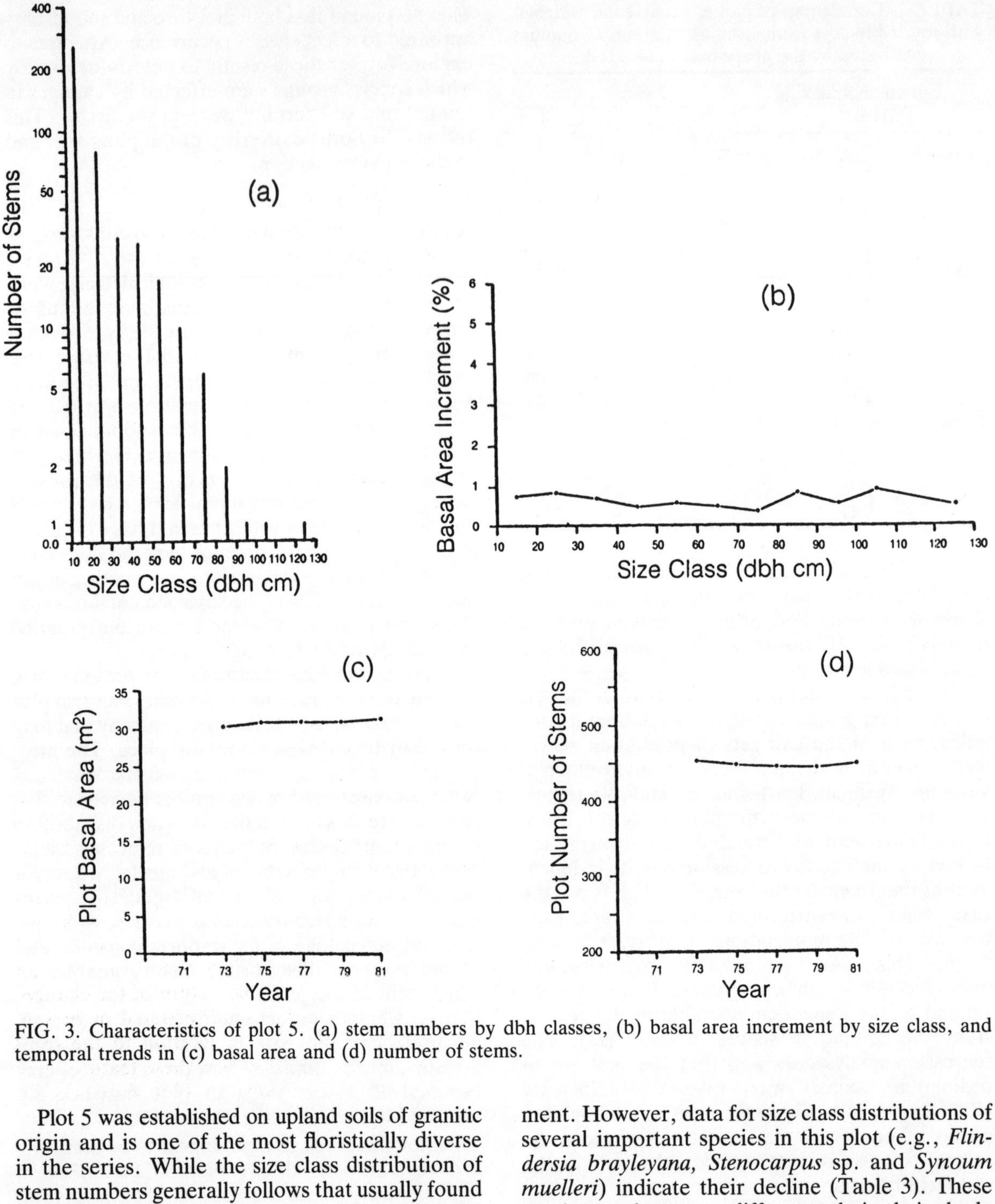

FIG. 3. Characteristics of plot 5. (a) stem numbers by dbh classes, (b) basal area increment by size class, and temporal trends in (c) basal area and (d) number of stems.

Plot 5 was established on upland soils of granitic origin and is one of the most floristically diverse in the series. While the size class distribution of stem numbers generally follows that usually found in rainforest (Fig. 3a), there are a few additional large individuals. However, the relative uniformity of growth rates within size classes (Fig. 3b) and the temporal consistency of basal area and stem numbers (Figs. 3c and 3d) suggest an overall stable structure with mortality balanced by recruitment. However, data for size class distributions of several important species in this plot (e.g., *Flindersia brayleyana*, *Stenocarpus* sp. and *Synoum muelleri*) indicate their decline (Table 3). These species are known to differ greatly in their shade tolerance and thus have contrasting requirements for successful establishment. Stochastic elements in their reproductive biology and the prevailing regime of environmental disturbance may have led to their decline in the plot, but not necessarily

TABLE 3. Plot 5 – Initial stem numbers, mortality and basal area increment by dbh size classes (cm) with recruitment to the 10-20 cm class and an indication of regeneration stocking in subplots for important species. Data from 1973-81.

Species	Regen.[a] stock.		Initial stem numbers										Init. tot.	Recr.[b]	Mortality										Basal area increment[c]									
	A	B	10-20	20-30	30-40	40-50	50-60	60-70	70-80	80-90	90-100	100			10-20	20-30	30-40	40-50	50-60	60-70	70-80	80-90	90-100	100	10-20	20-30	30-40	40-50	50-60	60-70	70-80	80-90	90-100	100
Flindersia bourjotiana	14	7	1	2	2	3	1	1	3	–	–	–	13	–	–	–	1	–	–	–	–	–	–	–	D	C	A	D	A	E	C	C	–	–
Chrysophyllum sp. RFK 3144	15	13	26	22	6	1	–	–	–	–	–	–	35	2	1	–	–	–	–	–	–	–	–	–	D	D	D	E	–	–	–	–	–	–
Stenocarpus sp. RFK 3174	1	2	–	–	–	–	–	1	1	–	1	1	4	–	–	–	–	–	–	–	–	–	–	–	–	–	–	–	E	D	–	E	*	–
Planchonella euphlebia	9	7	6	4	2	3	–	1	1	–	–	–	17	1	–	–	–	–	–	–	–	–	–	–	C	C	D	E	–	E	C	E	–	–
Syzygium sp. (WESA) RFRK 3030	7	7	–	–	–	–	–	1	–	–	–	1	2	–	–	–	–	–	–	–	–	–	–	–	–	–	–	–	–	A	–	–	–	C
Flindersia brayleyana	8	4	–	–	–	1	1	–	–	–	–	1	3	–	–	–	–	–	–	–	–	–	–	–	–	–	–	A	B	–	–	–	–	C
Synoum muelleri	6	–	–	–	–	–	3	1	–	–	–	–	4	–	–	–	–	–	1	–	–	–	–	–	–	–	–	–	E	E	–	–	–	–
Darlingia darlingiana	12	10	2	3	6	–	–	1	–	–	–	–	12	–	–	–	–	–	–	–	–	–	–	–	E	C	B	–	–	E	–	–	–	–
Garciniia sp. aff. *G. hunsteinii*	11	10	2	6	4	2	–	–	–	–	–	–	14	–	–	–	–	–	–	–	–	–	–	–	C	E	E	D	–	–	–	–	–	–
Cardwellia sublimis	14	11	1	3	1	1	1	1	–	–	–	–	8	1	–	–	–	–	–	–	–	–	–	–	B	C	A	E	C	C	–	–	–	–
Placospermum coriaceum	10	5	2	–	1	2	2	–	–	–	–	–	7	–	–	–	–	–	–	–	–	–	–	–	B	–	D	B	C	–	–	–	–	–
Beilschaiedia sp. aff. *B. obtusifolia*	9	11	13	5	–	1	–	–	–	–	–	–	19	1	–	–	–	–	–	–	–	–	–	–	B	B	A	A	–	–	–	–	–	–
Xylopia sp. RFK 2013	11	14	16	5	–	–	–	–	–	–	–	–	21	–	1	–	–	–	–	–	–	–	–	–	D	D	–	–	–	–	–	–	–	–
Cryptocarya sp. RFK 2436	9	6	24	–	–	–	–	–	–	–	–	–	24	–	–	–	–	–	–	–	–	–	–	–	D	–	–	–	–	–	–	–	–	–
Synima cardieri	16	14	17	2	–	–	–	–	–	–	–	–	19	1	–	–	–	–	–	–	–	–	–	–	C	C	–	–	–	–	–	–	–	–
Cryptocarya sp. aff. *C. hypospodia*	12	13	13	1	–	–	–	–	–	–	–	–	14	–	1	–	–	–	–	–	–	–	–	–	B	D	–	–	–	–	–	–	–	–

[a]The number of subplots of whole plot (max. 16) containing 1 or more small trees 0-3 m high (A) and 3 m high to 10 cm dbh (B).
[b]The number of small stems entering the 10-20 cm dbh class.
[c]Basal area increment – relative rates are indicated by A=100% over, B=25-100% over, C=within 25% of, D=50-100% below and E=over 100% below, plot mean for that size class.
*Only one tree in that size class.

in the surrounding forest. Overall the changes in this plot seem to reflect plot size in relation to the scale of the local floristic unit.

The type of change observed in plot 5 probably occurs to some degree in all plots. However, there also seem to be plots in which cyclical and perhaps unidirectional changes are taking place. Plots 11 and 13 contain changes which are probably cyclic. Plot 13 was established in lowland rainforest on relatively poor, clayey soils derived from shales. The distribution of stem numbers by size classes in this plot is suddenly truncated with very few individuals in excess of 50 cm dbh (Fig. 4a). Although the basal area is relatively constant (Fig. 4c) and the tree numbers appear to be falling (Fig. 4d), growth rates are relatively rapid among trees of intermediate size (Fig. 4b). Structurally, the plot appears to be in the late stages of recovery from a catastrophe which, at this site, would have almost certainly been severe cyclone

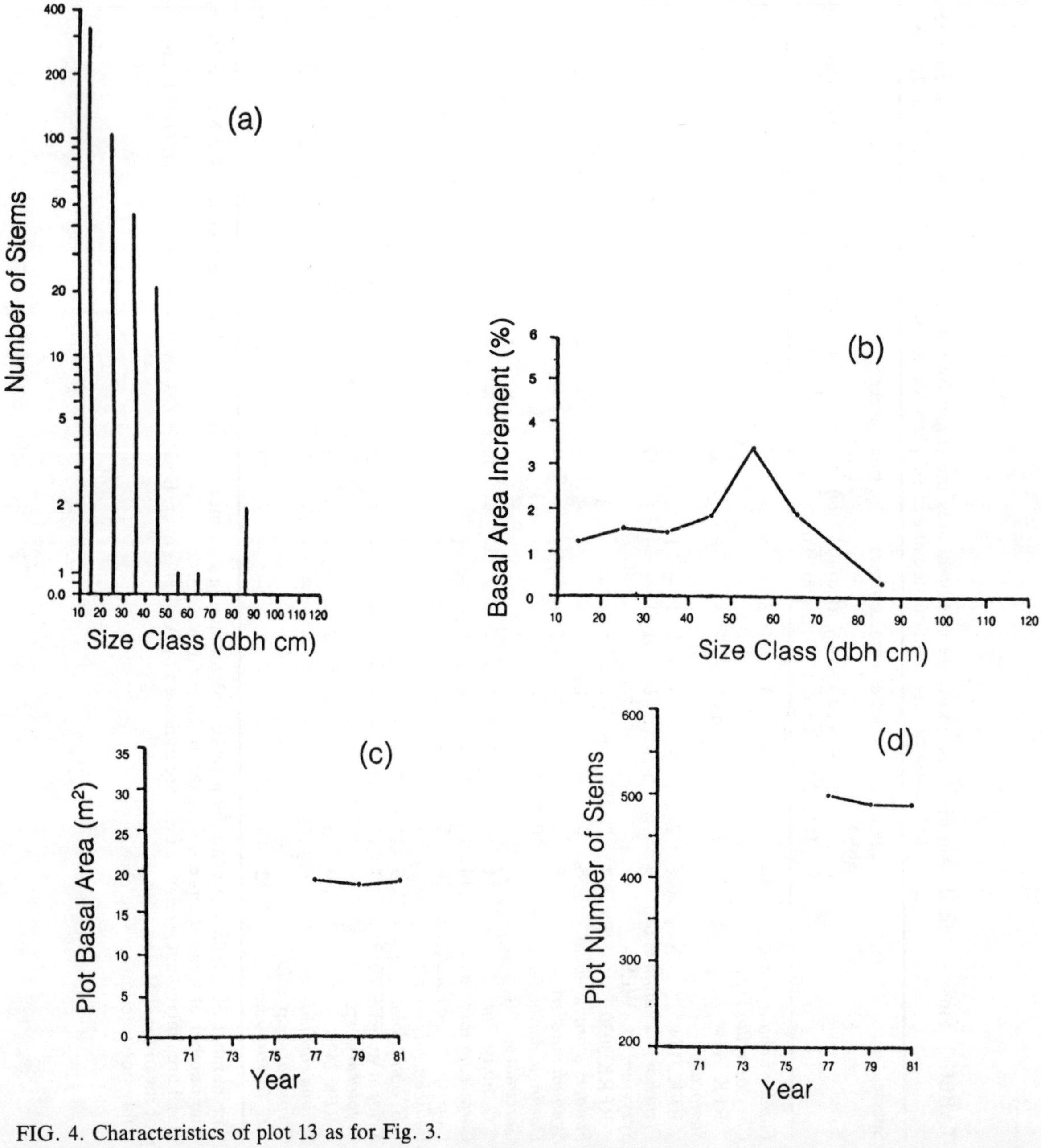

FIG. 4. Characteristics of plot 13 as for Fig. 3.

damage. Plot floristics support this view. The decline of the relatively shade intolerant species, *Acacia aulacocarpa,* appears to be balanced by the development of more shade tolerant species especially *Elaeocarpus, Syzygium* and *Crypto carya* spp. (Table 4). The very large individual of *Backhousia hughesii* in this stand is almost certainly a relic which survived this catastrophe. Cyclones are sufficiently frequent in this region to ensure that vegetation on lowland sites such as this will never reach the measure of stability apparent in the stand characterized by plot 5. The next severe cyclone affecting plot 13 will probably restart the cycle by allowing *Acacia* trees to regenerate and dominate the site until they reach senescene.

Plot 11 was established on the Atherton Tableland near Yungaburra. Although the soil is fertile (from a recent basalt), the average annual rainfall is only about 1 400 mm. While the site is far enough from the coast to avoid the worst effects of most cyclones, there is a report (T. D. McGreehan unpublished) that forests in this area were severely damaged by a cyclone in March 1918. There are also Queensland Department of Forestry records (Anon 1916) of severe drought and subsequent fires in the rainforests of this area during 1915 (nearby Atherton only had 498 mm of rain compared with its annual average of 1 425 mm). These catastrophes are not now strongly reflected in the structural features of the plot. Although there are a few more large stems than might be expected (Fig. 5a), basal area shows only a slight tendency to increase (Fig. 4c) and stem numbers are slowly falling (Fig. 5d). However, these events may have exerted an important influence on plot floristics favouring more of the relatively fast growing, shade intolerant species than would have occurred had the usual pattern of regeneration in single tree gaps been maintained. Since there appears to be an inverse relationship between shade tolerance and longevity in the rainforest flora of this region (G. C. Stocker, unpublished) it is most probable that the most intolerant species of this plot have already died out. Conspicuous among those remaining are *Aleurites moluccana, Flindersia brayleyana* and *Toona australis* (Table 5). Trees of *A. Moluccana* within the plot are virtually moribund; many of those just outside the plot have died within the past 10 years. Individuals of the other two species are much longer lived and, except for a few overtopped stems, most can be expected to survive for another 100-200 years.

As well as the changes illustrated by these three plots, observations elsewhere in the region suggest that another source of long term change in these forests is due to the prevailing fire regime (Unwin *et al.* 1988). Probable mechanisms involving fire, and in some areas other environmental catastrophes, were outlined by Stocker and Unwin (1986). Briefly, increased probability of fire following the arrival of Aborigines appears to have converted much of what was closed forest to open forest. Since European settlement and associated activities (especially grazing), there has been a trend for rainforest expansion in some areas. Observations to date indicate that the structure and floristics of the advancing rainforest ecotones depend on the dispersal, growth and survival characteristics of the available species particularly in relation to prevailing fire regimes (Unwin 1983). Shade tolerant species with limited capacity to disperse beyond the rainforest edge are poorly represented.

Discussion

At the biogeographical level, substantial advances have been made in the understanding of the ecology of Australian rainforest (Webb & Tracey 1981; Webb *et al.* 1984). The present work has presented some preliminary findings which emphasize the complexity of examining ecological relationships within ecofloristic provinces. The set of plots studied here comes from the ecologically most diverse of Australian rainforest (Webb *et al.* 1984). About one fifth of the rainforest of the region has been cleared (J. W. Winter, F. C. Bell, L. I. Pahl & R. G. Atherton, unpublished report) and much of the remainder that is accessible has been logged. This limits opportunity to study rainforest floristics in the native state. The opportunistic way in which the plot sites were selected compounds the problem of natural floristic diveristy that exists in the region.

Although the plots used in the present work could not be grouped on the basis of their floristic composition, it was found that their floristic composition was related to rainfall and soil fertility. Although such relationships might normally be expected, they do indicate that in undisturbed forests a deterministic framework exerts at least some controls on floristics at scales which are of importance to forest managers. While an increase in plot size might have clarified these relationships, this would normally have been counter-balanced by the introduction of additional topographic and edaphic variation to the plot sites.

TABLE 4. Plot 11 – Data from 1976-82 as for Table 3.

Species	Regen. stock. A	Regen. stock. B	Initial stem numbers 10/20	20/30	30/40	40/50	50/60	60/70	70/80	80/90	90/100	100	Init. tot.	Recr.	Mortality 10/20	20/30	30/40	40/50	50/60	60/70	70/80	80/90	90/100	100	Basal area increment 10/20	20/30	30/40	40/50	50/60	60/70	70/80	80/90	90/100	100
Dendrocnide photinophylla	3	7	23	16	6	1	1	1	2	1	–	–	51	–	1	–	–	–	–	–	–	–	–	–	E	C	C	E	E	E	E	E	–	–
Dysoxylum pettigrewianum	11	10	22	2	2	1	1	–	–	–	1	2	31	–	–	–	–	–	–	–	–	–	–	–	C	B	B	–	B	C	–	–	*	D
Argyodendron peralatum	13	14	22	2	1	3	1	2	1	–	–	1	33	2	–	–	–	–	–	–	–	–	–	–	C	A	A	A	D	C	E	D	–	E
Toona australia	–	2	4	6	3	3	5	1	1	–	–	–	23	–	–	1	–	–	–	–	–	–	–	–	C	B	A	B	A	B	B	–	–	–
Ficus watkinsiana	3	–	–	–	–	–	–	–	–	–	–	1	1	–	–	–	–	–	–	–	–	–	–	–	–	–	–	–	–	–	–	–	–	A
Aleurites moluccana	–	–	–	–	–	1	3	3	1	1	–	–	9	–	–	–	–	–	–	–	–	–	–	–	–	–	–	D	D	E	C	E	–	–
Flindersia brayleyana	–	–	–	2	2	2	3	1	–	–	–	–	10	–	–	–	–	–	–	–	–	–	–	–	–	C	C	B	B	B	–	–	–	–
Castanospermum australe	15	15	–	–	–	1	2	1	1	–	–	–	5	–	–	–	–	–	–	–	–	–	–	–	–	–	–	B	E	E	–	A	–	–
Endiandra cowleyana	–	1	1	4	3	–	2	–	–	–	–	–	10	–	–	–	–	–	–	–	–	–	–	–	A	B	B	–	E	–	–	–	–	–
Dephnandra repandula	1	3	14	6	3	1	–	–	–	–	–	–	24	–	–	–	–	–	–	–	–	–	–	–	C	E	E	E	–	–	–	–	–	–
Tetrasynandra laxiflora	7	2	2	5	6	–	–	–	–	–	–	–	13	–	1	–	–	–	–	–	–	–	–	–	E	E	E	–	–	–	–	–	–	–
Endiandra pubens	–	2	5	8	1	1	–	–	–	–	–	–	15	1	–	1	–	–	–	–	–	–	–	–	D	E	–	E	–	–	–	–	–	–
Litsea leefeana	5	5	6	2	3	1	–	–	–	–	–	–	12	–	–	–	1	–	–	–	–	–	–	–	A	A	B	C	–	–	–	–	–	–
Mallotus polyadenos	2	–	7	2	1	–	–	–	–	–	–	–	10	–	–	–	–	–	–	–	–	–	–	–	E	B	E	–	–	–	–	–	–	–

*Only one tree in this size class.

TABLE 5. Plot 13 – data from 1977-81 as for Table 3.

Species	Regen. stock. A	Regen. stock. B	Initial stem numbers 10/20	20/30	30/40	40/50	50/60	60/70	70/80	80/90	90/100	100	Init. tot.	Recr.	Mortality 10/20	20/30	30/40	40/50	50/60	60/70	70/80	80/90	90/100	100	Basal area increment 10/20	20/30	30/40	40/50	50/60	60/70	70/80	80/90	90/100	100
Acacia aulacocarpa	–	–	–	1	3	8	–	1	–	1	–	–	14	–	–	–	–	–	–	–	–	1	–	–	–	A	A	A	B	*	–	–	–	–
Flindersia bourjotiana	13	9	12	8	8	1	–	–	–	–	–	–	29	1	1	–	–	–	–	–	–	–	–	–	B	A	B	D	–	–	–	–	–	–
Eugenia kuranda	10	13	11	6	7	1	–	–	–	–	–	–	25	–	–	–	–	–	–	–	–	–	–	–	B	B	D	C	–	–	–	–	–	–
Eugenia sp. RFK 1101	16	15	34	9	3	–	–	–	–	–	–	–	46	–	–	–	1	–	–	–	–	–	–	–	C	E	E	–	–	–	–	–	–	–
Flindersia pimenteliana	2	1	5	4	5	1	–	–	–	–	–	–	15	–	1	–	1	–	–	–	–	–	–	–	A	A	B	C	–	–	–	–	–	–
Endiandra sp. aff. *E. hypotephra*	14	16	46	5	–	–	–	–	–	–	–	–	51	1	1	1	–	–	–	–	–	–	–	–	E	E	–	–	–	–	–	–	–	–
Eugenia johnsonii	4	7	8	6	2	1	–	–	–	–	–	–	17	–	–	–	–	–	–	–	–	–	–	–	C	C	C	E	–	–	–	–	–	–
Musgravea heterophylla	13	11	10	4	3	1	–	–	–	–	–	–	18	–	–	–	–	–	–	–	–	–	–	–	B	D	D	E	–	–	–	–	–	–
Backhousia hughesii	–	–	–	–	–	1	–	–	–	1	–	–	2	–	–	–	–	–	–	–	–	–	–	–	–	–	–	E	–	–	–	*	–	–
Xanthophyllum octandrum	12	13	9	7	1	–	–	–	–	–	–	–	17	–	–	–	–	–	–	–	–	–	–	–	C	E	D	–	–	–	–	–	–	–
Elaeocarpus sp. aff. *E. ferruginiflorus*	5	4	11	5	–	–	–	–	–	–	–	–	16	–	–	1	–	–	–	–	–	–	–	–	A	A	A	–	–	–	–	–	–	–
Cryptocarya sp. RKF 2423	15	14	22	2	–	–	–	–	–	–	–	–	24	–	–	–	–	–	–	–	–	–	–	–	C	D	–	–	–	–	–	–	–	–

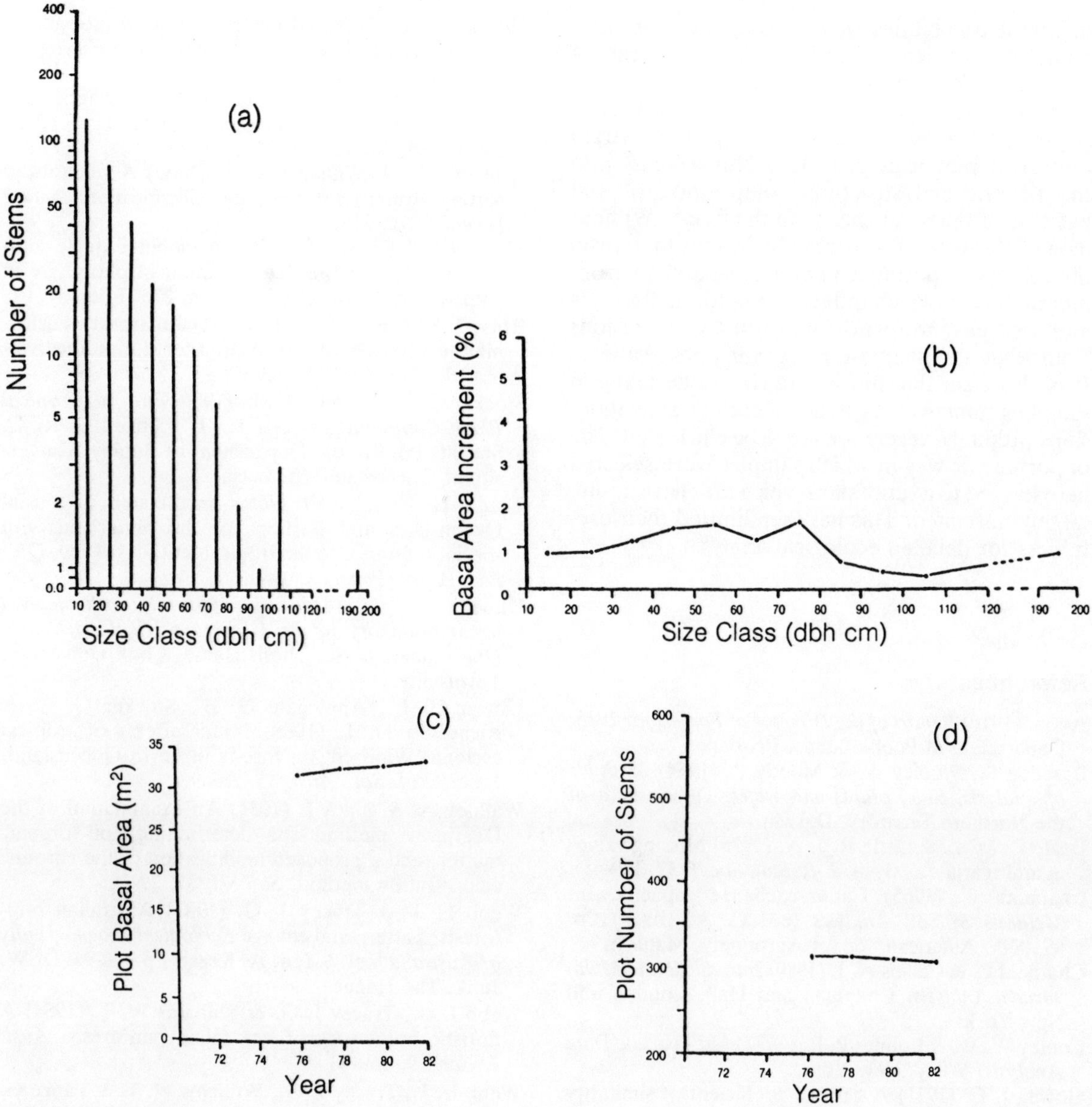

FIG. 5. Characteristics of plot 11 as for Fig. 3.

Indeed, it is difficult to find many sites in the region larger than 0.5 ha which do not include major physical or floristic discontinuities.

The problems of sampling forest in the region cause difficulties in assessing how fast floristic change is occurring. However, the plot data indicate that there are changes in site floristics over time scales of the order of the life of a forest tree. Large floristic changes at single sites are known to have occurred over relatively short periods (several maximum life-spans of typical forest trees) in the Recent (Kershaw 1983). The changes observed in the plots probably obscure more subtle shifts in floristics due to climatic fluctuations. It has been suggested (e.g., Webb & Tracey 1981; Kershaw 1983) that rainforest species were confined to small refuge areas within the region or even as far afield as New Guinea (G. C. Stocker & G. L. Unwin, unpublished), during drier periods of the Pleistocene. While the floristic composition of many of the plots more distant from obvious refuge areas appears to reflect the dispersal capabilities of the component species, additional evidence, perhaps in the form of genetic markers, needs to be gathered.

In conclusion, the results reported here have suggested that repeated measurements of virgin rainforest plots can provide useful insights into the floristic and structural composition of and patterns of temporal change in the forest. Without data of this type, it is extremely difficult to unravel the relative importance of biological and environmental factors which influence rainforest floristics and very easy to form inappropriate conclusions from what are otherwise just passing observations. It is also clear that much care should be taken in sampling rainforest vegetation if ecological relationships within the vegetation are to be elucidated. The opportunistic way in which samples were selected here has led to a set of plots which are, floristically, largely unrelated. This has then limited their usefulness for detailed ecological analysis.

References

Anon. (1916) *Report of the Director of Forests for 1915.* Department of Public Lands, Brisbane.

Baseden S., Wooley A. & Moody P. (1974) *Methods of analysis, soils, plants and water.* Department of the Northern Territory, Darwin.

Beale E. M. L. & Little R. J. A. (1975) Missing values in multivariate analysis. *J. R. Stat. Soc.* **B 37**, 129-45.

Chapman C. (1965) Cation-exchange capacity. In: *Methods of Soil Analysis* (ed. C. A. Black) Pp. 891-895. American Soc. of Agronomy, Madison.

Chatfield C. & Collins A. J. (1980) *Introduction to Multivariate Analysis.* Chapman and Hall, London and New York.

Cooley W. W. & Lohnes P. R. (1971) *Multivariate Data Analysis.* Wiley, New York.

Gower J. C. (1971) A general coefficient of similarity and some of its properties. *Biometrics* **27**, 857-71.

Hill M. O. & Gauch H. G. (1980) Detrended correspondence analysis: an improved ordination technique. *Vegetatio* **42**, 47-58.

Honda C. (1962) Rapid determination of nitrogen in soil by Kjeldahl method. *Nippon Dojo-Hiryogaku Zasshi* **33**, 195-200.

Kershaw A. P. (1983) A holocene pollen diagram for Lynch's crater, north-eastern Queensland. *New Phytol.* **94**, 669-82.

Lance G. N. & Williams W. T. (1966) A generalized sorting strategy for computer classification. *Nature (Lond.)* **212**, 218.

Murphy J. & Riley J. P. (1962) A modified single solution method for the determination of phosphate in natural waters. *Anal. Chim. Acta* **27**, 31-36.

Roise J. P. & Betters D. R. (1981) An aspect transformation with regard to elevation for site productivity models. *For. Sci.* **27**, 483-86.

Stocker G. C. & Unwin G. L. (1986) Fire. In: *Tropical Plant Communitites* (eds H. T. Clifford & R. L. Specht) Pp. 91-103. Department of Botany, University of Queensland, St. Lucia.

Teakle L. J. H. (1950) Notes on the soils of coastal Queensland and portions of the hinterland with special reference to the tropical latitudes. *Univ. Qld. Fac. Agric. Papers* **1**, 1-40.

Unwin G. L. (1983) *Dynamics of the rainforest-eucalypt forest boundary in the Herberton Highland, North Queensland.* M.Sc. Thesis, James Cook University, Townsville.

Unwin G. L., Applegate G. B., Stocker G. C. & Nicholson D. I. (1988) Initial effects of tropical cyclone 'Winifred' on forests in north Queensland. *Proc. Ecol. Soc. Aust.* **15**, 283-96.

Walkley A. & Black I. (1934) An examination of the Degtjarev method for determining soil organic matter, and a proposed modification of the chromic acid titration method. *Soil Sci.* **37**, 29-38.

Webb L. J. & Tracey J. G. (1981) Australian rainforests: pattern and change *Ecological Biogeography of Australia* Vol. 1. (ed. A. Keast) Pp.606-94. Dr W. Junk, The Hague.

Webb L. J., Tracey J. G. & Williams W. T. (1984) A floristic framework of Australian rainforests. *Aust. J. Ecol.* **9**, 169-98.

Webb L. J., Tracey J. G., Williams W. T. & Lance G. N. (1967) Studies in the numerical analysis of complex rain-forest communities. I. A comparison of methods applicable to site/species data. *J. Ecol.* **55**, 171-91.

Proc. Ecol. Soc. Aust. 15:61-80

Stand changes in north Queensland rainforests

D. I. NICHOLSON, N. B. HENRY and J. RUDDER

Queensland Department of Forestry, P.O. Box 210, Atherton, Qld. 4883

Abstract

This paper discusses some characteristics of virgin rainforest in north Queensland, the effects of logging and the responses that follow logging and logging with silvicultural treatment. We find that virgin rainforests generally have higher levels of tree stocking and basal area than Asian and African rainforests but are less species rich than the former. Mortality is generally less than 1% but exceeds the rate of recruitment so that there is a trend to lower species richness and basal area, with time. Such declines, it is suggested, could result from environmental stability at the scale of the samples and from climatic trends and short-term drought events. There is generally a fall in the Shannon Wiener index of diversity while evenness and uniformity generally rise. These changes are largely a consequence of the fall in species numbers.

The immediate effects of the logging operation are the reduction in the numbers of trees, reductions in basal area and an apparent reduction in the number of species. The latter effect is largely a consequence of small sample size and the constraint of the lower diameter limit in plot assessment. It does, however, have the effect of lowering diversity and raising evenness. These changes are correlated with the severity of logging as measured by basal area loss.

The after-logging response is to increase species numbers and the diversity index on all but lightly logged plots. There is a consistent rise in basal area resulting from lessened mortality and enhanced recruitment and tree growth.

The response following silvicultural treatment is much more marked with significant rises (from low levels) in species numbers and diversity and generally a rise in uniformity but a fall in evenness. Again there is a marked rise in basal area indicating the good response to treatment.

Some natural perturbations affecting structure and composition are discussed briefly.

All data indicate that a certain level of disturbance in the rainforest is necessary for maintenance of a high level of diversity. It is shown that logging generally provides this disturbance to a sufficient degree and leaves the rainforest as an intact and viable ecosystem with enhanced potential for production.

Introduction

The Queensland Department of Forestry maintains a series of yield plots to monitor growth and other changes in north Queensland rainforest. Details of these plots and their measurement can be found in Volck (1968) and Nicholson *et al.* (1983). The Department's main concern has been to monitor growth and stand changes in logged forest so that sound management prescriptions can be formulated. Thus fewer plots have been established in virgin forest but, two of these are of long standing and provide valuable data from undisturbed rainforests while the value of other plots in virgin rainforest is increasing with time. Other data come from investigations established in virgin rainforest before a logging or a treatment and so are of a static nature. They provide useful measures of species occurrence and stand structure.

Ninety plots have been established in logged rainforests, some of which have been treated silviculturally subsequently or enrichment planted. Data for a selection of plots typical of logged areas covering as far as possible the wide variation in sites and logging methods are presented in this paper. Responses to silvicultural treatments are also presented even though treatment of rainforests ceased in 1972.

In all cases a lower limit of 10 cm diameter breast height (DBH) is used in analysing the data. This imposes certain artificial constraints, especially in relation to species number, but it has certain advantages in making all data uniform throughout the measuring period and makes the recording of recruits depend on growth and not on the vigilance of the measure team. (Most plots were measured to smaller sizes for much of their lifetime.) It also makes the data consistent with most published work.

Before studying the data it should be mentioned that two types of "plot" are used. All the earlier work used conventional fixed area plots with defined boundaries. These continue to be used. The second type is based on a number of point samples using a simple wedge prism to select the sample trees. There is a wide literature on this method (e.g., Beers & Miller 1964; Bitterlich 1984).

The use of the wedge prism has the advantage that it allows a large sample for large trees and conversly a small sample for small trees, thus saving much measurement time. However from a purely ecological view point this may be a disadvantage since it will be shown that the species richness of any rainforest area owes much to the smaller tree fraction.

Most of the data in this paper are presented by individual plots since the range of variation is very high, both in the recorded data as well as the intrinsic factors of plot size, type, period of measure and soil parent material, though plots are grouped by these factors where possible. Means have been given in some cases, especially for parameters not seriously effected by plot size, as the "per hectare" data and diversity measures. These latter lend themselves to this operation and statistical analysis as they are normally distributed (Odum 1971).

Some Characteristics of Virgin Rainforest

Data base

The data base for virgin rainforest is shown in Table 1 with the plots arranged in order of increasing altitude. For most of the following discussion these are grouped by plot type, parent material and by forest type where distinguishable, as follows.

Soils developed on granite

Conventional plots — 591, 625/2, 408, 207.
Prism "plots"
  Mature rainforest — 679/1, 679/2, 582
  Kauri rainforest — 577

Soils developed on metamorphic rock

Conventional plot — 626/2

Soils developed on basalt

Conventional plot — 594, a stand dominated by rose butternut
Prism "plots"
  Mature rainforest — 701, 576

Plot 591 forms part of the data base used by Connell (1978, 1979) and Connell *et al.* (1984) which contains much extra useful information.

Mean stand tables

These are of utmost significance from a forester's point of view since so many management decisions are influenced by tree size. For our present purposes, however, only mean data will be given (Table 2). The size distribution follows the classical reversed J curve (Dawkins 1958) of rapidly decreasing numbers with increasing size.

Apart from the rose butternut (*Blepharocarya involucrigera* F. Muell) stand (Plot 594) and to a lesser degree, the kauri (*Agathis microstachya* Bailey and C. White) stand, it is difficult to note real differences in these data. Taken together it is seen that they are generally higher than most overseas data, for example those given by Dawkins (1958), where they are comparable to his upper limit; more comparable but still higher than those of Paijmans (1970) for Papua New Guinea; or Malaysia generally (Nicholson 1979; Brunig 1983) and the Philippines (Nicholson 1979). The Philippine stands often differ in having a higher large tree representation (Nicholson 1985) but a lower overall tree stocking and basal area.

TABLE 1. Data Base for Virgin Forest in north Queensland.

Plot No.	Type	Parent Material	Altitude m a.s.l.	Rainfall mm (approx.)	Plot Area or No. of Prism Pts.	Years of Measure
626/2	Prism	Metam.	380	2 030	.2023	12.0
701	Prism	Basalt	400	2 000	60	—
594	Plot	Basalt	720	2 000	{ 4857 { .2838	{ 12.0 { 19.9
591	Plot	Granite	730	1 800	.4087	32.2
576	Prism	Basalt	760	3 000	106	—
625/2	Plot	Granite	945	1 750	.2023	15.7
577	Prism	Granite	1 006	1 800	60	—
679/1	Prism	Granite	1 036	1 680	60	—
679/2	Prism	Granite	1 036	1 680	62	5.3
582	Prism	Granite	1 070	1 800	62	—
408	Plot	Granite	1 100	1 650	1.6187	—
207	Plot	Granite	1 130	1 650	.1012	10.9

The rose butternut stand is outstandingly different but is typical of a relatively small total area (less than 800 ha, H. E. Volck, pers. comm.) of gregarious occurrences of this species. Sixty percent of the trees and 58% of the basal area of trees over 10 cm diameter are made up by this species. The reasons for its occurrence and why the stand can grow to such high levels of tree density and basal area are not known. It is thought to be a relatively young stand originating from a catastrophic event such as a cyclone or a fire.

TABLE 2. Stocking by Nos. ha^{-1} in 10 cm diameter classes and basal area ha^{-1} over 10 cm diameter, in north Queensland rainforest.

Data Group	Diameter Class Lower Limit											
	10	20	30	40	50	60	70	80	90	100+	Total	B.A.
GRANITE SOILS												
Conv. Plots (4)	495	155	99	49	26	13	10	5	5	5	862	59.4
Prism Data (3)	468	171	69	54	31	14	9	5	2	3	826	53.6
Kauri Type (1)	570	115	54	25	18	10	5	3	2	3	804	37.9
METAMORPHIC SOILS												
Conv. Plot (1)	460	138	94	44	30	15	10	–	–	5	796	50.1
BASALT SOILS — TYPICAL												
Prism Data (2)	442	142	73	45	28	13	6	5	3	4	760	48.5
BASALT SOIL — ROSE BUTTERNUT STAND												
Conv. Plot (1)	581	375	208	97	49	8	2	2	2	–	1 324	79.8

The species grows in typical mature rainforest as well as these gregarious stands and there it behaves as a "normal" member of the stand. It is suggested that more detailed study of these gregarious stands should be very rewarding as they must be approaching limiting levels of stocking.

Species composition

Even when restricting the discussion to tree forms, our northern rainforests are very rich in species. This richness has rightly excited both the scientific and popular mind. How such richness can maintain itself in an apparently constant environment has exercised the minds of many ecologists (e.g., Connell 1979; Connell *et al.* 1984). We believe with Bourgeron (1983) that no answer has been suggested that satisfactorily meets all situations.

The present estimate of tree species in north Queensland is about 900 (B. P. M. Hyland, pers. comm.). This richness is spread over about 830 000 ha of rainforest (Nicholson 1985). It represents the extreme of the species/area curve and it is instructive to study this curve at the other extreme — small areas of uniform habitat.

TABLE 3. Some species/area data.

Area (ha)	Expected No. of Species (10 cm+) Plot 408	Sabah (1)	West Malaysia (2)	N.G. (3)	Nigeria (4)
0.10	35	45			
0.25	60	71	—	80	—
0.50	83	113	—	115	43
0.75	97	138	—	138	—
1.00	108	155	—	—	60
1.25	117	171	—	—	—
1.50	124	184	—	—	70
1.62 (4 acres)	127	191	227	—	—
1.75	130	196	—	—	—
2.00	136	204	251	—	—

(1) Nicholson (1965).
(2) Wyatt-Smith (1949) — Bukit Lagong Plot.
(3) Paijmans (1970) — Plot 2.
(4) Richards (1952) — Page 235.

Data used in this discussion come from areas varying from 0.10 ha to 1.62 ha as shown in Table 1, which carry from 28 to 125 species. Only Plot 408 lends itself to determining this species/area curve and even here, because of the design, we

TABLE 4. Comparative ecological measures in virgin rainforest in north Queensland.

Plot	Area (ha)	Date	BA/ha (m^2)	Trees/ ha	No. Spp/ Plot	No. Fam./ Plot	H_s (2)	H_s (e)	Even- ness	Uni- formity
CONVENTIONAL PLOTS										
591	.4087	1952	69.6	957	77	29	5.351	3.709	.834	5.08
625/2	.2023	1969	61.4	934	61	28	5.342	3.703	.901	3.10
408	1.6187	1967	56.0	827	125	39	5.752	3.987	.826	10.71
207	.1012	1956	69.0	870	28	18	4.008	2.778	.834	3.14
626/2	.2023	1969	50.1	796	48	23	5.013	3.475	.898	3.35
594	.4857	1951	79.8	1 324	52	20	2.618	1.815	.459	12.37
PRISM PLOTS										
679/1	.470*	1980	53.1	863	47	18	4.543	3.149	.818	8.63
679/2	.633*	1980	51.4	716	66	19	5.168	3.582	.855	6.87
582	.605*	1977	56.3	912	64+	24+	**	**	**	**
577	.314*	1977	37.9	804	55+	22+	**	**	**	**
701	.527*	1985	48.5	707	62	29	4.982	3.453	.840	6.01

* Area spot $m_2 = \frac{\pi \bar{d}^2}{9 \times \text{B.A.F.}}$ (from Zeide 1985)

** Not known. Plot 582 had 31 sample trees not identified.
Plot 577 had 21 sample trees not identified.

may have an overestimate of species numbers. The data were collected in 40 0.040 ha sub-plots equally allocated at random within four 4.047 ha contiguous strata. Thus the data may not be strictly permissable (Whitmore 1984) but they must be very close to the true relationship.

The form of the curve is given by the equation:

$$\text{No. of species} = 428.6 \exp[-1.37568 * \text{PLOT AREA}^{-0.25982}]$$
$$(R = 0.9986)$$

Table 3 shows some species numbers for areas up to and just outside the range of data for this experiment and some other published data for comparison.

With respect to Plot 408, it is clear that the species limit has not been reached even though the curve is rapidly levelling off. It is expected that rarer species would continue to be added with further increases in area. But this must not be carried to extremes as changes in habitat would soon negate the usefulness of the species/area curve.

The effect of plot area is further discussed below in relation to logging. Papua New Guinea and South-East Asian rainforests are considerably richer than our sample, whereas the African rainforest is poorer.

Table 4 shows other comparative measures of species occurrence for each plot. Plot areas for the prism estimates are calculated from a formula derived from Zeide (1985).

Species richness is best indicated by number of species, and diversity by the Shannon-Wiener index, H_s (Stocker *et al.* 1985). In Table 4 this latter index is given as log base 2 following Lloyd and Ghelardi (1964) and Cox (1967) as well as log base e (Stocker *et al.* 1985 and others), but in other tables only base 2 is used. Base e figures may be derived from base 2 figures through a multiplier 0.69315. Evenness uses the commonly applied ratio of H_s to $H_{max.}$ (Lloyd & Ghelardi 1964). Another useful measure that seems not to have been used much is simply the mean number of stems per species (Richards 1952; Wyatt-Smith 1949). For plots of near equal size or over time in the one plot it gives an easily visualized measure of "uniformity" which is not the case with the other measures used.

Despite the low dependence of richness and diversity on plot size (Stocker *et al.* 1985) it should be taken into account with plots of small size. The species area relationship has already been mentioned and using the same data, the Shannon-Wiener index and evenness vary as in Table 5.

Assuming that other sets of data behave in a similar fashion the information given in Table 4 can be better understood.

TABLE 5. Relationship of diversity and evenness with area (from Plot 408)

Area	Shannon Wiener Index, Base 2	Evenness
0.10	4.60	.90
0.25	5.13	.87
0.50	5.42	.85
0.75	5.56	.84
1.00	5.64	.83
1.25	5.70	.83
1.50	5.74	.83
1.62	5.76	.83
1.75	5.78	.83
2.00	5.80	.82

TABLE 6. Examples of species distribution in north Queensland rainforests.

Frequency of Individuals on Plot	No. of Corresponding Species	
	Plot 591	Plot 594
1	31	28
2	12	7
3	6	3
4	3	5
5	6	2
6	3	1
7	3	—
8	1	—
9	1	—
10	1	—
11	1	—
13	1	—
15	1	—
16	1	1
19	—	1
22	1	—
23	1	—
24	1	—
28	2	—
30	1	—
33	—	1
39	—	1
64	—	1
385	—	1
No. of Trees	391	643
No. of Species	77	52
$H_s(2)$	5.351	2.614
Evenness	.854	.459
Uniformity	5.08	12.37

When the diversity of Plot 207 is adjusted for the area relationship its value rises and the difference between it and the other three granite plots are not as great as indicated in Table 4. It is probable then that most granite soils (and probably metamorphic soils as well, Plot 626/2) carry forest of similar diversity. Plot 594 is markedly less diverse and this agrees with its visual appearance due to the preponderance of rose butternut. As indicated above, evenness is not easily conceptualized since in fact maximum evenness occurs with maximum diversity. Thus Plot 594 has low evenness also, that is, there is a large disparity in the numbers of individuals in each species. Uniformity on the other hand is a simple measure that reflects relative richness in species occurring on a plot. In Table 4 it can be seen that Plot 594 is much more uniform than Plot 591 (areas similar) whereas 626/2 is similar or only slightly more uniform than 625/2. Diversity as measured by the Shannon-Wiener index shows the same but inverse relationship — but has the advantage of being insensitive to plot area above about 0.5 ha (Table 5).

Taking two plots of nearly equal size as examples, it is helpful to show the distribution of species numbers from which these measures were calculated and to indicate what they actually represent in the forest (Table 6).

Despite their great value in assessment work the prism "plots" may not lend themselves well to these ecological measures. This is because with the high basal area factors found useful in rainforest work, small trees of about 10 or 20 cm D.B.H. are sampled on very small areas and it is just in these sizes (and smaller) that we find the greatest concentration of species. For example, in Plot 591, 22 of the 58 species occurring in the 10-20 cm D.B.H. size class grow no bigger. (See also Whitmore 1984). Therefore the slightly lower diversities and higher uniformities may be an artifact of the method. Despite this possibility, the relatively low diversity of Plot 701 is pointed out. This is in an area of lowland basalt (see Table 1)

TABLE 7. Mortality data — virgin rainforest in north Queensland.

	Size Distribution in 10 cm Size Classes											
	10	20	30	40	50	60	70	80	90	100+	TOTAL	B.A.
591 — .4087 ha; 32.2 years												
No. ha^{-1}y^{-1}	3.19	0.68	1.06	0.38	0.30	0.08	0	0	0	0.08	5.78	0.478
% of Initial	0.52	0.33	0.84	0.62	1.24	0.78	0	0	0	0.62%	0.54%	0.68%
625/2 — .2023 ha; 15.7 years												
No. ha^{-1}y^{-1}	6.30	0.63	0.94	0.31	0	0	0.31	0	0	0.31	8.82	0.718
% of Initial	0.88	0.53	1.19	0.45	0	0	1.06	0	0	6.37%	0.83%	1.15%
% with slip*	1.14	1.06									1.07%	1.27%
207 — .1012 ha; 10.9 years												
No. ha^{-1}y^{-1}	0.91	0.91	0.91	0	0	0	0	0	0	0	2.72	0.145
% of Initial	0.15	0.61	0.71	0	0	0	0	0	0	0	0.28%	0.21%
MEAN OF PLOTS ON GRANITE — .7122 ha; 58.8 plot years												
No. ha^{-1}y^{-1}	3.74	0.69	1.00	0.34	0.22	0.06	0.06	0	0	0.12	6.21	0.501
% of Initial	0.57	0.38	0.42	0.55	0.94	0.59	0.19	0	0	1.63%	0.59%	0.74%
626/2 — .2023 ha; 12.0 years												
No. ha^{-1}y^{-1}	5.36	1.64	0	0.41	0	0.41	0	0.41	0	0	8.24	0.601
% of Initial	1.10	1.19	0	0.92	0	2.77	0	8.37	0	0	1.00%	1.19%
594 — .4857 ha; 12.0 years + .2838 ha; 19.9 years												
No. ha^{-1}y^{-1}	7.41	3.05	0.96	0.26	0.17	0.09	0	0	0	0	11.94	0.481
% of Initial	1.18	0.87	0.46	0.26	0.35	1.05	0	0	0	0	0.88%	0.559%

*See text.

regarded by many people as the acme of rainforest diversity (Anon 1985?).

Further details of family representation and numbers of species at the initial measure of these plots is given in Appendix 1. The full specific composition is available on request. One must guard against the impression, gained from such lists, of a static rainforest community. As will be seen later considerable changes occur in even quite short time spans. For this reason the date has been included in Table 4. It must also be stressed that this sample does not include all families in north Queensland rainforests.

Mortality and recruitment in virgin rainforest

Using only the five plots with more than 10 years of record and weighting the mean for the plots on granite for plot area and length of record, we find death rates as given in Table 7.

The percentage figures include recruits as they participate in deaths. They effect only the first size class and the total figures. (Migration from class to class has been ignored but it does not affect the total figures.)

It is of interest to note that in Plot 625/2 a landslip occurred in 1979. In all, eight trees (40 trees/ha) were affected, all below 30 cm D.B.H. Though they are natural deaths they have not been included in the overall mean for the granite plots. Catastrophes of this nature are a real factor in the environment of the plots but are not adequately sampled by such a small number and area of short term plots, so are best handled separately, noting that they are real and must be important in the long term.

The mortality data were sorted into five year periods to determine any historical correlation among themselves or with climatic factors (Table 8). Because all plots are not measured at the same time there is some inaccuracy in this grouping but major trends would still be evident.

The striking feature of this breakdown is the high death rate in 1975-80 and that in this period only, it extends to the larger trees. All four plots, measured over this period, show the same trend. Rainfall records (for Atherton, Mareeba and Cairns) show very dry periods in 1978 and 1980 and the Bureau of Meteorology reports widespread drought conditions in 1977-79 (Anon 1985). It is tempting and reasonable to link the extra deaths to this dry period but other dry periods do not show up so clearly, for example, those of 1960/61 and 1965/66.

Foley (1957) reports a long term trend of rainfall decline in the north-east Australian tropics, extending from 1896 up to 1947. Such a trend could have important consequences for the composition and structure of rainforest especially when superimposed on short term drought effects, as mentioned above.

The specific composition of deaths reveals no clear trends. Trees commonly regarded as pioneers, such as *Polyscias*, and those of more mature rainforest, such as *Beilschmiedia*, are represented. In the five plots, deaths occurred in 72 species of which 27 are included in species "lost" to a particular plot. Appendix II lists these species. By comparing with Appendix I it can be seen that deaths occur in 54% of the families represented. The absence of deaths in the remainder is felt to be more an effect of small sample size rather than of specific factors.

Recruitment data are given in Table 9. They include 55 species in the five virgin plots under discussion of which 15 were new to a particular plot. Full particulars are in Appendix III. Pioneer species were noticeably absent as were many of the "intermediate" type species. Since pioneer species are included in the mortality data this is a clear indication that they are being lost from the stand. These species require a massive crown disturbance (Whitmore 1984, Chap. 8) or a large

TABLE 8. Weighted mean numbers per hectare, of deaths in five-year periods (Plots 591, 625/2*, 207, 626/2, 594)

Year	No. of Plots	10	20	330	40	50	60	70	80	90	100+	TOTAL	B.A.
1953-55	2	7.8	3.4	—	—	—	—	—	—	—	—	11.2	0.257
1955-60	3	22.1	9.0	10.0	1.0	—	—	—	—	—	—	42.2	1.989
1960-65	3	29.1	9.0	4.0	3.0	1.0	1.0	—	—	—	—	47.2	2.381
1965-70	2	21.5	2.5	6.3	1.3	1.3	—	—	—	—	—	32.8	1.577
1970-75	4	35.6	4.6	1.8	1.8	1.8	—	—	—	—	—	45.6	1.759
1975-80	4	32.8	18.2	5.5	1.8	0.9	1.8	0.9	0.9	0	1.8	64.7	6.462
1980-84	3	12.3	3.4	2.2	1.1	1.1	—	—	—	—	—	20.1	1.062

*Landslip deaths in 1979 are excluded.

TABLE 9. Summary of temporal changes in north Queensland virgin rainforest.

Plot	591	625/2*	207	626/2	MEAN	594
Area (ha)	0.4087	0.2023	0.1012	0.2023	(Previous 4 Plots)	0.4857 / 0.2838
No. of Years	32.2	15.7	10.9	12.0		31.9
Annual Recruitment						
No. of Species Plot^{-1}	0.65	1.15	0.46	0.67	N/A	0.99
No. of New Spp Plot^{-1}	0.03	0.32	0	0.08	N/A	0.29
No. of Trees ha^{-1}	3.27	7.87	8.16	4.94	6.06±3.75	4.01
B.A.m^2 ha^{-1}	0.027	0.066	0.073	0.042	0.052±0.033	0.038
Annual Mortality						
No. of Species Plot^{-1}	1.06	1.59	0.18	1.33	N/A	0.71
No. of "Lost" Spp Plot^{-1}	0.31	0.32	0.09	0.67	N/A	0.18
No. of Trees ha^{-1}	5.78	11.34	2.72	8.24	7.02±5.82	11.95
B.A.m^{2}ha^{-1}	0.478	0.789	0.445	0.601	0.503±0.431	0.481
No. Species Plot^{-1}						
Start	77	61	28	48	N/A	52
End	68	61	27	41	N/A	56
Diversity H_s (base 2)						
Start	5.351	5.342	4.008	5.013	4.929±1.008	2.614
End	5.278	5.364	3.974	4.924	4.885±1.013	2.809
Evenness						
Start	0.834	0.901	0.834	0.898	0.867±0.060	0.477
End	0.867	0.904	0.836	0.919	0.882±0.060	0.506
Uniformity						
Start	5.08	3.10	3.14	3.35	N/A	12.37
End	5.27	2.92	3.48	3.71	N/A	9.47
Annual % B.A. Change	—0.16%	—0.43%	+0.59%	—0.15%	—0.04%±.70%	+0.09%

*Including landslip effects. N/A: Not applicable due to various plot sizes.

gap (Whitmore 1984, Chap. 7) so this loss is a natural consequence of their ecology but it also implies the presence of such disturbances in virgin rainforest. In fact, all species require some such opportunity before their seedlings can make positive growth (Anon 1986a). This partly explains the gap in the size class distribution from seedlings, present in many thousands per hectare (Anon 1986a), to the first tree size class (400-600/ha).

A summary of these changes in time is given in Table 9 where species figuring in recruitment and mortality are reported on an annual basis to account for different periods of measurement.

Plots 591, 207 and 626/2 show a decrease in species numbers and diversity with time, (the size of the latter measure is not stressed here, only the direction of change) and as a consequence uniformity increases. Evenness increases in all plots. This indicates a rise towards maximum diversity with the lesser number of species, but it is not enough to counter the downward "species effect" on the diversity measure. This over-riding effect of species richness in this non-rigorous environment, supports the suggestion of Tramer (1969).

Plot 625/2 shows a slight rise in diversity and decrease in uniformity, probably due to the absence of the "species effect" (species numbers do not fall) as well as to the higher net decrease in stems.

The figures not dependent on plot size for these four plots have been averaged, though the very wide confidence bands must be noted.

Plot 594 is very different in many respects, especially in that species numbers, diversity and evenness rise and uniformity falls. Such movements support the earlier suggestion that this is a young stand, as also does its higher death rate (59% of which is made up by rose butternut).

Effects of Logging

The data from a number of plots that have been logged within their lifetime have been brought together in Table 10 so that the effects of the logging operation itself, apart from later recovery, can be quantified. Some of these data have already been published (Nicholson 1985) with the emphasis on logging damage to the commercial stand. Here, all species are considered as being of equal value ecologically. Though the plots are grouped into "virgin and second cut" and "granite and basalt soil types", the major effects of logging depend on its intensity and this is borne out by all studies both here and overseas (Nicholson 1958; Redhead 1960). Considering all plots, then, we see that removals vary from seven to 43 as logged stems and from zero to 268 as smashed stems amounting to a reduction of 1-34% of stems. Similarly basal area is reduced by between 8% and 43% (mostly by the logged component). For forest management these figures have to be supplemented with information on the commercial component of the stand as well as the consequent site damage. It has been shown (Nicholson 1985; Gillman *et al.* 1985) that these have been acceptable in the past and should improve with the current strict logging and environmental controls. These aspects are being kept under review.

With regard to the species composition, it apears that logging reduces species numbers. However, there are a number of considerations that negate this apparent loss. Firstly, the artificial lower assessment limit of 10 cm diameter neglects the store of smaller plants. For example, in both Plots 408 and 625/5 three of the "lost" species occur on the plots as plants between 6 m high and 10 cm diameter and in Plot 615/3 two of the three "lost" species are similarly found.

Second, there is a strong relationship between both plot size and severity of logging with the percentage species lost. The data in Table 10 were used to develop the following equation:

$$\text{\% species lost} = 0.2344 - 10.402 \text{ Plot area} + 0.5450 \text{ \% B.A. removed.}$$
$$(R^2 = 0.816)$$

This equation adequately explains the apparent anomalies of Table 10 where, for example, Plots 408 and 207 with similar basal area removals show such dissimilar species losses. When solved, as in Figure 1, it shows very clearly that even with heavy logging no species are lost when an adequate area is sampled. (The intercepts along the x axis).

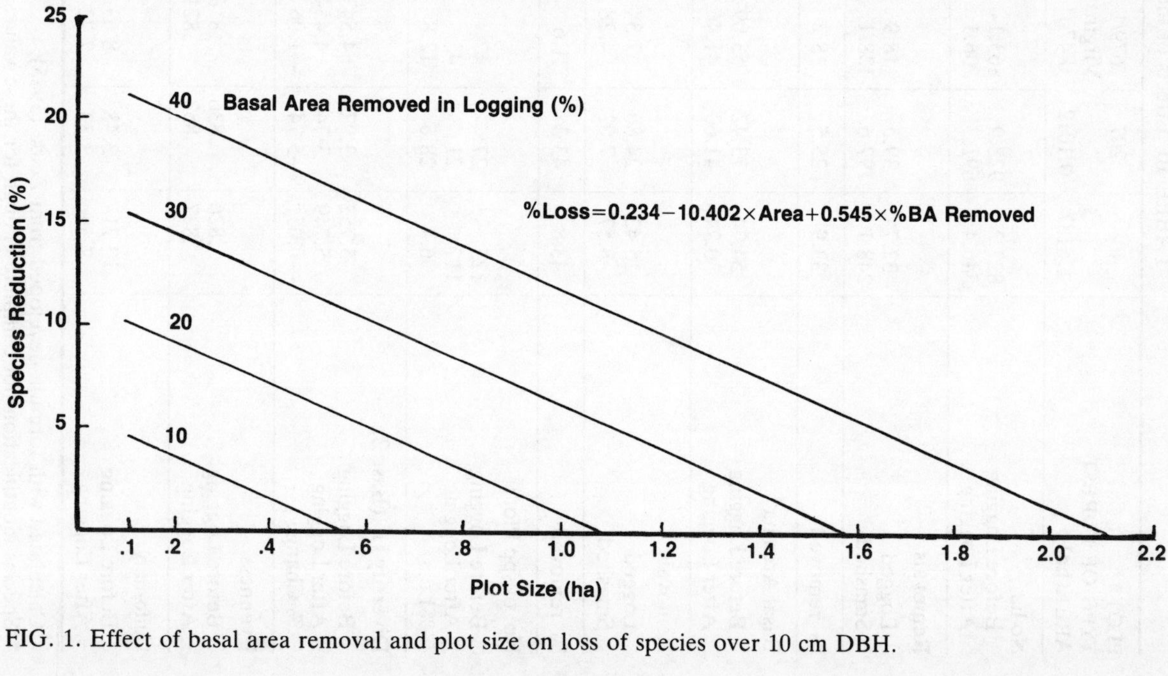

FIG. 1. Effect of basal area removal and plot size on loss of species over 10 cm DBH.

TABLE 10. Effects of the logging operations in north Queensland rainforests.

PLOT	408	207	679/1	582	577	625/5	615/3	615/6	616/1	619/1	618/1	640
TYPE OF FOREST			Virgin on Granite					2nd Cut on Granite			2nd Basalt	
AREA(ha)	1.6187	0.1012	0.470*	0.605*	0.314*	0.2023	0.2185	0.2064	0.2084	0.3966	0.2003	0.335*
No ha^{-1}												
Before Logging	827.2	928.9	863.1	911.9	804.3	716.8	814.6	939.9	734.2	746.3	723.9	742.3
After Logging	546.4	691.7	706.1	618.6	553.5	440.0	759.7	930.2	604.6	726.2	619.1	637.1
Removals												
Logged	32.7	39.5	18.9	25.6	18.2	19.8	9.2	9.7	43.2	7.6	25.0	9.4
Smashed	248.1	197.6	138.1	267.7	232.7	257.0	45.8	0	86.4	12.6	79.9	95.8
% removed	33.9	25.5	18.2	32.2	31.2	38.6	6.7	1.0	17.7	9.3	14.5	14.2
Basal Area ha^{-1}												
Before Logging	54.74	73.42	53.09	56.26	37.94	48.05	61.51	80.62	55.23	57.59	50.16	37.50
After Logging	30.74	41.60	41.46	37.02	25.86	27.24	51.85	73.93	40.59	51.24	36.20	32.33
Removals												
Logged	15.42	24.80	7.34	11.40	8.41	9.00	6.91	6.69	13.44	5.12	11.08	3.33
Smashed	8.57	7.02	4.28	7.85	3.67	11.81	2.74	0	1.20	1.24	2.88	1.83
% removed	43.8	43.3	21.6	34.2	31.8	43.3	15.7	8.3	26.5	11.0	27.8	13.8
No. of spp Plot^{-1}												
Before Logging	125	27	47	65**	55**	53	54	57	62	75	50	69
After Logging	117	21	43	61	54**	44	51	57	55	73	45	65
%Loss	6.4	28.6	12.8	—	—	17.0	5.6	0.0	11.3	2.7	10.0	5.8
Diversity H_s (Base 2)												
Before Logging	5.752	3.974	4.543	**	**	5.123	5.076	5.162	5.311	5.519	4.893	5.263
After Logging	5.749	3.746	4.454	**	**	4.969	5.012	5.172	5.184	5.473	4.937	5.159
% Change	—0.05	—5.74	—1.96	—	—	—3.00	—1.26	0.19	—2.39	—0.83	0.89	—1.98
Evenness												
Before Logging	.826	.836	.818	**	**	.894	.882	.885	.892	.886	.896	.862
After Logging	.837	.853	.821	**	**	.910	.883	.887	.897	.884	.900	.857
Uniformity												
Before Logging	10.71	3.48	8.63	**	**	2.74	3.30	3.40	2.47	3.95	2.90	3.60
After Logging	7.56	3.33	7.72	**	**	2.18	3.25	3.37	2.29	3.95	2.82	3.28

*Calculated with formula developed from Zeide (1985).
**Species identification not complete enough for these values.

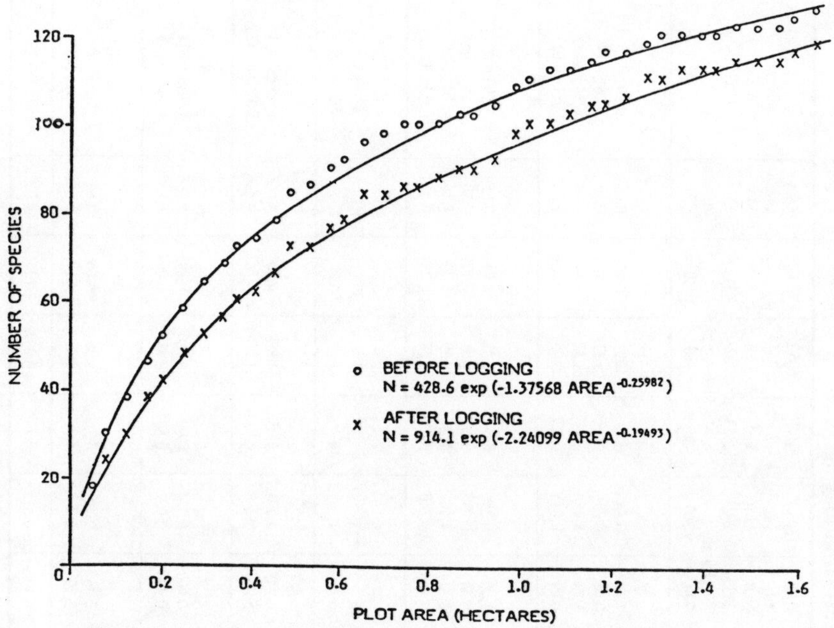

FIG. 2. Effect of lower limit of measurement on species numbers (from Plot 408).

It should be noted that this result is obtained without appeal to the store of smaller plants, which would only have the effect of reducing the area required to register a no species loss. It is acknowledged that a wider range of data or a logging area with a very rare species, may require larger sample areas. In our largest data set (Plot 408), when both the before and after logging species/area curves are plotted on the same graph, as in Figure 2, it is clear that the two curves are approaching each other even though the area sampled is only 1.6 ha. This is independent confirmation that in general, losses recorded in small plots are of little consequence in real life logging situations.

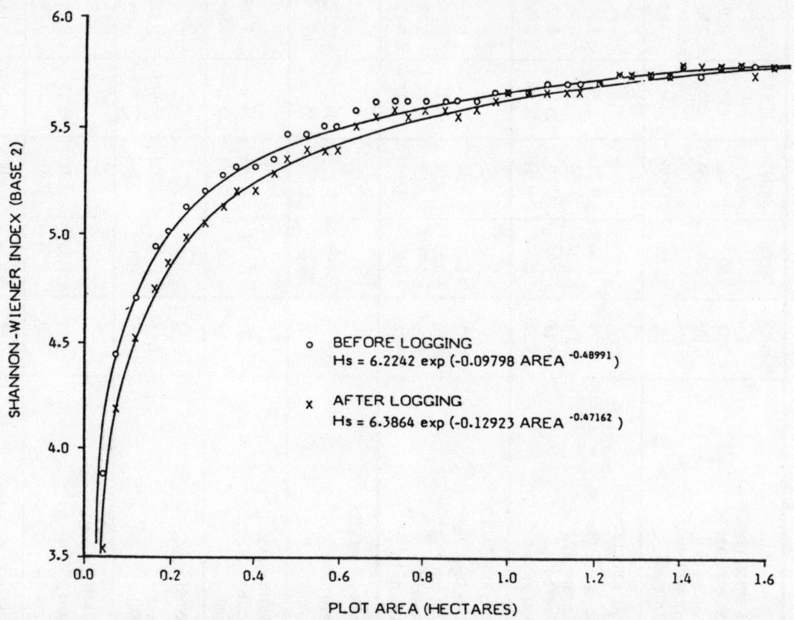

FIG. 3. Effect of logging and plot size on diversity (from Plot 408).

TABLE 11. Response of north Queensland rainforest from logging.

PLOT	615/3	615/6	616/1	625/5	625/3	625/1	625/4	Mean of Granite Plots	618/1	608/1	610/1	626/1	614/1	609/1	Overall Mean 13 Plots
YEARS SINCE LOGGING*	1	1	2	0	4	17	17		0	2	2	15	3	1	
YEARS OF MEASURE	30.2	30.2	23.8	15.8	15.9	15.8	15.9		16.0	31.0	30.5	12.0	29.9	30.9	
PLOT AREA (ha)	.2185	.2064	.2084	.2023	.2023	.2023	.2023		.2003	.2023	.2023	.2023	.2129	.1518	
PARENT MATERIAL	Granite								Basalt		Metamorphic			Rhyol.	
ALTITUDE (m)	1 100	1 100	1 130	730	1 065	700	790		740	760	440	360	30	770	
ANNUAL RECRUITMENT															
No. of Spp. Plot^{-1}	0.33	0.60	1.39	1.39	1.01	0.70	1.70	1.02±0.46	1.13	0.84	0.49	0.92	0.50	0.65	0.90±0.25
New Spp. Plot^{-1}	0.03	0.07	0.76	0.57	0.06	0.32	0.44	0.32±0.26	0.44	0.29	0.16	0.17	0.03	0.10	0.26±0.14
Trees ha^{-1}	3.18	4.97	11.69	15.02	7.77	4.38	16.79	9.11±5.02	8.74	8.45	6.32	8.24	5.81	7.04	8.34±2.43
B.A. m^2 ha^{-1}	0.026	0.042	0.106	0.161	0.066	0.038	0.147	0.084±0.051	0.091	0.086	0.052	0.068	0.050	0.063	0.077±0.025
ANNUAL MORTALITY															
No. of Spp. Plot^{-1}	0.53	0.70	0.84	0.63	0.57	0.76	0.57	0.66±0.11	1.06	0.55	0.59	0.83	0.74	0.65	0.69±0.09
Lost Spp. Plot^{-1}	0.10	0.23	0.29	0.13	0	0.13	0.06	0.13±0.09	0.38	0.03	0.20	0.08	0.17	0.10	0.15±0.06
Trees ha^{-1}	5.00	5.77	6.25	6.88	3.73	5.94	4.97	5.51±0.95	8.43	4.46	4.70	6.59	7.70	8.10	6.04±0.88
B.A. m^2 ha^{-1}	0.246	0.324	0.597	0.277	0.120	0.437	0.143	0.306±0.154	0.738	0.240	0.030	0.287	0.333	0.310	0.314±0.115
NUMBER OF SPECIES															
Start	51	57	51	44	62	53	53	53.0±5.2	45	43	47	47	39	42	48.7±3.9
End	49	52	62	50	63	56	59	55.9±5.3	46	51	46	48	35	42	50.7±4.8
DIVERSITY (BASE 2)															
Start	5.014	5.170	5.066	4.969	5.489	4.841	5.063	5.087±0.188	4.937	4.994	5.052	4.888	4.708	4.666	4.989±0.124
End	4.897	5.067	5.311	5.139	5.464	4.933	5.166	5.140±0.186	5.003	5.128	4.925	4.934	4.587	4.636	5.015±0.147
EVENNESS															
Start	0.884	0.886	0.893	0.910	0.922	0.845	0.884	0.889±0.023	0.900	0.920	0.909	0.880	0.891	0.865	0.891±0.014
End	0.872	0.889	0.892	0.911	.914	0.849	0.878	0.886±0.021	0.906	0.904	0.892	0.883	0.894	0.860	0.888±0.011
UNIFORMITY															
Start	3.31	3.33	2.49	2.18	3.03	3.55	3.58	3.07±0.50	2.76	2.70	3.70	4.26	4.05	3.64	3.28±0.37
End	3.20	3.54	2.47	2.28	3.19	3.27	3.68	3.09±0.49	2.72	2.76	4.00	4.23	4.17	3.52	3.31±0.37
ANNUAL B.A. % CHANGE	+0.62%	+0.37%	+0.38%	+0.78%	+1.08%	+0.25%	+1.45%	+0.70%±0.40%	+0.40%	+0.81%	+0.74%	+0.96%	+0.17%	+0.29%	+0.63%±0.23%

*To start of measure period.

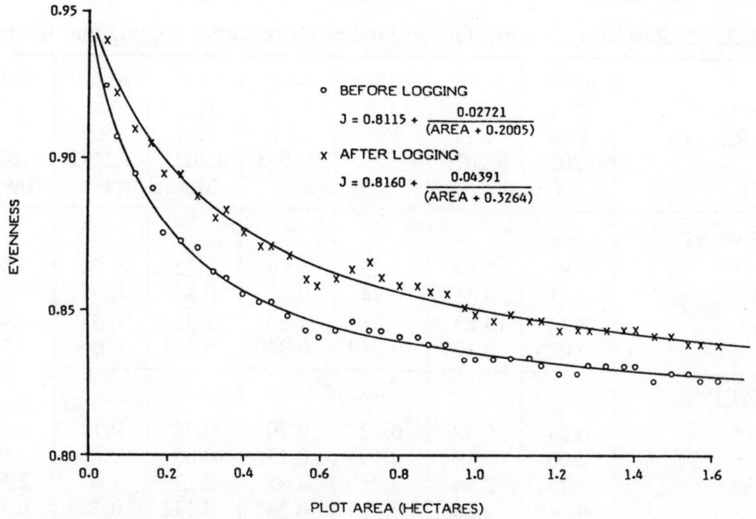

FIG. 4. Effect of logging and plot size on evenness (from Plot 408).

Percentage change in diversity is significantly correlated with both plot area and basal area removed, although Plot 618/1 is anomalous. Percentage change in evenness is significantly correlated with basal area removed, but not with plot area.

The effects of plot size on diversity and evenness have been examined in Plot 408, and equations fitted as shown in Figures 3 and 4. In this case the effects of logging on diversity is slight, but evenness is obviously increased, at least in part because of the reduction in number of species.

After Logging Response

Table 11 gives the after logging responses for 13 plots. Means have been calculated for the granite plots and all plots as individual plot areas are approximately equal.

Most plots show an increase in species numbers with time after logging, with a mean increase of three for the granite plots and two for all plots. However, Plots 615/3, 615/6, 610/1 and 614/1 show a decrease. Apart from the sampling limit mentioned earlier which could nullify this loss, it could be a continuation of the general downward trend noted in the virgin plots. The fact that the decrease occurs in lightly logged plots with a long record, tends to support this view as does the fact that these and all plots have a basal area increase.

Again there is a close positive correlation of diversity with species numbers, only Plot 625/3 having an inverse relationship, but here the movements are small and could be explained by the plot's low death rate, allowing stem numbers to increase. Thus the mean diversity for granite plots and all plots increases with time after logging, though the increase is not significant. Evenness and uniformity change little, with means falling and rising respectively. The values for diversity and evenness in these logged forests both at the start or end of the period, compare more than favourably with the virgin data (Table 9).

It is interesting to note the general similarity of diversity on granites, basalts and metamorphics which confirms the observation made earlier about low altitude basalts not necessarily carrying the most diverse rainforest.

Silvicultural Treatment Response

Volck (1968) describes the silvicultural treatments applied to selectively logged rainforest from about 1946. The basic aim was to remove all non-commercial trees and saplings to allow the remaining commercial species to better utilize the site. The effects of treatment have been covered in a number of papers, e.g., Volck (1968), Nicholson (1985), Anon (1983, 1986a) which show that always there is a marked growth response. In actual volume terms the yield is still low, however, and because the cost was high, no routine silvicultural treatment is currently practiced by the Department (Nicholson 1985). However, there are limited situations where it could be applied effectively (e.g., in privately managed rainforests.)

Even though only about 5 000 ha have been treated, the response will give important clues to the reaction from other severe perturbations. The

TABLE 12. Response of north Queensland rainforest after logging and treatment.

PLOT YEARS SINCE LOG/TR. YEARS OF MEASURE PLOT AREA (ha) PARENT MATERIAL	615/4 20, 2 17.0 0.2023 Granite	615/5 14, 2 16.8 0.2023	618/2 4, 0 26.7 0.2023 Basalt	608/2 17, 0 15.8 0.1955	610/2 19, 0 13.1 0.2023 Metamorphic	614/2 3, 0 29.9 0.2193	609/2 0, 0 29.6 0.2023 Rhyol.	Mean (7 Plots)
ANNUAL RECRUITMENT								
No. of Spp. Plot^{-1}	0.82	0.89	1.23	1.96	1.22	0.74	0.91	1.11 ± 0.39
New Spp. Plot^{-1}	0.41	0.36	0.94	0.82	0.84	0.37	0.27	0.57 ± 0.26
No. of Trees ha^{-1}	8.14	13.83	27.58	34.96	51.32	10.37	11.47	22.52 ±14.92
B.A. ha^{-1}	0.073	0.125	0.289	0.369	0.560	0.096	0.116	0.233± 0.168
ANNUAL MORTALITY								
No. of Spp. Plot^{-1}	0.24	0.42	0.52	0.70	0.38	0.17	0.41	0.41 ± 0.16
Lost Spp Plot^{-1}	0.06	0.12	0.19	0.13	0.08	0.0	0.07	0.09 ± 0.06
No. of trees ha^{-1}	1.75	2.94	3.33	4.85	2.27	1.07	2.53	2.68 ± 1.12
B.A. ha^{-1}	0.048	0.050	0.213	0.335	0.172	0.109	0.104	0.147± 0.194
NO. OF SPECIES								
Start	20	13	13	34	11	15	27	19.0 ± 7.94
End	28	17	33	45	21	28	33	29.3 ± 8.42
DIVERSITY (BASE 2)								
Start	3.367	3.100	3.123	4.774	2.846	3.231	4.286	3.532± 0.660
End	3.794	3.296	4.334	4.807	2.971	3.895	4.592	3.956± 0.621
EVENNESS								
Start	0.779	0.838	0.844	0.938	0.823	0.827	0.901	0.850± 0.049
End	0.789	0.806	0.859	0.875	0.676	0.810	0.910	0.818± 0.070
UNIFORMITY								
Start	5.30	6.23	2.38	1.94	3.45	3.87	2.96	3.73 ± 1.43
End	4.57	6.94	4.91	3.51	8.00	4.25	4.00	5.17 ± 1.54
ANNUAL B.A. % CHANGE	+1.93%	+2.20%	+6.61%	+4.42%	+4.82%	+2.68%	+1.76%	+3.49 ± 1.69

effect of the treatment itself has not been fully studied but as each of the plots in Table 12 has a pair in Table 11 comparisons can be made. Species numbers and diversity are much less, while evenness and uniformity move to levels indicating a slightly more uniform and less even composition.

The severe loss in species during treatment is a designed result which aims at better production on fewer species. (See also Terborgh 1973 and Plot 594 in this paper, which show that low diversity can lead to high productivity.)

Table 12 shows that treatment induces much more marked responses than logging. There is an increase in recruitment, less mortality and a marked increase in species numbers and diversity. There is less consistency in the other ecological measures but the trend is to less even species composition with a more uniform appearance. Following treatment there is often a tendency for small numbers of aggressive species to become better represented. This is the case in Plot 610/2 where *Alphitonia whitei* forms a high proportion of the recruitment, thus lowering evenness and increasing uniformity.

The very high and consistently positive basal area increase is worthy of notice indicating the growth response to treatment.

Observations on Natural Changes

Mention has been made earlier of drought, fire and landslides having effects on the stand structure and composition. Another important natural event of great current interest is the occurrence of cyclones. Cyclone "Winifred" is being discussed elsewhere in this volume (Unwin *et al.* 1986). Mention is made here of its effect on diversity, evenness and uniformity in two plots at Mission Beach that were severely damaged. Before and after the cyclone, Plots 614/1 and 614/2 had the following measures:

	Plot 1		*Plot 2*	
	Before	*After*	*Before*	*After*
% of B.A. Destroyed	—	36.4	—	28.4
No. of Spp. per Plot	35	35	28	27
Diversity (base 2)	4.594	4.606	3.874	3.958
Evenness	0.896	0.898	0.805	0.832
Uniformity	4.23	3.29	4.32	3.74

The loss of one species in Plot 2 is apparent only, because it occurs as 50 stems per hectare below 10 cm diameter. With such a severe effect, as shown by the loss of basal area, it is perhaps surprising that diversity and evenness rise while uniformity falls. Such movements do indicate, however, that the cyclone left the forest in a sound regenerative condition.

Plot 3 of the Mission Beach series which, because it was in the lee of a ridge (but only about 50 m from Plot 2), suffered only one cyclone death but has apparently suffered severely from some unknown but probably pathological cause. Of 31 recruited stems of *Elaeocarpus angustifolius* Blume resulting from a 1955 treatment, 20 died between 1982 and 1986. Another similar event has been reported in which many trees of *Podocarpus neriifolius* D. Don. growing on the MacIllraith Range in north Queensland died in a period from about 1970 to 1977. It was estimated that half the stock, and in all sizes, died. No cause could be pinpointed (G. C. Stocker pers. comm.).

These are just some examples of other perturbations that can occur in nature and that alter the composition and structure of the rainforest.

Discussion

Any significant use of virgin rainforest will cause some change in the total ecosystem. For the study of this system in its natural state, the Queensland Department of Forestry has set aside a number of scientific areas in strategic locations which will only be used for non-destructive scientific studies (see also Davidson 1985). To date 17 such reference areas, containing rainforest, and totalling 9 039 hectares have been established in Queensland. In addition there are very large areas of protected forest, the National Parks system and numerous plots within State Forests maintained by CSIRO.

This paper has shown that in the case of some virgin rainforest there is a decline in species richness and diversity with time. This conflicts with often made statements that environmental stability leads to greater diversity (see refs. in Bourgeron 1983). Rather it agrees with others, for example, Connell (1978, 1979), Middleton and Merriam (1985) who hold that a certain level of disturbance is necessary to maintain a high level of diversity. Connell (1979) proposes that highly diverse communities are usually in a state of non-equilibrium in a process moving towards an equilibrium of lower diversity. Connell *et al.* (1984) gives some supportive data.

The data presented in this paper show that logging tends to reverse this natural decline in diversity, though not always completely, while silvicultural treatment always does. This supports Connell's "intermediate hypothesis" and also that the compensatory mechanisms postulated are external to the system (Connell 1979). However, there still remain indications that within this intermediate disturbance regime chance plays a large role in determining which species are advantaged by having seedlings already established and able to respond quickly. This plays a big part in Malaysian forests (Nicholson 1965) and allows one to postulate groups of species (perhaps the three groups Connell proposes) with similar ecological requirements.

Management practices specified by the Department are consistent with the goals of the National Conservation Strategy for Australia (Anon 1986b) and the objectives of multiple use management as described in Anon (1984) and Shea (1986). We believe they clearly maintain the rainforest as an intact and viable ecosystem allowing all the ecological process to proceed while allowing the use of their valuable resources.

Acknowledgements

Many members of the staff of the Queensland Department of Forestry have had an input to the work described in this paper. In recent times particular mention must be made of H. E. Volck, A. L. Bragg, M. G. Keys and A. F. Wright who have established, measured and maintained these plots over many years. Their work is gratefully acknowledged along with many others who have helped from time to time. The Department's permission to publish is acknowledged.

References

Anon. (1983) *Rainforest Research in North Queensland*. Queensland Department of Forestry, Brisbane.

Anon. (1984) *Forest Management in Queensland*. Queensland Department of Forestry, Brisbane.

Anon. (1985) *Queensland Year Book*. Queensland Government, Brisbane.

Anon. (1985?) Downey Creek. *Tropical Rainforest Society Newsletter*. **1**.

Anon. (1986a) *Research Report 1985*. Queensland Department of Forestry, Brisbane.

Anon. (1986b) Foreword. In: *Tropical Plant Communities* (eds H. T. Clifford and R. L. Specht). Department of Botany, University of Queensland, St Lucia.

Beers T. W. & Miller C. I. (1964) Point sampling, research results, theory and application. *Purdue University, Research Bulletin*, **786**.

Bitterlich W. (1984) *The Relascope Idea. Relative Measurements in Forestry*. Commonwealth Agricultural Bureau, Farnham Royal, Slough.

Bourgeron P. S. (1983) Spatial Aspects of Vegetation Structure. In: *Ecosystems of the World 14A, Tropical Rain Forest Ecosystems* (ed. F. B. G. Golley) pp. 29-47. Elsevier, Amsterdam.

Brunig E. F. (1983) Vegetation Structure and Growth. In: *Ecosystems of the World 14A, Tropical Rain Forest Ecosystems*. (ed. F. B. Golley) pp. 49-75. Elsevier, Amsterdam.

Connell J. H. (1978) Diversity in Tropical rainforests and Coral Reefs. *Science (N.Y.)* **199**, 1302-10.

Connell J. H. (1979) Tropical rainforests and coral reefs as open non-equilibrium systems. In: *Population Dynamics* (eds R. M. Anderson, B. D. Turner & L. R. Taylor) pp. 141-63. Blackwell Scientific, Oxford.

Connell J. H., Tracey J. G. & Webb L. J. (1984) Compensatory recruitment, growth and mortality as factors in maintaining rainforest tree diversity. *Ecol. Mono*. **54**, 141-64.

Cox G. W. (1967) *Laboratory Manual of General Ecology*. W. M. C. Brown Company, Dubuque, Iowa.

Davidson J. (1985) Economic use of tropical moist forests. *Commission on Ecology Papers, No. 9*. IUCN.

Dawkins H. C. (1958) The management of natural tropical high-forest with special reference to Uganda. *Imp. For. Inst. Paper*, **34**.

Foley J. C. (1957) Droughts in Australia. *Bureau of Meteorology, Bulletin* **43**.

Gillman G. P., Sinclair D. F., Knowlton R. & Keys M. G. (1985) The Effect on some soil chemical properties of the selective logging of a north Queensland rainforest. *For. Ecol. Manag*. **12**, 195-214.

Lloyd M., & Ghelardi R. J. (1964) A table for calculating the "Equitability" component of species diversity. *J. Anim. Ecol*. **33**, 217-25.

Middleton J. & Merriam G. (1985) The rationale for conservation. Problems from a virgin forest. *Biol. Cons*. **33**, 133-45.

Nicholson D. I. (1958) An analysis of logging damage in a tropical rain forest, North Borneo. *Mal. For*. **21**, 235-45.

Nicholson D. I. (1965) A study of virgin forest near Sandakan, North Borneo. In: Ecological Research in Humid Tropics Vegetation, Kuching 1963, pp. 67-87.

Nicholson D. I. (1979) *The Effects of Logging and Treatment on the Mixed Dipterocarp Forests of South East Asia*. Food and Agriculture Organization of the United Nations, FO: MISC/79/8.

Nicholson D. I. (1985) The Development of Silviculture in North Queensland Rainforests. In: *Managing the Tropical Forest* (eds K. R. Shepherd & H. V. Richter) pp. 203-23. Development Studies Centre, The Australian National University, Canberra.

Nicholson D. I., Henry N. B., Rudder J. & Anderson T. M. (1983) The Research Basis of Rainforest Management in North Queensland. Paper to the XV Pacific Science Congress, Dunedin, New Zealand.

Odum E. P. (1971) *Fundamentals of Ecology. W. B. Saunders, Philadelphia*.

Paijmans, K. (1970) An Analysis for Four Tropical Rain Forest Sites in New Guinea. *J. Ecol*. **58**, 77-101.

Redhead J. F. (1960) An Analysis of Logging Damage in Lowland Rainforest, Western Nigeria. *Inform. Bull. Dept. For. Res. Nigeria*, **10**, 5-16.

Richards P. W. (1952) *The Tropical Rain Forest*. Cambridge University Press, Cambridge.

Shea G. M. (1986) Forestry Management. In: *Tropical Plant Communities* (eds H. T. Clifford & R. L. Specht) pp. 113-21. Department of Botany, University of Queensland, St Lucia

Stocker G. C., Unwin G. L. & West P. W. (1985) Measures of richness, evenness and diversity in tropical rainforest. *Aust. J. Bot*. **33**, 131-37.

Terborgh J. (1973) On the notion of favourableness in plant ecology. *Am. Nat*. **107**, 481-501.

Tramer E. J. (1969) Bird species diversity: components of Shannon's Formula. *Ecology* **50**, 927-29.

Unwin G. L., Applegate G. B., Stocker G. C. & Nicholson D. I. (1986) Effects of cyclone "Winifred" on forests in north Queensland. *Proc. ecol. Soc. Aust*. **15**, 283-96.

Volck H. E. (1968) Silvicultural Research and Management in North Queensland Rain Forests. Paper to Ninth Comm. For. Conference.

Whitmore T. C. (1984) *Tropical Rain Forests of the Far East*. Clarendon Press, Oxford.

Wyatt-Smith J. (1949) A note on tropical lowland evergreen rainforest in Malaya. *Mal. For*. **12**, 58-64.

Zeide, B. (1985). Distance to point-sampled trees and their diameters. *For. Ecol. Manag*. **11**, 131-37.

APPENDIX 1. Composition of samples of north Queensland virgin rainforest by families, number of trees hectare^{-1} and number of species.

Plot No.+	591		625/2		408		207		626/2		594		679/1		679/2		582		577		701	
Area (Ha)	.4087		.2023		1.6187		.1012		.2023		.4857		.470*		.633*		.605*		.314*		.527*	
Family	No./ha	No. Sp.	No./ha	No. Sp.	No./ha	No. Sp.	No./ha	No. Sp.	No./ha	No.	No./ha	No. Sp.	No./ha	No. Sp.	No./ha	No. Sp.	No./ha	No. Sp.	No./ha	No. Sp.	No./ha	No. Sp.
Alangiaceae	2.4	1	4.9	1	6.8	1	–	–	–	–	–	–	–	–	–	–	–	–	–	–	17.9	1
Annonaceae	–	–	–	–	–	–	–	–	4.9	1	2.1	1	–	–	–	–	–	–	–	–	2.4	1
Apocynaceae	–	–	–	–	.6	1	–	–	4.9	1	12.4	1	–	–	–	–	–	–	–	–	1.6	1
Araliaceae	4.8	2	14.8	1	2.4	2	–	–	4.9	1	8.2	1	–	–	–	–	24.7	1	24.3	2	6.3	2
Arucariaceae	22.0	1	24.7	2	1.2	1	–	–	–	–	–	–	–	–	–	–	.3	1	13.2	1	–	–
Atherospermataceae	17.1	1	–	–	26.6	2	–	–	–	–	2.1	1	–	–	10.8	1	–	–	15.5	1	40.5	2
Balanopaceae	4.9	1	9.9	1	2.5	1	9.9	1	–	–	–	–	.3	1	–	–	.8	1	–	–	–	–
Blepharocaryaceae	–	–	–	–	–	–	–	–	14.8	1	792.7	1	–	–	–	–	–	–	–	–	–	–
Burseraceae	–	–	19.8	2	32.7	2	9.9	1	9.9	1	–	–	.4	1	.5	1	5.2	1	10.0	2	–	–
Corynocarpaceae	–	–	–	–	–	–	–	–	–	–	–	–	1.2	1	–	–	–	–	–	–	–	–
Cunoniaceae	36.6	3	49.4	2	80.2	4	59.3	2	–	–	–	–	147.2	1	94.1	1	116.1	3	54.6	1	19.0	3
Davidsoniaceae	–	–	–	–	–	–	–	–	–	–	–	–	–	–	17.1	1	–	–	–	–	–	–
Ebenaceae	2.4	1	–	–	3.7	1	–	–	–	–	–	–	–	–	–	–	9.1	1	26.1	1	–	–
Elaeocarpaceae	51.7	4	24.7	2	36.5	9	9.9	1	60.3	5	6.2	2	32.2	5	38.2	8	20.9	5	12.9	3	42.3	3
Escalloniaceae	9.8	1	9.9	1	12.3	2	–	–	–	–	–	–	–	–	11.1	1	–	–	–	–	–	–
Euphorbiaceae	26.8	3	9.8	2	22.3	3	–	–	103.8	2	49.4	3	–	–	43.6	1	33.1	1	38.3	4	58.2	3
Flacourtiaceae	–	–	–	–	3.7	2	–	–	–	–	–	–	–	–	.6	1	–	–	–	–	–	–
Flindersiaceae	90.5	3	34.6	3	101.3	3	29.7	2	69.1	4	111.2	3	108.2	4	57.5	4	31.6	4	49.7	4	8.2	1
Guttiferae	73.4	1	–	–	–	–	–	–	–	–	–	–	–	–	–	–	–	–	–	–	–	–
Himantandaceae	2.4	1	–	–	–	–	–	–	–	–	–	–	.3	1	–	–	.1	1	–	–	.3	1
Icacinaceae	26.9	2	39.6	2	65.5	3	88.9	2	9.9	1	–	–	–	–	6.5	2	18.2	1	.4	1	141.0	3
Lauraceae	102.4	12	158.2	8	107.6	19	237.2	1	148.3	6	27.0	8	93.9	7	86.5	7	52.1	8	40.5	5	111.0	14
Leguminosae	–	–	9.9	1	7.4	1	9.9	1	4.9	1	–	–	31.0	1	–	–	.7	1	.8	1	4.2	1
Meliaceae	–	–	–	–	3.0	3	–	–	–	–	2.1	1	15.2	1	4.6	1	14.8	1	5.3	2	14.3	2
Monimiaceae	–	–	9.9	1	.6	1	–	–	–	–	–	–	–	–	–	–	6.5	1	–	–	7.3	1
Moraceae	–	–	–	–	3.7	2	29.6	1	4.9	1	2.1	1	.2	1	1.4	1	1.0	1	2.4	1	–	–
Myristicaceae	–	–	–	–	–	–	–	–	–	–	–	–	–	–	–	–	–	–	–	–	26.6	1
Myrsinaceae	19.6	1	–	–	3.1	2	9.9	1	–	–	–	–	–	–	–	–	–	–	–	–	–	–
Myrtaceae	43.7	8	79.0	7	37.1	14	59.3	4	39.5	3	16.6	5	239.7	7	119.4	16	84.2	8	79.9	8	12.4	3
Ochnaceae	53.8	1	128.5	1	22.8	1	59.3	1	–	–	–	–	–	–	–	–	–	–	–	–	–	–
Oleaceae	4.9	1	–	–	25.3	1	39.5	1	–	–	–	–	–	–	–	–	–	–	–	–	8.7	1
Pittosporaceae	2.4	1	4.9	1	.6	1	–	–	–	–	–	–	–	–	–	–	–	–	–	–	–	–
Podocarpaceae	–	–	4.9	1	.6	1	–	–	–	–	–	–	–	–	–	–	–	–	–	–	–	–
Potaliaceae	–	–	–	–	.6	1	–	–	–	–	–	–	–	–	–	–	–	–	–	–	–	–
Proteaceae	34.0	8	64.1	8	43.7	10	69.2	4	69.1	6	96.9	7	25.3	5	61.3	8	91.2	9	13.5	2	41.9	3
Rhamnaceae	7.3	1	4.9	1	9.2	2	–	–	14.8	1	6.2	1	25.4	2	9.8	1	6.9	2	5.3	1	1.6	1
Rosaceae	–	–	–	–	–	–	–	–	–	–	–	–	–	–	–	–	–	–	–	–	3.8	1
Rubiaceae	80.7	4	88.9	3	24.1	3	49.4	2	19.7	3	–	–	–	–	–	–	–	–	–	–	2.0	1
Rutaceae	14.6	4	4.9	1	21.6	8	–	–	74.1	3	135.9	2	59.3	2	37.8	4	84.8	5	29.6	4	51.6	2
Sapindaceae	–	–	44.4	4	20.3	7	9.9	1	24.7	2	39.2	8	35.2	3	79.0	5	16.4	2	26.6	2	17.5	2
Sapotaceae	83.1	5	19.8	1	1.2	2	–	–	–	–	2.1	1	18.1	2	36.1	2	21.9	3	38.0	2	3.8	1
Siphonodontaceae	–	–	–	–	–	–	–	–	–	–	–	–	–	–	–	–	–	–	4.7	1	7.9	1
Sphenostemonaceae	–	–	4.9	1	–	–	9.9	1	–	–	–	–	–	–	–	–	–	–	–	–	–	–
Sterculiaceae	56.3	1	24.7	1	76.5	3	–	–	64.3	1	6.2	2	8.0	1	–	–	91.1	2	106.2	5	46.8	2
Stilaginaceae	7.3	2	–	–	.6	1	–	–	14.8	1	–	–	–	–	–	–	–	–	–	–	–	–
Strychnaceae	–	–	–	–	.6	1	–	–	–	–	–	–	–	–	–	–	–	–	–	–	–	–
Symplocaceae	–	–	–	–	1.9	1	–	–	–	–	–	–	–	–	–	–	6.1	1	–	–	–	–
Thymelaeaceae	9.8	1	–	–	–	–	–	–	4.9	1	–	–	–	–	–	–	–	–	–	–	–	–
Urticaceae	–	–	–	–	–	–	–	–	–	–	2.1	1	–	–	–	–	–	–	–	–	–	–
Verbenaceae	–	–	9.9	1	–	–	–	–	–	–	–	–	–	–	–	–	–	–	–	–	2.4	1
Winteraceae	–	–	–	–	6.2	1	–	–	24.7	1	–	–	–	–	–	–	–	–	–	–	–	–
Xanthophyllaceae	58.7	1	29.7	1	7.4	1	79.1	1	19.8	1	2.1	1	–	–	–	–	–	–	–	–	3.1	1
Unknown	4.9	2 Trees	–	–	3.1	5 Trees	–	–	–	–	2.1	1 Trees	22.1	2 Trees	–	–	173.9	31 Trees	206.0	21 Trees	1.9	1
TOTALS	956.7	77	934.3	61	827.2	125	869.6	28	795.8	48	1 323.8	52	863.0	47	716.0	66	911.9	64+	804.3	55+	706.9	60

+See Table 1 for plot type and site details. *Areas calculated with Zeide's (1985) formula.

APPENDIX II. Species over 10 cm with deaths — Virgin Rainforest Plots 591, 625/2*, 207, 626/2, 594.

Species	Lost to Plot Shown
APOCYNACEAE	
Alstonia muellerana	
ARALIACEAE	
Polyscias australiana	591, 626/2
Polyscias elegans	591
ARAUCARIACEAE	
Agathis microstachya	
ATHEROSPERMATECEAE	
Doryphora aromatica	
BLEPHAROCARYACEAE	
Blepharocarya involucrigera	
CUNONICEAE	
Ceratopetalum succirubrum	
Gillbeea adenopetala	591
EBENACEAE	
Diospyros sp.	591
ELAEOCARPACEAE	
Aceratium concinnum	626/2
Elaeocarpus angustifolius	591
Elaeocarpus arnhemicus	
Elaeocarpus largiflorens	591
Elaeocarpus sericopetalus	626/2
Elaeocarpus sp.	
Sloanea macbrydei	626/2
ESCALLONIACEAE	
Polyosma alangiancea	
EUPHORBIACEAE	
Macaranga subdentata	
FLINDERSIACEAE	
Flindersia bourjotiana	
Flindersia brayleyana	
Flindersia pimenteliana	
GUTTIFERAE	
Garcinia sp. aff. *G. hunsteinii*	
ICACINACEAE	
Apodytes brachystylis	
Citronella smythii	
LAURACEAE	
Beilschmiedia bancroftii	594
Beilschmiedia obtusifolia	
Beilschmiedia sp.	
Cryptocarya angulata	
Cryptocarya cinnamomifolia	
Cryptocarya mackinnoniana	
Cryptocarya murrayi	
Cryptocarya rigida	
Endiandra hypotephra	
Endiandra sp. aff. *E. muelleri*	
LEGUMINOSAE	
Archidendron vaillantii	626/2
MYRSINACEAE	
Rapanea achradifolia	
MYRTACEAE	
Austromyrtus lucida	625/2
Syzygium canicortex	625/2
Syzigium cormiflorum	
Syzygium endophloim	
Syzygium papyraceum	591
Syzygium wesa	
Syzygium sp.	594
OCHNACEAE	
Brackenridgea nitida	
OLEACEAE	
Chionanthus axillaris	591
PROTEACEAE	
Cardwellia sublimis	
Darlingia darlingiana	207
Lomatia fraxinifolius	625/2
Opisthiolepis heterophylla	
Oreocalis wickhamii	594
Placospermum coriaceum	626/2
RHAMNACEAE	
Alphitonia whitei	
RUBIACEAE	
Antirhea myrtoides	
Antirhea tenuifolia	
Canthium coprosmoides	625/2
Canthium odoratum	626/2
Randia fitzalanii	
RUTACEAE	
Acronychia acidula	
Brombya platynema	
Euodia bonwickii	591
SAPINDACEAE	
Guioa montana	
Mischocarpus lachnocarpus	
Mischocarpus sundaicus	594

*625/2 without landslip.

APPENDIX II — *continued*

	Lost to Plot Shown		Lost to Plot Shown
SAPINDACEAE — *cont'd*		WINTERACEAE	
Rhysotoechia mortoniana	625/2	*Bubbia semecarpoides*	
Synima cordieri	626/2		
Toechima lanceolatum		XANTHOPYLLACEAE	
SAPOTACEAE		*Xanthophyllum octandrum*	
Chrysophyllum sp.		MISCELLANEOUS SPP.	591,594
STERCULIACEAE			
Argyrodendron trifoliolatum			
Sterculia laurifolia		(A TOTAL of 72 SPECIES)	

APPENDIX III. Species with Recruits to 10 cm+ — Virgin Rainforest Plots 591, 625/2, 207, 626/2, 594.

	New to Plot Shown
ARALIACEAE	
Polyscias australiana	
ATHEROSPERMATACEAE	
Doryphora aromatica	
BALANOPACEAE	
Balanops australiana	
BLEPHAROCARYACEAE	
Blepharocarya involucrigera	
CUNONIACEAE	
Ceratopetalum succirubbum	594
ELAEOCARPACEAE	
Elaeocarpus sericopetalus	
Elaeocarpus sp.	
ESCALLONIACEAE	
Polyosma rhytophloia	594,626/2
EUPHORBIACEAE	
Macaranga subdentata	
Mallotus polyadenos	
Rockinghamia angustifolia	
FLINDERSIACEAE	
Flindersia bourjotiana	
Flindersia brayleyana	
Flindersia laevicarpa	
Flindersia pimenteliana	594
GUTTIFERAE	
Garcinia sp. aff. *G. hunsteinii*	
ICACINACEAE	
Apodytes brachystylis	
LAURACEAE	
Beilschmiedia obtusifolia	
Cryptocarya cinnamomifolia	
Cryptocarya corrugata	625/2
Cryptocarya hypospodia	
Cryptocarya mackinnoniana	
Cryptocarya murrayi	
Cryptocarya rigida	
Endiandra hypotephra	594
Endiandra palmerstonii	594
Endiandra pubens	594
Endiandra sp. aff. *E. muelleri*	

	New to Plot Shown
MONIMIACEAE	
Hedycarya loxocarya	594
MYRSINACEAE	
Rapanea achradifolia	
MYRTACEAE	
Austromyrtus sp.	
Syzygium cormiflorum	625/2
Syzygium endophloium	
Syzygium wilsonii ssp. *cryptophlebium*	
OCHNACEAE	
Brackenridgea nitida	
OLEACEAE	
Chionanthus axillaris	625/2
PROTEACEAE	
Cardwellia sublimis	
Darlingia darlingiana	
Helicia nortoniana	
RUBIACEAE	
Antirhea myrtoides	
Randia fitzalanii	
Timonius singularis	625/2
RUTACEAE	
Acronychia acidula	
Acronychia acronychioides	
Meliocope fareana	
SAPINDACEAE	
Guioa montana	
Sapindaceae sp.	591
Synima cordieri	594
Toechima erythrocarpum	594
Toechima lanceolatum	
SAPOTACEAE	
Chrysophyllum sp.	
Planchonella pohlmaniana	625/2
STERCULIACEAE	
Argyrodendron peralatum	
STILAGINACEAE	
Antidesma erostre	
XANTHOPHYLLACEAE	
Xanthophyllum octandrum	
(A TOTAL of 55 SPECIES)	

Proc. Ecol. Soc. Aust. 15:81-87

Soil seed stores in a rainforest succession

R. ABDULHADI and D. LAMB

Botany Department, University of Queensland, St. Lucia, Qld. 4067

Abstract

The soil seed populations in a successional sequence in subtropical rainforest in south-east Queensland were assessed. The six sites used included regrowth forests aged two years, 20 years, 50 years and 59 years. Two undisturbed forest sites were also sampled.

The density of seeds in the two undisturbed forests was similar (566 and 603 m^{-2}). The density of seeds in the two-year-old forest was significantly ($P < 0.05$) lower than these whereas the 20, 50 and 59-year-old sites had significantly ($P < 0.05$) high densities (2 082 m^{-2}, 1 694 m^{-2} and 947 m^{-2} respectively). The total species richness and species density followed a similar trend with maxima being recorded in the older regrowth forests. Changes in the composition of the soil seed pool are discussed and the reasons for these patterns reviewed. It is concluded that a single disturbance in these forests can affect the soil seed stores for over 60 years.

Introduction

It is well known that the soil seed store in undisturbed rainforests is dominated by secondary species (Guevara & Gomez-Pompa 1972; Liew 1973; Cheke *et al.* 1979; Hall & Swaine 1980; Whitmore 1983). It is also known that these soil seed stores strongly influence the secondary successions that occur when the forest is disturbed (Hopkins & Graham 1984).

What is less well known is how the soil seed store itself is affected by disturbance and how long such effects may persist. We have examined these changes in a series of regrowth forests in sub-tropical rainforest in south-east Queensland.

Study Sites and Methods

The study area was in the vicinity of O'Reilly's Guest House in the Lamington National Park (28°14'S, 153°7'E). Rainfall in the area is about 1 880 mm annually with most falling in the summer months between November and April. There are no temperature measurements available for the area but the environment is probably similar to the nearby Tambourine Mountain where the temperature range varies from a mean daily minimum of 8.3C in July to a mean daily maximum of 26.8 in January. The vegetation in the area has been mapped and described by McDonald & Whiteman (1979) as a tall closed forest alliance dominated, above 800 m, by a *Caldcluvia paniculosa* (F. Muell) Hoogland — *Cryptocarya erythroxlyon* Maiden & Betche — *Dysoxylum fraseranum* (Adr. Juss) Benth. association. This can be described as cool subtropical rainforest or complex notophyll vine forest using Webb's classification (Webb 1968). Within this area a number of study sites were chosen in regrowth forests of different ages:

Site 1. Two-year-old regrowth which has developed after a severe storm in September 1983. The regeneration is composed of seedlings and vegetative regrowth. The site is 50 m wide and 100 m long and is surrounded by intact forest. The elevation is 920 m.

Site 2. Twenty-year-old regrowth which has developed on abandoned farmland. The original forest was probably burnt or partially burnt at the time of clearing. The site is about 40 ha in area and 500 m wide at the narrowest point. It is surrounded on three sides by intact forest and on the fourth by pasture. The elevation is 860 m.

Site 3. Fifty-year-old regrowth which has developed after an attempt to clear the original forest failed. The forest was cleared and partially burnt but weeds colonized and regrowth prevented pasture establishment. The area covers about 16 ha and is 400 m wide at its narrowest point. It is bounded on two sides by intact forest and by pasture and disturbed forest on the other two sides. The elevation is 800 m.

Site 4. Fifty-nine-year-old forest regrowth which has developed after the original forest was cleared, burnt and converted to pasture for several years before being abandoned. The area is about 16 ha and about 300 m wide at its narrowest point. It is surrounded on three sides by undisturbed forest and bounded on the fourth by a road and repeatedly disturbed vegetation dominated by *Acacia melanoxylon* R.Br. The elevation is 900 m.

Site 5. Undisturbed forest immediately adjacent to site 1. Elevation is 920 m.

Site 6. Undisturbed forest close to site 2. Elevation is 860 m.

All the sites lie at elevations between 860 m and 920 m and all have kraznozem soils derived from tertiary basalts.

At each site ten 25 cm × 50 cm × 5 cm soil samples were taken at random along five transects each of 100 metres using a metal sampling frame. In the case of the regrowth forests all samples were taken from the centre of the site (Site 1) or at least 100 m from the site margin (Sites 2, 3, 4). The soils were spread over vermiculite in germination trays in a glasshuse and watered daily. Seedling emergence was monitored weekly for 39 weeks and the soil was disturbed twice (week 15 and week 30) to help promote the germination of buried seed. Six control trays containing steam sterilized soil were placed among the samples to monitor contamination. At the end of the period five of the 10 samples from each site were examined using direct sieving and a flotation technique (Jones & Bunch 1977), to count the number of and identify ungerminated seed.

The floristic composition of seeds at the six sites was compared using the Bray-Curtis dissimilarity measure and a group average clustering strategy (Clifford & Stephenson 1975).

Results

Two species, *Portulaca oleracea* L. and *Cardamine hirsuta* L., were found in the control trays. Both species are common weeds and have been removed from the records discussed below. Nor do the data below include seed extracted by the direct sieving and flotation methods. These were less than 1% of the total seed number at any site and were almost entirely dominated by *Acacia melanoxylon*.

In the results reported below the numbers of species at each site is represented by the total number of species recorded from the 10 soil samples at each site. The size of the soil seed population is expressed on a per m^2 basis and is the mean of the 10 soil samples at each site.

Species composition of seedbank

A total of 143 species belonging to 106 genera and 51 families were recorded in the soil seed banks at the six study sites. Of these, 95 species were native and 48 were introduced species. Most of the native species were trees (32%) or vines (20%) while most of the introduced species were herbs (74%).

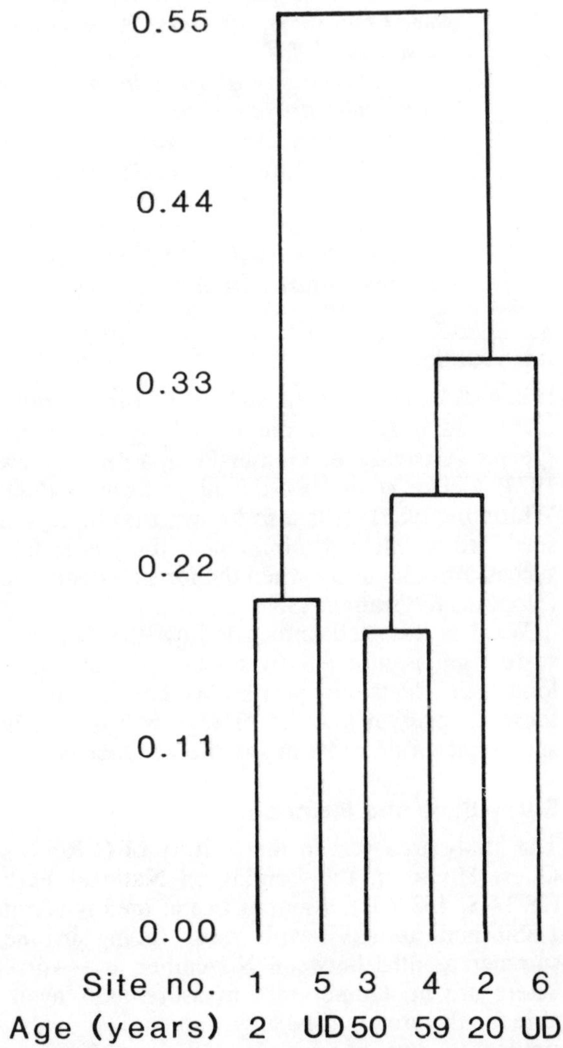

FIG. 1. Dendrogram resulting from the classification of the soil seed bank composition of the six sites using the Bray-Curtis dissimilarity index and group average sorting.

TABLE 1. Shannon diversity indices and numbers of species with the corresponding percentage in parentheses.

Site	Ages (years)	Number of species			Shannon diversity
		Introduced	Native	Total	
5	?	14 (31.11)	31 (68.89)	45	2.67
1	2	9 (26.47)	25 (73.53)	34	2.33
2	20	31 (42.46)	42 (57.54)	73	2.63
3	50	29 (34.11)	56 (65.89)	85	3.35
4	59	26 (35.13)	48 (64.87)	74	3.33
6	?	18 (29.50)	43 (70.50)	61	2.76

Most species were capable of rapid germination and 115 of the 143 species germinated within the first six weeks. Most pioneer species such as *Conyza* spp., *Eupatorium* spp., *Crassocephalum crepidioides* (Benth.) S. Moore *Gnaphalium* spp., *Rubus* spp., and *Solanum* spp. and secondary tree species *(Caldcluvia paniculosa, Dendrocnide excelsa* (Wedd.) Chew, *Duboisia myoporoides* R.Br., *Omalanthus populilfolius* Grah. and *Trema aspera* (Bronga). Bl. germinated quickly. However, *Acacia melanoxylon* R.Br. and *Alphitonia excelsa* (Fenzl.) Benth. continued to germinate over 36 weeks.

Cluster analysis showed the six sites formed two distinct groups. One contained the youngest regrowth site (Site 1) and the adjacent undisturbed forest (Site 5). The second group contained the older regrowth forests and the other undisturbed forest (Site 6). This second group was further divided into regrowth forests and undisturbed forest (Fig. 1). The initial two groups probably reflect the slight altitudinal difference between the various sites, 1 and 5 being at a higher elevation than the others. In this respect it is interesting that a small patch of *Nothofagus moorei* occurs not far from Sites 1 and 5 and it may be that both are near the upper margin of the cool subtropical rainforest type. For this reason Site 1 is probably best regarded as the sequel to a disturbance at Site 5 while Sites 2, 3 and 4 are sequels to and precursors of disturbance at Site 6. This interpretation of the relationship between the sites will be used throughout the remainder of this paper.

The initial effect of a disturbance was a reduction in the number of species in the soil seed bank (Table 1). Thus the undisturbed forest at Site 5 had 45 species but this declined to 34 species in the two-year-old forest (Site 1). With further time however, the forest reached a maximum of 85 species in the soil store at the 50-year-old forest. With further successional development the species richness began to decline. The diversity index followed a similar trend.

At least 26% of the total species number at any site were introduced species. The proportion and absolute number was lowest in the two-year-old site (Site 1) and highest in the three old regrowth forests (Sites 2, 3, 4). The two undisturbed forests (Sites 5, 6) had an intermediate number of introduced species.

Numbers of species in various life forms

The most common life form at most sites were herbs (Fig. 2). There were 11 herb species at the undisturbed forest at Site 5 and 10 species at the adjacent two-year-old site. The number of herb species increased sharply as the succession developed and reached 38 species at the 50-year-old site. The number subsequently declined to 20 species in the nearby undisturbed site (Site 6).

No other life form was as rich in species as the herbs. However, most of the other life forms had increased their species richness during the succession to reach a maximum in the 50 or 60-year-old regrowth. Thus the number of secondary tree species increased from 12 species in the undisturbed forest to 15 species in 60-year-old regrowth. Similar changes occurred among the vines, shrubs and grasses. By contrast the number of primary forest tree species declined from four species in the two undisturbed forests to only one species in the successional forests.

Soil seed-bank size

The size of the soil seed banks varied considerably between the different sites (Table 2). The undisturbed forest at Site 5 contained 566 seeds m^{-2}, the adjacent two-year-old site contained only 345 seeds m^{-2} but the 20-year-old site contained 2 082 seeds m^{-2}. With further successional development the number declined from this peak to 603 seeds m^{-2} in the other undisturbed forest (Site 6). All three of the older regrowth forests (i.e. Sites 2, 3 and 4) had significantly larger seed populations than the undisturbed forest ($P < 0.05$).

Both undisturbed forests had similar numbers of exotic seeds (about 90 seeds per m^2) making

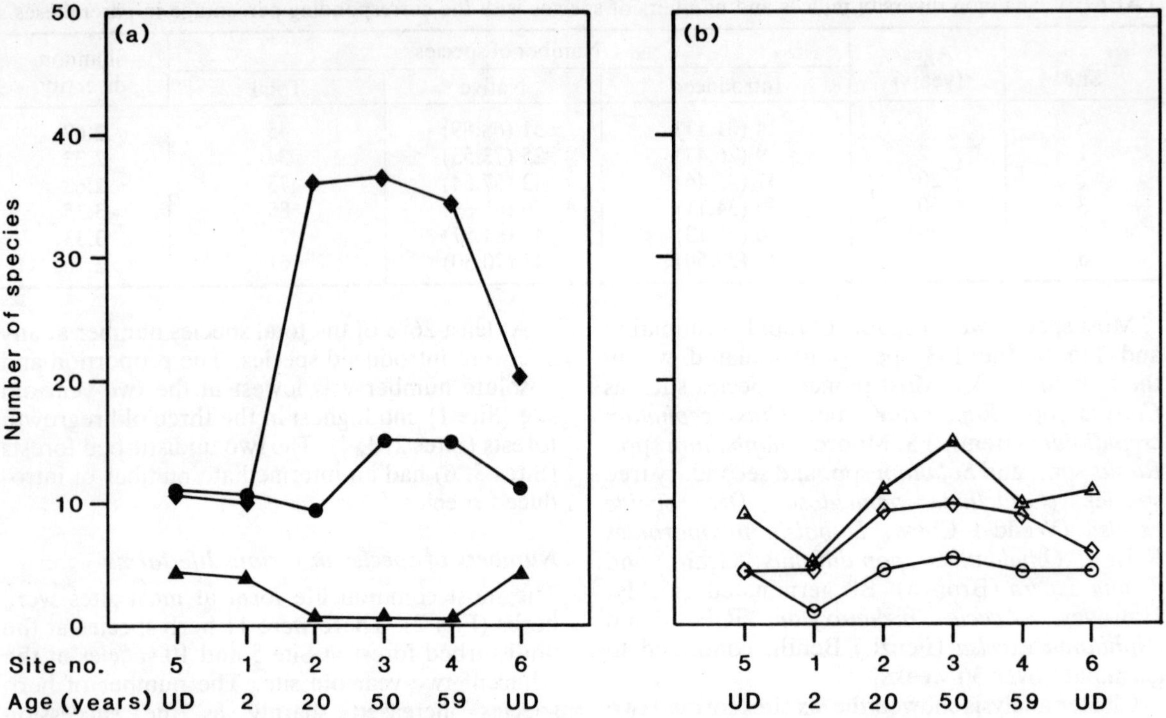

FIG. 2. The total number of species found in 10 0.125 m^2 samples in various life forms at each site. (a) Herbs ◆, secondary forest trees ●, primary forest trees ▲; (b) Vines △, shrubs ◇, grasses ○.

up, in both cases, about 14% of the total seed population. The number and proportion of exotic seed increased during the succession to reach a maximum at the 50-year-old forest. Thereafter the importance of exotic seed declined sharply.

Number of seed in various forms

Herb and vine species dominated the soil seed stores at most sites and their population densities varied similarly with successional development. In the case of herb species there were 80-120 seeds m^{-2} at the two undisturbed forests and at the two-year-old forest. However, this seed bank increased sharply with successional development and reached over 700 seeds m^{-2} at the 20-year-old forest site. With further time the number declined and the older regrowth forests had much lower densities (Fig. 3). The number of vine species seed dropped after a disturbance from 70 seeds m^{-2} in the undisturbed forest at Site 5 to 20 seeds m^{-2} in the adjacent two-year-old forest. But then, like the herbs, the density increased sharply to reach a maximum at the 20-year-old forest. The population of vine species seeds here was 680 m^{-2}. As with the herbs, the seed populations of vine species then declined with further successional

TABLE 2. The numbers of seeds of introduced and native species with the corresponding percentage in parentheses.

Site	Ages (years)	Numbers of seeds per square meter		
		Introduced spp	Native spp.	Total (±SD)
5	?	85 (15.06)	481 (84.94)	566± 56.32
1	2	82 (26.67)	253 (73.33)	345± 66.87
2	20	319 (15.32)	1 763 (84.68)	2 082±498.51
3	50	367 (21.66)	1 327 (78.34)	1 694±438.34
4	59	124 (13.08)	823 (86.92)	947±241.38
6	?	96 (15.92)	507 (84.08)	603± 78.18

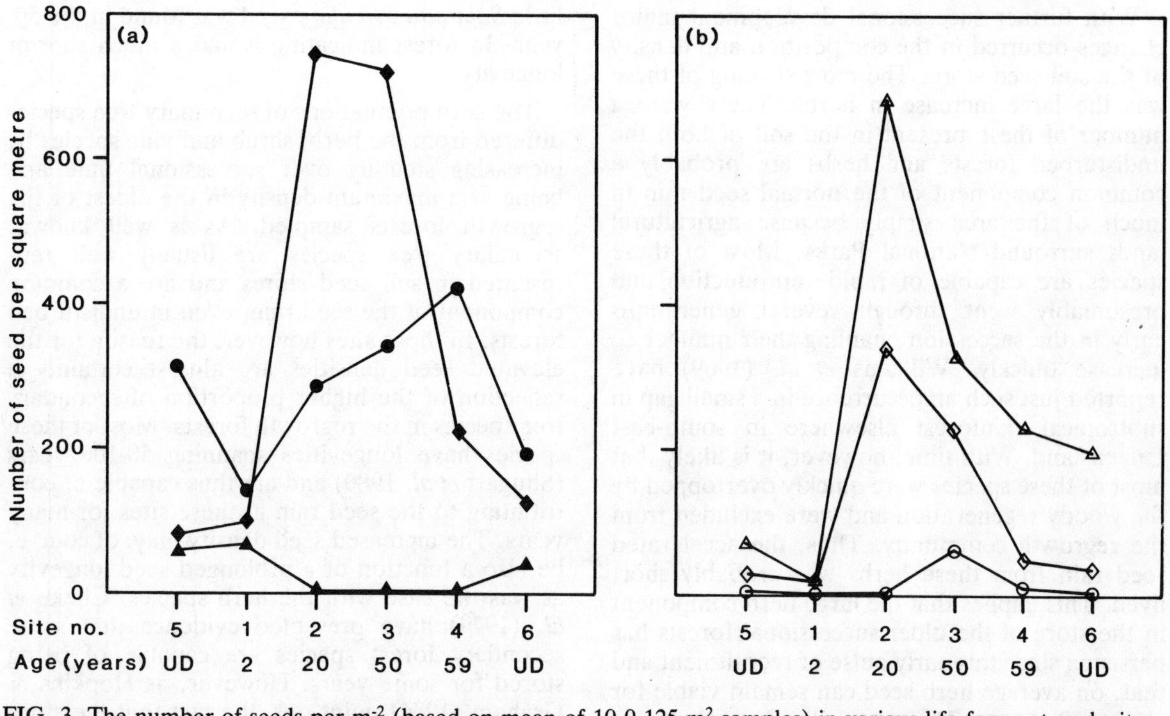

FIG. 3. The number of seeds per m^{-2} (based on mean of 10 0.125 m^2 samples) in various life forms at each site. (a) Herbs ◆, secondary forest trees ●, primary forest trees ▲; (b) Vines △, shrubs ◇, grasses ○.

development. The vine seed populations were dominated by *Rubus rosifolius* Sm. At the 20-year-old site, for example, it made up 86% of the total population.

Shrub species seed followed a similar pattern over time to that of the herbs and vine although the number were smaller. That is, seed densities increased in the younger regrowth forests and then declined as these forests aged. Like the vines the shrub species seed populations were dominated by a single species. In this case *Solanum aviculare* Forst. made up 82% of the population of shrub seed at the 20-year-old site.

The population of secondary tree species seed differed from these trends. The initial disturbance caused a decline in seed densities from 305 seeds m^{-2} in an undisturbed forest (Site 5) to only 136 seeds m^{-2} at the adjacent two-year-old site. However, the population then increased throughout the succession and the largest population (415 seeds m^{-2}) was found at the oldest regrowth forest (Site 4). The numbers of seed of primary forest species in the soil seed banks were much lower than all these life forms and varied between 42 and 70 seeds m^{-2} at the two undisturbed forests and the two-year-old regrowth forest. There were virtually no primary forest tree species seed present in the older regrowth forests. Grass species seed were mostly present in small numbers at most sites.

Discussion

The working assumption at the beginning of this study was that despite their different histories, the six sites could be arranged in a chronological sequence which would show how soil seed stores responded to a major disturbance. This assumption had to be slightly modified following a cluster analysis which suggested the youngest regrowth site (Site 1) and its adjoining undisturbed forest (Site 5) were floristically different from the other four sites. Collectively, however, the six sites still show the short term and long term effects of disturbance on soil seed stores. The short term effects were a sharp decline in species richness and in seed density. Most life forms lost species although, proportionally, the greatest loss occurred amongst the vines. The greatest loss in terms of seed numbers were the secondary tree species. Many of these species germinated after the storm of 1983 created a large canopy gap and are now present at the site in the young forest regrowth at Site 1 (Olsen & Lamb unpublished data).

With further successional development major changes occurred in the composition and density of the soil seed store. The most striking of these was the large increase in herbs. There were a number of these present in the soil of both the undisturbed forests and herbs are probably a common component of the normal seed rain in much of the area simply because agricultural lands surround National Parks. Most of these species are capable of rapid reproduction and presumably went through several generations early in the succession enabling their number to increase quickly. Williams *et al.* (1969) have reported just such an occurrence in a small gap in subtropical rainforest elsewhere in south-east Queensland. With time, however, it is likely that most of these species were quickly overtopped by the woody regeneration and were excluded from the regrowth community. Thus, the accelerated seed rain from these herbs was probably short lived. This implies that the large herb component in the store of the older successional forests has persisted since this early pulse of recruitment and that, on average herb seed can remain viable for about 50 years. This generalization obviously masks many important differences and not all herb seed may be capable of this long term storage.

Like the herbs, the numbers of many vine and shrub species were also temporarily increased by a disturbance. Unlike the herbs however, the numbers fell sharply after 20 years and were not matched by a large increase in species number. In both cases most of the increase was due to a single species (*Rubus rosifolius* and *Solanum aviculare* respectively). Both these species are present in the understorey of the 20-year-old forest where scattered gaps in the still incomplete canopy allows light to reach the forest floor. However, neither are present in the understorey of the better developed older regrowth forests with more complete canopy covers. If these two species are excluded then the increase in shrub and vine species seed populations are much more modest. Little is known about the longevity and reproductive behaviour of most of these species and it is not possible, therefore, to draw many conclusions about the long term viability of the seed. In the particular case of *Rubus rosifolius* however, substantial populations of seed remained present in the soil store of the 59-year-old regrowth forest suggesting much could persist for up to 40 years. In contrast little *Solanum aviculare* seed was found in the 50-year-old forest indicating it had a much shorter longevity.

The seed populations of secondary tree species differed from the herb, shrub and vine species by increasing steadily over successional time and being at a maximum density in the oldest of the regrowth forests sampled. As is well known, secondary tree species are usually well represented in soil seed stores and are a common component of the seed rain, even in undisturbed forests. In those sites however, the reason for the elevated seed densities are almost certainly a reflection of the higher proportion of secondary tree species in the regrowth forests. Most of these species have longevities spanning 50-100 years (Shugart *et al.* 1980) and are thus capable of contributing to the seed rain at these sites for many years. The increased seed density may, of course, be also a function of a prolonged seed longevity as was the case with the herb species. Cheke *et al.* (1979) have presented evidence that some secondary forest species are capable of being stored for some years. However, as Hopkins & Graham (1984) point out, the fact that the store of secondary species in primary forest is much lower than that in secondary forest suggests the life of much of this seed is probably limited to a few years.

The number of species and the seed densities found in this study are comparable with those found in studies of other rainforest soils (e.g., Hopkins & Graham 1983; Cheke *et al.* 1979; Saulei 1985). One noticeable difference from most of these other studies, however, was the high proportion of exotic species in the soil seed pool. This is probably a reflection of the proximity of agricultural ecosystems to the six study sites. Most of these exotic species have considerable powers of dispersal. The two undisturbed forest sites, for example, had similar seed densities and similar numbers of exotic species even though Site 6 was about 700 m from cleared forest land and Site 5 was over 2 km distant. Such exotic species are thus likely to be a permanent feature of most of the seed rain within much of the Lamington Park except perhaps for an inner core.

We conclude that a simple disturbance in these forests can have consequences for the soil seed populations which may last for over 60 years. In some cases this is because of the longevity of the seed involved. In others it is because the disturbance temporarily allows more secondary species to become established at a site and contribute to the seed rain.

References

Cheke A. S., Nanakorn W. & Yankoses C. (1979) Dormancy and dispersal of seeds of secondary forest species under the canopy of a primary tropical rainforest in Northern Thailand. *Biotropica* **11,** 88-95.

Clifford H. T. & Stephenson W. (1975) *An Introduction to Numerical Classification.* Academic Press, New York.

Guevara S. & Gomez-Pompa A. (1972) Seeds from surface soils in a tropical region of Vera Cruz, Mexico. *J. Arnold Arbor.* **53,** 312-35.

Hall J. B. & Swaine M. D. (1980) Seed stocks in Ghanaian forest soils. *Biotropica* **12,** 256-63.

Hopkins M. S. & Graham A. W. (1983) The species composition of viable soil seed banks beneath lowland tropical rainforests in north Queensland, Australia. *Biotropica* **15,** 90-99.

Hopkins M. S. & Graham A. W. (1984) Viable soil seed banks in disturbed lowland tropical rainforest sites in north Queensland. *Aust. J. Ecol.* **9,** 71-79.

Jones R. M. & Bunch G. A. (1977) Sampling and measuring the legume seed content of pasture soils and cattle faeces. *CSIRO Division of Tropical Crops and Pastures. Trop. Agronomy Technical Memorandum* **7,** 1-9.

Liew T. C. (1973) Occurrence of seeds in a virgin forest top soil with particular reference to secondary species in Sabah. *Malay. For.* **36,** 185-93.

McDonald W. J. F. & Whiteman W. G. (1979) *Moreton Region Vegetation Map Series: Explanatory Booklet for Murwillumbah Sheet.* Botany Branch, Queensland Dept. of Primary Industries, Brisbane.

Saulei S. (1985) *The recovery of tropical lowland rainforest after clear-fell logging in the Gogol Valley, Papua New Guinea.* Ph.D. University of Aberdeen.

Shugart H., Hopkins M. S., Burgess I. P. & Mortlock A. T. (1980) The development of a succession model for subtropical rainforest and its application to assess the effects of timber harvest at Wiangaree State Forest, New South Wales. *J. env. Manag.* **11,** 243-65.

Webb L. J. (1968) Environmental relationships of the structural types of Australian rainforest vegetation. *Ecology* **49,** 296-311.

Whitmore T. C. (1983) Secondary succession from seed in tropical rainforests. *Forestry Abstracts* **44,** 767-79.

Williams W. T., Lance G. N., Webb L. J., Tracey J. G. & Dale M. B. (1969) Studies in the numerical analysis of complex rainforest communities III. The analysis of successional data. *J. Ecol.* **57,** 515-35.

Proc. Ecol. Soc. Aust. 15:89-98

Origin and evolution of rainforest biogeographic diversity using evidence from the Proteaceae*

E. C. SAXON[1] and J. G. TRACEY[2]

[1]Q.N.P.W.S. Research and Planning Branch, Box 2066, Cairns, Qld. 4870
[2]C.S.I.R.O. Tropical Forest Research Centre, Box 780, Atherton, Qld. 4883

Abstract

Distribution data on rainforest trees of the family Proteaceae were examined to establish whether or not they were likely to be found in National Parks. Spatial patterns of the 48 taxa, most of which are endemic to the region, were correlated with a number of environmental gradients and disturbance. Some species associations were found to occur mainly in areas subject to logging. There are considered vulnerable to disturbance patterns which are different from those in which they evolved.

Introduction

In 1982, Michael Soulé told delegates to the 3rd World National Parks Conference that unless they were planning with a 200-year time frame their efforts were useless. We hope you will appreciate our taking a 200 million year time frame. Rainforest planning and management require this evolutionary perspective, because decisions are being made which foreclose opportunities for the future. It also requires a regional perspective. The core of our work is trying to determine the extent to which one may safely extrapolate from what little is known about a few particular places to what we would like to know about everywhere else.

Planners have various rules of thumb for determining the size and shape of reserves to sample effectively populations in the absence of comprehensive data on the distribution of species. Australian rainforests have been classified by structural and floristic types which are in turn related to the environment and to habitats for groups of species. Reserves have then been proposed with as much structural and floristic variation as possible (Webb 1966; Tracey 1981).

These rules work best in two situations. If species distributions are entirely chaotic, then the size of the sample determines the number of taxa included. If the distributions are completely orderly, then the location of the sample determines the number of taxa included. It is well known, however, from herbarium records and from surveys, that the distribution of species in the rainforests of north Queensland is neither chaotic nor orderly.

We recognize that if conservation is to be based on the maintenance of ecological processes, landscape management rather than land tenure control is called for. If conservation is to be based on representative ecosystems, these units, and their natural boundaries, must be mapped. If conservation is to be based on species, the effects of dispersal and fragmentation of populations must be considered. Unfortunately for conservation biologists, these three aims are not easy to achieve simultaneously. Evolutionary history, environmental variation, and ecological processes *are not necessarily correlated.*

The problem is made more intractable because evolutionary history must be examined on a global scale stretching over millions of years, and environmental gradients are examined on a regional scale, stable over millennia, yet ecological processes affect individual organisms within generations.

*Footnote: The opinions contained in this paper are not necessarily those of either QNPWS or CSIRO.

Proteaceae are being used as a test case for a number of hypotheses on the extent to which evolutionary history, environmental gradients and ecological processes are correlated. This determines just how compatible different conservation objectives are in practice. As a starting point, we wanted to determine whether the objective of reserving a comprehensive floristic sample was achieved by the preservation in national parks and the conservation in state forests of the bulk of the area of rainforests in north Queensland.

The choice of north Queensland as a study area requires no explanation, but reasons for the selection of Proteaceae may not be so obvious. The family is ancient and has many endemic taxa throughout the study area. Thirty-seven of the infra-generic taxa surveyed are local endemics, the remaining eleven occur elsewhere in Australia. The family is also widespread in the Southern Hemisphere. Not least important, to the field ecologist, are the facts that a thoroughly revised taxonomy has been published (Johnson & Briggs 1975) and that sterile individuals are readily identifiable. Looked at in a more mundane light, they include variable timber species (the silky oaks), species of horticultural interest (the macadamias), and, outside the region, the Proteoideae which are the basis of a cut flower industry.

Four questions are addressed in the following discussion:

1. What is the significance of the north Queensland rainforest Proteaceae in a global evolutionary perspective?
2. Where do individual species occur in north Queensland rainforests?
3. Can the local occurrence of species assemblages be predicted?
4. What are the future consequences of present land use?

The Evolution of the Proteaceae

Vicariance biogeography (Wiley 1981) holds, as its working hypothesis, that the evolution of a landscape matches the evolution of its phylogenetically related biota. By comparing the family tree of related taxa, the Proteaceae, with the physical history of a region, north Queensland as a fragment of Gondwanaland, it is possible to understand the taxonomic and biogeographic isolation characteristic of the region.

The biogeography of the family Proteaceae (Table 1) highlights the conservation significance of north Queensland as a remnant of the Gondwanaland supercontinent (Webb & Tracey 1981; Johnson & Briggs 1975; Schuster 1976; Smith, Hurley & Briden 1981). Major radiation within the family occurred there 130-65 myr ago. Today, Australia conserves biota from *two* provinces dating from that time: West Australia-East Antarctica-India-Africa; and East Australia-West Antarctica-New Zealand-South America. From the time of the origin of angiosperms until relatively recently, north Queensland remained a peninsula at the edge of the southern continent. The recent collision with the Sunda plate has allowed some communication with Asia, but the diffusion of Proteaceae there has been very limited.

North Queensland Proteaceae comprise a large, but unbalanced selection among the five, presumably Gondwanic, subfamilies. Proteoideae are found predominantly from New Caledonia across Australia to Africa (with only one provisionally attributed species in north Queensland rainforests); Persoonioideae are found in New Caledonia, New Zealand and Australia (also having only one species in north Queensland rainforests); Sphalmioideae (1 species) and Carnarvonioideae (2 species) are endemic to north Queensland rainforests. Grevilleoideae occur predominantly from Melanesia across Australia to South America (with 43 taxa in north Queensland rainforests).

It is highly improbable that the chance effects of numerous selective pressures and the ebb and flow of populations in refugia (Webb, Tracey & Williams 1984) would result in the isolation of five taxa in four subfamilies in north Queensland with virtually no close relatives surviving anywhere, while the 43 taxa in the remaining sub-family have congeneric or at least sub-tribal relatives as far away as South America. It is more likely that the ones that are currently isolated here were never particularly abundant and that they had not widely radiated in the Cretaceous assemblage in this area. Variation in abundance and isolation within the much more numerous Grevilleoideae may be the result of different selective pressures operating on different guilds within the taxonomic group.

Ecologically, the Proteaceae can be divided into three groups, or guilds, having markedly different growth habits and distributions. One group, the wind or insect pollinated and wind or gravity dispersed species, includes all the taxonomically isolated species. They are all rainforest canopy trees.

Members of a second group, in the sub-family Proteoidea or Grevilleoideae, bear indehiscent fruits which are presently animal dispersed. They can be understorey, sub-canopy or canopy trees in the rainforest. Many of these species have con-

TABLE 1. Biogeographical and taxonomic position of the Proteaceae revised after Johnson & Briggs (1975) (North Queensland endemics **Bold**). Numbers in [ ] identify taxa in Table 2.*

SUBFAMILY Tribe "Subtribe"	N.Q. genera locations of others	N.Q. species, varieties locations of others	
PERSOONIOIDEAE			
Persoonieae			
"Placosperminae"	***Placospermum***	***coriaceum***	[1]
"Persooniinae"	Australia (3) New Zealand New Caledonia		
Bellendeneae	Australia		
PROTEOIDEAE			
Conospermeae			
"Cenarrheninae"	***Gen. indet.*** Australia New Caledonia (2)		[2]
"Dilobeiinae"	Madagascar		
"Stirlingiinae"	Australia		
"Petrophilinae"	Australia (2)		
"Conosperminae"	Australia (2)		
Franklandieae			
"Franklandiinae"	Australia		
"Adenanthinae"	Australia		
Proteeae			
"Aulacinae"	South Africa (2)		
"Proteinae"	South Africa (11)		
SPHALMIOIDEAE	**Sphalmium**	**racemosum**	[3]
CARNARVONIOIDEAE	**Carnarvonia**	**araliifolia**	[4]
		sp. nov. *"Mt. Haig"*	[5]
GREVILLEOIDEAE			
Oriteae	*Orites*	***peltata***	[6]
		excelsa also eastern Australia	[7]
		spp. Australia & South America	
	Neorites	***kevediana***	[8]
Knightieae			
"Knightiinae"	***Darlingia***	***darlingiana***	[9]
		ferruginea	[10]
	New Caledonia New Zealand		
"Cardwelliinae"	***Cardwellia***	***sublimis***	[11]
Embothrieae			
"Stenocarpinae"	*Stenocarpus*	***reticulatus***	[12]
		sp. nov. *"fern leaf"*	[13]
		sp. nov. *"tripinnate"*	[14]
		sinuatus also eastern Australia	[15]
		salignus var. *salignus* also eastern Australia	[16]
		salignus var. *moorei* also northern Australia	[17]
		spp. Australia, New Caledonia	
	Australia		
"Buckinghamiinae"	***Opisthiolepis***	***heterophylla***	[18]
	Buckinghamia	***celcissima***	[19]
		sp. nov. *"Noah Creek"*	[20]
"Lomatiinae"	*Lomatia*	***fraxinifolia***	[21]
		spp. Australia, South America	

*Footnote to Table 1 see page 85.

SUBFAMILY Tribe "Subtribe"	N.Q. genera locations of others	N.Q. species, varieties locations of others	
GREVILLEOIDEAE – continued			
"Embothriinae"	*Oreocallis*	***wickhamii***	
		var. *"divided leaf"*	[22]
		var. *"simple leaf"*	[23]
		spp. Australia, South America	
	Australia		
	South America		
Grevilleeae	*Grevillea*	**sp. *indet.***	[24]
		pinnatifida	[25]
		also northern Australia	
		hilliana	[26]
		also eastern Australia	
		spp. Australia, Celebes, New Guinea, New Caledonia	
	Australia		
	Melanesia		
Helicieae			
"Hollandaeinae"	***Hollandaea***	***sayerana***	[27]
		sp. nov. *"Roaring Meg"*	[28]
"Heliciinae"	*Helicia*	***blakei***	[29]
		greyii	[30]
		lamingtoniana	[31]
		lewisensis	[32]
		recurva	[33]
		australasica	[34]
		also northern Australia	
		nortoniana	[35]
		also northern Australia	
		spp. south-east Asia	
	Australia		
"Triuiinae"	*Triunia*	*youngiana*	
		var. *montana "br"*	[36]
		var. *montana "wh"*	[37]
		var. youngiana	
		eastern Australia	
Macadamieae			
"Hicksbeachiinae"	*Virotia*	***heyana***	[38]
		spp. New Caledonia	
	Hicksbeachia	*pinnatifolia*	[39]
		also eastern Australia	
	Athertonia	***diversifolia***	[40]
	south-east Asia, Melanesia		
	Madagascar		
"Gevuininae"	*Bleasdalea*	*bleasdalei*	[41]
		also eastern Australia	
		spp. New Guinea, Melanesia	
	New Caledonia		
	South America (2)		
"Floydiinae"	Australia		
"Macadamiinae"	*Macadamia*	***whelani***	[42]
		aff. *hildebrandii*	[43]
		hildebrandii	[44]
		also northern Australia, Celebes	
		spp. eastern Australia	
	South America		
	South Africa		

SUBFAMILY Tribe "Subtribe"	N.Q. genera locations of others	N.Q. species, varieties locations of others	
"Macadamiinae" – continued			
"Roupalinae"	New Caledonia South America		
"Lambertiinae"	Australia		
Banksieae			
"Musgraveinae"	***Austromuellera***	***trinervia***	[45]
		sp. nov. *"Mt. Lewis"*	[46]
	Musgravea	***heterophylla***	[47]
		stenostachya	[48]
"Banksiinae"	Australia Australia, New Guinea		

*Revisions of Proteaceae have resulted in the following name changes. Numbers in [ ] as in Table 1: [13] ***Stenocarpus davallioides*** D. Foreman and B. Hyland; [14] ***Stenocarpus cryptocarpus*** D. Foreman and B. Hyland; [20] ***Buckinghamia ferruginiflora*** D. Foreman and B. Hyland; [25] ***Grevillea baileyana*** McGillivray; [36] ***Triunia montana* (C. White)** D. Foreman; [37] ***Triunia erythrocarpa*** D. Foreman.

generic relatives which may presently occur anywhere from South Africa to South America. This required either extremely slow radiation from a very early date or multiple parallel evolution after the breakup of Gondwanaland (Johnson & Briggs 1975). Radiation of what are now higher taxa into niches defined by canopy position and the presence of frugivores could be extremely ancient. Mammals now have to be included in the picture of Gondwanic tree evolution since the discovery of a badger-sized toothed monotreme in an early Cretaceous deposit at Lightning Ridge (Archer *et al.* 1985). Radiation of lower taxa into local species has probably gone on episodically ever since then.

The third guild, mostly shrubs, has radiated widely in cold or dry areas with parallel evolution being evident in three subfamilies at opposite ends of the former West Australia-South Africa province. As with the second group, taxonomic, biogeographic and ecological evidence support the hypothesis of very ancient radiation into particular habitats with subsequent, possibly recent, further radiation where a previously rare habitat (such as Gondwanic sand plains and mountain tops) becomes widespread. Where the habitat remains rare or shrinks (such as rainforests) the species numbers remain low, with any new ones probably replacing old ones.

Individual Species Distributions

It was well known that some species were widespread in rainforests of the region, while others had very narrow distributions. We are testing hypotheses that these differences would be correlated with:

1. topographic position in the landscape; and/or,
2. stratigraphic position in the forest canopy; and/or,
3. fruiting characteristics.

Three methods are used to map the distribution of individual species: herbarium data; presence/absence records for 217 half hectare sites between Cooktown and Townsville (Tracey 1982); and, gradient transects with population counts on quadrats, each 300 m^2.

Three gradient transects were laid out from east to west across the rainforest massif. From south to north, these are in Bellenden Kerr N.P. and S.F. 755, Mossman Gorge N.P. and S.F. 143, and Thornton's Peak N.P. and T.R. 165. Traverses were located at 10 points of major change in topography from base to peak to base on each transect. Individual quadrats were then arrayed at ten points from ridge to gully to ridge on each traverse. This arrangement means that samples are from comparable points in the landscape on each transect regardless of the actual distance that may separate quadrats.

Three types of distribution were observed (Table 2):

1. taxa with widespread populations, in similar topographic positions;
2. taxa with a few, disjunct, populations in different positions; and,
3. taxa with a single population each.

Among the species having the first type of distribution pattern (found in similar positions on two adjoining transects) are the most taxonomically isolated species: the monotypic subfamily, *Sphalmium;* the two species in the family Carnarvonioideae; the single rainforest member of Persoonioideae in the region, *Placospermum;* and the single member of the Proteoideae in the region, a species as yet undescribed. They are all canopy trees.

The distribution patterns within the other family, are independent of canopy position and fruit type. Within the Grevilleoideae, there are

TABLE 2. Taxonomic, environmental and ecological isolation among north Queensland rainforest Proteaceae [See Table 1 for Taxa Names].

Taxon	Taxonomic Links: species	genus	subtribe	tribe	sub-family	Forest Position: canopy	sub-canopy	understorey	Fruit Type: follicular	indehiscent	Clumping Pattern: widespread	few sites	same habitat	one site	sparse
1.	.	.	.	.	.		.	.	│	.	.	x	.	.	.
2.	.	.	.	.	.	│	.	.	.	│	.	.	x	x	.
3.	.	.	.	.	.	│	.	.	.	.	.	x	x	.	.
4.	.	│	│	│	│	│	.	.	│	.	x	.	.	.	.
5.	.	│	│	│	│	│	.	.	│	.	x	.	.	.	.
6.	.	│	│	│	│	│	.	.	│	.	.	x	.	.	.
7.	.	│	│	│	│	│	.	.	│	.	x	.	u	.	.
8.	.	.	│	│	│	│	.	.	│	.	.	x	.	.	.
9.	.	│	│	│	│	│	.	.	│	.	x	.	.	.	.
10.	.	│	│	│	│	│	.	.	│	.	.	x	x	.	.
11.	.	.	.	│	│	│	.	.	│	.	x	.	.	.	.
12.	.	│	│	│	│	│	.	.	│	.	x	.	u	.	.
13.	.	│	│	│	│	│	.	.	│	.	.	x	.	.	.
14.	.	│	│	│	│	│	.	.	│	.	.	x	.	.	.
15.	.	│	│	│	│	│	.	.	│	.	x	.	.	.	.
16.	.	│	│	│	│	│	.	.	│	.	.	x	u	.	.
17.	.	│	│	│	│	│	.	.	│	.	.	x	x	.	x
18.	.	.	.	│	│	│	.	.	│	.	x	.	.	.	.
19.	.	│	│	│	│	│	.	.	│	.	x	.	.	.	.
20.	.	│	│	│	│	│	.	.	│	.	.	x	.	.	.
21.	.	.	.	│	│	│	.	.	│	.	x	.	u	.	.
22.	│	│	│	│	│	│	.	.	│	.	.	x	.	.	.
23.	│	│	│	│	│	│	.	.	│	.	x	.	.	.	.
24.	.	│	│	│	│	│	.	.	│	.	.	x	u	.	.
25.	.	│	│	│	│	│	.	.	│	.	x	.	.	.	.
26.	.	│	│	│	│	│	.	.	│	.	.	x	l	.	.
27.	.	│	│	│	│	.	.	│	.	│	x	.	.	.	.
28.	.	│	│	│	│	.	.	│	.	│	.	.	x	x	x
29.	.	│	│	│	│	.	.	│	.	│	.	x	x	.	.
30.	.	│	│	│	│	.	.	│	.	│	.	x	.	.	.
31.	.	│	│	│	│	.	.	│	.	│	x	.	.	.	.
32.	.	│	│	│	│	.	.	│	.	│	.	x	u	.	.
33.	.	│	│	│	│	.	.	│	.	│	.	x	u	.	.
34.	.	│	│	│	│	.	.	│	.	│	x	.	.	.	.
35.	.	│	│	│	│	.	.	│	.	│	x	.	.	.	.
36.	│	│	│	│	│	.	.	│	.	│	x	.	u	.	.
37.	│	│	│	│	│	.	.	│	.	│	.	x	x	.	.
38.	.	.	│	│	│	.	│	.	.	│	x	.	x	.	.
39.	.	.	│	│	│	.	│	.	.	│	x	.	x	.	x
40.	.	.	│	│	│	.	│	.	.	│	.	x	x	.	.
41.	.	.	.	│	│	.	│	.	.	│	x	.	u	.	.
42.	.	│	│	│	│	│	.	.	.	│	x	.	.	.	.
43.	.	│	│	│	│	│	.	.	.	│	.	x	x	.	.
44.	.	│	│	│	│	│	.	.	.	│	.	x	x	.	.
45.	.	│	│	│	│	│	.	.	│	.	.	x	.	.	.
46.	.	│	│	│	│	│	.	.	│	.	.	.	x	x	.
47.	.	│	│	│	│	│	.	.	│	.	.	x	.	.	.
48.	.	│	│	│	│	│	.	.	│	.	x	.	u	.	.

u=upland l=lowland.

canopy, sub-canopy and understorey species, and both follicular and indehiscent fruiting types. These attributes are consistent within sub-tribes, but are not correlated with species distribution patterns. In eight of 20 genera of this subfamily there are species pairs, usually with non-over-lapping distributions.

Carnarvonia araliifolia occurs in several lowland sites while *Carnarvonia* "Mt. Haig" is found at almost all upland sites. *Musgravea heterophylla* is found at many sites in lowland areas north of the Barron River whereas *Musgravea stenostachya* is found in disjunct upland populations. *Helicia australasica* and *H. nortoniana* are widespread, while *H. recurva and H. lewisensis* are found on one mountaintop and *H. blakei* is found on another. This is consistent with the hypothesis that populations within north Queensland rainforests have been sufficiently isolated from each other to give rise, occasionally, to new, narrowly endemic species.

The Predictability of Species Assemblages

Having come to some understanding of the distribution of individual taxa, the next step was to consider whether there are predictable assemblages of species and whether they occur in predictable places in the landscape. The aim was to use the known distribution of taxa to predict their unknown distribution and then to compare this with present-day land use. It was suggested that where unique assemblages were presently known only from areas which were to be logged, that it would be possible to indentify replicate communities in national parks or on land unlikely to be logged.

In the numerical taxonomic analysis, 217 sites with complete tree species data were clustered according to their assemblages of Proteaceae. The groups generated in this way were then analysed in terms of variables associated with each site, including: structural classification, classification based on the total floristics, climate attributes derived from the AEGIS data base by Prof. Henry Nix' group, and logging history mapped by the Queensland Department of Forestry (Table 3).

Obviously there are many sites with only a few species and a few species occur at many sites. Consequently, there are many sites which are similar in terms of species assemblage because, whether they have many species or few, they include the most common species of the region. These statistically likely assemblages are not the most pressing conservation planning problem. They are the sites likely to be included in a random network of reserves. At the opposite extreme lie the sites with more rare species than their total species richness would predict. We wanted to know whether they are already in reserves, and, if not, whether similar floristic assemblages can be found in matching habitats.

The algorithm provided in the FOBEID subroutine of C.S.I.R.O.'s TAXON package was used to rank sites in order of number of species present and to rank species in order of number of occurrences in the region. The subroutine then calculates the probability of the species assemblage at each site arising as a random sample of the region. It is the improbable assemblages which would slip through the net if reserves were designated by random samples.

A statistically significant break separates the 31 sites with the least random samples of the region from the remaining 186. It may be stated, parenthetically, for those who lack confidence in mathematical algorithms, that sites which seem, subjectively, most "important" did turn up on the list.

Looking at the region as a whole and the floristics as a whole, there was no significant correlation between groups of sites with similar assemblages of Proteaceae (48 taxa) and the floristic classes assigned on the basis of 740 taxa (Williams & Tracey 1984) or 595 taxa (Tracey & Abel, unpublished results). Among the 31 sites with atypical assemblages, however, the central and the northern mountain top sites are distinctive from the others in both Proteaceae and the whole floristic suite.

Among the site climate attributes derived from the AEGIS data base by Prof. Henry Nix' group, there is a high degree of correlation between some precipitation and temperature estimates and the grouping of the 217 sites. There is a similar relationship between some precipitation and solar radiation estimates and the grouping of the 31 sites. The broad environmental parameters such as structural class, soil, altitude, longitude and latitude (Tracey & Webb 1975) do not distinguish all the clusters of sites from one another, though some groups can be tightly defined in these terms.

It remains to be seen whether combinations of these attributes have predictive value for finding additional sites with similar assemblages. It is one thing to combine attributes to define the habitat of a single species, but the habitat of a species assemblage cannot be mapped simply as the intersection of the constituent distributions. There is every reason to expect each taxon to react differently to any given environmental gradient.

TABLE 3. Cramer values for significant associations of attributes with sites clustered by Proteaceae species.

		217 Site Analysis (mean Cramer value for all attributes = .5598		31 Site Analysis (mean Cramer value for all attributes = .5226	
Attribute	Statistical test	Cramer value	P =	Cramer value	P =
Floristics					
Williams #3	Chi Square			.9618	.003
Abel #4	Chi Square			.5424	.276
Logging					
Logged/unlogged	Chi Square	.6693	.096	.8539	.050
Date of logging	Chi Square	.5614	.048		
Spatial					
Latitude	F			.6697	.008
Longitude	F			.5753	.060
Altitude	F	.6187	.000		
Climatic					
Precipitation					
Total	F			.5476	.093
Highest month	F	.5635	.000	.5684	.067
Lowest month	F	.5719	.000		
Range	F	.5677	.000	.5852	.050
Wettest quarter	F	.5694	.000	.5598	.077
Driest quarter	F	.5724	.000		
Temperature					
Annual mean	F	.6159	.000		
Highest month	F			.5389	.106
Lowest month	F	.6082	.000		
Seasonality	F	.6005	.000		
Wettest quarter	F	.5981	.000		
Driest quarter	F	.5883	.000		
Radiation					
Annual mean	F			.5719	.063
Lowest month	F			.5790	.056
Range	F			.5323	.117
Seasonality	F			.5640	.072
Wettest quarter	F			.5780	.057

There are pragmatic and theoretical approaches to solving this problem. A pragmatic approach would start with the observable patterns of forest structure and refine the structural types by reference to low level photographs and evidence of forest disturbance. A theoretical approach would combine models of forest dynamics with non-linear topological models, such as the "density of area above contour" function developed by Ian Briggs (pers. comm.). Both approaches would obviously require adequate ground controls. Only ground surveys will discern the results of recent historical factors.

Of all the ecological, climatic, structural, soil and topographic attributes with which we thought Proteaceae assemblages might be correlated, the simplest, most direct evidence turned out to be the most significant. Logging history turns out to be the single non-floristic variable most highly correlated with groupings of sites based on species assemblage. Some species are consistently found in logging areas while others are found in areas which are not subject to logging. The possibility of this being a chance association is less than 10% for the 217 sites and only 5% for the 31 sites. It is not that species are present or absent because of the logging, but rather that logging is present or absent because of the species.

Some species assemblages are found only in high altitude, structurally simple, rainforests which are regarded as unproductive and inaccessible from the forester's point of view. Some of the others are found mostly in rainforests which have been logged in the past, and others in rainforests which may be logged in the future.

Future Consequences of Present Land Use

Hopkins (1981) suggests that "the effect of different types of disturbance on rainforest will depend on the disturbance regime within which it has evolved. Disturbance which, because of type, periodicity or intensity, is unusual relative to a particular forest type will have a far greater impact than that which falls within the usual disturbance regime and *vice versa.*"

The question might be asked, whether a consequence of heavy predation on mature canopy trees may be the collapse of the exploited population to a level which can no longer sustain the exploiting population.

Such has been the experience of ocean fisheries, a similarly complex, multi-species "renewable resource" industry, when it went beyond an initial large harvest and repeatedly culled large individuals from depleted stocks. Some 50 years of research on the relationship between fish stock size, recruitment, environment and exploitation have seen a series of approaches tried and discarded. The most recent casualties have been predictive modelling (because none of the managers are prepared to accept the unpalatable results) and maximum sustainable yield (because nobody is interested simply in tonnes of fish, but in particular species and particular age classes).

What is currently being advocated among fishery researchers (May 1984) involves deliberate variation of rates of exploitation, elaborate monitoring of responses to disturbance, and rapid feedback of results into regulatory management, what Graeme Caughley (1985) calls "cautious fiddling, a technique dignified by the label 'adaptive management' ". There is genuine doubt as to whether such an approach is feasible at all, even though fish (unlike trees) have shorter generation times than fishermen. Still, the approach of staggering harvest among fragmented units within a single forest type is being tried in temperate forests (Harris 1984).

What then, of sustainable yield in rainforests? Rainforest structure, environment and disturbance can be mapped on a range of scales. That puts us one step ahead of the fishermen who have difficulty knowing where their stocks will actually be located at any moment in time. All the dynamic variables can be measured, including: timber species composition, size classes and population patterns, natural recruitment, post-harvest populations, and the intensity and frequency of culling. Models can be developed to predict the extent to which disturbance simplifies forest structure and determines species composition.

The patterns in north Queensland rainforests are so complex that computer technology is only just able to handle geographic data bases which would depict forest types and disturbance patterns in relation to environmental features on scales relevant to land use planning. Maps can be compiled which clearly show areas unlikely to be logged, irrespective of whether they are in national parks or state forests. Within the loggable areas of state forests, such maps could serve as the basis for quantifying and locating the effects predicted by the dynamic models.

Even if theoretically feasible, the problem becomes insoluble if the time required to obtain the information on which informed decisions can be made is actually longer than the time-frame in which the decisions have to be made. Furthermore, if individual species preservation is the aim, then there is no substitute for detailed knowledge of distribution patterns of the species with which we are concerned and a much better knowledge of their biology than presently exists or is practically obtainable. A conservation policy which relies on the impossible is an illusion.

While we cannot confidently predict the future consequences of our present action (because we are not yet measuring the present consequences of our past actions), we can conclude by considering the likely evolution of the Australian continent and its rainforests over the next 50 million years. Dougal Dixon does this in his book entitled, *After Man: A Zoology of the Future.* The present direction of plate tectonic movement will bring Australia further into the tropics. When the plate eventually collides with Asia, a new mountain chain will be pushed up between them. Over most of the world's terrestrial biota, those species presently dependent on man will be the only ones to survive his extinction. Australia, in Dixon's view, may be an exception. The native mammals and reptiles may radiate to fill vacated niches. If the rainforests of Australia survive the extinction of man, they are likely to reoccupy much of the continent, prosper for millions of years, and might even give rise to the next sapient creature, descended from a marsupial.

References

Archer M., Flannery T. F., Ritchie A., & Molnar R. E. (1985) First Mesozoic mammal from Australia – an early Cretaceous monotreme. *Nature* **318,** 363-66.

Caughley G. (1985) Problems in wildlife management. In: *The Study of Populations* (ed. H. Messel) pp. 129-35. Pergamon, Sydney.

Dixon D. (1981) *After Man: A Zoology of the Future.* Granada, London.

Harris L. D. (1984) *The Fragmented Forest: Island Biogeographic Theory and the Preservation of Biotic Diversity.* University of Chicago, Chicago.

Hopkins M. S. (1981) Disturbance and change in rainforests and the resulting problems of functional analysis. In: *Vegetation Classification in Australia* (eds A. N. Gillison and D. J. Anderson) pp. 42-52. C.S.I.R.O. and ANU Press, Melbourne.

Johnson L. A. S. & Briggs B. G. (1975) On the Proteaceae – the evolution of a southern family. *Bot. J. Linn. Soc.* **70,** 83-182.

May R. M. (ed) (1984) *Exploitation of Marine Communities.* Springer, Berlin.

Schuster R. M. (1976) Plate tectonics and its bearing on the geographic origin and dispersal of Angiosperms. In: *Origin and Evolution of the Angiosperms* (ed C. B. Beck) pp. 48-138. Columbia University Press, New York.

Smith A. G., Hurley A. M. & Briden J. C. (1981) *Phanerozoic Paleocontinental World Maps.* Cambridge University Press, Cambridge.

Tracey J. G. (1981) Australia's rainforests: where are the rare plants and how do we keep them? In: *The Biological Aspects of Rare Plant Conservation* (ed H. Synge) pp. 165-78. Wiley, Chichester.

Tracey J. G. (1982) *The Vegetation of the Humid Tropical Region of North Queensland.* C.S.I.R.O., Melbourne.

Tracey J. G. & Webb L. J. (1975) *Key to the Vegetation of the Humid Tropical Region of North Queensland, with 15 maps at 1:100 000 scale.* C.S.I.R.O., Indooroopilly.

Webb L. J. (1966) The identification and conservation of habitat types in the wet tropical lowlands of north Queensland. *Proc. Roy. Soc. Qld.* **78,** 59-86.

Webb L. J. & Tracey J. G. (1981) Australian rainforests: patterns and change. In: *Ecological Biogeography of Australia* (ed A. Keast) pp. 607-94. D. W. Junk, The Hague.

Webb L. J., Tracey J. G. & Williams W. T. (1984) A floristic framework of Australian rainforests. *Aust. J. Ecol.* **9,** 169-98.

Wiley E. O. (1981) *Phylogenetics: The Theory and Practice of Phylogenetic Systematics.* John Wiley & Sons, New York.

Williams W. T. & Tracey J. G. (1984) Network analysis of northern Queensland tropical rainforests. *Aust. J. Bot.* **32,** 109-16.

Proc. Ecol. Soc. Aust. 15:99-100

A database for Australian rainforest trees

T. WHIFFIN[1] and B. P. M. HYLAND[2]

[1]*Department of Botany, La Trobe University, Bundoora, Vic. 3083*
[2]*Australian National Herbarium, Division of Plant Industry, CSIRO, P.O. Box 780, Atherton, Qld. 4883*

Rainforest trees, while important scientifically, ecologically, and commercially, are one of the least known groups in the Australian flora. The last general rainforest tree flora was that of Francis (1951), although field keys for north Queensland (Hyland 1971, 1982) and New South Wales and southern Queensland (Floyd 1977-1983, Williams *et al.* 1984) have been produced subsequently. Increased collection and study have markedly increased the number of species recognized, to the extent that many of the existing revisions and floras are no longer satisfactory. Identification often proves difficult because of this, and the problem is compounded when dealing with sterile material only.

This paper outlines a long term project which aims to collect and describe all Australian rainforest tree species. This information will be placed within a computerised database, and hence corrections and additions can be made continually. Thus it will be possible at any stage to provide up-to-date information on any species or geographic area. As a consequence of this project, a number of different studies which rely on these data will be greatly facilitated, such as those involving study of the variation, evolution, phytogeography, ecology, silviculture and conservation of Australian rainforest trees.

Initially, then, the main aim is to provide a computerised database for Australian rainforest trees, to assist in the process of preparation of keys, descriptions, floras etc. This database will have three main sections:

(i) a taxonomic database, containing a continuously updated list of Australian rainforest tree species, with additional information on publication details, synonymy etc.;

(ii) a morphological database, containing data for each species on a wide range of characters, particularly including field characters, vegetative characters, reproductive characters, and seedling characters; and,

(iii) a geographical database, containing verified records of distributional data for each species.

The taxonomic database has four main functions. Firstly, it can be used to produce checklists, either of all Australian rainforests or, in conjunction with the geographical database, of specified geographic areas. Secondly, it can be used to find taxonomic or nomenclatural information on selected species, including the tracing of synonyms. Thirdly, it will contain information on codes or recognized species from previous publications, as in the RFK code numbers of Hyland (1982), so that links can be made to the information provided there. Finally, the taxonomic database serves as the point of entry into the morphological and geographical databases, links being provided from each species to the appropriate parts of these databases.

The morphological database will contain the information recorded for each species on all characters used. This will be used for the preparation of descriptions and keys. This database will be held in the DELTA system format, which is ideally suited to a database of this type (Dallwitz 1980, 1984). As all species represented in the database have an overall similarity of form (being trees), there will be few of the problems of homology noted by Pankhurst (1983). It is anticipated that while some of the information will be taken from the published literature, most will come from direct observation of herbarium specimens or living plants in the field. This will avoid errors, and produce a more consistent and reliable database.

Four character types will be incorporated into the morphological database — field characters, vegetative characters, reproductive characters, and seedling characters. Character lists for each of these sets are currently being refined and tested before the major data recording begins. The field characters, which comprise bark and leaf features, are derived from those used by Hyland (1982), to allow compatibility with that key. Descriptions and keys will be produced involving all or selected parts of the morphological database. Thus it will be possible to produce special purpose keys, such as keys based only on vegetative characters for identification of sterile material, or keys based only on seedling characters for use in ecological studies.

The geographical database will be the last to be undertaken, and the exact form and nature of this have yet to be determined. It bears an obvious relationship to computerised herbarium databases, and thus hopefully there will be a link between this database and current or planned herbarium databases.

Special programmes are being written to handle the taxonomic and geographical databases. A trial run, involving approximately fifty species from a range of different families, will shortly be undertaken, and this will be used to identify and correct any problems in the system. The subsequent stages will involve the compiling and recording of information from herbarium specimens. Any identified gaps in the database, particularly those relating to field characters, will be filled on field collecting trips to rainforest areas.

Once the database has been set up, it will be possible to make the data available in a wide variety of forms, including:

(i) a comprehensive, printed flora, including keys, descriptions and distribution maps;
(ii) keys based on selected characters, such as field keys for foresters, ecologists or naturalists;
(iii) floras or checklists for specified geographical areas, or specified taxonomic groups; and,
(iv) subsections of the database on floppy disk, thus allowing, for example, online identification by any user with an appropriate computer.

This last use is perhaps a pointer to the future. To investigate the methodology, and as an example of potential use, a microcomputer-based version of the rainforest key of Hyland (1982) has been produced. The 45 leaf features and 48 stem features recorded by Hyland (1982) for 799 rainforest tree taxa from north Queensland have been incorporated into a single online key. The advantages over the printed version are that the data and the nomenclature can be (and in fact have been) updated. In addition, once the data are computer based, they can be searched or compared in new ways, to assist in the identification process. It should also be possible to produce keys of the same form for different groups of species, or with additional or different character sets.

References

Dallwitz M. J. (1980) A general system for coding taxonomic descriptions. *Taxon* **29**, 41-46.

Dallwitz M. J. (1984) User's guide to the DELTA system — a general system for coding taxonomic descriptions. *CSIRO Division of Entomology Report*, **13**.

Floyd A. G. (1977-1983) *N.S.W. Rainforest Trees, parts I-XIII.* Forestry Commission of N.S.W., Sydney.

Francis W. D. (1951) *Australian Rainforest Trees.* Second revised edition. Forestry and Timber Bureau, Canberra.

Hyland B. P. M. (1971) *A Key to the Common Rain Forest Trees between Townsville and Cooktown based on Leaf and Bark Features.* Queensland Department of Forestry, Brisbane.

Hyland B. P. M. (1982) *A Revised Card Key to Rainforest Trees of North Queensland.* CSIRO, Melbourne.

Pankhurst R. J. (1983) The construction of a floristic database. *Taxon* **32**, 193-202.

Williams J. B., Harden G. J. & McDonald W. J. F. (1984) *Trees and Shrubs in Rainforests of New South Wales and Southern Queensland.* University of New England, Armidale.

Proc. Ecol. Soc. Aust. 15:101-105

Solar radiation regimes in a north Queensland rainforest

S. M. TURTON

Department of Geography, James Cook University of North Queensland, Townsville, Qld. 4811

Introduction

Solar radiation penetrating forest canopies plays a key role in the response of understorey vegetation and in controlling the surrounding microclimate. It influences plant growth by providing visible radiation for photosynthesis and radiant energy for evapotranspiration and environmental heating (Ross 1975).

Hemispherical (fish-eye) canopy photographs provide a simple and quick method for estimating the spatial and temporal distribution of solar radiation beneath forests over intervals of a day or more (e.g., Evans & Coombe 1959; Anderson 1964; Madgwick & Brumfield 1969). They also provide a convenient record of canopy cover and structure, and are useful for comparisons with other more direct measurement techniques. However, Anderson (1964) has stressed the importance of analysing the effect of the canopy on diffuse-sky radiation and on direct-beam radiation separately. Her technique was substantiated by comparing the estimates from the fish-eye photographs with nearly three years data of daily totals of solar radiation at several sites under a deciduous forest.

The aim of this study was to use hemispherical canopy photographs to estimate the quantity of solar radiation (diffuse and direct) reaching the forest floor at various times of the year. A closed canopy (deep shade) site and a small gap site were selected for comparison of understorey radiation environments within the same forest type.

Methods

The study was conducted in the Curtain Fig State Forest, north Queensland (17°17′S., 145°34′E., c. 680 m a.s.l.). The forest type is complex notophyll vine forest with scattered deciduous and semi-evergreen trees (Tracey 1982).

Total irradiation (Q_T) in KJ m^{-2} day^{-1} above the canopy was estimated using an empirical equation based on data from the humid tropics (Turton 1987). The equation gave mean atmospheric transmission coefficients of 0.70 for a cloudless (clear) sky and 0.30 for an overcast sky. Diffuse irradiation (Q_d) in KJ m^{-2} day^{-1} above the canopy was estimated using an empirical formula developed by Sayigh (1977). For a cloudless sky $Q_d/Q_T \simeq 0.275$ and for an overcast day it was assumed that $Q_d/Q_T = 1.0$. Direct irradiation (Q_b) in KJ m^{-2} day^{-1} above the canopy was derived from the difference between Q_T and Q_d.

Hemispherical photographs of the canopy at the two sites were analysed to determine their diffuse, direct and total site factors (Anderson 1964). Anderson's (1971) spider's-web grid overlay was used to assess the relative gap frequency of each photograph (Fig. 1) and to estimate Q_d at the forest floor. Solar tracks for the solstices and equinox were calculated for 17°17′S. and superimposed over the photographs (Fig. 2). Q_b reaching the forest floor was then estimated for the selected times of the year following the method outlined in Anderson (1971). The procedure was simulated for cloudless and overcast skies, except that Q_b was assumed to be zero under an overcast sky.

Results and Discussion

Tables 1 to 3 illustrate estimated daily values for Q_d, Q_b, Q_T above the canopy and at the forest floor for the two sites under cloudless skies, together with diffuse, direct, and total site factors. Table 4 shows estimates for Q_T under overcast skies.

Research in temperate forests has generally emphasized the importance of Q_d (shade light) in the understorey relative to Q_b (sunflecks), particularly beneath dense canopies (Anderson 1964). However, the consensus from research in tropical forests is that Q_b provides a significant contribution to Q_T in the understorey for most of the year. In this study, at the summer solstice Q_b represented 50.5% of the daily total reaching the

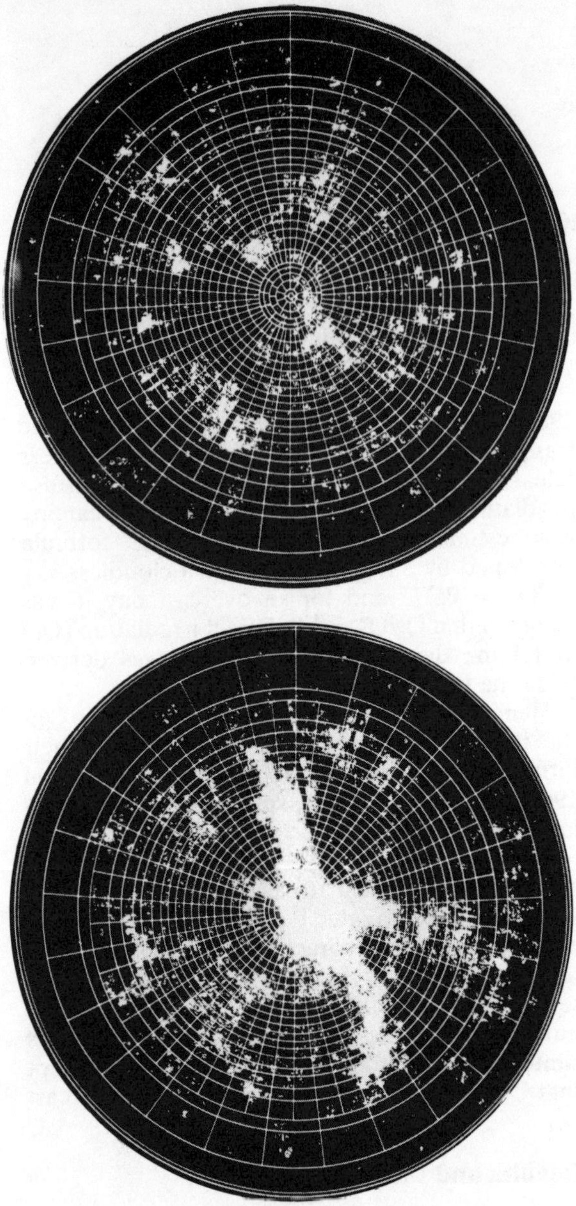

FIG. 1. Hemispherical photographs of the two rainforest canopies. The photographs are overlaid with the spiders'-web grid, designed so that each segment contributes 0.1% of the total radiant flux on a horizontal surface from a standard overcast sky: (a) closed canopy site, (b) small gap site.

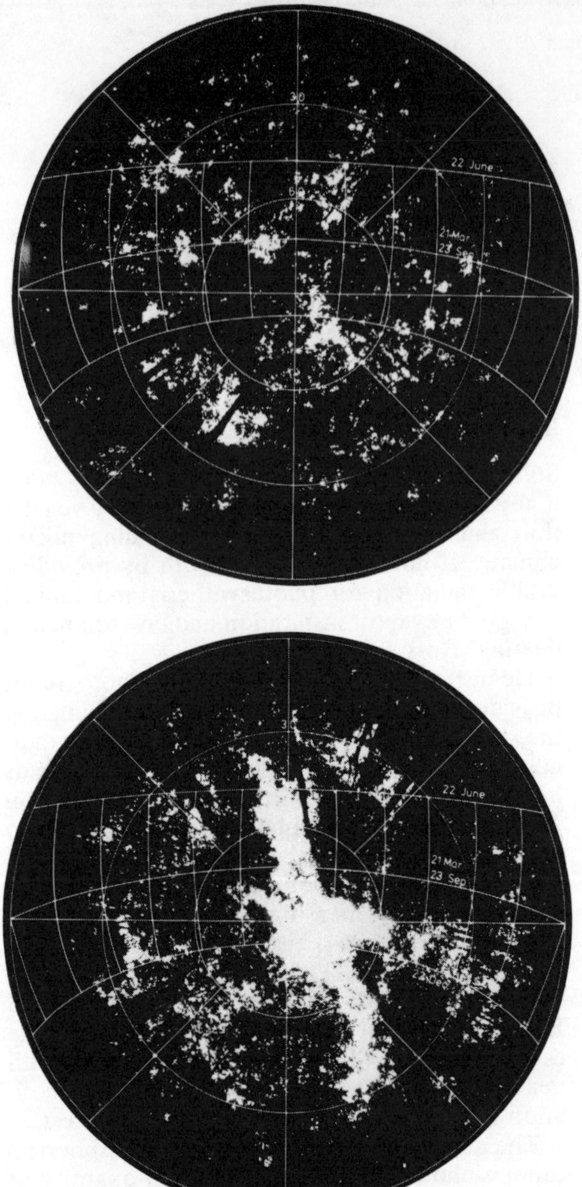

FIG. 2. Hemispherical photographs of the two rainforest canopies. The photographs are overlaid with a polar diagram for 17°17′S showing solar tracks for the solstices and equinox: (a) closed canopy site, (b) small gap site.

forest floor (Q_T) at the closed canopy site and 71.3% at the small gap site. At the equinox Q_b was 50.8% and 60.9% respectively, decreasing to 16.3% and 25.9% for the respective sites by the winter solstice. Evans (1956) found that sunflecks contributed about 70% of Q_T on the forest floor during a 3-month measurement period in a Nigerian rainforest. Similarly, in a lowland rainforest in Ecuador, Grubb and Whitmore (1967) concluded that 70% of Q_T from 1000-1400 hrs under sunny conditions was contributed by sunflecks. Whitmore and Wong (1959), in a rainforest near Singapore, estimated that over an entire year 50% of Q_T was derived from sunflecks; this is to be compared

TABLE 1. Estimated diffuse irradiation, Q_d (KJ m^{-2} day^{-1}) above and beneath the two rainforest canopies under cloudless skies (Q_d/Q_T = 0.275 after Sayigh 1977).

	Diffuse Irradiation (KJ m^{-2} day^{-1})		
	Above canopy	At forest floor closed canopy*	At forest floor small gap†
Summer solstice (Dec 22)	7 995	448	1 423
Equinoxes (Mar 21/Sept 23)	6 989	391	1 244
Winter solstice (June 22)	5 221	292	929

*Diffuse site factor = 5.6%.
†Diffuse site factor = 17.8%.

TABLE 2. Estimated direct irradiation, Q_b (KJ m^{-2} day^{-1}) above and beneath the two rainforest canopies under cloudless skies.

	Direct Irradiation (KJ m^{-2} day^{-1})		
	Above canopy	At forest floor closed canopy	At forest floor small gap
Summer solstice (Dec 22)	21 076	457 (2.2)*	3 532 (16.8)
Equinoxes (Mar 21/Sept 23)	18 424	404 (2.1)	1 934 (10.5)
Winter solstice (June 22)	13 763	57 (0.4)	325 (2.4)

*The values in parentheses are the direct site factors (%).

TABLE 3. Estimated total irradiation, Q_T (KJ m^{-2} day^{-1}) above and beneath the two rainforest canopies under cloudless skies (atmospheric transmission coefficient = 0.70 after Turton 1987).

	Total Irradiation (KJ m^{-2} day^{-1})		
	Above canopy	At forest floor closed canopy	At forest floor small gap
Summer solstice (Dec 22)	29 071	905 (3.1)*	4 955 (17.0)
Equinoxes (Mar 21/Sept 23)	25 413	845 (3.0)	3 335 (12.5)
Winter solstice (June 22)	18 984	349 (1.8)	1 254 (6.6)

*The values in parentheses are the total site factors (%).

TABLE 4. Estimated total irradiation, Q_T (KJ m^{-2} day^{-1}) above and beneath the two rainforest canopies under overcast skies (atmospheric transmission coefficient = 0.30 after Turton 1987).

	Total Irradiation (KJ m^{-2} day^{-1})		
	Above canopy	At forest floor closed canopy*	At forest floor small gap†
Summer solstice (Dec 22)	12 459	698	2 218
Equinoxes (Mar 21/Sept 23)	10 891	610	1 939
Winter solstice (June 22)	8 136	456	1 448

*Total site factor = 5.6%.
†Total site factor = 17.8%.

with a yearly mean of 46.0% for both sites presented in this paper.

The direct site factors (Table 2) also compare well with those given in Pearcy (1983) for the understorey of a Hawaiian evergreen forest at similar latitudes. The direct site factors ranged from 0 to 17.1%, and he estimated that on relatively clear days, the contribution of sunflecks may be as great as 80% of the daily total (Pearcy 1983). Bjorkman & Ludlow (1972) and Chazdon & Fetcher (1984a) have reported similar maximum contributions by sunflecks of 62% and 77% respectively, beneath subtropical and tropical rainforests.

The ecological and physiological significance of sunflecks is the focus of recent studies in tropical rainforests (e.g., Pearcy & Calkin 1983; Chazdon & Fetcher 1984a, 1984b; Pearcy *et al.* 1985). However, the daily values for Q_b under cloudless skies (Table 2) do not show individual sunfleck events which may account for a large fraction of the daily carbon gain beneath a closed canopy (Pearcy & Calkin 1983).

Although Q_b is obviously important in the understorey, one should not neglect the predominance of cloud cover in the humid tropics and its possible effects on forest productivity (Longman & Jenik 1974). A recent study by Aylett (1985) in an upper montane rainforest in Jamaica concluded that cloudiness is almost certainly important in reducing productivity. A similar effect probably occurs on the summits of the upland mountains of north Queensland. By comparing the estimated values for Q_T beneath cloudless and overcast skies (Tables 3 and 4), it is evident that the differences are most marked during the summer half of the year, which also happens to be the most cloudy period.

The size and orientation of gaps in the canopy undoubtedly play an important part in the understorey vegetation dynamics (Hartshorn 1980) and in controlling the surrounding microclimate. The small gap analysed in this study (Fig. 2) has a north-south orientation which limits the quantity of Q_b reaching the forest floor in comparison with a similar sized gap with an east-west orientation.

Furthermore, in the seasonally-wet tropics of north Queensland, understorey moisture stress is likely to occur beneath quite small gaps during the dry-season (August-November) because of the large quantities of Q_b reaching the ground (Table 2). In forest types containing deciduous or semi-evergreen trees, such as this site, the effect is enhanced still further. Drying out of the soil and leaf litter would also promote fire regimes.

Acknowledgements

Funding for this project was made possible through a Special Research Grant from the James Cook University of North Queensland, Townsville. I would like to thank Mr Geoff Stocker (CSIRO, Forest Research, Atherton) for ideas concerning the content of this paper.

References

Anderson M. C. (1964) Studies of the woodland light climate I: the photographic computation of light conditions. *J. Ecol.* **52**, 27-41.

Anderson M. C. (1971) Radiation and crop structure. In: *Plant Photosynthetic Production: Manual of Methods* (eds Z. Sestak, T. Catsky & P. G. Jarvis) pp. 412-67. Dr W. Junk, The Hague.

Aylett G. P. (1985) Irradiance interception, leaf conductance and photosynthesis in Jamaican upper montane rain forest trees. *Photosynthetica.* **19**, 323-37.

Bjorkman O. & Ludlow M. M. (1972) Characterization of the light climate on the floor of a Queensland rainforest. *Carnegie Inst. Wash. Yearbook.* **71**, 85-94.

Chazdon R. L. & Fetcher N. (1984a) Photosynthetic light environments in a lowland tropical rain forest in Costa Rica. *J. Ecol.* **72**, 553-64.

Chazdon R. L. & Fetcher N. (1984b) Light environments of tropical rain forests. In: *Physiological Ecology of Plants of the Wet Tropics* (eds E. Medina, H. A. Mooney & C. Vazquez-Yanes) pp. 27-36. Dr W. Junk, The Hague.

Evans G. C. (1956) An area survey method of investigating the distribution of light intensity in woodlands, with particular reference to sunflecks, including an analysis of data from rain forest in Southern Nigeria. *J. Ecol.* **44**, 391-428.

Evans G. C. & Coombe D. E. (1959) Hemispherical and woodland canopy photography and the light environment. *J. Ecol.* **47**, 103-13.

Grubb P. J. & Whitmore T. C. (1967) A comparison of montane and lowland forest in Ecuador III: the light reaching the ground vegetation. *J. Ecol.* **55**, 33-57.

Hartshorn G. S. (1980) Neotropical forest dynamics. *Biotropica* **12**, (Suppl.), 23-30.

Longman K. A. & Jenik J. (1974) *Tropical Forest and its Environment.* Longman, London.

Madgwick H. A. I. & Brumfield G. L. (1969) The use of hemispherical photographs to assess light climate in the forest. *J. Ecol.* **57**, 537-42.

Pearcy R. W. (1983) The light environment and growth of C_3 and C_4 tree species in the understorey of a Hawaiian forest. *Oecologia (Berl.)* **58**, 19-25.

Pearcy R. W. & Calkin H. W. (1983) Carbon dioxide exchange of C_3 and C_4 tree species in the understorey of a Hawaiian forest. *Oecologia (Berl.)* **58**, 26-32.

Pearcy R. W., Osteryoung K. & Calkin H. W. (1985) Photosynthetic responses to dynamic light environments by Hawaiian trees. *Plant Physiol.* **79**, 896-902.

Ross J. (1975) Radiative transfer in plant communities. In: *Vegetation and the Atmosphere Vol. 1: Principles* (ed. J. L. Monteith) pp. 13-55. Academic Press, London.

Sayigh A. A. M. (1977) *Solar Energy Engineering*. Academic Press, New York.

Tracey J. G. (1982) *The Vegetation of the Humid Tropical Region of north Queensland*. CSIRO Publ., Melbourne.

Turton S. M. (1987) The relationship between total irradiation and sunshine duration in the humid tropics. *Solar Energy* **38**, 353-54.

Whitmore T. C. & Wong Y. K. (1959) Patterns of sunfleck and shade light in tropical rain forest. *Malay. For.* **22**, 50-62.

Proc. Ecol. Soc. Aust. 15:107-113

Beetle pollination in Australian tropical rainforests

A. K. IRVINE[1] and J. ARMSTRONG[2]

[1]*Tropical Forest Research Centre, C.S.I.R.O Atherton, Qld. 4883*

[2]*Illinois State University, Normal, Illinois, U.S.A.*

Abstract

There are few in-depth studies on entomophily and cantharophily, particularly in the putative bird/bat pollination systems in Australia. A review on cantharophily is presented. Examples of beetle pollination in the rainforests of north Queensland are given and described in terms of cantharophilous syndromes. Beetle pollination occurs throughout the rainforest structure, from ground-storey plants to upper canopy trees and also in pioneer species. In some pollination systems, the nectar/pollen-eating beetles predominate, in others the food-body eaters or herbivore beetles are dominant. The pollination associations range from flexible systems where cantharophily varies in importance, to specialized entirely cantharophilous systems. In the latter more than one species of beetle may be involved or the system may be so specialized, that only one species of beetle is the pollinator. Some implications of these pollination systems for management and conservation of the species described are given.

Introduction

There are few studies highlighting cantharophily (beetle pollination) in Australia other than that presented by Hamilton (1897). The review by Armstrong (1979) of biotic pollination mechanisms in Australia confirms this situation. In fact entomophily itself has seldom been studied in depth in Australia apart from a very recent work on *Eucalyptus regnans* F. Muell. by Griffin *et al.* (in press). Most workers recently have concentrated on the more conspicuous bird pollination, the impact-effect of marsupial pollination and to a lesser extent bat and rodent pollination, without really comparing the influence of insects in these systems. Griffin (1982) stated "It is difficult to believe that one could find any species (eucalypts) without some type of insect visitor." We concur with this view, but would extend the observation to more groups than eucalypts. Crome & Irvine (1986) demonstrated that in a rainforest tree (*Syzygium cormiflorum,* (F. Muell.) B. Hyland, Myrtaceae) with typical bat/bird-syndrome flowers, the low, steady occurrence of insects, produced fertilization results equal to that of the frequent, conspicuous, honeyeater visitations during the study period.

Hawkeswood (1981) recorded 12 species of beetles pollinating *Angophora woodsiana* F. M. Bail. (Myrtaceae) in S.E. Queensland and buprestid beetles pollinating *Leptospermum, Angophora* and *Eucalyptus* in the Blue Mountains, New South Wales (1978). Griffin *et al.* (in press) listed beetles as part of the pollinator spectrum in *E. regnans.* The prominence of beetles visiting myrtaceous species in temperate Australia, is interesting to compare with their scarcity as flower visitors in the northern hemisphere as claimed by Faegri & van der Pijl (1976), who state that beetle pollination is characteristic of the tropical zone. If so, then prominence of beetles as flower visitors in Australia, could provide speculative support to theories *re* the Gondwanan-Austral origin of angiosperms (Webb *et al.* 1986). The prominence of dipterans and thysanopterans as flower visitors in Australia provides further fuel as both groups were prominent with coleopterans in the fossil record (Riek 1970), at the time of the appearance of angiosperms.

In this paper, we discuss cantharophily and give examples in north Queensland rainforests. We will also briefly focus on implications of these studies for land-use management and conservation.

Cantharophily and Beetle-flower Syndromes

There is much confusion in the literature about the nature of the beetle flower syndrome. Percival (1965) stated, "A great many beetle species visit flowers, but in the majority of cases the plant reaps no benefit. They eat pollen, anthers, lick nectar, gnaw flower parts, and ovaries. Pollen does not adhere to their smooth bodies. They tend to move little between flowers and their flight is erratic." Ashton (1969) stated "beetle pollination is not likely to be oligolectic, at least in tree families." Faegri & van der Pijl (1976) described the cantharophilous syndrome as:
"Pollination units with few visual attractions, no special or definite shape, no depth effect, no nectar guide, generally large, flat, cylindric or shallow, bowl-shaped, — sometimes closed — easy of access (beetles being poor fliers). Colours dull, frequently greenish or off-white. Odour strong, fruity or aminoid. Attractants open, easily accessible, pollen, food bodies or nectar. Sexual organs exposed."

In their text, they stated that due to lack of specialization, beetle flowers are often overlooked, but later describe two types of beetle flowers: "the big single flower, frequently amorphic or haplomorphic *sensu* Leppik (1956) and an advanced type the inflorescence". They did not mention the inflorescence in their syndrome.

Armstrong (1979) adapting from Faegri & van der Pijl (1976) stated "compared with other insects, most anthophilous beetles have gnawing mouthparts and seemingly primitive instincts, which generally preclude their effective participation in the pollination process." "Beetles are unable to land precisely on blossoms, hence most anthophilous species are restricted . . . to the simplest flowers, chiefly of the 'dish and bowl type.' " In the same text Grant (1950) is quoted, suggesting that flowers regularly probed by beetles "have evolved more advanced means of ovule protection than flowers sucked by . . . bee, . . .". Rickson (1979) analysing the food bodies of the protein-rich tissue on the tips of staminodes, stamens and inner tepals in *Calycanthus* stated "these cells represent a very specialized type of plant cell". House (1986) quoting from Radford *et al.* (1974), ". . . flowers actinomorphic, pollen presented in large quantities, flower parts many . . ." and from Carlquist (1969) ". . . stamens and staminodes are broad and leaf-like."

Faegri & van der Pijl (1976) emphasized the term *"mess and soil pollination"* for beetle pollination, with implications of primitiveness of both beetles and the pollination system. Although they mentioned that it occurred in more evolved pollination systems, it was not highlighted when dealing with other groups. It is a very effective process of pollination and very widespread. However it is hardly confined to beetles, and the "mess" of pollen widely occurs in the absence of biotic vectors. There is the implication of damage in the term and if such is intended, then it could be applied to beetle pollination systems where feeding on food bodies and flower parts occurs. Perhaps "mess and *spoil*" would be a more appropriate term for such pollination systems, as Thien (1974) had inadvertently used.

Gottsberger (1974, 1977) appears to be the first worker to attempt to sort out the different types of cantharophily. His work added brush blossoms, the inflorescence and a sweet fragrant odour to Faegri & van der Pijls' syndrome but more importantly, separated the syndrome into three types.

1. An open, unprotected, less specialized, medium-sized flower, with sexual organs exposed, hermaphroditic, protogynous, having a sweet fragrant odour, relatively few stamens, attracting only a few species of beetles ((1977) *cf.* with a wide spectrum of unspecialized beetles (1974)). Gottsberger based this type on *Drimys brasiliensis* which has cymose inflorescences aggregated at the end of branches. He regarded it as the most primitive type of cantharophily, due to the especially primitive position of *Drimys*. Thien (1980) queried the relative roles of thrips and flies, that Gottsberger mentioned visiting *Drimys*. Thien stated that Diptera, Hymenoptera and Coleoptera visit *Drimys granadensis* var. *mexicana*. Due to the open nature of the flower, the syndrome is likely to attract a wide range of insects, probably with odour being the prime sorter of potential visitors. It is actually an entomophilous syndrome with cantharophily varying in degree of importance.

2. Flowers specialized into enlarged, more robust structures, with a proliferation of petals and/or sex organs (i.e., food bodies). Closure of the perianth at some stage, providing a protective chamber for feeding, copulating, ovipositing and breeding. Hermaphroditic, protogynous with strong fruity or aminoid, imitative or deceitful odours, acting directly on the instincts of visitors. Ovary semi-inferior. Flowers of this type attract a crowd of beetles.

3. Flowers specialized into an inflorescence. Gottsberger focused on cymose, corymbose and capitulate forms in this type but probably also meant to include expanded inflorescences (panicles), e.g., Arecaceae (Henderson 1984; Beach 1984) and many rainforest groups.

This type of specialization overlaps the other two syndromes in function and lends itself to nectar-pollen-eating beetles, or to chewing food-body beetles. The latter feed on trichomes, rachillae, petals, anthers, stigmas and pollen. In the paniculate inflorescence visited by nectar/pollen eaters, flowers are often actinomorphic, with an open, simplistic form. Flower size may influence the degree of entomophily. Odours may be sweet smells or fruity or aminoid smells.

The unifying theme amongst flowers visited commonly by beetles, tends to be a strong odour. Faegri & van der Pijl (1976) state that odour, probably together with texture, is a very important factor for short distance recognition (i.e., within 1 cm or less). Odour can also be used for long distance attraction as shown by Williams & Dodson (1972) *ex* Williams (1983) for male euglossine bees, for beetles both in an arum lily (Monteith 1973) and in *Eupomatia* (Armstrong & Irvine, unpublished).

Beetle Pollination in North Queensland

Two of the examples presented, *Alphitonia petriei* Braid & C. T. White, *Flindersia brayleyana* F. Muell. are based on studies by Irvine (unpublished). *Balanophora fungosa* Forster & G. Forster is based on work by Cox & Irvine (unpublished) whereas *Myristica insipida* R.Br. and *Eupomatia laurina* R.Br. are based on work done by Armstrong & Irvine (unpublished). The pollination systems are compared mainly with Gottsberger's syndromes.

The pioneer tree *Alphitonia petriei* (Rhamnaceae) is a sequentially developing, mass-flowering tree with numerous, shallow, dish-shaped, actinomorphic flowers in numerous axillary corymbose-like inflorescences. Hermaphrodite flowers are about 10 mm diameter when petals are fully extended. Perianth, greeny cream, and the shiny green open disk is covered with a clear sticky nectar. The five stamens, initially hooded by the petals dehisce a few hours before the stigmas appear receptive (protandrous). Odour is a heavy sweet smell. Flowering occurs in the driest months (Sept.-Oct.) in upland rainforest, curtailing as the storm season nears in November. The tree has a flexible, entomophilous pollination system. Stamen size tends to negate bird and mammal contact and these animals have not been recorded to date. Flies tend to predominate particularly in warmer parts of the day, but beetles are prominent in early mid-morn and late afternoon. Wasps are also active visitors, but occur in lower numbers. The beetles feed on nectar, which may have shed pollen-grains present. Ten beetle species have been recorded from limited observations to date. None of them physically damage the flower.

In terms of cantharophily, it is the specialized inflorescence type, attracting predominantly medium-sized, non-flower-damaging, nectar/pollen-feeding beetles, whose awkward flight is quite suitable for landing on the rough outlines of inflorescences, but like *Drimys,* the open nature of the flowers attracts other insects and cantharophily is only part of the system.

Flindersia brayleyana (Flindersiaceae) is an upper canopy, late secondary, sequentially developing, mass-flowering tree. The inflorescences form showy white, terminal, paniculate heads superficially resembling cauliflowers. Flowers are hermaphrodite, actinomorphic, similar in size and shape to *A. petriei* flowers. Petals (five) are bright white. A bright orange sticky disk surrounds the greenish ovary and style, when the latter emerges. Stamens and staminodes (five of each) curve over the central part of the flower. Stamens dehisce 18-24 hours before the stigma appears receptive (protandrous). Odour is a heavy sweet smell. Flowering occurs in the storm season (Nov.-Dec.) prior to the wet season. Like *A. petriei* it has a flexible entomophilous system in which birds and mammals are similarly negated and have not been observed pollinating. In this case beetles (2-12 mm) tend to dominate, but flies are also very prominent. Lepidopterans (both night and day species) rank next, whereas wasps are few in kind and number. *Apis mellifera* does not visit this species but visits *A. petriei* actively.

If storms occur in the months when *F. brayleyana* is flowering, many scarab beetles emerge from the soil and swarm into the air. One of these, *Phyllotocus apicalis* Macleay is well adapted for crawling over inflorescence surfaces. It is helpless on leaf and bare ground surfaces. In favourable storm seasons, masses of these beetles form an active cloud, about one metre above the flowers between 8.00 and 11.00 a.m. Female beetles actively forage over flowers for nectar, whilst males eagerly pursue them for mating. The flying cloud consits of dislodged males and females, as well as newly arrived beetles, bouncing back and forth, attempting to gain access to flowers and mates. In the afternoon all is quiet. Inspection reveals virtually each open flower, occupied by a beetle quietly foraging on the sticky disk or resting. Next morning, cloud activity occurs again with the arrival of newly-emerged beetles. This results in some insects leaving the tree and seeking other flowers, which may not necessarily be *Flindersia,* but the spatial relationships of *F. brayleyana* are generally close enough to ensure inter-tree movement.

The work of Frankie (1976), Stephenson (1982) and Frankie & Haber (1983) suggested reasons why inter-tree movement occurs amongs pollinators of mass-flowering trees. Intra specific competition for food and mates, certainly applies to the behaviour of *P. apicalis* and a response to nectar depletion may also occur. In terms of cantharophily, *F. brayleyana* is similar to *A. petriei,* but with a shift in some seasons at least to beetle dominance. There appears to be some sorting of insects between two species with flowers of similar shape and size. This may be due to the difference in odour, which seems unattractive to at least several hymenopteran groups in *F. brayleyana*. Again beetle visitors are non damaging and protandry, not protogyny occurs.

On the forest floor, the achlorophyllous, root parasite *Balanophora fungosa* (Balanaphoraceae) is pollinated by a wide spectrum of visitors. Inflorescences usually occur in clumps, 8-12 cm high. It is monoecious, with male flowers forming a ring around the base of the female head, which contains thousands of microscopic apetalous female flowers. Each female flower possesses at least one vesicle, that secretes mouse-like smelling nectar in the dawn-early morning period and possibly at night. By 10.00 a.m. the nectar is usually evaporated or consumed but the mousey odour persists throughout the day and re-secretion occurs next morning. This process lasts for at least four or five days, persisting probably whilst the female flowers are receptive. Anther dehiscence may be slightly protandrous or homogamous. Visitors recorded feeding on pollen and nectar are ants (seven species) some flies, and a springtail. In some rainforest communities *Rattus fuscipes,* has been observed feeding on male flowers. (Cox & Irvine, unpublished).

Four species of insects have been recorded using the basal bracts below the male flowers, as brood chambers, two species of *Lasiodactylius* (Nitidulidae-Coleoptera), one tipulid and one pyralid. Adult and larval *Lasiodactylus* may be present in the bracts at the same time. All have been found with pollen on their bodies. As the larval stages develop, they commence feeding on the central fleshy stalk that supports the flower head, tunneling in between and parting the female flowers. They do not appear to damage these flowers but rather break them up into small islets, that peel or shed as the stalk succumbs to foraging. This occurs when the ovary/fruit appears to be mature and ready for dispersal, probably via water through the soil.

The pollination system is very flexible, inviting not only pollen/nectar-eating animals, but also provides a protective chamber (basal bracts) and food bodies (bracts and central stalk) for plant-eating insects, that participate in the pollination and dispersal process. The latter system may have some minor mechanical loss of ovules via tunnelling between the female flowers. The genetic differences between clumps of *B. fungosa* on the same tree root are unknown, but the activity of adult *Lasiodactylus* suggests that they would readily move between clumps, as pollen is the adult food.

In terms of cantharophily, *B. fungosa* has a compound inflorescence head and monoecy occurs. Food bodies are not provided by multiplication of flower organs or flowers, but rather by the central supporting tissue, basal bracts and pollen. The odour is not deceitful, as *Lasiodactylus* find food, shelter and a successful breeding site. The system is also very suitable for ant, moth and rodent pollination. Govindappa & Shivamurthy (1975) only noted hymenopteran species visiting *Balanophora abbreviata* Blume (also monoecious) for nectar and pollen.

The next two cases are examples of micro-cantharophily. They are both more exclusively cantharophilous. *Myristica insipida* (Myristicaceae) is a dioecious, (occasionally slightly monoecious) mid to upper-canopy tree with inconspicuous, aggregated, axillary inflorescences. Flowers are small (5-6 mm long, 2-3 mm wide), cream-green, urn-like, with male flowers appearing slightly more elongate. Central sexual parts are easily accessible to a visitor, small enough to enter the flower. Both sexes produce the same sweet floral fragrance, which appears to be the initial attractant. The flower visitors recorded to date, are a minute parasitic fly (whose larvae destroy both male and female flowers), thrips (which use flowers as domiciles, browsing on the internal wall of the receptacle and ovary surface) and small beetles (eight species, in mid-upland rainforest and only two of these species in coastal lowland rainforest). There are no observations as yet of thrips foraging on or making deliberate contact with the stigma, but they do browse on pollen. After being initially attracted by the fragrant odour of the flowers, pollen seems to be of greater interest to the beetles, as their numbers are much higher at staminate trees, than at pistillate trees. But they do show interest in the seemingly sticky surface of the stigma and do not immediately fly away from female flowers.

This system fits Gottsberger's less specialized-flower/beetle syndrome, with the difference that its flowers are much smaller (markedly limiting the size of potential visitors), dioecy occurs, and it appears exclusively cantharophilous. This is supported by Armstrong's (1986) work on *Myristica fragrans.*

Eupomatia laurina is a shrub to small understorey tree, which occurs in rainforest on the tropical lowlands, but occurs in rainforest and tall, moist, eucalypt, open forest in the tropical uplands and subtropical areas. The pollination system of this species has become very specialized. Only one species of weevil, *Elleschodes hamiltoni* T. Blackburn (Hamilton 1897) is known to visit the flower throughout its geographic range, from east Victoria to north Queensland. We have not recorded any other animal visitor to the flowers, other than predatory green ants waiting on the edge for weevils to arrive. They do not enter the flower. The flower provides shelter, specialized food bodies and a brood place which provides sufficient nutrition to enable weevil larvae to develop to maturity, before pupating in the soil. They can emerge as adults in 14-16 days, ready to track more flowers. The many stamens and staminodes form a fused structure (synandrium) which sheds as one unit, neatly removing weevil eggs as it falls. The eggs hatch and the larvae develop on the fallen synandrium. As the flowers only last one day, adult weevils leave each flower at dusk, passing through dehiscing anthers (protogynous) and arrive at flowers opening at dawn the next day. The beetles fertilize stigmas as they move around on the floor of the chamber provided by the closed staminodes. The mostly solitary flowers emit a strong fruity-musky scent, which attracts the weevils from some distance.

The system conforms to the specialized larger flower beetle syndrome of Gottsberger's. It differs in that the odour is not deceitful, it is very specific, having only one pollinator and there is no loss of ovules in the pollination system.

Implications for Management and Conservation

Knowledge of these pollination relationships provides information on plant/animal associations in rainforest communities and information on the flexibility or specialization of a pollination system. Together, these factors indicate whether a breeding system could be upset by disturbance within a rainforest community and also indicate how a breeding system may behave outside the community. Combining this knowledge with the ecophysiological behaviour of a species, provides a sound basis for management decisions. For instance both *A. petriei* and *F. brayleyana* stand up to disturbance very well. In fact *A. petriei* depends upon disburbance and seedlings of *F. brayleyana* require light gaps for regeneration. Both species can establish in cleared land beyond the rainforest edge and their flexible pollination systems ensure reproductive success in this situation. This indicates that both species would be very suitable for plantation and revegetation projects in cleared areas of land, which were once their former habitat.

The reaction of *B. fungosa* to disturbance and increased light is unknown. It seems to have a wide range of hosts (Irvine, unpublished) and if plants are able to survive disturbance, then its flexible pollination system would also ensure reproductive success. With *M. insipida* and *E. laurina,* the situation becomes more tentative. Both species are micro-beetle specialists and in some rainforest communities it seems that the number of species available for pollination of *M. insipida* are few, whilst there is only one available for *E. laurina.* In this situation the effects of disturbance on pollinators become very important. Seedlings of both species are sensitive to light and need at least partial shade to establish in north Queensland. Unless apomictic, *E. laurina* is not likely to reproduce sexually in areas remote from its natural habitat. In rainforest patches where these two species exist, the impact of pesticide use in adjacent land could affect their survival and reproduction. These two species, whose specialized pollination system has enabled survival for a long period may be highly vulnerable under the impact of man's disturbance.

Conclusion

Beetle pollination in north Queensland's rainforests occurs in all layers of the forests, from ground layer plants to upper canopy trees, as well as occurring in pioneer trees. The plant/pollinator association may be a flexible system where beetles feature prominently in the spectrum together with other insects such as flies, wasps, lepidopterans and thrips. Some of these groups together with beetles are associated with bird, bat, marsupial and rodent flower visitors. In some cases, the cantharophily is associated predominantly with flies and some lepidopterans whereas hymenopterans are few in kind and number. The plant/pollinator association may be quite specialized and be entirely cantharophilous with more than one species of beetle involved or be as specialized as a one to one relationship.

Acknowledgements

We thank Tom Weir, Entomology Division, C.S.I.R.O., for beetle identifications and colleagues Andrew Graham and Frank Crome for comments.

References

Armstrong J. (1979) Biotic pollination mechanisms in the Australian flora — a review. *N.Z. J. Bot.* **17**, 467-508.

Armstrong J. (1986) Floral biology of *Myristica fragrans* Houtt. (Myristicaceae), the nutmeg of commerce. *Biotropica* **18**, 32-38.

Ashton P. S. (1969) Speciation among tropical forest trees: some deductions in the light of recent evidence. *Biol. J. Linn. Soc.* **1**, 155-96.

Beach J. A. (1984) The reproductive biology of the peach or "pejibaye" palm *(Bactris gasipaes)* and a wild congener *(B. porschiana)* in the Atlantic lowlands of Costa Rica. *Principes* **28**, 107-19.

Carlquist S. (1969) Toward acceptable evolutionary interpretations of floral anatomy. *Phytomorphology* **19**, 332-62.

Crome F. H. J. & Irvine A. K. (1986) "Two bob each way". The pollination and breeding system of the Australian rainforest tree, *Syzygium cormiflorum. Biotropica* **18**, 115-25.

Faegri K. & Van Der Pijl L. (1976) *The Principles of Pollination Ecology*. Second revised edition. Pergamon Press, Oxford, U.K.

Frankie G. W. (1976) Pollination of widely dispersed trees by animals in Central America, with an emphasis on bee pollination systems. In: *Variation, Breeding and Conservation of Tropical Trees* (eds J. Burley & B. Styles) pp. 151-59. Academic Press, London.

Frankie G. W. & Haber W. A. (1983) Why bees move among mass-flowering neotropical trees. In: *Handbook of Experimental Pollination Biology* (eds C. E. Jones & R. J. Little) pp. 360-72. Van Nostrand Reinhold Company Inc. New York.

Gottsberger G. (1974) The structure and function of the primitive angiosperm flower — a discussion. *Acta Bot. Neerl.* **23**, 461-71.

Gottsberger G. (1977) Some aspects of beetle pollination in the evolution of flowering plants. *Plant Syst. Evol. Suppl.* **1**, 211-26.

Govindappa D. A. & Shivamurthy G. R. (1975) The pollination mechanism in *Balanophora abbreviata* Blume. *Ann. Bot.* **39**, 977-78.

Grant V. (1950) The pollination of *Calycanthus occidentalis. Amer. J. Bot.* **37**, 294-97.

Griffin A. R. (1982) Pollination ecology of eucalypts — a framework for study. In: Proc. of Symp. Pollination '82 (eds E. G. Williams, R. B. Knox, J. H. Gilbert, & P. Bernhardt) pp. 42-56. School of Botany, Univ. of Melbourne.

Griffin A. R., Ohmart C. P. & Hand F. C. (1986) Estimation of the effective pollinators of *Eucalyptus regnans* F. Muell., a mass-flowering tree species. *Oecologia (Berl)*.

Hamilton A. G. (1897) On the fertilization of *Eupomatia laurina,* R.Br. *Proc. Linn. Soc. N.S.W.* **22**, 48-56.

Hawkeswood T. J. (1978) Observations on some Buprestidae (Coleoptera) from the Blue Mountains, N.S.W. *Aust. Zool.* **19**, 257-75.

Hawkeswood T. J. (1981) Insect pollination of *Angophora woodsiania* F. M. Bail. (Myrtaceae) at Burbank, South-east Queensland. *Vict. Nat.* **98**, 120-29.

Henderson A. (1984) Observations on pollination of *Cryosophila albida. Principes.* **28**, 120-26.

House S. M. (1986) *Relationships between Breeding and Spatial Pattern in some Dioecious Tropical Rainforest Trees*. Ph.D. Thesis, Australian National University, Canberra.

Leppik E. E. (1956) The form and function of numerical patterns in flowers. *Amer. J. Bot.* **43**, 445-55.

Monteith G. B. (1973) Entomological notes (Dung beetles as pollinators of an arum lily). *News Bull., Ent. Soc. Qld.* **97**, 13.

Percival M. S. (1965) *Floral Biology*. Pergamon Press, Oxford.

Radford A. E., Dickson W. C., Massey J. R. & Bell C. R. (1974) *Vascular Plant Systematics*. Harper & Row.

Rickson F. R. (1979) Ultrastructural development of the beetle food tissue of *Calycanthus* flowers. *Amer. J. Bot.* **66**, 80-86.

Riek E. F. (1970) Fossil History. In: *Insects of Australia*, pp. 168-86. C.S.I.R.O. Div. Ent., Melbourne Univ. Press, Melbourne.

Stephenson A. G. (1982) When does outcrossing occur in a mass flowering plant? *Evolution* **36**, 762-67.

Thien L. B. (1974) Floral biology of magnolia. *Amer. J. Bot.* **61**, 1037-45.

Thien L. B. (1980) Patterns of pollination in the primitive angiosperms. *Biotropica* **12**, 1-13.

Webb L. J., Tracey J. G. & Jessup L. W. (1986) Recent evidence for the autochthony of Australian tropical and subtropical rainforest elements. *Telopea* **2**, 575-89.

Williams N. H. (1983) Floral fragrances as cues in animal behaviour. In: *Handbook of Experimental Biology* (eds E. G. Jones & R. J. Little) pp. 50-72. Van Nostrand Reinhold Company Inc., New York.

Williams N. H. & Dodson C. H. (1972) Selective attraction of male euglossine bees to orchid floral fragrances and its importance in long distance pollen flow. *Evolution* **26**, 84-95.

Proc. Ecol. Soc. Aust. 15:115-126

The temporal distribution of butterfly species richness in a subtropical region of Australia

C. J. HILL

*School of Australian Environmental Studies, Griffith University, Nathan, Qld. 4111**

Abstract

The species richness of a given area or habitat is a simple but widely used community index with several applications. Many studies of species richness ignore the temporal distribution of species but this component may be an important factor in the species richness of insects because of their short generation times.

The results of monthly surveys in Brisbane showed that the greatest butterfly species richness occurred in mid-summer — the season in which the highest temperatures and rainfall are recorded. A comparison with studies from other tropical localities found that the peak in butterfly species richness also occurred in the wet season, although there were some contradictory results. A more detailed appraisal of the Brisbane results found that there was a second peak in species richness in spring. Several factors were examined in relation to this double peak in species richness. Voltinism and the availability of adult food resources were suggested as possible causal factors.

The results of this study suggest that the temporal distribution of species is indeed an important component of butterfly species richness and that, in tropical regions, the temporal distribution of butterfly species is largely determined by the pattern of rainfall. Furthermore, it highlights the need for additional studies of this type, particularly in the tropics.

Introduction

The species richness of a given area or habitat is perhaps the simplest and most widely used community index. It is often used as the biological measure of a habitat when decisions are to be made concerning conservation and management or for environmental impact statements. In addition the factors which affect species richness (such as area, habitat heterogeneity or disturbance) are currently the subject of much theoretical debate (e.g., Simberloff & Gotelli 1984; Jarvinen 1984; Willis 1984). Species richness has three components: the species composition (i.e., the particular species which are present); the temporal distribution of these species and finally the total number of species. Both Shapiro (1975) and Gilbert (1984) observe that most of the work in community ecology (and therefore species richness) has focussed on birds and mammals. Furthermore, these studies often ignore time as a resource whereas for insects, which often have relatively short generation times, the temporal distribution of species (with respect to each other and the environment) may be both important and revealing (Shapiro 1975). This aspect of insect species richness may be particularly relevant in the tropics where species can have several generations within one year.

Unlike many insects, butterflies are taxonomically well-known and, generally, recognizable in the field without capture. Accordingly, they are an obvious choice for a study concerning the temporal distribution of insect species richness. In this paper I describe the results of the first detailed temporal survey of an Australian butterfly fauna in the subtropics or tropics. The results of this study of within-year variation in butterfly species richness in the Brisbane region are compared with those from other tropical localities, with an emphasis on the effect of

**Present Address:* Dept. of Zoology, James Cook University, Townsville, Qld. 4811.

climate on the temporal distribution of butterfly species richness. Finally the Brisbane results are analyzed in greater detail and the influence of factors in addition to climate are discussed.

Methods

Study sites

Four study sites, with well defined boundaries, were selected within the Moreton Bay region of south-east Queensland (Fig. 1). The climate throughout the area is subtropical with a hot, wet summer from October to March and a mild, dry winter from April to September. The average annual temperature is 25.4°C and there is an average annual rainfall of 1 148.1 mm. A description of each site is summarized in Table 1.

Site 1: Toohey Forest. Toohey Forest, with an area of some 400 hectares, is an island of high ground rising from the plains of south Brisbane. In spite of the pressure of surrounding suburban development, the vegetation is in a relatively natural condition, although the area is used extensively for recreational purposes and is subject to regular bushfires.

The vegetation is open forest (Specht 1970) and usually consists of three layers — trees, shrubs and herbs. The tree layer is dominated by members of the family Myrtaceae, the most abundant and diverse genus being *Eucalyptus* but *Tristania conferta* and *Angophora woodsiana* are also common in places. The shrub layer is comprised of a number of species, including members of the families Fabaceae,

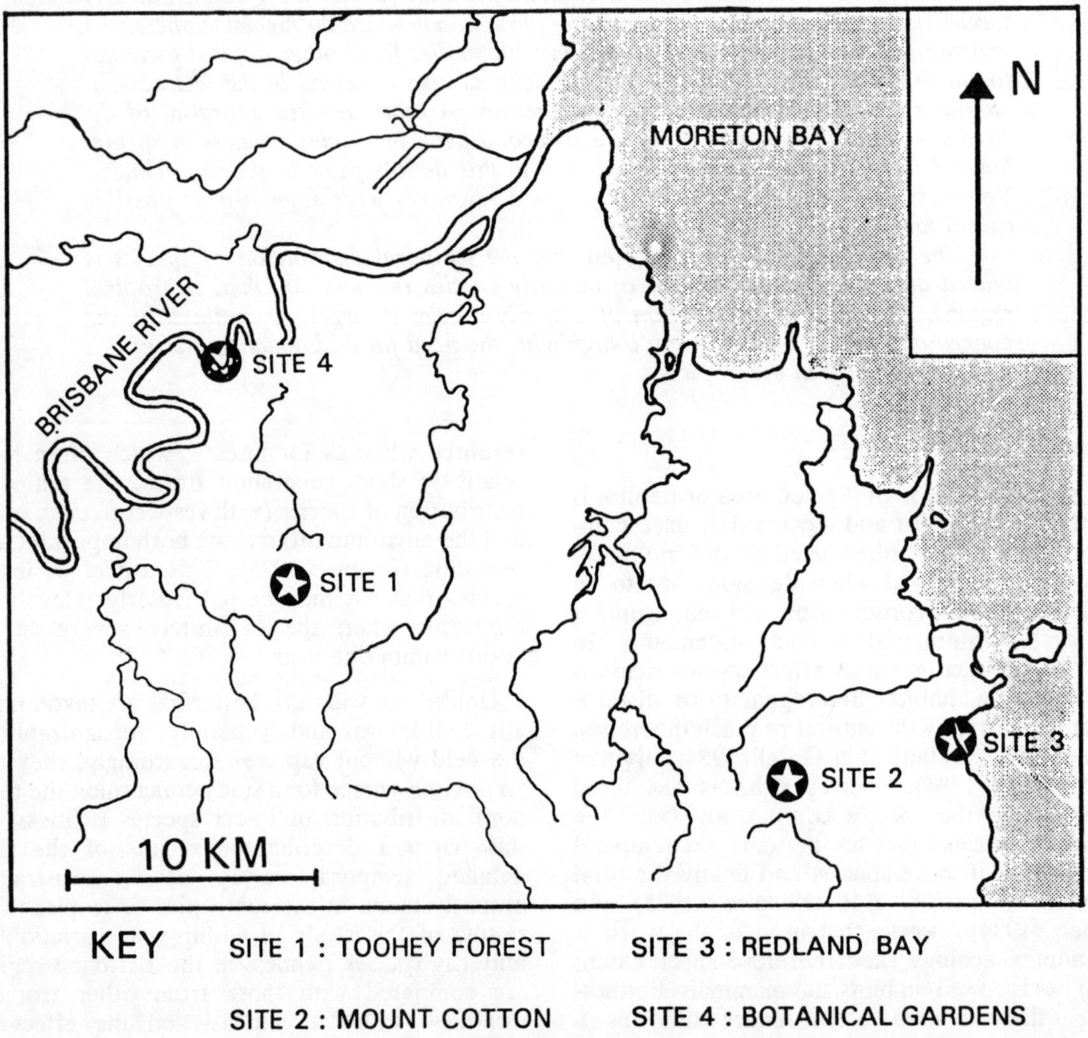

FIG. 1. The location of the study sites in the Moreton Bay region of south-east Queensland.

TABLE 1. Descriptive details of each study site.

Site No.	Location (Lat.+Long.)	Altitude (metres)	Area (ha)	Transect Length (m)	Sample Time (SD) (min)	Sampling Period
Site 1	153°3′30″E 27°32′45″S	35-194	398.3	3 825	119.4(5.2)	Sept. 1982-Aug. 1983
Site 2	153°14′17″E 27°36′35″S	24-60	80.0	2 447	95.5(4.5)	Sept. 1982-Aug. 1983
Site 3	153°17′48″E 27°35′50″S	0-2.5	14.7	2 100	94.8(9.0)	Sept. 1982-Aug. 1983
Site 4	153°10′44″E 27°28′41″S	5-18	14.6	2 110	85.9(7.3)	Nov. 1982-Oct. 1983

Mimosaceae and Xanthhorroeaceae. The herb layer contains several families but four species are probably the most widespread and abundant, these being: *Imperata cylindrica* and *Themeda australis* (Poaceae), *Goodenia rotundifolia* (Goodeniaceae) and *Passiflora suberosa* (Passisifloraceae).

Site 2: Mount Cotton. The study site at Mount Cotton is an agricultural research station 26 km south-east of Brisbane. The station has a total area of 100 hectares, 20 hectares of which are set aside as a eucalypt forest nature reserve while the remainder is used for experimental purposes.

The site was originally covered by open forest but, at present, consists of grass-covered paddocks and fields used for crop trials. The area is subject to regular disturbance in the form of grazing, trampling, mowing, crop planting and harvesting. The high intensity of disturbance has led to the establishment of diverse and abundant populations of weedy plant species. The Verbenaceae are well represented by *Lantana camara* and *Verbena bonariensis,* and other abundant species include *Ageratum houstonianum* (Asteraceae) and *Asclepias fruticosa* (Asclepiadaceae).

Site 3: Redland Bay. The Redland Bay site, with an area of 14.7 hectares, is situated on the coast of Moreton Bay. It contains areas of mudflat, saltmarsh and mangroves which are subject to regular tidal innundation and most of the site is covered by water at high tide. The area has little disturbance apart from lepidopterists in search of specimens.

The vegetation consists almost entirely of mangroves, of which the dominant species is *Avicennia marina* (Avicenniaceae). However, there are several distinct zones of vegetation running more or less parallel with the shore line. The seaward fringe consists of a thin strip of *A. marina* trees, inland and adjacent to which is an intermittent band of *Rhizophora stylosa* (Rhizophoraceae). The next zone landwards contains shrub-like *A. marina* mixed with *Ceriops tagal* (Rhizophoraceae) and some clumps of *Aegiceras corniculatum* (Myrsinaceae). The landward side is bordered by small *A. marina* and between these and the high tide line is an area of saltmarsh and mudflat.

Site 4: Botanical Gardens. The Botanical Gardens are situated in the centre of the city of Brisbane and have an area of 14.6 hectares. They were first established in 1828 in order to accelerate and expand the production of food and commercial plants, however their present function combines horticulture with recreation.

The vegetation could by no means be described as natural but it is representative of that in the suburban areas of Brisbane. Lawns, herbaceous borders, shrubs and trees are the four major vegetational types found at the site. The trees are a mixture of native and exotic species, *Delonix regia* (Caesalpinaceae), *Jacaranda mimosaefolia* (Bignoniaceae) and *Cinnamomom camphora* (Lauraceae) being exotic tree species which occur at the site. Native tree species include *Castanospermum australe* (Fabaceae), *Geijera parviflora* (Rutaceae) and various *Ficus* sp. (Moraceae). The shrubs and herbs present usually have large colourful flowers such as *Euphorbia pulcherrima* (Euphorbiaceae), *Clerodendrum* spp. (Verbenaceae), *Pentas lanceolata* (Rubiaceae) and *Impatiens sultanii* (Balsaminaceae), all of which are exotic species. However the Australian flora are well represented by various species of *Acacia* (Mimosaceae), *Callistemon, Leptospermum, Melaleuca* (Myrtaceae) and *Grevillia* (Proteaceae).

Sampling Methods

A non-overlapping transect route was selected for each study site (Table 1). The route was designed to incorporate as much intra-habitat variation in vegetation, topography and disturbance history as possible. The same transect was used to sample both butterflies and plants in each site at one monthly intervals for one year. At Sites 1, 2 and 3 sampling began in September, 1982 and ended in August, 1983. At site 4 the equivalent dates were November, 1982 and October, 1983.

Butterflies

Butterflies were sampled using a modified version of Pollard's method (Pollard 1977). This technique takes the form of an observer walking a transect at a uniform pace and identifying and counting the butterflies which are observed within a predetermined area. For the purposes of this study the sampling area was set at up to ten metres in front, five metres either side and directly above the observer. These distances were found to be the limits at which most species could be identified or, when necessary, captured. All samples were carried out between 1030 and 1430 so as to reduce sampling bias due to interspecific differences in activity patterns. This time period was found to include the period of greatest activity for butterfly species in the Brisbane area (C. Hill *pers obs.*). Climatic conditions also affect butterfly activity (Kingsolver 1983), therefore habitats were only censussed on sunny days with little or no wind.

As an area was sampled over a period of time it became increasingly clear that some of the butterflies observed were emigrants from other habitats. A set of criteria was chosen to determine which species were resident within the study sites. A resident species was defined as one for which oviposition, mating or immature stages were observed. For some species oviposition and/or immature stages were difficult to observe due to the inaccessibility of their larval foodplains. Therefore a second category, called possible resident, was used. The criteria for this category were the observation of the adult butterflies for two or more months of the sampling period and the presence of the larval foodplant at the site. Transient species were those for which the adult butterflies were observed but no larval foodplants were known to be present at the site. For the purpose of this paper the resident and possible resident species are grouped into one category, 'RPR' species.

Flowering Plants

The plant species in flower were identified and recorded for a width of five metres along the total length of the transect. Because the primary aim of this sampling effort was to assess the diversity of potential nectar-source plants, those species with flowers which were not easily visible (such as grasses) were not included in the sample. Using this method an estimate was made of the species richness of flowering plants for each month.

Only a proportion of the flowering plant species available in a habitat is visited by adult butterflies. In order to determine which flowers were being utilized a record was kept of any visits by butterflies which were observed. With only a finite amount of time to be spent in any one study site and with some parts of the habitat being difficult to observe, this method may not be sufficiently comprehensive. Therefore plant species for which records existed from butterfly visits observed outside the study site, personal communications from other researchers or observations in the literature were also included. By incorporating all of these sources of information a fairly comprehensive list of the plant species utilized as adult food resources by butterflies was obtained for each study site.

Results

The mean monthly values of four climatic factors are shown in Fig. 2. It is apparent that the Brisbane region experiences its lowest temperatures and rainfull during the winter months of June to August. Conversely the highest temperatures and rainfall occur in mid-summer (January and February). The evaporation rate is at its highest in December and the lowest humidity is recorded during November. Figure 3 shows the amount of rainfall which fell during each month of the sampling period at the study sites. There is no positive relationship between the distribution of rainfall during the sampling period and the temporal distribution of butterfly species richness at the sites (Table 2), although at Site 3 there is a negative correlation between monthly rainfall

TABLE 2. The correlation between the temporal distributions of rainfall during the sampling period and the total number of butterfly species at each site.

Site	r_s	Sig.
1	0.0594	NS
2	-0.4000	NS
3	-0.6454	*
4	0.1246	NS

n = 12; NS: not significant; $p < 0.05$.
r_s: Spearman's Rank Correlation Coefficient.

during the sampling period and the species richness of butterflies.

In Fig. 4 the records of all butterfly species observed at the four sites have been combined. The lowest values of butterfly species richness occur in June and July but there are two peaks of species richness, the first in September and October and the second from January to March. An alternative viewpoint is that there are unexpectedly low values of species richness in November and December. Figure 5 uses data collected by Prof. R. L. Kitching from ten lepidopterists in the Brisbane region. The participants in the survey were asked to record (on the basis of their observations and collections), the months in which adults of each butterfly species occurred. The data plotted are the total number of records for each month for those species which occurred in this study. The number

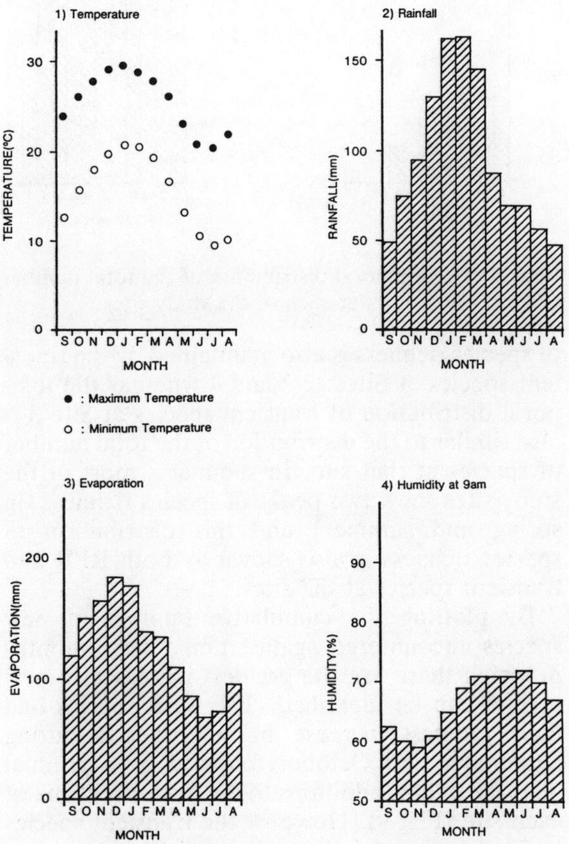

FIG. 2. The Brisbane climate. Mean monthly values of temperature, rainfall, evaporation and humidity for all years of record to the end of 1972. (Source: Commonwealth Bureau of Meterology Regional Office for Queensland).

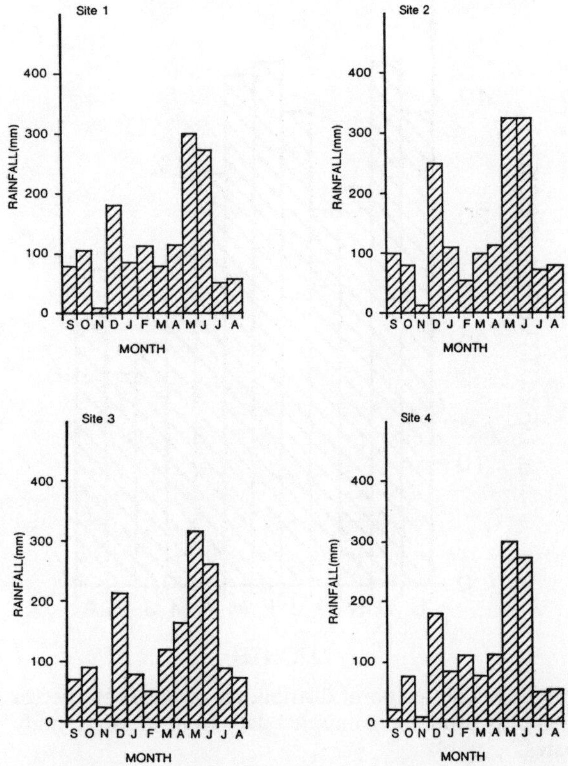

FIG. 3. The monthly rainfall at the study sites during the sampling period.

of records, rather than the number of species, was used because this measure places less emphasis on the records of individual lepidopterists and gives a better indication of the months in which a species is most likely to occur. The data show a close concordance with the results of this study, with the lowest species richness in winter, the highest in summer and a secondary peak of species richness in spring.

When the study sites are considered separately (Fig. 6) the total number of species at Sites 1, 2 and 4 show the same pattern of temporal distribution to that observed for the combined data. At Site 3 however, there are low values of species richness during winter but no clear peaks during spring or summer. Figures 7 and 8 show the temporal distribution of RPR and transient butterfly species at the sites. The pattern observed for the total number of species is again shown by the RPR species at Sites 1 and 2 but less clearly at Site 4. In addition the RPR species at Site 3 show a similar pattern of temporal distribution to the other sites. Interestingly the pattern of two peaks

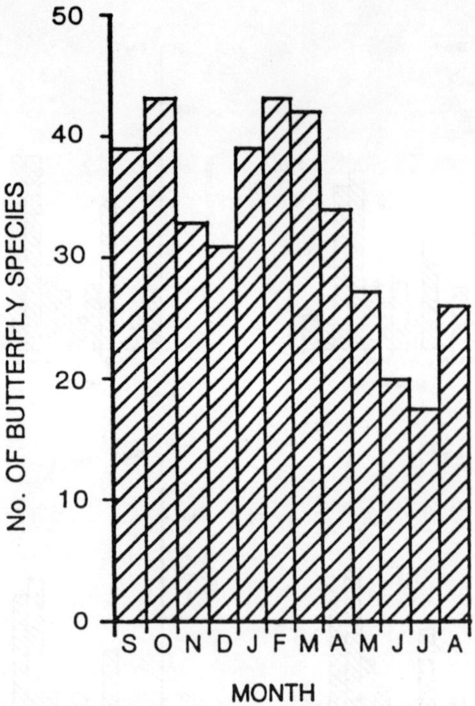

FIG. 4. The temporal distribution of butterfly species richness using the combined data from the four study sites.

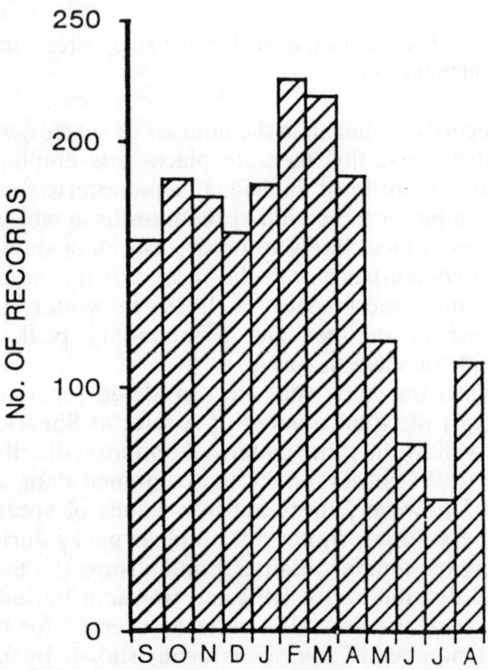

FIG. 5. The temporal distribution of monthly records from 10 Brisbane lepidopterists.

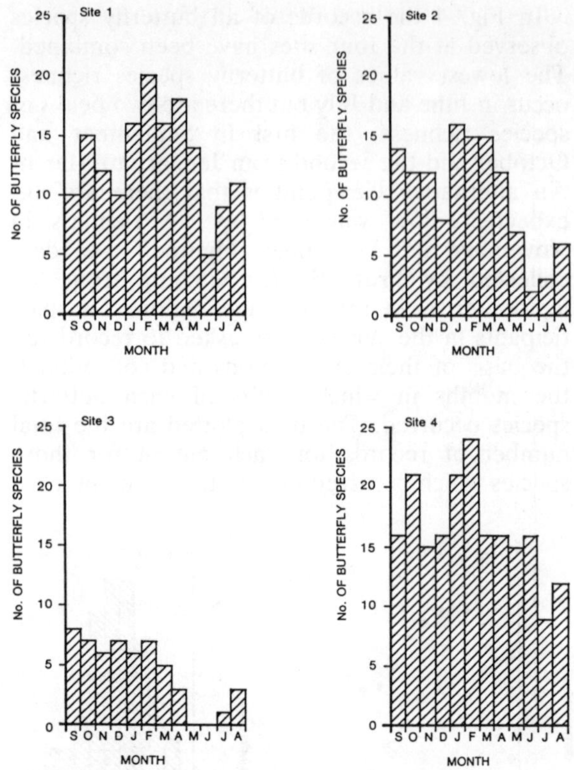

FIG. 6. The temporal distribution of the total number of butterfly species at each of the study sites.

of species richness is also maintained by the transient species at Sites 1, 2 and 4 whereas the temporal distribution of transient species at Site 3 is also similar to the distribution of the total number of species at that site. In summary, most of the study sites show two peaks of species richness (in spring and summer) and this distribution of species richness is also shown by both RPR and transient species at the sites.

By plotting the cumulative number of new species encountered against time, those months in which there are the greatest increases in new species can be identified. The RPR species had their greatest increase in new species during September and October followed by a gradual increase in new additions to the butterfly fauna of each site (Fig. 9). However the transient species at Sites 1, 2 and 4 showed two distinct increases in new species, the first of these occurred during September and October and the second from January to February (Fig. 10). The transient species in Site 3 showed a gradual increase throughout the summer months (Fig. 10).

Table 3 shows the results of tests for correlations (Spearman's Rank Correlation) between the temporal distribution of butterfly species richness and the number of plant species in flower. At Sites 1, 2 and 4 neither the RPR nor the transient butterfly species were correlated with the total number of plant species in flower. In fact the transient species at these sites were not correlated with any category of flowering plant species. However the temporal distribution of RPR species *was* correlated with the number of plant species in flower at these sites when the analysis was confined to those plant species which were utilized as adult food resources by the butterflies. At Site 3 both RPR and transient species were correlated with the utilized plant species in flower. The RPR species were also correlated with the total number of plant species in flower.

Discussion

Comparison with other tropical regions

The flight period of butterflies (and hence their temporal distribution of species richness) is

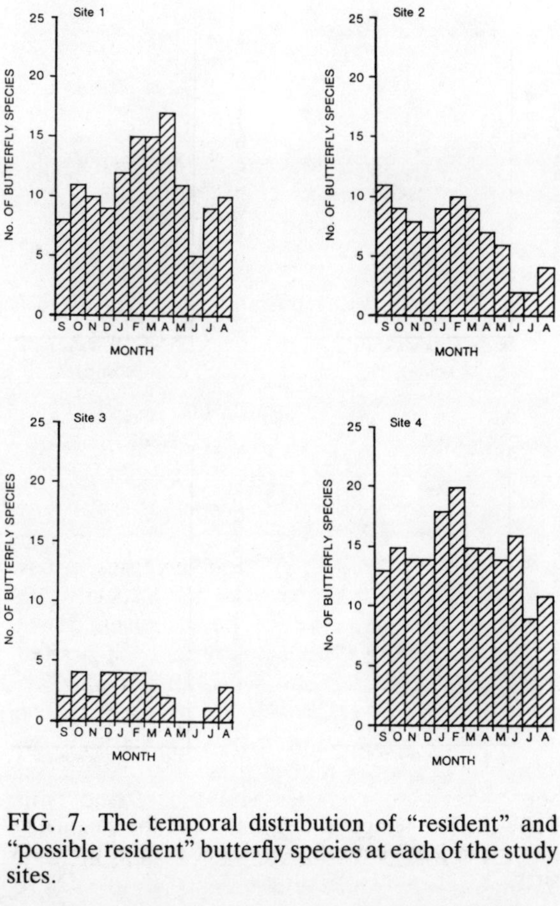

FIG. 7. The temporal distribution of "resident" and "possible resident" butterfly species at each of the study sites.

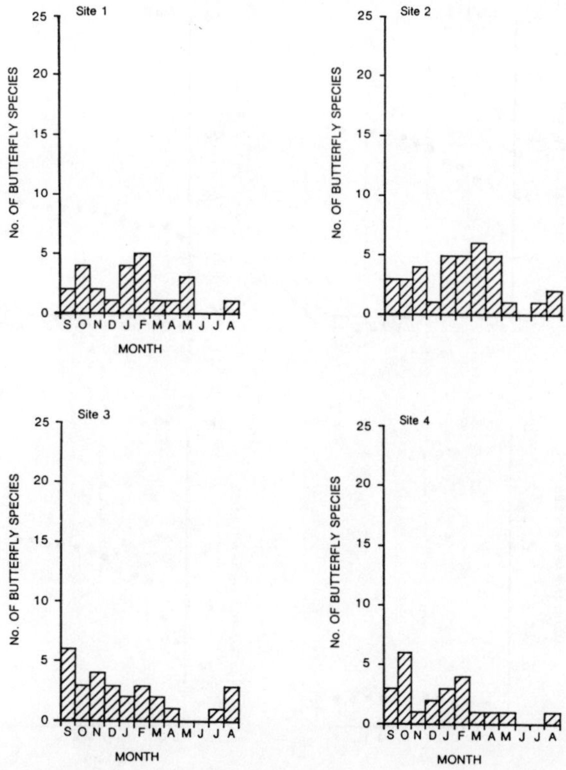

FIG. 8. The temporal distribution of "transient" butterfly species at each of the study siters.

influenced by several factors of which climate is probably the most important. Climate may be broken down into two major components — temperature and rainfall. Temperature affects butterflies directly since they are poikilothermic and must raise their body temperature to a certain level before they can fly (Brussard & Ehrlich 1970). In addition it governs the rate of larval development (Gossard & Jones 1977). Rainfall has an indirect effect inasmuch as it determines the growing season for many plant species and, since most butterfly larval require young foliage especially in their early instars, it can determine the availability of suitable oviposition sites. Cloud cover (and hence rainfall) may act directly to suppress activity in butterflies (Brussard & Ehrlich 1970; Gossard & Jones 1977; Zalucki 1981). Given these important effects, it is not surprising that the overall temporal pattern of butterfly species richness in this study corresponded strongly with climate. Generally the highest butterfly species richnesses corresponded with the season in which the highest temperature and rainfall occurred and conversely the lowest butterfly

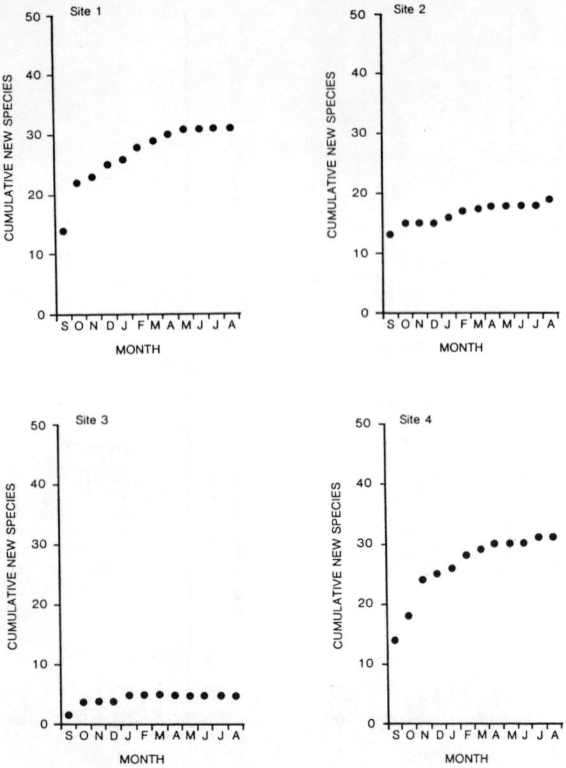

FIG. 9. The temporal distribution of the cumulative new "resident" and "possible resident" butterfly species at each of the study sites.

species richness occurred during the months with the lowest temperatures and rainfall.

The relative importance of temperature and rainfall and their effects (positive or negative) on butterfly species richness varies according to the latitude of the region in question. For instance, in those regions with a temperate or mediterranean climate the greatest species richness coincides with the period of highest temperature and, usually, lowest rainfall (e.g., Shapiro (1975), Opler & Langston (1968) in North America; Kitching *et al.* (1978) in Australia). In this, the only sub-tropical region studied so far, the peak in butterfly species richness coincides with high temperatures and rainfall (this study) but the fact that temperatures are sufficiently high to allow some species to breed throughout the year suggests that rainfall is of the greatest importance in these regions. In the tropics temperature has little effect in determining the temporal distribution of butterfly species richness since it remains at a relatively constant high level. Rainfall however becomes of great importance although its effects may be varied (Wolda 1983). Because the wet season is the period in which most new plant growth occurs (Wolda 1978; Frith & Frith 1985), one might expect that the highest butterfly species richness would also occur at this time. Indeed, some studies have shown that this is the case, for example Heithaus (1974) in Costa Rica, Owen (1971) in Sierra Leone and Percival (1974) in Jamaica. Studies of other tropical insects have also found that the greatest species richness or abundance occurs in the wet season (Ackerman 1983; Janzen 1973; Frith & Frith 1985; Broadhead & Wolda 1985).

In contrast to these studies, others have found the maximum butterfly species richness during the dry season (e.g., Emmel & Leck (1970) in Panama and Fox *et al.* (1965) in Liberia). In both of these studies it was suggested that the inhibitory effect of rainfall on the flight activity of butterflies was responsible for the result. Janzen (1973) has also found that, in areas with a very mild dry season, the greatest number and species of insects occurs in the dry season.

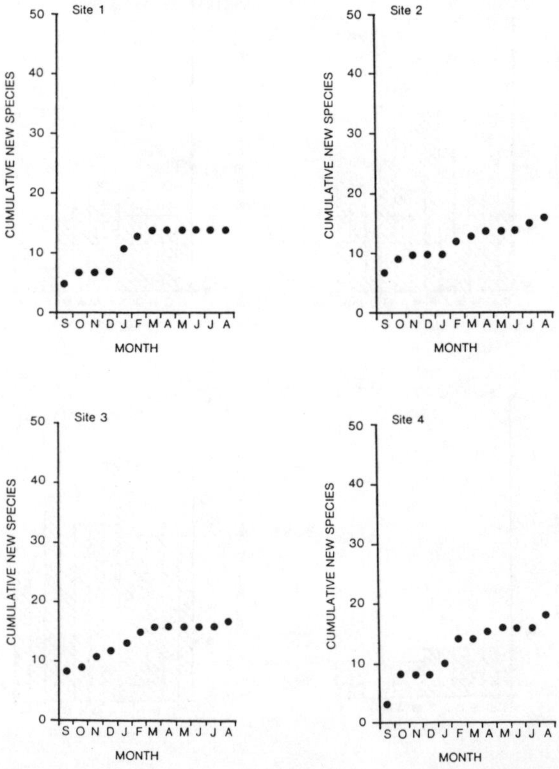

FIG. 10. The temporal distribution of the cumulative new "transient" butterfly species at each of the study sites.

TABLE 3. The correlation between the temporal distributions of the number of butterfly species and the number of plant species in flower.

Site	Butterfly species type	Plant species in flower			
		Total		Utilized[a]	
		r_s	sig	r_s	sig
1	RPR	0.3982	NS	0.7016	*
	TRANS	0.1727	NS	0.2441	NS
2	RPR	0.4884	NS	0.6506	*
	TRANS	0.1541	NS	0.2441	NS
3	RPR	0.9019	**	0.8516	**
	TRANS	0.3642	NS	0.5684	*
4	RPR	0.4050	NS	0.6096	*
	TRANS	0.4448	NS	0.2588	NS

n = 12; NS: not significant; *: $p<0.05$; **: $p<0.01$; r_s: Spearman's Coefficient of Rank Correlation; [a]: The maximum number of plant species in flower which are utilized as adult food resources by butterflies.

To complicate matters further, Emmel & Leck (at another site in Panama) found that the highest number of butterfly species occurred during the wet-dry season transition period. This result was attributed to a combination of an overlap in wet and dry season faunas and favourable climatic conditions.

One explanation for these contradictory results might be that there is a threshold level in the number of rain-days in a month which occur during the wet season. If the rainfall pattern of the region lies above this threshold level then it might have a negative effect on the flight activity of adult butterflies. Therefore the butterfly fauna switch their adult activity towards the dry season. However this explanation appears to be unlikely, given the opposite results that were obtained in Sierre Leone (Owen 1971) and Liberia (Fox *et al.* 1965), these being adjacent countries on the coast of West Africa. In order to resolve this problem further studies of the temporal distribution of butterfly species richness are required and, perhaps of equal importance, similar sampling techniques should be adopted.

The Brisbane region in detail

Although Sites 1, 2 and 4 had their highest species richness in mid-summer, the results show that they also had a peak in spring (September or October), as did the RPR species at Site 3. This double peak in butterfly species richness may simply be a unique event which happened to occur during the sampling period, or it might be a true representation of the temporal distribution of butterfly species richness in this region. The fact that the pattern is similar throughout most of the study sites suggests that the event occurs over the whole region rather than a particular site. Furthermore, the results from Site 4 also show the pattern even though the sampling commenced two months later and continued for two months longer at this site. In addition data from local lepidopterists collected over a number of years also supports the observation that there are spring and summer peaks of butterfly species richness in the Brisbane region. If the low values of butterfly species richness in November and December were due to an event which occurred over the whole region then rainfall experienced during the sampling period (rather than expected rainfall) seems to be the most likely factor responsible. The results show that the rainfall experienced during the sampling period was not typical of the average pattern of rainfall in the Brisbane region. There was lower than average rainfall in November but this was followed by high rainfall in December. In general there was no relationship between rainfall during the sampling period and the temporal distribution of butterfly species richness.

Assuming that the pattern is a true representation of the temporal distribution of butterfly species richness in this region then to what causal factors may it be attributed? In Shapiro's (1975) comprehensive study on the temporal distribution of butterfly richness in North America, double or multiple peaks in species richness were also observed at some sites. They were attributed to several factors including: (i) transient species, (ii); voltinism, (iii) climate and (iv) flowering phenology.

(i) It is possible that one or other of the two peaks of species richness observed in this study was caused by an influx of transient species at a particular time of the year. However an analysis of temporal distribution of RPR and transient species separately shows that both categories had the two peaks of species richness at the majority of sites. It therefore seems unlikely that either peak in species richness was caused by the arrival of transient species.

(ii) Another explanation for the observed pattern of temporal distribution may lie in the life histories of the species themselves. The peaks of species richness may represent successive flight periods of the same species (i.e., bi- or multi-voltine species) or a separate spring and summer fauna (uni-voltine species). If most of the species involved are multi-voltine then a plot of cumulative new species with time would show an initial increase in new species corresponding to the first peak of species richness and little or no increase with the second peak of species richness. In other words, the species which occur in the first peak of species richness also make up the majority of species in the second peak. Alternatively, for uni-voltine species one would predict two increases in new species corresponding to the two peaks of species richness. Such plots for RPR species suggest a mostly bi- or multi-voltine species composition.

The low values in the number of species observed at the sites in November and December may indicate a time period which is unfavourable for adult activity and is therefore avoided by many of the species in this region. Alternatively the low values may simply be due to the time required for successive generations of adults to develop, i.e. eggs laid by adults in September and October do not develop into adults until January or February, although it would be surprising if such synchrony occurred, given that butterfly families tend to diapause at different developmental stages of their life histories (see Hayes 1982 for review).

In the case of the transient species both peaks of richness correspond to increases in new species and thus to some extent represent a separate spring and summer fauna. It is possible that the transient spring fauna may be derived from local habitats (other than those at the study sites) and the summer fauna may originate from geographically distant localities which have expanded their range earlier. Assuming that this were the case, then the two peaks of species richness found for transient species would be a consequence of range expansion and contraction during the year. Jones & Rienks (1987) have suggested that this sort of seasonal geographic range expansion and contraction occurs in three butterfly species of the genus *Eurema* (Pieridae) in Australia.

(iii) There are two climatic variables which could account for the unexpectedly low values of species richness observed in November and December. These are evaporation rate and humidity. Evaporation rate is at its highest and humidity at its lowest at this time of the year. Therefore there is the potential for adult butterflies to be under stress from desiccation during this period. However most butterflies have reasonably sophisticated physiological and behavioural thermo-regulatory mechanisms (Clench 1967a; Kingsolver 1985) and, moreover, they are able to obtain water both from their adult food source (usually floral nectar) and more directly by drinking (Norris 1936; Arms *et al.* 1974). Accordingly it seems unlikely that the high evaporation and the low humidity would affect them deleteriously, given that supplies of floral nectar and or water were available.

(iv) Finally it has been suggested that the availability of floral nectar may influence the timing of the flight period for some butterfly species (Clench 1967b; Owen 1971; Shapiro 1975). The results show that (for three of the sites) the temporal ditribution of RPR species was not correlated with the total number of plants in flower but (for all sites) was correlated with the plant species in flower which they utilized as adult food resources. However, the possibility exists that these correlations may be due to both butterfly and flowering plant species responding in a similar fashion to an environmental factor (e.g., climate).

It seems unlikely that the observed pattern of temporal distribution of butterfly species richness can be solely attributed to the arrival of transient species. But the possibility that the low values of species richness in November and December are simply due to the time required for adult development between flight periods cannot be dismissed. Alternatively, the pattern may result from a period of the year which is unfavourable for flight activity. If this is so, the correlation between the temporal distribution of butterfly species richness and the temporal distribution of the maximum number of utilized flower species suggests that adult food resources may be an important factor in determining flight activity patterns. Floral nectar is not only a source of sugar and amino-acids but also a source of water for butterflies. When this fact is coupled with the observation that there is the potential for adult butterflies to be under stress from desiccation at this time of the year, a possible explanation for the low values of species richness in November and December emerges.

Conclusions

The life-cycle of butterflies is closely tied to the local vegetation (since their larvae feed on a variety of specific plant species and most adults require nectar from flowers) therefore any changes to the vegetation of a habitat are likely to have an immediate effect on the butterfly community at present. As a result of this association, they are excellent candidates as indicator species in the assessment of environmental quality, change or impact of conserved area. If butterflies are to fulfill their potential in this role, then it is important that the factors which affect their species richness are understood. This study has shown that the temporal distribution of butterfly species richness is an important component of their overall species richness in both the subtropics and tropics. It has found that although the overall pattern of temporal distribution of butterfly species richness is largely determined by the cycles of temperature and rainfall of the region, other factors such as voltinism and the availability of adult food resources may also be important. Finally, this study has highlighted the need for further research in this area both in Australia and elsewhere.

Acknowledgements

I am most grateful to Roger Kitching and Jane Hughes for their help and encouragement in their capacity as supervisors of my Ph.D., from which this work was derived. Carla Catterall, Roger Kitching, Jane Hughes and Carole Wallace provided many useful criticisms of an earlier draft of this manuscript. This work was funded by a Griffith University Postgraduate Scholarship.

References

Ackerman J. D. (1983) Diversity and seasonality of male euglossine bees (Hymenoptera: Apidae) in Central Panama. *Ecology* **64**, 274-83.

Arms K., Feeny P. & Lederhouse R. C. (1974) Sodium: Stimulus for puddling behaviour by tiger swallowtail butterflies, *Papilio glaucus Science, N.Y.* **185**, 372-74.

Broadhead E. & Wolda H. (1985) The diversity of Psocoptera in two tropical forests in Panama. *J. Anim. Ecol.* **54**, 739-58.

Brussard P. F. & Ehrlich P. R. (1970) Adult behaviour and population structure in *Erebia epipodea* (Lep.: Satyrinae). *Ecology* **51**, 880-85.

Clench H. K. (1967a) Behavioural thermoregulation in butterflies. *Ecology* **47**, 1021-34.

Clench H. K. (1967b) Temporal dissociation and population regulation in certain hesperiine butterflies. *Ecology* **48**, 1000-06.

Emmel T. C. & Leck C. F. (1970) Seasonal changes in organization of tropical rain forest butterfly populations in Panama. *J. Res. Lepid.* **8**, 133-52.

Fox R. M., Lindsey A. W., Clench H. K. & Miller L. D. (1965) The butterflies of Liberia. *Mem. Amer. ent. Soc.* No. 19.

Frith C. B. & Frith D. W. (1985) Seasonality of insect abundance in an Australian upland tropical rainforest. *Aust. J. Ecol.* **10**, 237-48.

Gilbert L. E. (1984) The biology of butterfly communities. In: *The biology of butterflies.* Symp. R. ent. Soc. Lond. No. 11 (eds R. I. Vane-Wright & P. R. Ackery) pp 41-52. Academic Press, London.

Gossard T. W. and Jones R. E. (1977) The effects of age and weather on egg laying in *Pieris rapae* L. *J. appl. Ecol.* **14**, 65-71.

Hayes J. L. (1982) A study of the relationships of diapause phenomena and other life history characters in temperate butterflies. *Am. Nat.* **120**, 160-70.

Heithaus E. R. (1974) The role of plant pollinator interactions in determining community structure. *Ann. Mo. Bot. Gdn.* **61**, 675-91.

Janzen D. H. (1973) Sweep samples of tropical foliage insects: effects of seasons, vegetation types, elevation, time of day, and insularity. *Ecology* **54**, 687-701.

Jarvinen O. (1984) Dismemberment of facts: a reply to Willis on subdivision of reserves. *Oikos* **42**, 402-03.

Jones R. E. & Rienks J. (1987) Reproductive seasonality in the tropical genus *Eurema* (Lepidoptera: Pieridae). *Biotropica* **19**, 7-16.

Kingsolver J. G. (1985) Thermal ecology of *Pieris* butterflies (Lep.: Pieridae): a new mechanism of behavioural thermoregulation. *Oecologia (Berl.)* **66**, 540-45.

Kitching R. L., Edwards E. D., Ferguson D., Fletcher M. B. & Walker J. M. (1978) The butterflies of the Australian Capital Territory. *J. Aust. ent. Soc.* **17**, 125-33.

Norris M. J. (1936). The feeding habits of the adult Lepidoptera Heteroneura. *Trans. R. entomol. Soc. London.* **85,** 61-90.

Opler P. A. & Langston R. L. (1968) A distribution analysis of the butterflies of Contra Costa County, California. *J. Lepid. Soc.* **22,** 89-107.

Owen D. F. (1971) *Tropical Butterflies: The Ecology and Behaviour of Butterflies in the Tropics with special reference to African Species.* Clarendon Press, Oxford.

Percival M. (1974) Floral ecology of coastal scrub in south-east Jamaica. *Biotropica* **6,** 104-29.

Pollard E. (1977) A method for assessing changes in abundance of butterflies. *Biol. Conserv.* **12,** 115-34.

Shapiro A. M. (1975) The temporal component of butterfly species diversity. In: *Ecology and Evolution of Communities* (eds M. L. Cody & J. M. Diamond) pp. 181-95. Harvard Univ. Press, Cambridge, Massachusetts.

Simberloff D. & Gotelli N. (1984) Effects of insularization on plant species richness in the prairie-forest ecotone. *Biol. Conserv.* **29,** 27-46.

Specht R. L. (1970) Vegetation. In: *The Australian Environment.* (ed. G. W. Leeper) pp. 43-67. CSIRO and Univ. Melb. Press, Melbourne.

Willis E. O. (1984) Conservation, subdivision of reserves, and the anti-dismemberment hypothesis. *Oikos* **42,** 396-98.

Wolda H. (1978) Seasonal fluctuations in rainfall, food and abundance of tropical insects. *J. Anim. Ecol.* **47,** 369-81.

Wolda H. (1983) Spatial and temporal variation in abundance in tropical animals. In: *Tropical Rain Forest: Ecology and Management* (eds S. L. Sutton, T. C. Whitmore & A. C. Chadwick) pp. 93-106. Blackwell, Oxford.

Zalucki M. P. (1981) The effects of age and weather on egg laying in *Danaus plexippus* L. (Lepidoptera: Danaidae). *Res. Popul. Ecol.* **23,** 318-27.

Proc. Ecol. Soc. Aust. 15:127-138

Ecological specialization of mammals in Australian tropical and sub-tropical rainforest: refugial or ecological determinism?

J. W. WINTER

Queensland National Parks and Wildlife Service, Pallarenda, Townsville, Qld. 4810

Abstract

Rainforest mammals (excluding bats) of eastern Australia are classified by ecological category into rainforest specialists, forest generalists, rainforest ecotone species, ubiquitous and independent species. When grouped according to biogeographical regions several patterns are evident. Forest generalists occur in all biogeographical regions, but rainforest specialists are restricted to north-eastern Australia, primarily in the Wet Tropical Region. Ecotone species are apparently restricted to the South-eastern Australian Region.

It is postulated that the size of rainforest refugia 25 000-15 000 years B.P., when climatic conditions were at their driest since the Miocene, was a major influence in these patterns. The lack of rainforest specialists in the South-eastern Region is attributed to refugia contracting below the minimum critical size for rainforest specialist mammals. In the Wet Tropical and Cape York Peninsula Regions some refugia were always sufficiently large during this dry phase to support a community of rainforest specialist mammalian species, with the Australian endemics representing relict species of a once more extensive community. Ecological factors may be important determinants in (1) the degree of specialization in the Cape York Peninsula Region because of climatic contrasts between rainforest and sclerophyll habitats (2) the distribution of ecoton especies, and (3) the greater degree of rainforest specialization of some forest generalist species at lower latitudes. Dispersal from a centre of origin, New Guinea, is the most plausible explanation for rainforest specialist species also occurring in lowland New Guinea rainforest.

Introduction

Discontinuous distribution of rainforest in eastern Australia is primarily regarded as a refugial pattern of a forest that once extended overmuch of eastern and central Australia in the Miocene, with present day ecological influences (Barlow 1981; Kershaw 1981; Walker & Singh 1981; Webb & Tracey 1981). Distribution of fauna within the rainforest is seen as relict endemic elements with vicariant relationships within some taxa between rainforest isolates; elements that have dispersed either north or south particularly in relation to New Guinea; and elements that have invaded the rainforest from the adjacent sclerophyll forests (Archer 1981, 1984; Brereton & Kikkawa 1963; Darlington 1961; Kikkawa *et al.* 1981; Schodde & Calaby 1972; Taylor 1972). Compositions of faunal communities have been used as evidence that some tracts of rainforest have in the past disappeared and reappeared with reconstituted faunas (Darlington 1961; Kikkawa *et al.* 1981).

Darlington (1961) drew attention to a trend in which some rainforest Carabidae are more strictly limited to rainforest in the tropics than in the south temperate zone. He interpreted this as the result of more intense ecological factors in the tropics with regard to temperature and evaporation rates. A similar trend is apparent within the mammals with several species having distributions entirely restricted to rainforest in north-eastern Australia, but none in south-eastern Australia.

In this paper I document the ecological and bigeographical relationships of the mammals (other than the bats) of tropical and sub-tropical rainforests of eastern Australia, and explore historical and present day ecological factors determining biogeographical patterns, using the concepts of the dispersive, vicariant and ecological deterministic models to explain the biogeographic distribution of species (Endler 1982a, 1982b).

Methods

Tropical and sub-tropical rainforest extends in a discontinuous band from Cape York in the north to the Illawarra in the south (Fig. 1). It comprises the cool forests of floristic region A (excluding A3) and the warm forests of florestic region B of Webb & Tracey (1981). Three biogeographical regions are recognized for mammals (Darlington 1961; Schodde & Calaby 1972; Kikkawa *et al.* 1981; modified by Winter unpublished data). They are Cape York Peninsula (includes Darlington's Tip-of- and Mid-peninsula areas), the Wet Tropics from Townsville to Cooktown (Darlington's Base-of-peninsula), and south-eastern Australia which includes the Eungella area (unlike Shodde & Calaby 1972, who include Eungella in the Wet Tropical Region). The three regions are separated by major dry corridors (Webb & Tracey 1981; Kikkawa *et al.* 1981), acting as secondary disjunctions in mammal distributions (Winter 1984a).

Rainforest mammals can be divided into several ecological categories according to their degree of dependence of rainforest.

* Rainforest specialist species — strictly confined to the rainforest over the whole of their geographical range.
* Forest generalist species — occurring in rainforest and other types of woody vegetation e.g. open forest or shrubland.
* Rainforest ecotone species — confined to the interface (often broad) between rainforest and more open habitats.
* Ubiquitous species — occurring in a wide range of habitats from rainforest to open grassland.
* Independent species — those that are responding to a feature of the environment independently of the general vegetation type e.g., semi-aquatic species.

In the assignation of a species to one of the above categories the following criteria were used:

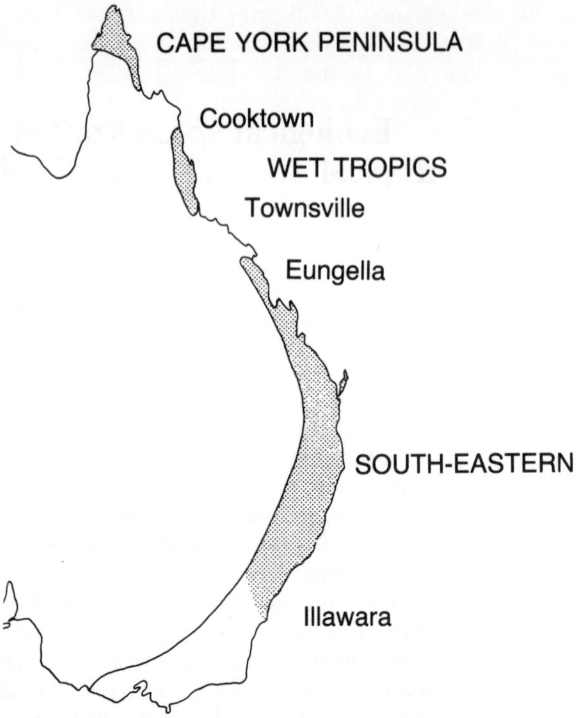

FIG. 1. Distribution limits of tropical and subtropical rainforest (shaded) in eastern Australia (after Hitchcock 1976; Webb & Tracey 1981) and rainforest biogeographical regions (modified from Darlington 1961, Kikkawa *et al.* 1981; Schodde & Calaby 1972).

* A species was considered to occur in a particular habitat only when it was known to or inferred to exist in that habitat as a stable reproducing population. The occasional occurrence of an individual, even one that reproduced, was not necessarily taken to constitute a viable population.
* The habitat use by a species was considered to be that recorded over its whole geographical range, irrespective of whether it was restricted to rainforest in part of its range.
* Ecotone species are difficult to define in the present context. For the purposes of this paper they are taken as those which are associated with rainforest, yet not confined to it nor found completely dissociated from it like forest generalist species. It is a fine, primarily subjective distinction which may not stand more rigorous examination.

* To counteract observer bias through wanting to reject the null hypothesis — "There is no difference in ecological categories between biogeographical regions" — borderline species were assigned to the category least likely to lead to a rejection of the null.

Results

Ecological categories

Forty-one species of nonvolant mammal have been recorded from tropical and sub-tropical rainforest in eastern Australia. Fourteen are classified as rainforest specialists, 18 as forest generalists, five as ecotone species, and two each for ubiquitous and independent species (Table 1).

Several species presented ecological classification problems. *Thylogale stigmatica* is primarily a rainforest species with fallen leaves and fruit of rainforest trees as major items of its diet (Cook 1979; P. Johnson 1983). It will however graze on the edges of rainforest (Johnson 1983), though apparently not to the extent of *T. thetis* (Johnson 1980) and it is recorded as occurring in fringe habitats (Calaby 1966; Winter & Atherton unpublished). I have classified it as a forest generalist, rather than an ecotone species, because of its diet and sightings of it deep within rainforest (pers. obs.). Whilst *Echymipera rufescens* (Gordon &

TABLE 1. Nonvolant rainforest mammals of eastern Australia in relation to their ecological category and biogeographic region.

		BIOGEOGRAPHIC REGION			
		CY	WT	SE	
RAINFOREST SPECIALIST SPECIES					
Phalanger orientalis	#	*	–	–	
Antechinus leo		*	–	–	
Uromys caudimaculatus	#	*	*	–	
Rattus leucopus		*	*	–	
Pogonomys mollipilosus		*	*	–	
Cercartetus caudatus		–	*	–	
Antechinus godmani		–	*	–	
Pseudocheirus herbertensis	#	–	*	–	
Pseudocheirus archeri	#	–	*	–	
Hemibelideus lemuroides	#	–	*	–	
Hypsiprymnodon moschatus	#	–	*	–	
Dendrolagus lumholtzi	#	–	*	–	
Dendrolagus bennettianus	#	–	*	–	
Melomys hadrourus		–	*	–	
FOREST GENERALIST *SPECIES*					
Echymipera rufescens		*	–	–	
Phalanger maculatus	#	*	–	–	
Melomys capensis		*	–	–	
Dactylopsila trivirgata		*	*	–	
Petaurus breviceps		*	*	*	
Thylogale stigmatica	#	*	*	*	
Perameles nasuta		*	*	*	
Trichosurus vulpecula	#	–	*	–	
Antechinus flavipes		–	*	*	
Pseudocheirus peregrinus	#	–	*	*	
Melomys cervinipes		–	*	*	
Dasyurus maculatus		–	*	*	
Antechinus stuartii		–	*	*	
Sminthopsis leucopus		–	*	*	
Rattus fuscipes		–	*	*	
Antechinus swainsonii		–	–	*	
Trichosurus caninus	#	–	–	*	
Cercartetus nanus		–	–	*	
RAINFOREST ECOTONE SPECIES					
Potorous tridactylus		–	–	*	
Thylogale thetis		–	–	*	
Macropus parma		–	–	*	
Macropus dorsalis		–	*	*	
Petrogale persephone		–	*	(*)	
GENERALIST SPECIES					
Tachyglossus aculeatus		*	*	*	
Canis familiaris dingo		*	*	*	
INDEPENDENT SPECIES					
Ornithorhynchus anatinus		—	*	*	
Hydromys chrysogaster		*	*	*	
TOTAL		15	29	21	= 41

NG = New Guinea, CY = Cape York Peninsula
WT = Wet Tropics, SE = South-eastern Australia,
# = primarily folivores and/or frugivores.

Sources: Barry 1984; Boscacci 1982; Calaby 1966; Fox 1982; Gordon & Lawrie 1977; Hockings 1981; Johnson 1980; Maynes 1977, 1982; Morton *et al.* 1980; Russell 1980 & pers. comm.; Press 1986; Seebeck 1981; Smith 1984; Strahan 1983; Taylor & Horner 1973; Van Dyck 1980, 1982a, 1982b, 1985; Van Dyck & Oglivie 1977; Warneke 1971; Watts & Aslin 1981; Wellesley-Whitehouse 1981; Winter 1978, 1984b; Winter & Atherton (unpublished data); Winter, Russell & Winter (unpublished report).

Lawrie 1977), *Dactylopsila trivirgata* (Russell 1980), *Melomys capensis* (Winter & Atherton unpublished data) and *Phalanger maculatus* (Winter 1983) have been recorded from habitats other than rainforest, the records are not necessarily indicative of viable populations of these poorly known species. For the present I have classified them as forest generalists on the grounds of not biasing the results towards the rejection of the null hypothesis. *Pseudocheirus herbertensis* has also been recorded in open forest, raised young to the newly independent stage, eaten the leaves of *Eucalyptus intermedia* and *E. torrelaina* (Russell 1980; Winter pers. obs.), and is capable of constructing a drey (Cairn & Grant 1890; N. Goudberg pers. comm.); all indicative of a forest generalist species. Nevertheless, I have classified it as a rainforest specialist because I am familiar with the species and do not consider the above observations indicative of viable populations outside the rainforest; the female and young recorded by Russell (1980) soon disappeared, probably predated by a rufous owl (Russell pers. comm.). *Isoodon macrourus* and *Melomys burtoni* have been recorded from rainforest on Cape York Peninsula, but only in narrow riverine strips and small rainforest isolates (Winter & Atherton unpublished data). I consider them to be examples of open habitat, not rainforest, species capable of penetrating short distances into rainforest, particularly in the absence of the species usually occupying the niche, namely *Echymipera rufescens* and *Melomys capensis*. *Phascogale tapoatafa* is a possible forest generalist, but has been excluded because of the paucity of information on its use of rainforest.

Two factors affect the probability of finding and recognizing viable populations of rainforest species in tall open forest (includes wet sclerophyll, the closest ecologically to rainforest) in north-eastern Australia. First, the extent of tall open forest, particularly of the wet sclerophyll type, is less in Cape York Peninsula and the Wet Tropics than in the South-eastern Region. An indication of this is the width of the rainforest plus tall to medium eucalypt open forest band in eastern Australia (Fig. 2). In the Wet Tropics tall to medium open forest is primarily restricted to a narrow band no more than five kilometres wide on the western edge of the rainforest (Types 14 and 15b, Tracey & Webb 1975; Tracey 1982). By comparison tall to medium open forest communities in the South-eastern Region were more extensive (e.g., McDonald & Elsol 1984; McDonald & Whiteman 1979). On Cape York Peninsula the extensive areas of open forest differ significantly from those further south in that the understorey and shrub layers are almost entirely of relatively xeromorphic species (Pedley & Isbell 1971). This is probably a result of the hot climate with a marked dry season in the winter (Koppen type Aw, Fig. 2), associated with regular fire, in contrast to the temperate moist climatic types (Koppen type C) which extend into the Wet Tropical Region through the effects of elevation. Thus there is a greater contrast between open forest and rainforest habitats on Cape York Peninsula than in the more southerly regions.

Secondly, there have been many fewer ecological studies in north-eastern than in south-eastern Australia and these have tended to concentrate on rainforest rather than on adjacent habitats in the ecological continuum with rainforest. However, in the Wet Tropics there are sufficient studies (Boscacci 1982; Russell 1980, 1984, pers comm.; Winter 1978; Winter *et al.*

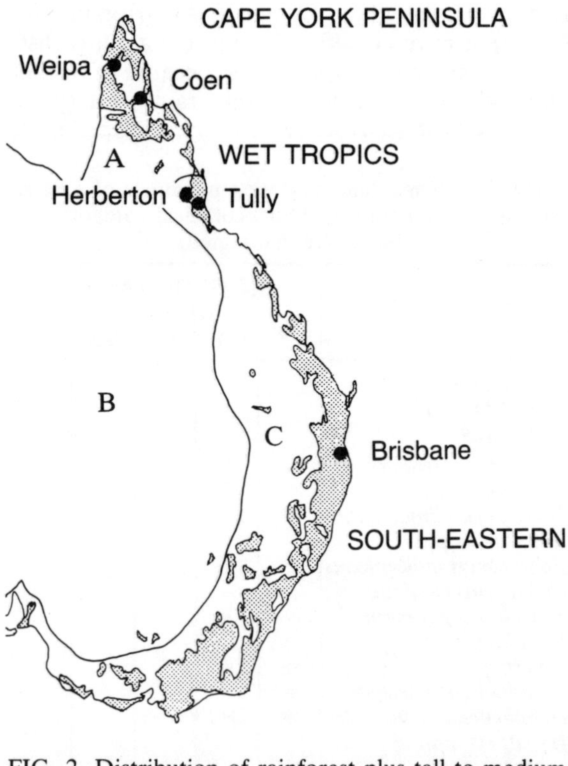

FIG. 2. Distribution of rainforest plus tall to medium eucalypt open forest (shaded) in eastern Australia and Koppen climatic types, A = hot moist, B = dry hot to warm, C = temperate moist (Division of National Mapping 1977).

1979; Winter and Atherton unpublished data) to provide an acceptable degree of confidence in the ecological classification of the species. The degree of confidence is lower for the Cape York Peninsula Region where information is based mainly on a limited number of short term survey studies (e.g., Gordon & Lawrie 1977; Taylor & Horner 1973) and on two major fauna surveys (Winter & Atherton, unpublished data and reports). Hence the problems with the classification of several Peninsula species.

Ecological patterns

When rainforest mammals are tabulated according to their ecological classification and biogeographic region the following major relationship is evident.

* Forest generalist species occur in all three biogeographical regions, though perhaps proportionally fewer in Cape York Peninsula bearing in mind the classification problems for some species already discussed, but
* all rainforest specialist species are restricted to the Wet Tropical and Cape York Peninsula Regions.
* Ubiquitous and independent species are widespread and occur in all three regions, except for *Ornithorhynchus* which does not occur in Cape York Peninsula Region.

Less certain, because of the problems of ecological classification, is the relationship between ecotone species and distribution. Within these limitations an apparent relationship is one in which ecotone species are restricted to the South-eastern Region. The two apparent exceptions, *Macropus dorsalis* and *Petrogale persephone,* are unusual but cannot be considered as belonging to either of the two northerly regions; the former has a range extending into the latitudes of the wet tropics, but in the northern part of its distribution is restricted to dry vine scrubs within Webb & Tracey's (1981) floristic province C1 to the west of the Wet Tropics proper, and is therefore treated as a northern extension of the South-eastern Region; the latter has a very restricted range in dry vine forest on rocky hillsides in the Proserpine area. Whilst well into tropical latitudes it is part of the Eungella area which is taken as the northernmost extent of the South-eastern Region. It is an anomalous species because of its very restricted range and close relationship to *Petrogale xanthopus* of the dry hinterland (Maynes 1982).

Discussion

Three deterministic models are used to explain the biogeographic distribution of species. They are the dispersive, vicariant and ecological models which are not necessarily exclusive and more than one may be entertained to provide explanations for biogeographical patterns (Endler 1982a, 1982b, 1982c). The existing biogeographical patterns of rainforest mammals in eastern Australia could reflect, (1) ecological differences in operation now through climate and diversity of habitat, (2) historical differences imposed by past events in geological time, or (3) a combination of both. The latter is the most likely and it is a matter of attempting to determine the relative level of influence of historical and present ecology on the determination of present distribution patterns, something which is difficult to do particularly when the data has not been designed to test for such differences. A problem which is applicable to the present data set (Endler 1982a, 1982c).

Ecological factors

Climatic differences: There are significant climatic differences between the Cape York Peninsula region and the more southerly regions. Cape York Peninsula lies within Koppen's hot moist climate with a marked winter dry season (less than 3% of the annual rainfall — Table 2), that is, a monsoonal climate. Conversely the Wet Tropic and South-eastern Regions both come within Koppen's temperate moist climatic types (Fig. 2); extending into the Wet Tropical Region through the effects of elevation which is over 600 m for much of the region. Like Cape York Peninsula, the Wet Tropical and South-eastern Regions have a winter dry season, but it is less pronounced as even on the drier western edge of the Wet Tropics (Herberton), dry season rainfall accounts for 11.4% of the annual total, and Brisbane in the South-eastern Region receives 25.4% of its total in the dry season (Table 2). On the assumption that the buffering effect of rainforest (other than totally deciduous types) on high temperatures and evaporation rates is greater in Cape York Peninsula, the contrast between rainforest and more open habitats will be greater with less ecological continuity between the habitats. Climate may therefore have an overriding influence on the ability of mammals to use both the rainforest and open habitats in Cape York Peninsula, but less so in the other two regions. Climatic similarities

TABLE 2. Regional mean annual rainfall (mm) and dry season (May to September inclusive) rainfall as a percentage of the annual mean (Bureau of Meteorology 1977).

Region	Site	Rainfall		
		Annual Mean	Dry Mean	Season %
Cape York Peninsula	Weipa	1648	30	1.8
	Coen	1144	33	2.9
Wet Tropics	Herberton	1132	129	11.4
	Tully	4321	977	22.6
South-eastern	Brisbane	1146	291	25.4

between the South-eastern Region, the upland Wet Tropical Region and upland New Guinea are suggested by their faunal affinities which has led to the proposal of an upland Tumbunan biogeographical division covering all three (Schodde & Calaby 1972).

A species which may be affected by the less pronounced climatic differences between the South-eastern and upland Wet Tropical Regions is *Antechinus stuartii* with a latitudinal change in its degree of confinement to rainforest. In New South Wales, towards the southern end of its range, *A. stuartii* occupies a wide range of habitats other than rainforest (Fox 1982; Statham & Harden 1982). In south-eastern Queensland it is primarily a rainforest species, though it can be found in humid gullies in sclerophyll forest (Van Dyck 1982a; Barry 1984). In the Wet Tropical Region it appears to be confined to upland rainforest (Van Dyck 1982a; Winter & Atherton unpublished data). *Sminthopsis leucopus* may show a similar trend though its presence in the Wet Tropical Region is known from only two records, one in rainforest the other in the transition zone with tall open forest (Van Dyck 1985). Southern populations occur in a wide range of habitats, including beech forest, eucalypt woodland and heath (Ahern 1983; Morton *et al.* 1980).

Species diversity: As a general rule species diversity is higher in the tropics than in more temperate latitudes (Krebs 1972). Thus the higher number of rainforest specialist mammals in the Wet Tropics compared with the South-eastern Region may be directly proportional to a higher species diversity. There are more rainforest mammals (29) in the Wet Tropics than in the South-eastern Region (22) in accordance with general theory, but on the assumption that the proportion of specialist to total rainforest species remains constant for the two regions, the number of specialists from the regions (12 and 0 respectively) differs significantly from that expected ($X^2 = 7.6215$, $p < 0.01$). Cape York Peninsula, on the other hand, with only 15 rainforest species is lower in species diversity than the South-eastern Region, but has five rainforest specialists. This indicates that rainforest specialization of mammals is not a simple relationship between species diversity and latitude.

Species diversity of rainforest mammals may be a function of rainforest area in each region. There is a trend, though not significant, for mammal numbers to be higher in the larger areas of rainforest, but this trend does not hold for the specialist rainforest species (Table 3). The lack of specialist species in the South-eastern Region may result from the more fragmented distribution of rainforest (see Map 1 & 2 in Williams *et al.* 1984) in comparison to the two northern regions. Nevertheless two of the largest contiguous rainforest areas in the South-eastern Region at the advent of the European were the "Big Scrub" near Lismore and Eungella with areas of 75 000 ha (A. G. Floyd *pers. comm.*) and 73 000 ha (Winter & McDonald 1986) respectively. Both are considerably larger than the area (unaffected by clearing) presently occupied by several population isolates of rainforest specialist species in the Wet Tropical Region (Winter 1984a; Winter, Bell, Pahl & Atherton unpublished report). For example *Melomys hadrourus* is restricted to the Thornton Peak uplands with a rainforest area of 24 000 ha (Winter 1984b), and the population isolate of *Hemibelideus lemuroides* on the Mt. Carbine Tableland is restricted to about 22 000 ha of upland rainforest though both are contiguous with large areas of lowland rainforest.

Rainforest diversity, as a measure of habitat diversity, may determine the ability of specialist

TABLE 3. Rainforest mammals in relation to rainforest area.

Region	Area (ha)	Rainforest mammals	
		Total	Specialists
Cape York Peninsula	260 260[2]	15	5
South-eastern	609 000[1]	21	0
Wet Tropics	791 300[2]	29	12

[1] = from Webb & Tracey (1981); [2] = from Winter, Bell, Pahl & Atherton (unpublished report).

rainforest species to exist. A high proportion (9:12) of the specialist species are folivores and or frugivores (Table 1). Thus one measure of habitat diversity is the number of rainforest trees for each region (Table 4). Again there is a trend for the highest number of rainforest mammals to occur in the region with the highest number of tree species, that is, in the Wet Tropics. No such trend is evident with regard to specialist species in that there are no specialists in the South-eastern Region even though it has a higher tree species diversity than Cape York Peninsula, which has at least one folivore/frugivore rainforest specialist *(Phalanger orientalis)* and one frugivore *(Uromys coudimaculatus)*.

Ecotone species: These are apparently restricted to the South-eastern Region. A possible explanation is that a mosaic of small rainforest patches, too small to support rainforest specialist species, within an extensive tall open sclerophyll forest, has led to the evolution of the ecotone species *Thylogale thetis, Macropus parma* and *Potorous tridactylus* in the South-eastern Region. The lack of ecotone species in the Wet Tropics (the status of *Bettongia penicillata* has yet to be determined) can be attributed to the smaller proportion of ecotones associated with relatively fewer but larger refugia. The evolution of *Macropus dorsalis* and *Petrogale persephone,* both present day ecotone species, may have a weaker association with the rainforest than the previous ecotone species: *M. dorsalis* because it may be associated with a habitat with a closed canopy, rather than with rainforest *per se,* and *P. persephone* with rocky refuge areas. Evolution of ecotone species is more readily explained by ecological factors related to present rainforest distribution, than by the past refugial model, though it is not inconsistent with the latter.

TABLE 4. Rainforest mammals in relation to the number of tree species in each region.

Region	Tree species	Rainforest mammals	
		Total	Specialists
Cape York Peninsula	418[1]	15	5
South-eastern	423[2]	21	0
Wet Tropics	724[1]	29	12

[1] = extracted from Hyland (1982); [2] = extracted from Williams *et al.* (1984).

Historical factors

Late Quaternary refugial effects: Between the middle Miocene and the present there has been intermitent drying of the climate with rainforest faunas and floras contracting in distribution and diversity (Galloway & Kemp 1981; Archer 1981). Pollen profiles provide a more detailed Quaternary history of the past 350 000 years with evidence of a climatic waxing and waning favourable to the existence of rainforest (Kershaw 1981). Conditions 25 000-15 000 years before the present (B.P.) were drier and cooler than at any other time in the last 50 000 years. There was a general lack of forest and woodland vegetation in south-eastern Australia (Kershaw 1981), and rainforest was replaced by sclerophyll woodland on the Atherton Tableland in the Wet Tropics (Kershaw 1978). In more recent times optimal conditions for rainforest growth were achieved between 7 000 and 5 000 years before the present with a slight reduction in rainfall within the last few thousand years (Kershaw 1978, 1981).

Botanists have postulated rainforest refugia, the largest of which correlate with climatic and topographic factors that favour mesic communities, e.g., summits and gullies on upper slopes of cloudy wet mountains, very wet lowlands, deep moist gorges of coastal lowlands and less commonly of subcoastal lowlands (Webb & Tracey 1981). Climatically controlled refugia form part of a continuum with ecologically related vegetation, and the flora within them is free to expand or contract with changes in climate (Webb & Tracey 1981). Faunal and floral communities within these refugia persist as relict assemblages representative of a once more widely distributed rainforest.

On the assumption that rainforest specialist mammals have not evolved from extant forest generalist species or genera (discussed below), since the driest conditions 25 000-15 000 years ago, then specialist species can be considered relict species.

If the species-area relationship (predicted from theoretical island biogeography — MacArthur & Wilson 1967), holds then the diversity of these relict communities will be dependent on the minimum areas attained by the refugia. Total extinction of rainforest specialist species will occur if the refugia fall below a critical size capable of supporting such a community.

Taking the above assumptions to be valid, then the presence or absence of rainforest specialists in the Wet Tropical and South-eastern regions can be attributed to differences in the size of past refugia.

It is postulated that major rainforest refugia in the Wet Tropics never contracted below the critical minimum size required to support a community of rainforest specialist mammals. In the South-eastern Australian Region, on the other hand, the total lack of rainforest specialist (relict) species indicates that all refugia shrank below the critical size required to maintain this community. Only forest generalists, capable of existing in rainforest and sclerophyll communities, either survived in the small rainforest refugia of the South-eastern Region, or reinvaded the rainforest as the latter expanded during subsequent wetter climates. For example the two South-eastern Regional rainforest folivores, *Pseudocheirus peregrinus* and *Trichosurus caninus,* can also live on a diet primarily of eucalypt leaves (Owen & Thomson 1965; Pahl 1984), enabling them to use the full spectrum of habitats within a rainforest/sclerophyll mosaic.

The argument, that the lack of rainforest specialists from the South-eastern Region is a result of past refugia being smaller than at the present, is supported by present-day patches of rainforest apparently being sufficiently large to support specialist species as already discussed.

Speciation: If the lack of rainforest specialist mammals in the South-eastern Region is the result of past rainforest refugia being smaller than at present, why then have none evolved as the rainforest expanded since the driest period 25 000-15 000 years B.P.? Conversely, could specialist species in the two northern regions be the result of speciation since 25 000-15 000 years B.P.?

Once a rainforest has lost its rainforest specialist species it may be recolonized by sclerophyll adapted species in addition to forest generalist species already present. Evolution of rainforest specialists will be dependent on the time elapsed since recolonization and on selection pressures. Brereton & Kikkawa (1963), in a comparison of species diversity of birds in rainforest and sclerophyll forests, suggest that sclerophyll adapted species do not easily colonize rainforest. With the mammals several species may be relatively recent colonizers of rainforest. *Petaurus breviceps* and *Trichosurus vulpecula* are well adapted to sclerophyll open forests and woodland and have the most extensive geographical ranges of any Australian arboreal mammal. *T. vulpecula* has colonized the rainforest, but only where the apparently better adapted *T. caninus* is absent, and in the Wet Tropical Region a distinct subspecies, *T. v. johnstoni* is restricted to the western half of the region suggesting only a marginal colonizing ability. The gliding ability of *P. breviceps* suggests an open forest rather than rainforest adaptation and the indistinguishable forms in the rainforest is indicative of a recent colonization.

Evidence that most rainforest specialist mammals are not the result of recent speciation from forest generalists is provided by the degree of restricted endemism and lack of vicariance with generalist species. Two, *Hemibelideus* and *Hypsiprymnodon,* are monotypic genera, four, *Dendrolagus, Phalanger, Pogonomys* and *Uromys,* are genera restricted to rainforest and one, *Pseudocheirus archeri,* is in a subgenus restricted to rainforest. Of the genera which have non-rainforest or forest generalist representatives, *Melomys hadrourus* is not closely related to any extant species and is therefore not a vicariant of recent origin (Winter 1984b), *Antechinus godmani* has its closest affinities with New Guinea montane rainforest species and is considered a relict (Van Dyck 1982b). *Pseudocheirus herbertensis,* though in the same subgenus as the forest generalist *P. peregrinus,* is more strongly divergent from typical *peregrinus* than the several subspecies of *peregrinus* (Tate 1945), some of which (in Tasmania and Western Australia) will have had histories of isolation much greater than 25 000-15 000 years. Though *Antechinus leo* has close affinities with *A. flavipes,* and may therefore be a vicariant of relatively recent origin, Van Dyck (1980) suggests that *A. leo* was the ancestral form.

There is some evidence of relatively recent incipient speciation occurring in the Wet Tropical Region. *Hemibelideus lemuroides* has two population isolates with a genetic divergence exhibited by significantly different numbers of white individuals in each population (Winter 1984a). The Black Mountain Corridor forms a tenuous rainforest link between the two populations, susceptable to disruption by relatively minor climatic changes. It probably did not exist 25 000-15 000 years ago and its loss may have been recurrent in more recent times. The Black Mountain Corridor is also an altitudinal barrier between the two populations, but may not have been if *H. lemuroides* populations migrated altitudinally in response to climatic changes. *Pseudocheirus herbertensis* is separated into two allopatric races by the Black Mountain Corridor, but speciation has progressed further than in *H. lemuroides* because the races are quite distinct colour morphs and have different chromosome numbers of 2N = 12 and 16 for *P. h. herbertensis* and *P. h. cinereus* respectively (McKay 1984). Each race is further divided into two major population isolates by secondary disjunctions (Winter 1984a), but with no overt signs of divergence within the races. Speciation to distinct allopatric forms (with different chromosome numbers) has apparently progressed further than in the equivalent population isolates of *H. lemuroides,* either through a more rapid rate of divergence, or through an earlier disjunction. The two Wet Tropical allopatric *Dendrolagus* are sufficiently dissimilar to indicate a divergence at an even earlier time (Groves 1982).

In the South-eastern Region none of the endemic species, species that might be expected to have speciated *in situ,* are rainforest specialists. The three macropods, *Thylogale thetis, Macropus parma* and *Potorous tridactylus,* may have evolved as ecotone specialists in response to the development of a rainforest-sclerophyll mosaic. *Antechinus swainsonii, Trichosurus caninus* and *Cercartetus nanus* are forest generalists, and *Petrogale persephone* is an anomolous species. Other forest generalists, except the more arid-adapted *Petaurus breviceps* and *Trichosurus vulpecula,* of the South-eastern Region have their northern geographical limit in the Wet Tropical Region, which is also the northern limit of temperate tall open forest (Tracey & Webb 1975; Tracey 1982). This suggests a group of species adapted to cool moist sclerophyll forest, most with a wide distribution, that have reinvaded the rainforest, with the south-eastern endemics restricted by either ecological factors, or different dispersal abilities from a southern centre of origin.

Dispersive influences: A group of rainforest specialist species in the two northern regions — *Phalanger orientalis, Uromys caudimaculatus, Rattus leucopus* and *Pogonomys molipillosus* — occur in lowland rainforest in New Guinea. Genetic divergence to the subspecific level at the most is indicative of relatively recent isolation of the populations. They may have colonized, or recolonized, the two northern Australian regions across a postulated land bridge and almost continuous lowland rainforest from New Guinea, the most recent being about 8 000 years B.P. (Kikkawa *et al.* 1981; Nix & Kalma 1972; Schodde & Calaby 1972). For this group of species the dispersal model is as good as the refugial model, if not better, in explaining their biogeographical distribution. *Thylogale stigmatica* may be an example of reverse dispersal (Kikkawa *et al.* 1981; Schodde & Calaby 1972). *Cercartetus caudatus,* though occurring in New Guinea, fits the vicariant refugial model better than the above species because its presence in New Guinea is restricted to the uplands and *C. nanus* is in rainforest in the South-eastern Region.

Conclusions

In conclusion, there is a group of rainforest specialist mammals restricted to the Wet Tropical and Cape York Peninsula Regions, with no specialists in the South-eastern Region. It is postulated that the lack of specialists in the South-eastern Region and the presence of Australian endemic rainforest specialists in the northern region is an historical refugial effect, possibly dating back to 25 000-15 000 years B.P. The endemic specialists are relicts of an earlier, much more extensive rainforest fauna, absent from the South-eastern Region because rainforest refugia contracted below the critical minimum level able to support rainforest specialists. Rainforest specialist species shared with lowland New Guinea fit the dispersive model, with New Guinea as the centre of origin, better than the refugial model. The present day ecological model may be a partial explanation for rainforest specialist species in the Cape York Peninsula Region, South-eastern Regional ecotone species, and some forest generalists being more restricted to rainforest in lower latitudes.

Finally, in the words of Endler (1982c) "Biogeography could profit by more attempts to examine the relative and joint effects of both history and ecology on distributions."

Acknowledgements

R. G. Atherton and R. Russell helped and inspired whilst in the field, and K. Plowman, P. Ogilvie and W. J. Freeland gave helpful critical comments on the manuscript.

References

Ahern L. D. (1983) White-footed Dunnart *Sminthopsis leucopus.* In: *Complete book of Australian mammals* (ed. R. Strahan) pp 55. Angus and Robertson, Sydney.

Archer M. (1981) A review of the origins and radiations of Australian mammals. In: *Ecological Biogeography of Australia* (ed. A. Keast) pp. 1435-88. W. Junk, The Hague.

Archer M. (1984) The Australian marsupial radiation. In: *Vertebrate Zoogeography and Evolution in Australasia* (eds M. Archer and G. Clayton) pp. 633-808. Hesperian Press, Carlisle.

Barlow B. A. (1981) The Australian flora: its origin and evolution. In: *The Flora of Australia, Volume 1, Introduction* (ed. A. S. George) pp 25-75. Australian Government Publishing Service, Canberra.

Barry S. J. (1984) Small mammals in a south-east Queensland rainforest: the effects of soil fertility and past logging disturbance. *Aust. Wildl. Res.* **11**, 31-39.

Boscacci L. J. (1982) *Ecology of* Rattus lutreolus lacus *in tall open forest, north-east Queensland, with emphasis on habitat use and reference to coexistant small mammals.* B.Sc. Hon. Thesis, James Cook University, Townsville.

Brereton J. LeG. & Kikkawa J. (1963) Diversity of avian species. *Aust. J. Sci.* **26**, 12-14.

Bureau of Meteorology (1977) *Rainfall Statistics, Australia.* Commonwealth Department of Science, Canberra.

Calaby J. H. (1966) Mammals of the Upper Richmond and Clarence Rivers, New South Wales. *CSIRO Div. Wildl. Tech. Tech. Paper* No. 10.

Cairn E. J. & Grant R. (1890) Report of a collection trip to north-eastern Queensland during April to September, 1889. *Aust. Mus. Rec.* **1**, 27-30.

Cook B. N. (1979) *Field observations of the behaviour of the macropod marsupial* Thylogale stigmatica *(Gould).* M.Sc. Thesis, University of Queensland, Brisbane.

Darlington P. J. (1961) Australian carabid beetles V. Transition of wet forest faunas from New Guinea to Tasmania. *Psyche.* **68**, 1-24.

Division of National Mapping (1977) *Atlas of Australian Resources.* Department of National Resources, Canberra.

Endler J. A. (1982a) Pleistocene forest refuges: fact or fancy? In: *Biological diversification in the tropics* (ed. G. T. Prance) pp. 641-57. Columbia University Press, New York.

Endler J. A. (1982b) Alternative hypotheses in biogeography: introduction and synopsis of the symposium. *Amer. Zool.* **22**, 349-54.

Endler J. A. (1982c) Problems in distinguishing historical from ecological factors in biogeography. *Amer. Zool.* **22**, 441-52.

Fox B. J. (1982) Ecological separation and coexistence of *Sminthopsis murina* and *Antechinus stuartii* (Dasyuridae, Marsupialia): a regeneration niche? In: *Carnivorous Marsupials* (ed. M. Archer) pp. 187-97, Roy. Soc. N.S.W., Mosman.

Galloway R. W. & Kemp E. M. (1981) Late Cainozoic environments in Australia. In: *Ecological biogeography of Australia* (ed. A. Keast) pp. 51-80. W. Junk, The Hague.

Gordon G. & Lawrie B. C. (1977) The rufescent bandicoot, *Echymipera rufescens* (Peters & Doria) on Cape York Peninsula. *Aust. Wildl. Res.* **5**, 41-45.

Groves C. P. (1982) The systematics of tree kangaroos (*Dendrolagus;* Marsupialia, Macropodidae). *Aust. Mammal.* **5**, 157-86.

Hitchcock P. (1976) Rain forest types. In: *Rain forests* (ed. W. Goldstein) pp. 32-49. N.S.W. National Parks and Wildlife Service, Sydney.

Hockings M. (1981) Habitat distribution and species diversity of small mammals in south-east Queensland in relation to vegetation structure. *Aust. Wildl. Res.* **8**, 97-108.

Hyland B. P. M. (1982) *A Revised Card Key to Rainforest Trees of North Queensland.* CSIRO, Melbourne.

Johnson K. A. (1980) Spatial and temporal use of habitat by the Red-necked Pademelon, *Thylogale thetis* (Marsupialia: Macropodidae). *Aust. Wildl. Res.* **7**, 157-66.

Johnson P. M. (1983) Red-legged Pademelon *Thylogale stigmatica.* In: *Complete Book of Australian mammals* (ed. R. Strahan) pp. 225. Angus and Robertson, Sydney.

Kershaw A. P. (1978) Record of last interglacial-glacial cycle from north-eastern Queensland. *Nature (Lond.).* **272**, 159-61.

Kershaw A. P. (1981) Quaternary vegetation and environments. In: *Ecological biogeography of Australia* (ed. A. Keast) pp. 81-101. W. Junk, The Hague.

Kikkawa J., Monteith G. B. & Ingram G. (1981) Cape York Peninsula. Major region of faunal interchange. In: *Ecological biogeography of Australia* (ed. A. Keast) pp. 1695-742. W. Junk, The Hague.

Krebs C. J. (1972) *Ecology, the Experimental Analysis of Distribution and Abundance.* Harper and Row, New York.

MacArthur R. H. & Wilson E. O. (1967). *The Theory of Island Biogeography.* Princeton University Press, Princeton.

McDonald W. J. F. & Elsol J. A. (1984) *Moreton Region Vegetation Map series, Summary Report and Species Checklist for Caloundra, Brisbane, Beenleigh and Murwillumbah sheets.* Queensland Department of Primary Industries, Brisbane.

McDonald W. J. F. & Whiteman W. G. (1979) *Moreton Region Vegetation Map series, Explanatory Booklet for Murwillumbah Sheet.* Queensland Department of Primaries Industries, Brisbane.

McKay G. M. (1984) Cytogenetic relationships of possums and gliders. In: *Possums and Gliders* (eds A. P. Smith and I. D. Hume) pp. 9-16. Australian Mammal Society, Sydney.

Maynes G. M. (1977) Distribution and aspects of the biology of the Parma Wallaby, *Macropus parma,* in New South Wales. *Aust. J. Wildl. Res.* **4**, 109-25.

Maynes G. M. (1982) A new species of rock wallaby, *Petrogale persephone* (Marsupialia: Macropodidae), from Proserpine Central Queensland. *Aust. Mammal.* **5**, 47-58.

Morton S. R., Wainer H. W. & Thwaites T. P. (1980) Distributions and habitats of *Sminthopsis leucopus* and *S. murina* (Marsupialia: Dasyuridae) in south-eastern Australia. *Aust. Mammal.* **3**, 19-30.

Nix H. A. & Kalma J. D. (1972) Climate as a dominant control in the biogeography of northern Australia and New Guinea. In: *Bridge and Barrier: the Natural and Cultural History of Torres Strait* (ed. D. Walker) pp. 61-91. Australian National University, Canberra.

Owen W. H. & Thomson J. A. (1965) Notes on the comparative ecology of the Common Brushtail and Mountain Possums in eastern Australia. *Vict. Nat.* **82**, 216-17.

Pahl L. T. (1984) Diet preference, diet composition and population density of the ringtail possum *(Pseudocheirus peregrinus cooki)* in several plant communities in southern Victoria. In: *Possums and Gliders* (eds A. P. Smith and I. D. Hume) pp. 253-60. Australian Mammal Society, Sydney.

Pedley L. & Isbell R. F. (1971) Plant communities of Cape York Peninsula. *Proc. R. Soc. Qd.* **82**, 51-74.

Press A. J. (1986) Comparison of numbers of *Rattus fuscipes* living in cool temperate rainforests and dry sclerophyll forests. *Aust. Wildl. Res.* **13**, 419-26.

Russell R. (1980) *Spotlight on Possums.* University of Queensland Press, Brisbane.

Russell R. (1984) Social behaviour of the Yellow-bellied Glider, *Petaurus australis reginae,* in North Queensland. In: *Possums and Gliders* (eds A. P. Smith and I. D. Hume) pp. 343-53. Australian Mammal Society, Sydney.

Schodde R. & Calaby J. H. (1972) The biogeography of the Australopapuan bird and mammal faunas in relation to Torres Strait. In: *Bridge and Barrier: the Natural and Cultural History of Torres Strait* (ed. D. Walker) pp. 257-300. Australian National Univerity, Canberra.

Seebeck J. H. (1981) *Potorous tridactylus* (Kerr) (Marsupialia: Macropodidae): its distribution, status and habitat preferences in Victoria. *Aust. Wildl. Res.* **8**, 285-306.

Smith G. C. (1984) The biology of the Yellow-footed Antechinus, *Antechinus flavipes* (Marsupialia: Dasyuridae), in a swamp forest on Kinaba Island, Cooloola, Queensland. *Aust. Wildl. Res.* **11**, 465-80.

Statham H. L. & Harden R. H. (1982) Habitat utilization of *Antechinus stuartii* (Marsupialia) at Petroi, Northern New South Wales. In: *Carnivorous Marsupials* (ed. M. Archer) pp. 165-85. Roy. Soc. N.S.W., Mosman.

Strahan R. (ed.) (1983) *Complete Book of Australian Mammals.* Angus and Robertson, Sydney.

Tate G. H. H. (1945) Results of the Archbold expeditions. No. 54, The marsupial genus *Pseudocheirus* and its subgenera. *Amer. Mus. Novitates* **1287**, 1-30.

Taylor J. M. & Horner B. E. (1973) Results of the Archbold Expeditions. No. 98. Systematics of native Australian *Rattus* (Rodentia, Muridae). *Bull. Am. Mus. Nat. Hist.* **150**, 1-130.

Taylor R. W. (1972) Biogeography of insects of New Guinea and Cape York Peninsula. In: *Bridge and Barrier: the Natural and Cultural History of Torres Strait* (ed. D. Walker) pp. 213-30. Australian National University, Canberra.

Tracey J. G. (1982) *The Vegetation of the Humid Tropical Region of North Queensland.* CSIRO, Melbourne.

Tracey J. G. & Webb L. J. (1975) *Vegetation of the Humid Tropical Region of North Queensland (15 maps at 1:100 000 scale + key).* CSIRO Aust. Long Pocket Labs., Indooroopilly, Queensland.

Van Dyck S. (1980) The cinnamon antechinus, *Antechinus leo* (Marsupialia: Dasyuridae), a new species from the vine-forests of Cape York Peninsula. *Aust. Mammal.* **3**, 5-17.

Van Dyck S. (1982a) The relationships of *Antechinus stuartii* and *A. flavipes* (Dasyuridae, Marsupialia) with special reference to Queensland. In: *Carnivorous Marsupials* (ed. M. Archer) pp. 723-66. Roy. Soc. N.S.W., Mosman.

Van Dyck S. (1982b) The status and relationships of the Atherton Antechinus, *Antechinus godmani* (Marsupialia: Dasyuridae). *Aust. Mammal.* **5**, 195-210.

Van Dyck S. (1985) *Sminthopsis leucopus* (Marsupialia: Dasyuridae) in North Queensland rainforest. *Aust. Mammal.* **8**, 53-60.

Van Dyck S. & Ogilvie P. (1977) *Antechinus swainsonii* (Waterhouse 1840), the dusky marsupial mouse, an addition to the mammal fauna of Queensland. *Mem. Qd Mus.* **18**, 69-73.

Walker D. & Singh G. (1981) Vegetation history. In: *Australian Vegetation* (ed. R. H. Groves) pp. 26-43. Cambridge University Press, Cambridge.

Warneke R. M. (1971) Field study of the Bush Rat, *Rattus fuscipes* Waterhouse (Rodentia, Muridae). *Wildl. Contributions, Vict.* **14.**

Watts C. H. S. & Aslin H. J. (1981) *The Rodents of Australia.* Angus and Robertson, Sydney.

Webb L. J. & Tracey J. G. (1981) Australian rainforests: patterns and change. In: *Ecological Biogeography of Australia* (ed. A. Keast) pp. 605-94. W. Junk, The Hague.

Wellesley-Whitehouse H. L. (1981) *A study on the comparative ecologies of two rainforest murid rodents,* Uromys caudimaculatus *and* Rattus fuscipes coracius, *in north Queensland.* B.Sc. Hons. Thesis, James Cook University of North Queensland, Townsville.

Williams J. B., Harden G. J. & McDonald W. J. F. (1984) *Trees and Shrubs in Rainforests of New South Wales and Southern Queensland.* Botany Department, University of New England, Armidale.

Winter J. W. (1978) The rainforest: where strange things live. In: *Exploration North* (ed. H. J. Lavery) pp. 113-46. Richmond Hill Press, Richmond.

Winter J. W. (1983) Spotted Cuscus *Phalanger maculatus*. In: *Complete Book of Australian Mammals* (ed. R. Strahan) pp. 154. Angus and Robertson, Sydney.

Winter J. W. (1984a) Conservation studies of tropical rainforest possums. In: *Possums and Gliders* (eds A. P. Smith and I. D. Hume) pp. 469-81. Australian Mammal Society, Sydney.

Winter J. W. (1984b) The thornton peak melomys, *Melomys hadrourus* (Rodentia, Muridae); a new rainforest species from north eastern Queensland, Australia. *Mem. Qd. Mus.* **21**, 519-39.

Winter J. W. & McDonald K. R. (1986) Eungella, the land of cloud. *Aust. Nat. Hist.* **22**, 39-43.

Winter J. W., Russell A. W. & Winter E. M. (1979) A sighting of the Yellow-bellied Glider *(Petaurus australis)* in north-eastern Queensland. *N. Qd. Nat.* **45**, 2-4.

Proc. Ecol. Soc. Aust. 15:139-148

A comparison of Australian tropical rainforest and Tertiary fossil leaf-beds

D. C. CHRISTOPHEL and D. R. GREENWOOD

Botany Department, University of Adelaide, G.P.O. Box 498, Adelaide, S.A. 5001

Abstract

The fossil record is revealing the nature of Australian vegetation prior to the Miocene contact between Australia and Asia. There are divergent views on whether fossil leaf beds always give an accurate picture of past vegetation. Leaf litter from sites in modern Australian tropical rainforests, however, reflect the physiognomic character of the surrounding forest. Complex Mesophyll (CMVF), Complex Notophyll (CNVF), and Simple Notophyll Vine Forests (SNVF) have characteristic foliar physiognomic signatures. The physiognomic signature of stream bed deposits of SNVF leaf litter was within the range of variation observed for forest floor leaf litter from SNVF.

These results suggest that Tertiary fossil leaf-beds can be used in some instances to reconstruct past vegetation based on the physiognomic characteristics of the fossil leaves and the overall representation of taxa where modern affinities can be determined. The physiognomic character of the Eocene leaf assemblages from Anglesea in Victoria appears to match tropical montane and sub-tropical rainforest (Simple Notophyll Vine Forest) from Australia in both physiognomic character and the presence of some taxa.

The Anglesea flora has many elements in common with the modern wet tropical flora of northern Queensland. The Podocarpaceae is represented by at least six species from five modern genera at Anglesea (Dacrycarpus, Decussocarpus, Falcatifolium, Podocarpus *and* Prumnopitys), *whereas only two genera are present in the modern wet tropical forests of Australia* (Podocarpus *and* Prumnopitys).

Many angiosperm taxa characteristic of the modern Australian wet tropical forests are also found as significant components of the Anglesea fossil flora, including Gymnostoma *(Casuarinaceae),* Musgravinaeanthus *and* Orites *(Proteaceae),* Austrodiospyros *(Ebenaceae),* Brachychiton *(Sterculiaceae) and* Quintinia *(Escalloniaceae). In the case of* Gymnostoma, *and species of the Musgraveinae, the modern taxa are restricted to a small region in northern Queensland. The close match to the Anglesea* Orites *species is also restricted to the same region.*

At Noah Creek in N.E. Queensland, the streamside vegetation is dominated by the only extant Australian species of Gymnostoma *and an undescribed species of* Orites *which is highly similar to the Anglesea fossil species.* Podocarpus *and* Prumnopitys *are also present.*

These facts, along with the rarity of some of the plant associations found at Anglesea and Noah Creek (e.g., the Gymnostoma/Orites/*Musgraveinae association) suggest that Noah Creek may provide as good a comparative model as possible for the Anglesea flora.*

The demonstration of the correlation between rainforest type and physiognomic analyses of leaf litter from modern Australian rainforests, coupled with the evidence that similar types of plant associations can be noted from the modern forests and the Eocene floras, suggests that a more accurate interpretation than previously expected of Australian vegetation prior to the Miocene may be possible. The existing evidence predicts that forests very similar to those found in the modern wet tropics of Australia were present in southern Australia in the Eocene, prior to Australia's contact with Asia.

Introduction

The widespread acceptance of the plate tectonics theory of continental drift has promoted the idea that much of the Australian flora has either been inherited from the pan-Gondwanan flora, or is autochthonous, evolving from the more labile members of the ancient Gondwanan flora. There is, however, debate as to the extent to which the composition of Australia's tropical rainforests reflect immigration by northern taxa into Australia (mainly from the extensive rainforest flora of the Malesian region), or the persistence of Gondwanan taxa (Barlow 1981; Whiffin & Hyland 1986).

The post-Miocene northern invasive flora corresponds largely to the "Indo-Malayan" element espoused by Burbidge (1960) and her predecessors, and is based on the observation that for many taxa found in the tropical rainforests of Australia, a much higher diversity is present in Malesia, often with only a single species present in Australia (Martin 1982; Whiffin & Hyland 1986).

There are contradictory views however. Webb *et al.* (1986) suggest that the integrated nature of tropical rainforest, with its highly interdependent synusiae, makes the idea of individual rainforest taxa colonizing the Australian tropics *en masse* implausible.

An ever-growing catalogue of early Tertiary pollen records of taxa with affinities to modern genera in the Australian tropical rainforests is demonstrating that some so-called Indo-Malayan or Malesian taxa were present in Australia well before the establishment of the link between Asia and Australia (Martin 1977, 1982). Similar finds of plant megafossils [*Parafatsia,* (Blackburn 1982) and *Austrodiospyros,* (Basinger & Christophel 1985)] have reinforced the expectation amongst some botanists that analysis of early Tertiary vegetation may now be able to provide insight into the origins of the Australian tropical rainforest flora.

Methods of Comparing Fossil and Modern Vegetation

There are two main methods used to reconstruct past vegetation. The more traditional approach has been to identify the modern affinities of most of the fossil taxa from a deposit and infer from the ecological preferences of the closest modern relatives the nature of the palaeoenvironment. This method has given contrasting interpretations of the early Tertiary vegetation of southern Australia, with palynologists initially predicting a cool to warm temperate vegetation dominated by *Nothofagus* and Podocarpaceae (Kemp 1981), whereas the megafossil record suggests a vegetation of sub-tropical to tropical character (Christophel 1981). This is explained by the fact that fossil pollen often reflects the regional flora with prolific producers such as *Nothofagus* swamping the samples, whereas fossil leaf beds generally reflect the local flora (Martin 1982).

Numerous studies on modern plant communities have demonstrated a link between leaf form and size, or physiognomy, and the environment (Dolph 1978; Davis & Taylor 1980; Givnish 1984). In Australia, Webb (1959, 1968) classified Australian rainforest on physiognomic criteria, primarily leaf size and margin type, using canopy leaves. These studies have given rise to the idea that the physiognomic characteristics of fossil leaf beds may yield significant palaeoecological information (Dolph & Dilcher 1979). This constitutes the second main method of reconstructing past vegetation and is usually referred to as leaf or foliar physiognomic analysis.

Foliar Physiognomic Analysis of Fossil and Modern Tropical Rainforest Leaf Beds

Previous attempts to apply leaf physiognomic analysis to Tertiary floras assumed a direct correspondence between the physiognomic characteristics of the forest community and the leaves found in a fossil deposit (Dolph & Dilcher 1979; Christophel 1981). This assumption has been questioned on the grounds that most of the studies on modern forest communities have dealt with regional floras (Roth & Dilcher 1978) or have been based exclusively on canopy leaves (Martin 1982). The physiognomic analysis of three Eocene floras by Christophel (1981) was further questioned by Nix (1982) who suggested that as southern Australia lay in high southern latitudes in the Eocene, a tropical (megathermal) or subtropical (mesothermal) climate was not possible for Anglesea, Nerriga, or Maslin Bay.

In order to further refine the foliar analysis of Australian Tertiary floras, a study of the physiognomic characteristics of leaf litter from the types of tropical rainforest recognized by Webb (1968) was initiated. Leaf litter on the forest floor represents the leaves of all the members of the plant community in the immediate vicinity, and not just a single stratum (Ferguson 1985), whereas most foliar physiognomic studies are based on either herbarium specimens or field collections of living leaves. The leaves in leaf litter better reflect a fossil leaf bed than herbarium collections, as recent studies (Ferguson 1985) have found that leaves do not travel in air much further than the height of the parent tree, and that larger leaves tend to be screened out by intervening vegetation, producing a different physiognomic character than the original forest (Drake & Burrows 1980).

A fossil leaf bed is usually formed in a stream or lake where water flow was reduced to allow settling of both the leaves and a layer of sediment. Several studies have suggested that where the point of leaf deposition is some distance from the source of the leaves, such as in a lake where the leaves have been brought in by a stream, the resultant leaf bed will be dominated by the lakeside and streamside flora with the smaller, more coriaceous canopy leaves distorting the physiognomic interpretation (Roth & Dilcher 1978; Drake & Burrows 1980). These changes are the net result of the preferential destruction of larger, more fragile leaves from the interior of the forest due to hydrodynamic forces in the stream, and the low input of leaves from the interior of the forest due to the screening effect of the streamside and lakeside vegetation (Ferguson 1985). These same studies have also found that these effects become more pronounced with increasing distance from shore, and hence source.

When leaves are deposited directly into a potential fossil deposit, such as by forest on a flood plain with numerous ox-bow lakes or ephemeral channels in a braided stream system, the resultant leaf bed will closely resemble the leaf litter from the same forest. The study on modern leaf litter from Australian tropical rainforest has found that each of the three forest types examined has leaf litter with a characteristic physiognomic signature. The implication is that where fossil leaf beds come from situations like that described above, it is possible to make a direct comparison between the forest types and the fossil vegetation.

Three rainforest types were examined ranging from the tropical lowland (Complex Mesophyll Vine Forest or CMVF), to the montane tropical rainforest (Simple Notophyll Vine Forest or SNVF), and the forest physiognomically intermediate between these two (Complex Notophyll Vine Forest or CNVF). Samples were taken from several localities in north-east Queensland in the region of Cairns and Atherton (Fig. 1), with each sample representing all of the litter scraped from a square metre of ground. Approximately 200 leaves from each sample were measured for length, width (as a fraction of length), and the position of maximum width (as a fraction of length). The type of margin and the taxonomic identity of each leaf was also recorded but will not be considered here. A frequency histogram of leaf length for the three forest types indicates the clear partitioning of the leaf litter from different forest types (Fig. 2). The mean value for each character for each sample has been plotted on a box diagram (Fig. 3) to illustrate the characteristic physiognomic signature of the three forest types from the region studied.

The SNVF leaf litter was characterized by microphylls (< 75 mm length) with an average width of 30-40% of length (narrow elliptic), and were generally symmetric about the midpoint of the leaf. The CNVF leaf litter was composed of a mixture of microphylls and notophylls (75-125 mm length) with an average width of 40-45% of length and distinctly ovate (position of maximum

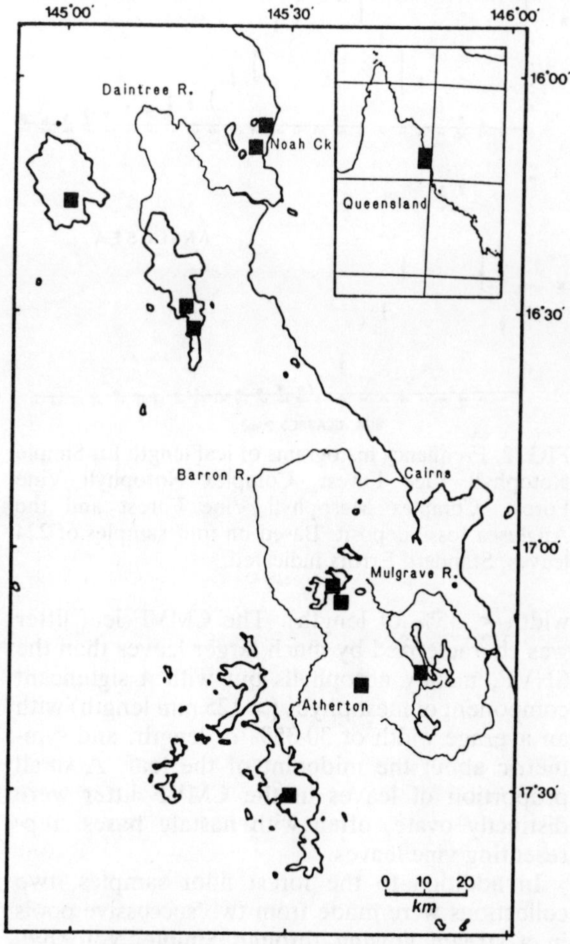

FIG. 1. Location of leaf litter collection sites (black squares). 1 000 metre contour indicated.

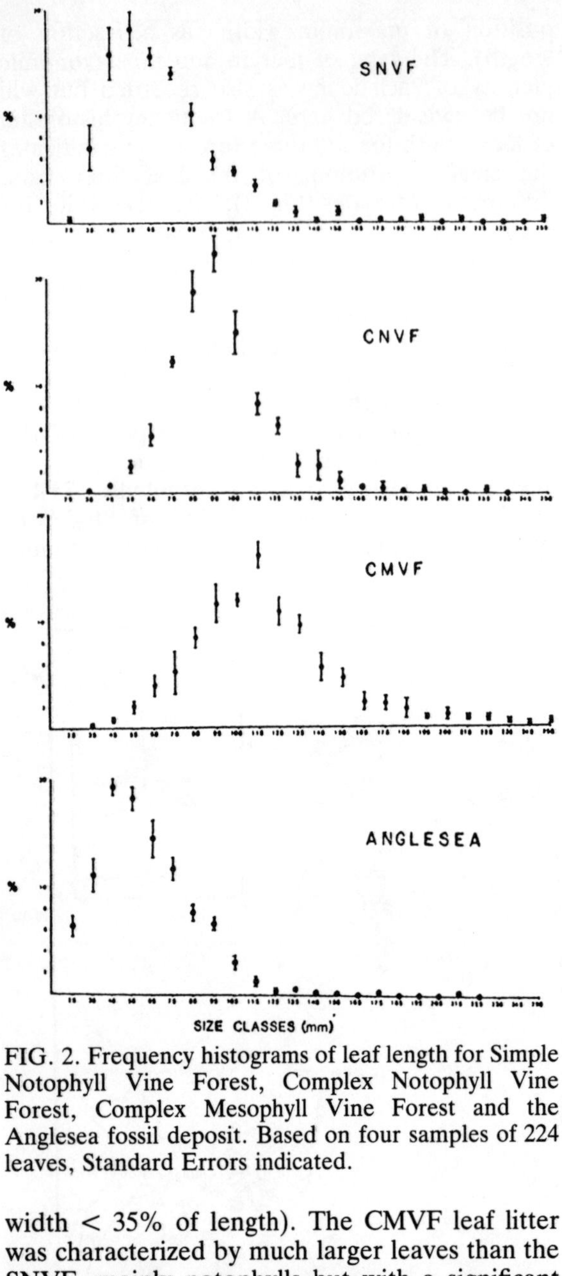

FIG. 2. Frequency histograms of leaf length for Simple Notophyll Vine Forest, Complex Notophyll Vine Forest, Complex Mesophyll Vine Forest and the Anglesea fossil deposit. Based on four samples of 224 leaves, Standard Errors indicated.

width < 35% of length). The CMVF leaf litter was characterized by much larger leaves than the SNVF, mainly notophylls but with a significant component of mesophylls (> 125 mm length) with an average width of 30-35% of length, and symmetric about the midpoint of the leaf. A small proportion of leaves in the CMVF litter were distinctly ovate, often with hastate bases, representing vine leaves.

In addition to the forest floor samples, two collections were made from two successive pools in a stream flowing through Simple Notophyll Vine Forest on the Windsor Tableland. The results suggest that these two samples have similar physiognomic characteristics to leaf litter from the surrounding forest type. Based on this data, it is suggested that where leaves have been deposited with little transport after leaf fall, the physiognomic characteristics (the physiognomic signature) can be used to compare leaf populations from unknown forest type to leaf populations from known forest types. On this premise, collections of fossil leaves from two Eocene localities from southern Australia, Anglesea in Victoria, and a newly discovered deposit near Golden Grove in South Australia (Fig. 4a), were measured in the same manner as the leaf litter. These results are also plotted on the box diagram (Fig. 3).

The Anglesea samples were characterized by microphylls with an average width of 30-35% of length (narrow elliptic), and the position of maximum width was on average 40-45% of length, reflecting a small group of ovate leaves in some samples. This situation is to a lesser extent found in some SNVF litter samples. Overall, the

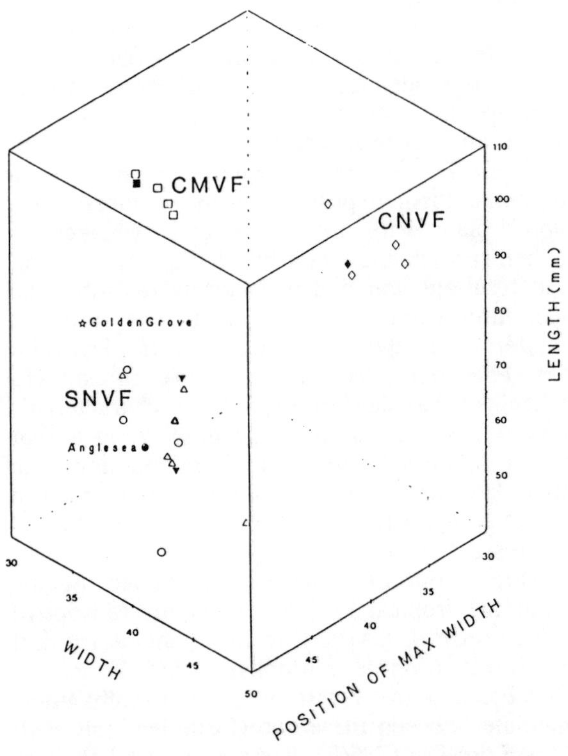

FIG. 3. 'Box Diagram' displaying the physiognomic signatures of SNVF (open △), CNVF (open ◊), CMVF (open □) the Anglesea flora (open ○) and the Golden Grove flora (☆). The stream samples of SNVF are indicated by the inverted solid▼'s. Means of each are indicated by solid symbols.

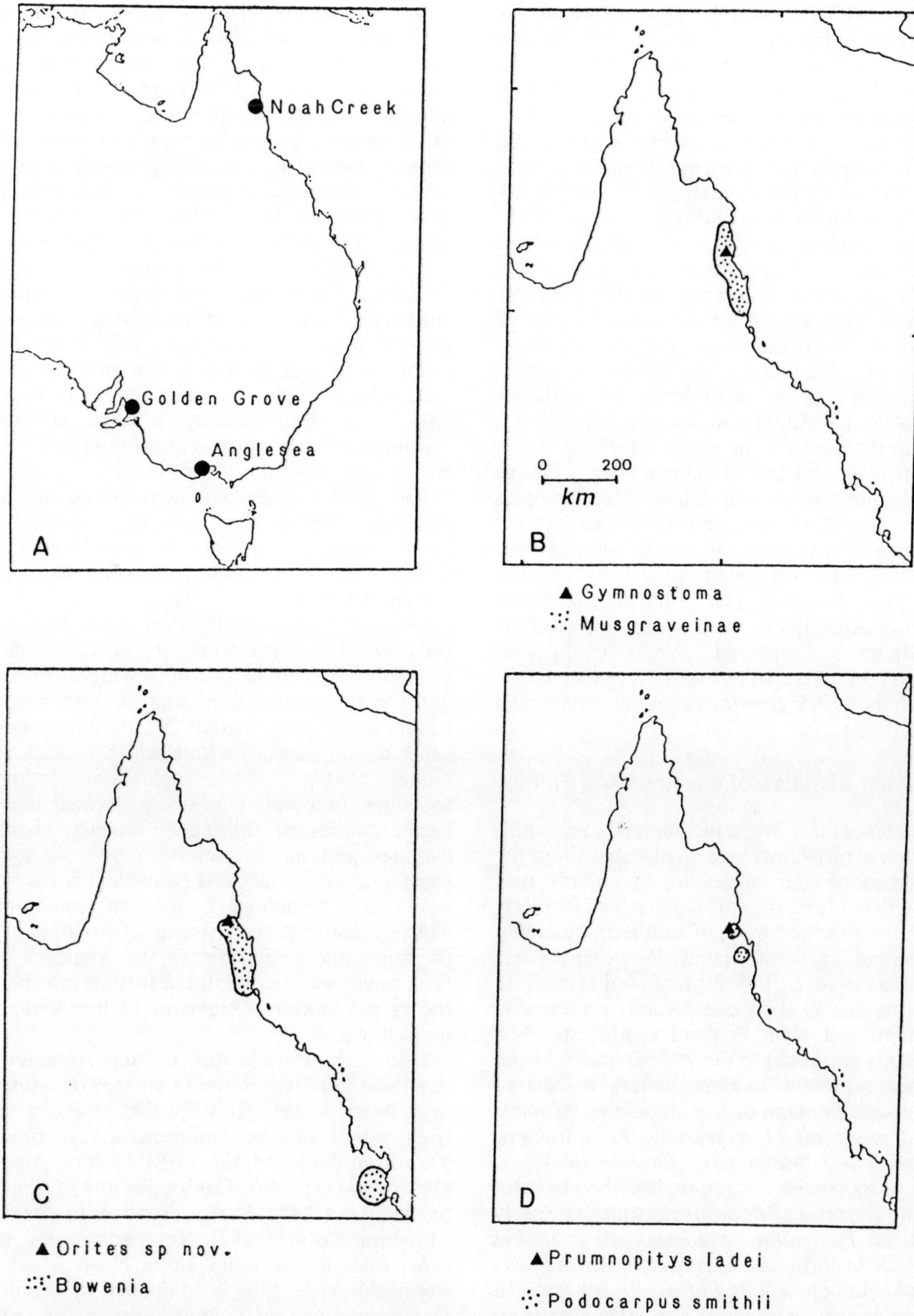

FIG. 4. Maps showing location of fossil localities, Noah Creek and distribution of selected taxa. (a) Fossil localities and Noah Creek. (b) Distribution of extant *Gymnostoma* and Musgraveinae. (c) Distribution of extant *Orites* sp. nov. and *Bowenia.* (d) Distribution of *Prumnopitys ladei* and *Podocarpus smithii.*

Anglesea and SNVF samples appear to have the same physiognomic signature (Fig. 3). The Golden Grove sample has a physiognomic signature at the upper limit of the length character for SNVF (Fig. 3). Field observation of oxidized material on fresh clay blocks from Golden Grove suggest that further collections may give a much greater average length, increasing the similarity with the physiognomic signature of CMVF litter.

In the original study of the physiognomic characteristics of the Anglesea flora, Christophel (1981) suggested a similarity to either Simple Notophyll Vine Forest or Complex Notophyll Vine Forest. The present study supports a similarity to SNVF, although the fossil leaves are generally narrower than the SNVF litter. The variation between the individual samples of fossil leaves is similar to that seen for the SNVF litter, which represents two distinct sites. The Golden Grove sample appears quite distinct from the Anglesea samples and is intermediate between CMVF and SNVF.

It is clear that some fossil samples are not close matches to the suggested analogous modern forest types; however, given the high southern latitude of Australia in the Eocene, some differences must be expected. What is apparent though, is that there were forests very like SNVF and perhaps CNVF growing in southern Australia in the Eocene.

Taxonomic Affinities of the Anglesea Flora

The Anglesea flora, while incompletely known, is nevertheless revealing a wealth of detail about the past vegetation and suggesting that many taxa now restricted to north-east Queensland formerly grew in the Eocene forests of southern Australia. For example, the conifer family Podocarpaceae is often considered an indicator of cool temperate conditions due to their conspicuous presence in Tasmanian and New Zealand rainforests. The megafossils attributed to the Podocarpaceae from Anglesea represent modern genera associated with tropical montane or less commonly, tropical lowland rainforest (*Dacrycarpus, Falcatifolium, Podocarpus* and *Prumnopitys;* Greenwood 1987). A total of six species have been described to date, including a species of *Podocarpus* which is closely related to *P. smithii,* a species of restricted occurrence in north-east Queensland which grows in Simple Notophyll Vine Forest. *Prumnopitys* in Australia is also restricted to the same area (Fig. 4b). The *Decussocarpus* species described has also been found at a new Eocene deposit at Golden Grove, South Australia.

Several angiosperm taxa described from Anglesea also have their closest extant relatives growing in the north Queensland rainforests. *Musgraveinanthus* (Christophel 1984) has affinities with the Musgraveinae (Proteaceae), all of which have a distribution restricted to north-eastern Queensland (Fig. 4b). Remains of *Gymnostoma* (Casuarinaceae) have also been described from Anglesea (Christophel 1980). Only one species of *Gymnostoma* occurs in Australia, and it is also restricted to this same region (Fig. 4b). A lobed Proteaceae leaf type is frequently encountered in some lenses of the Anglesea locality, and this fossil has been found to be most similar to an extant species of *Orites* from north-eastern Queensland which is as yet undescribed (Figs. 4c, 5d). This extant species has an auriculate, clasping leaf base which is also found in the fossil and is unique within the tribe.

Hill (1978) described two species of fossil *Bowenia.* One species from Anglesea has since been collected as a complete pinnule (Fig. 5a). The restricted distribution of extant *Bowenia* is shown in Fig. 4.

Using the positive identification of five of the taxa described above from Anglesea — *Podocarpus, Bowenia, Gymnostoma, Musgraveinanthus* and *Orites* — and considering the restricted distributions of their extant counterparts, a very small region remains within which to look for a modern analogue of the Anglesea vegetation. In so doing, however, it must be stressed that the basic realities of shifting continents, changing climates and natural selection over 40 million years guarantee that exact equivalent floras would not exist. Nonetheless, modern communities which demonstrate strong taxonomic and physiognomic similarities to the Anglesea fossil flora might well be helpful in further interpreting the as yet unknown elements of that and other fossil floras.

One such possible direct comparison for the Anglesea flora was found in an area of rainforest near Noah Creek (16°07′S, 145°26′E), a small river north of the Daintree River, draining Thornton Peak on the north-eastern coast of Queensland (Fig. 4a). This locality and its similarity to Anglesea have been described in detail by Christophel *et al.* (1987). It is worth noting, however, that at one point along Noah Creek the streamside vegetation is dominated by one of the rare populations of *Gymnostoma* in the region, and by the undescribed species of *Orites.* Several hundred metres downstream from this spot, the vegetation is a typical lowland tropical forest

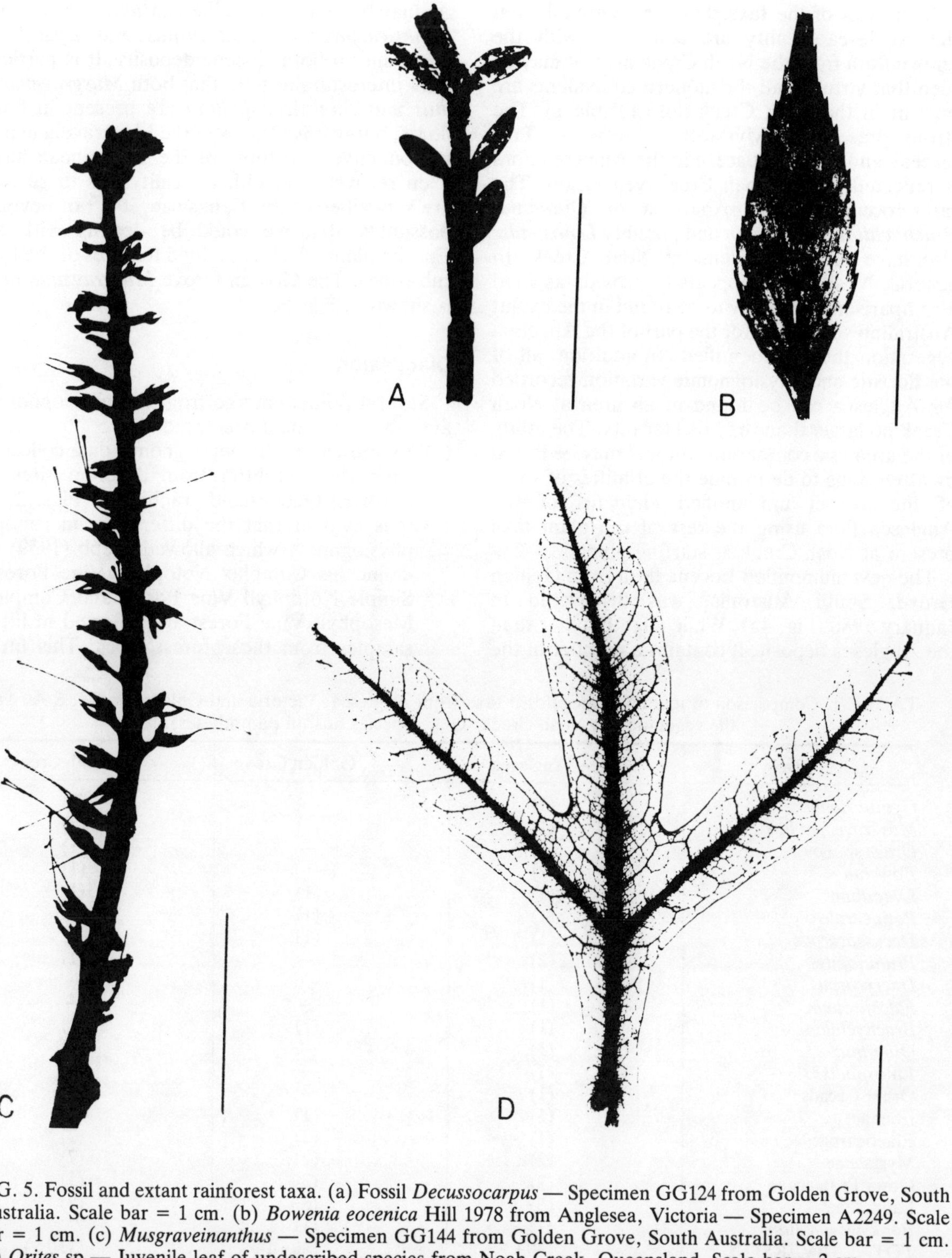

FIG. 5. Fossil and extant rainforest taxa. (a) Fossil *Decussocarpus* — Specimen GG124 from Golden Grove, South Australia. Scale bar = 1 cm. (b) *Bowenia eocenica* Hill 1978 from Anglesea, Victoria — Specimen A2249. Scale bar = 1 cm. (c) *Musgraveinanthus* — Specimen GG144 from Golden Grove, South Australia. Scale bar = 1 cm. (d) *Orites* sp — Juvenile leaf of undescribed species from Noah Creek, Queensland. Scale bar = 2 cm.

(CMVF) whose component taxa include two of the three described species of the Musgraveinae, and *Bowenia*.

If the rest of the taxa thus far identified from the Anglesea locality are compared with the known flora from the Noah Creek area, it may be seen that virtually all the modern equivalents are present in the Noah Creek flora (Table 1). The strong presence of Myrtaceae, Lauraceae, Proteaceae and Elaeocarpaceae in the Anglesea flora is reflected in the Noah Creek vegetation. The rarer occurrences at Anglesea of *Quintinia, Brachychiton, Lygodium* and possibly *Linospadix* also have their equivalents at Noah Creek. In general, Noah Creek appears to provide as good a comparison as is likely to be found in the extant Australian vegetation for the part of the Angelsea vegetation thus far identified. In addition, all of the floristic and physiognomic variation recorded for Anglesea can be found in an area at Noah Creek no larger than the fossil locality. The utility of the area as a comparative model may be tested by attempting to determine the affinities of some of the as yet unidentified elements of the Anglesea flora using the rest of the plant taxa present at Noah Creek as starting points.

The new mummified Eocene flora from Golden Grove, South Australia, was discovered in January 1986 (Fig. 4a). While slightly older than the Anglesea deposit, it contains a number of the key taxa found at Anglesea, and hence provides a possibility of studying the evolution of some of the taxa over reasonably short periods of time. As may be seen from Table 1, *Musgraveinanthus, Banksieaephyllum, Podocarpus* and *Lygodium* are found in both Eocene deposits. It is particularly interesting to note that both *Musgraveinanthus* and *Banksieaephyllum* are present in both floras, but neither leaves of the Musgraveinae nor reproductive structures of the Banksineae have been recovered at either locality. As these two are sub-tribes of the Banksieae, it is not beyond possibility that we could be dealing with an Eocene plant which combined features of the two sub-tribes. The Golden Grove *Musgraveinanthus* is shown in Fig. 5c.

Discussion

Several points emerge from the physiognomic and taxonomic data presented above.

(1) By presenting the physiognomic data collected from the leaf litter from different sites in northern Queensland graphically (Figs. 2, 3) it is evident that the differences in canopy physiognomy which allowed Webb (1959) to define his Complex Notophyll Vine Forest, Simple Notophyll Vine Forest and Complex Mesophyll Vine Forest are reflected in litter samples from those forest types. This litter

TABLE 1. Comparison of identified megafossil taxa from Anglesea, Victoria and Golden Grove, S.A. with the vegetation at Noah Creek, Qld. (Species nos. in parentheses).

Taxa	Anglesea	Golden Grove	Noah Creek
Gymnostoma	(1)	—	(1)
Musgraveinanthus	(1)	(1)	(3)
Orites sp. nov.	(1)	—	(1)
Bowenia	(2)	—	(1)
Lygodium	(1)	(1)	(1)
Podocarpus	(1)	(1)	(2)
Decussocarpus	(1)	(1)	—
Prumnopitys	(2)	—	(1)
Dacrycarpus	(1)	—	—
Falcatifolium	(1)	—	—
Brachychiton	(1)	(1)	(1)
Quintinia	(1)	—	(1)
Linospadix?	(1)	—	(2)
Other Cycads	(1)	—	(1)
Ebenaceae	(1)	(1)	(2)
Elaeocarpaceae	(1)	(2)	(2)
Myrtaceae	(2)	(2)	(7)
Other Proteaceae	(2)	(3+)	(5)
Lauraceae	(5+)	(3+)	(9)
Banksieaephyllum	(1)	(1)	?
TOTAL DIVERSITY	(80+)	(40+)	(110+)

data can then be compared directly with physiognomic data collected from fossil leaf deposits, which are interpreted as being litter from fossil vegetation. This was done with the result that strong physiognomic similarities were shown between three Eocene deposits from southern Australia and Webb's three forest types mentioned above. This then provided one piece of evidence for the hypothesis that the Eocene floras of southern Australia were most similar to the wet tropical vegetation of northern Australia.

(2) Examination of the taxa thus far identified from the Eocene Anglesea Locality showed that their nearest living relatives in Australia are found in these same wet tropical forests. Not only are specific (and rare) taxa such as members of the Musgraveinae, *Podocarpus smithii, Bowenia* and *Gymnostoma* found there, but the general floristic composition when examined at the family level is similar. A preliminary examination of the Golden Grove Eocene Flora suggests similar comparisons.

(3) By examining closely one locality in north-eastern Queensland (Noah Creek) where the rarer extant relatives of the Anglesea flora are found, it can be seen that both floristically and physiognomically all tested elements of the Anglesea flora can be accounted for within a small area similar in size to the fossil deposit. This demonstrates the viability of direct comparisons between the southern Australian Eocene Floras and the Wet Tropical Forests of northern Australia.

This paper began with the question of the origin of the Australian Tropical Rainforests. We believe that the fossil evidence shows clearly that the ancient relatives of a growing number of extant Australian rainforest taxa were to be found in the Eocene forests of southern Australia. Not only were these taxa present, but they were in similar associations, were in forests of similar general composition, and represented vegetations with similar physiognomic signatures to those of modern day northern Australia. Although some extant rainforest taxa (e.g., *Ficus, Saurauia*) as yet have no Australian fossil record and may possibly be Malesian in origin, the majority of northern Australian rainforest taxa, and perhaps their associations, appear to be autochthonous.

Acknowledgements

The authors thank B. P. M. Hyland, Bruce Gray and the staff of the CSIRO, Atherton, for assistance in the field. We thank the Queensland National Parks and Wildlife Service and the Department of Forestry for permission to collect material in the Atherton and Daintree areas. We also thank Sue Shaw for production of the photographic plates and Bronwyn Burns for preparation of the final draft of the manuscript.

References

Basinger J. F. & Christophel D. C. (1985) Fossil flowers and leaves of the Ebenaceae from the Eocene of southern Australia. *Can. J. Bot.* **63,** 1825-43.

Blackburn D. (1982) Tertiary megafossil flora of Maslin Bay, South Australia: Numerical taxonomic study of selected leaves. *Alcheringa* **5,** 9-28.

Burbidge N. T. (1960) The phytogeography of the Australian region. *Aust. J. Bot.* **8,** 75-212.

Christophel D. C. (1980) Occurrence of *Casuarina* megafossils in the Tertiary of south-eastern Australia. *Aust. J. Bot.* **28,** 299-59.

Christophel D. C. (1981) Tertiary megafossil floras of Australia as indicators of floristic associations and palaeoclimate. In: *Ecological Biogeography of Australia* (ed. A. Keast) pp. 379-390. Junk, The Hague.

Christophel D. C. (1984) Early Tertiary Proteaceae: The First Floral Evidence for the Musgraveinae. *Aust. J. Bot.* **32,** 177-86.

Christophel D. C., Harris W. K & Syber A. K. (1987) The Eocene Flora of the Anglesea Locality, Victoria. *Alcheringa* **11,** 303-23.

Davis J. M. & Taylor S. E. (1980) Leaf physiognomy and climate: a multivariate analysis. *Quaternary Res.* **14,** 337.

Dolph G. E. (1978) Variation in leaf size and margin type with respect to climate. *Cour. Forsch.-inst. Senck.* (W. Germany). **30,** 153-58.

Drake H. & Burrows C. J. (1980) The influx of potential macrofossils into Lady Lake, north Westland, New Zealand. *N.Z. J. Bot.* **18,** 257-74.

Ferguson D. K. (1985) The origin of leaf assemblages — new light on an old problem. *Rev. Palaeobot. Palynol.* **46,** 117-88.

Givnish T. J. (1984) Leaf and canopy adaptations in tropical forests. In: *Physiological ecology of plants of the Wet Tropics* (eds E. Medina, H. A. Mooney & C. Vazquez-Yanes) pp. 51-84. Junk, The Hague.

Greenwood D. R. (1987) Early Tertiary Podocarpaceae: megafossils from the Eocene Anglesea location, Victoria. *Aust. J. Bot.* **35,** 111-33.

Hill R. S. (1978) Two New Species of *Bowenia* Hook. ex Hook. f. from the Eocene of Eastern Australia. *Aust. J. Bot.* **26,** 837-46.

Kemp E. M. (1981) Tertiary palaeogeography and the evolution of the Australian climate. In: *Ecological Biogeography of Australia* (ed. A. Keast) pp. 31-50. Junk, The Hague.

Martin H. A. (1977) The history of *Ilex* (Aquifoliaceae) with special reference to Australia: Evidence from pollen. *Aust. J. Bot.* **25,** 655-73.

Martin H. A. (1982) Changing Cenozoic Barriers and the Australian Paleobotanical Record. *Ann. Missouri Bot. Gard.*, **69**, 625-67.

Nix H. (1982) Environmental determinants of biogeography and evolution in Terra Australis. In: *Evolution of the Flora and Fauna of Arid Australia* (eds W. R. Barker & P. J. M. Greenslade) pp. 47-66. Peacock Publications, Adelaide.

Roth J. L. & Dilcher D. L. (1978) Some considerations in leaf size and margin analysis of fossil leaves. *Cour. Forschung.-institut. Senck.* (W. Germany) **30**, 165-71.

Webb L. J. (1959) A physiognomic classification of Australian rainforests. *J. Ecol.* **47**, 551-70.

Webb L. J. (1968) Environmental relationships of the structural types of Australian rainforest vegetation. *Ecology*. **49**, 296-311.

Webb L. J., Tracey J. G. & Jessup L. W. (1986) Recent evidence for autochthony of Australian tropical and sub-tropical rainforest floristic elements. *Telopea* **2**, 575-90.

Whiffin T. & Hyland B. P. M. (1986) Taxonomic and Biogeographic evidence on the relationships of Australian plants. *Telopea* **2**, 591-610.

Proc. Ecol. Soc. Aust. 15:149-158

Stability amid turmoil?: towards an ecology of north Australian eucalypt forests

D. M. J. S. BOWMAN

Conservation Commission of the Northern Territory, P.O. Box 496, Palmerston, N.T. 0831

Abstract

The coastal regions of the Northern Territory support a considerable proportion of Australia's open eucalypt forest. These forests are restricted to areas which receive in excess of 1 200 mm of precipitation per annum. They are distributed according to edaphic factors such as degree of wet season inundation and dry season moisture deficit within this climatic zone. Understorey type varies from grassland to closed-scrub and is determined by dry season moisture supply. Fire occurrence is positively related to grass fuels. The intensity of fires increases with the time elapsed since the wet season rains. Experimental late season fires cause a slight depression in the density of tree and shrub saplings greater than 2 m in height and less than 10 cm DBH. Fire-free intervals greater than two years result in structural development of grassland understoreys which continues for at least 12 years. This change is accompanied by minor floristic replacement. Late season fires are capable of destroying Callitris *and monsoon forest boundaries, although the impact of fire is compounded by feral animal damage and subsequent invasion by herbaceous weeds. Expansion of* Callitris *and drought resistant monsoon forest species can occur in unburnt eucalypt forests but the high frequency of people- and lightning-caused fires makes succession unrealistic, especially given the tolerance and competitive abilities of the eucalypts. Naturally shrubby eucalypt forests are richer in vertebrate fauna than are grassy forests, but it is unclear whether this reflects the effect of structure, floristics or both. Ant species are sensitive to the structural change of vegetation induced by fire exclusion. Fire management may therefore have significant conservation implications not only for* Callitris *and monsoon forests but also for the fauna associated with the eucalypt forests.*

Introduction

The Northern Territory is blessed with a significant portion of Australia's *Eucalyptus* communities that range in height from 10 to 30 m and have a canopy cover of between 30 to 70% (Fig. 1). Until recently little was known about the northern eucalypt forest because most attention has been focused on the monsoon forests which occur as scattered patches of a few hectares throughout the coastal regions of northern Australia (Webb 1968; Stocker 1971; Beard 1976; Webb & Tracey 1981; Kikkawa *et al.* 1981; Beard *et al.* 1984; Clayton-Greene & Beard 1985; Russell-Smith 1985, 1986; Russell-Smith & Dunlop 1987; Bowman and Dunlop 1986). Floristic diversity is not the explanation for the disproportionate fascination with the geographically miniscule closed communities for Taylor and Dunlop (1985) have shown that eucalypt forests are floristically richer than either sandstone monsoon forest or lowland monsoon forest in Kakadu National Park. Biogeographic questions may account for the more detailed study of the monsoon forests, although the problem of why there are not eucalypt forest analogues elsewhere in the tropics has received scant consideration. Perhaps the major motive for research into closed forests is the thought that

FIG. 1. The distribution of open eucalypt forest in Australia. No account has been made of variation in understorey type, nor the subgeneric divisions of the genus *Eucalyptus*. (Source, Carnahan 1976).

eucalypt forests have expanded at the expense of the more "fragile" monsoon forests following indiscriminate burning by mankind.

The question of forest succession has not been directly tackled, although some workers assume that the dynamics between eucalypt and monsoon forest are similar to the interrelations between temperate rainforests and eucalypt forests (Stocker 1966; Webb 1968; Langkamp *et al.* 1981). This may reflect the origin of researchers as much as a prevailing scientific paradigm.

Concurrent with consideration of succession in the coastal regions of the Northern Territory was the development of the important Australian tradition of Land Systems mapping (Christian & Stewart 1953; Story *et al.* 1969; Story *et al.* 1976). This approach is based on the assumption that vegetation, landform and soils are highly interrelated and form "land systems" which can be mapped. Clearly such an approach argues for equilibrium of vegetation with physical factors in contrast to the disequilibrium generated by fire patterned vegetation.

This paper seeks to bring together the various strands of evidence to form a framework by which to manage the coastal eucalypt forests of the Northern Territory.

Spatial Distribution

Small scale mapping portrays eucalypt forests as widespread uniform communities in the coastal regions of the Northern Territory (Carnahan 1976), while larger scale maps show the complex patterning of eucalypt and non-eucalypt vegetation (Wilson & Bowman 1987). In reality eucalypt forests are sections of floristic and structural continua corresponding to underlying environmental gradients (Bowman 1986; Bowman & Minchin 1987).

Environmental correlates

The major environmental determinate of eucalypt forests in the Northern Territory is precipitation. However within the region which receives more than 1 200 mm per annum, edaphic factors determine the distribution and type of eucalypt forest. On low-lying sites with poor wet season drainage eucalypt forests are replaced by woodlands dominated by *E. confertiflora, E. papuana* on clayey soils or *Melaleuca* spp. on sandy soils (Pedley & Isbell 1971; Wilson & Bowman 1987; Bowman & Minchin 1987). Sites with good moisture supply throughout the dry season are dominated by monsoon forests which vary both structurally and floristically in relation to water availability (Specht *et al.* 1977; Russell-Smith & Dunlop 1987). On rocky ridges eucalypt forests grade into woodlands dominated by *E. bleeseri* and *E. dichromophloia* (Story 1976; Pedley & Isbell 1971). There is evidence that in addition to moisture supply soil fertility limits forest development, as eucalypt woodlands or *Callitris intratropica* forests occur on sandsheets or sandstone that has weathered *in situ* (Kirkpatrick *et al.* 1987; Bowman *et al.* 1988a).

The eucalypt forests are dominated by *E. tetrodonta, E. miniata,* and *E. nesophila* in northern coastal localities. They occur on deep well-drained red or yellow laterite soils (Pedley & Isbell 1971; Story 1976). Within these communities understorey type is controlled by dry season moisture supply, which varies in relation to the position of the water table (Bowman 1986; Wilson & Bowman 1987; Bowman & Minchin 1987). On elevated sites tall grasses, such as *Sorghum* and *Heteropogon,* form dense swards which die off in the dry season. On down-slope positions, closed-scrubs develop because broadleaf species such as *Planchonia careya, Alphitonia excelsa, Ficus opposita, Denhamia obscura* and *Terminalia ferdinandiana* are able to tap the water table during the seven month drought (Wilson & Bowman 1987; Bowman & Minchin 1987).

Competition

There has been little investigation of the ecophysiological processes in the northern eucalypt forests. Competition for water may be important in inter- and intra-specific relations, given the intense annual drought. Some evidence supports this view, such as the clumping of eucalypt saplings in forest gaps and the inverse relationship between grass and shrub cover (Bowman 1986). Thus sapling density and shrub cover may also be controlled by disturbance.

Disturbance

Termites and cyclonic storms have the greatest impact on trees while fire has more effect on the understorey.

Fox and Clark (1971) found that 92% of the *E. miniata* trees and 83% of the *E. tetrodonta* greater than 15 cm DBH, that they sampled were damaged by termites.

Extreme wind velocities associated with cyclones result in considerable canopy damage because eucalypt forests are restricted to well drained sites where the trees have firm root anchorage and are resistant to windthrow (Stocker 1976; Wilson & Bowman 1987). The regeneration that develops following cyclone damage to the canopy thus appears to be controlled by site quality, overwood competition and fire (Wilson & Bowman 1987).

Following 12 years of fire protection at the Munmarlary experimental site, Kakadu National Park, the density of tree and shrub saplings was found to increase (Fig. 2, Bowman *et al.* 1988b). Experimental early annual, late annual and early biennial fires had no significant impact on sapling density although late season fires caused a depression in sapling number (Fig. 2).

These results agree with the trends detected after five years of treatment by Hoare *et al.* (unpublished results) (Bell 1981). Braithwaite and Estbergs (1985) found that the vertical profile of the 20 most common shrub species differed between the woodland and forest communities that they studied elsewhere in Kakadu National

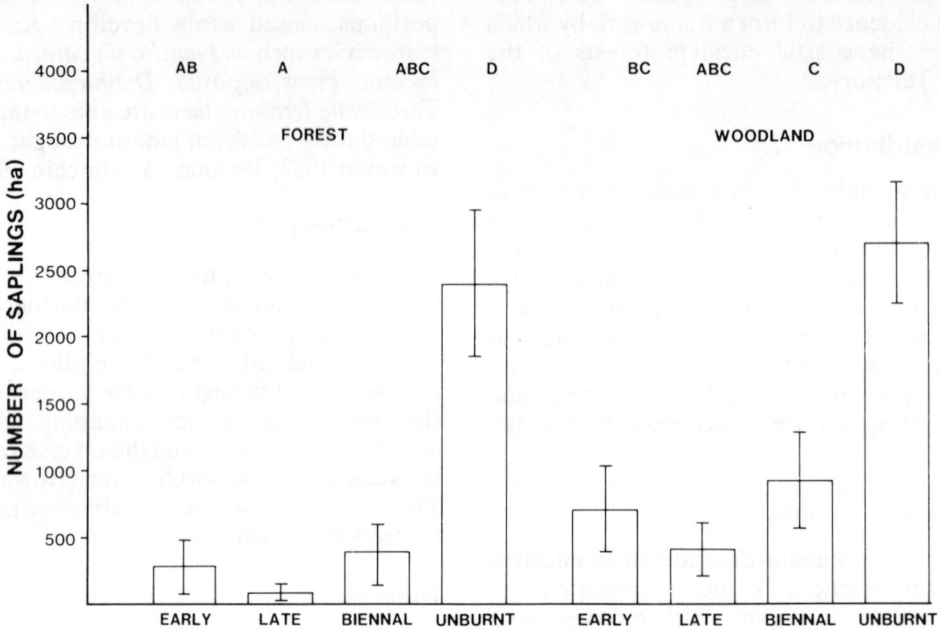

FIG. 2. Histogram showing the pooled density of tree and shrub saplings greater than 2 m high and less than 10 cm DBH for the early and late annual, early biennial fire regimes and the unburnt treatments at Munmarlary. Data are derived from five 10 × 10 m quadrats placed in three replicate 1 ha plots (n=15). Verticle bars indicate 95% confidence intervals. (From Bowman *et al.* 1986).

Park. They found that twice as many woodland species had "even vertical distributions" in the woodland than in the forest, and half as many woodland species had triangular vertical distributions. They attribute these trends to the earlier time of burning in the woodland than in the forest. Figure 2 shows that for all treatments at Munmarlary the woodland has a slightly greater density of saplings than the forest, possibly reflecting the greater competitive advantage of shrubs on these sites. Bowman (1986), Wilson & Bowman (1987) and Bowman & Minchin (1987) suggest that soils and fire control shrub and eucalypt regeneration growth rates and that these variables interact to determine the understorey type. Dry sites are dominated by grass, and the high probability of frequent fire and the slow growth of shrubs makes the development of a shrubby understorey unlikely unless the sites are artificially protected from fire.

Moisture content and gravel content of the surface soil best explains the differences in floristic composition of the 12 experimental plots (three one ha replicates of four different fire regimes) indicating the negligible effect of fire regime (Fig. 3). This conclusion is further supported by the small number of species positively or negatively associated with fire treatment, and the low number of species (37%) which showed major increases or decreases in abundance in the four fire regimes (Table 1). These results concur with Hoare *et al.* (unpublished ANPWS Report, 1980) who concluded that floristic composition of permanent quadrats varied little after five years of treatment. Thus the shrubby forests developed by fire protection are floristically distinct from the shrubby forests that occur naturally in the absence of fire management.

TABLE 1. Number of forest plant species which occur in more than five quadrats which are significantly positively or negatively associated with either of the four fire treatments at Munmarlary (n=98, $P<0.05$).

Fire Treatment	Positive Association	Negative Association
Early Annual	1	2
Late Annual	5	1
Early Biennial	1	3
Unburnt	2	2

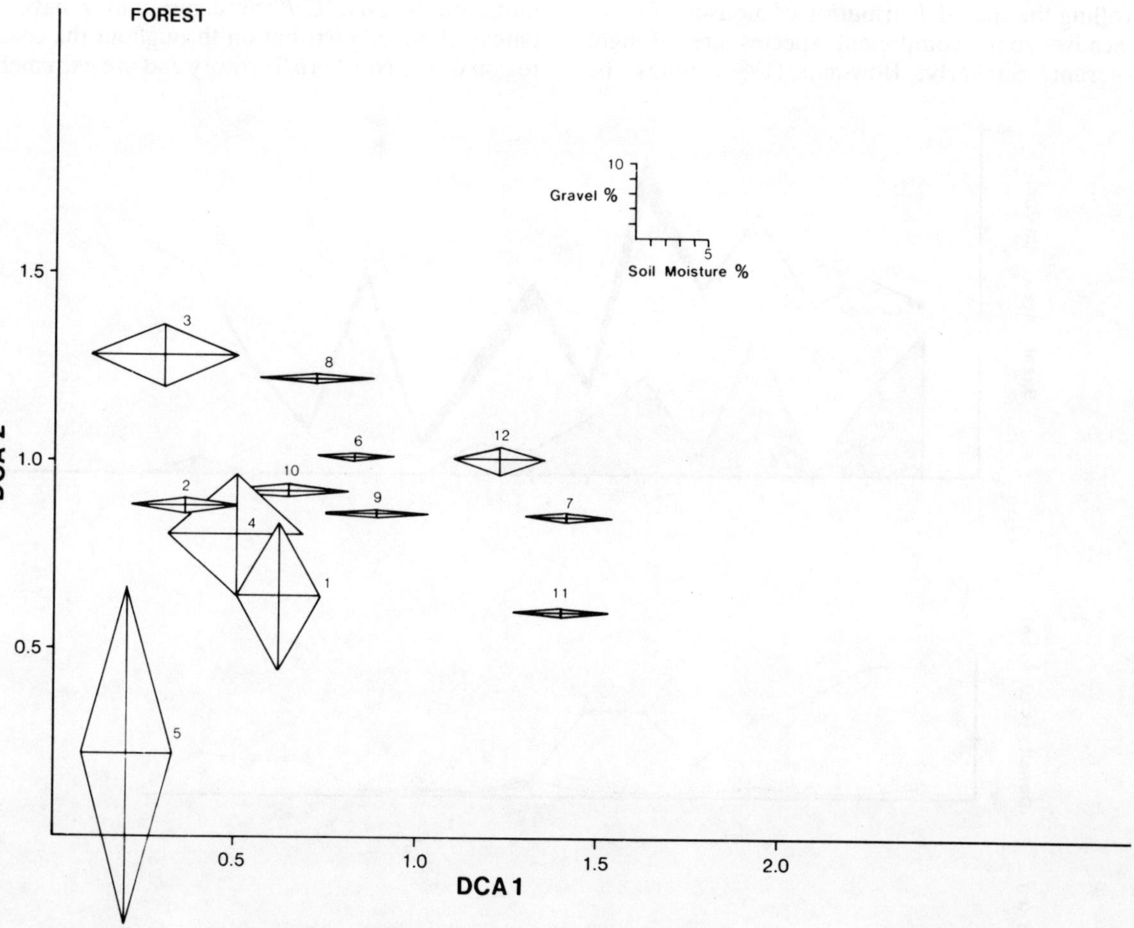

FIG. 3. Ordination showing the mean gravel and surface soil moisture values for each treatment cell plotted on the DCA ordination centroids. Numbers refer to fire treatments: Early annual = 4,5,11; Late annual = 1,6,9; Early biennial = 2,8,10; Unburnt = 3,7,12.

Boundaries and Succession

The application of the temperate eucalypt-rainforest succession model to northern eucalypt forests is considered inappropriate by Bowman (1986) because unlike the southern eucalypt species the monsoon species can regenerate in the absence of fire, and their stand structure is independent of understorey type. The mixed eucalypt-monsoon thicket studied by Bowman and Dunlop (1986) was found to be different from the seral mixed temperate eucalypt-rainforest communities of Tasmania (Gilbert 1959): the spatial distribution of the stand was stable over a 20-year period, it had evidence of wildfire within that time, and the dominant eucalypt, *E. polycarpa,* which is capable of regenerating beneath the dense canopy, was of a different species to the dominants of the surrounding open forest.

The boundary between the mixed thicket and eucalypt forest was found to relate to significant moisture difference, but the separation between the mixed and pure thicket appeared to be related to frequency of fire. Russell-Smith (1985) has demonstrated that monsoon forests have been lost in the recent past, and that fire can cause their boundaries to retreat. However the relationship between fire and closed communities is confounded by the effect of feral animal damage, and consequent invasion by herbaceous weeds.

Russell-Smith (1986) and Kirkpatrick *et al.* (1987) argue that fire may be important in controlling the spatial distribution of monsoon forests because some component species are drought tolerant. Similarly, Bowman (1986) notes the possibility that *Callitris* could be the climatic climax of the eucalypt forest sites for, like the monsoon forests, *Callitris* communities have a patchy clumped distribution throughout the coast region of the Northern Territory and are extremely

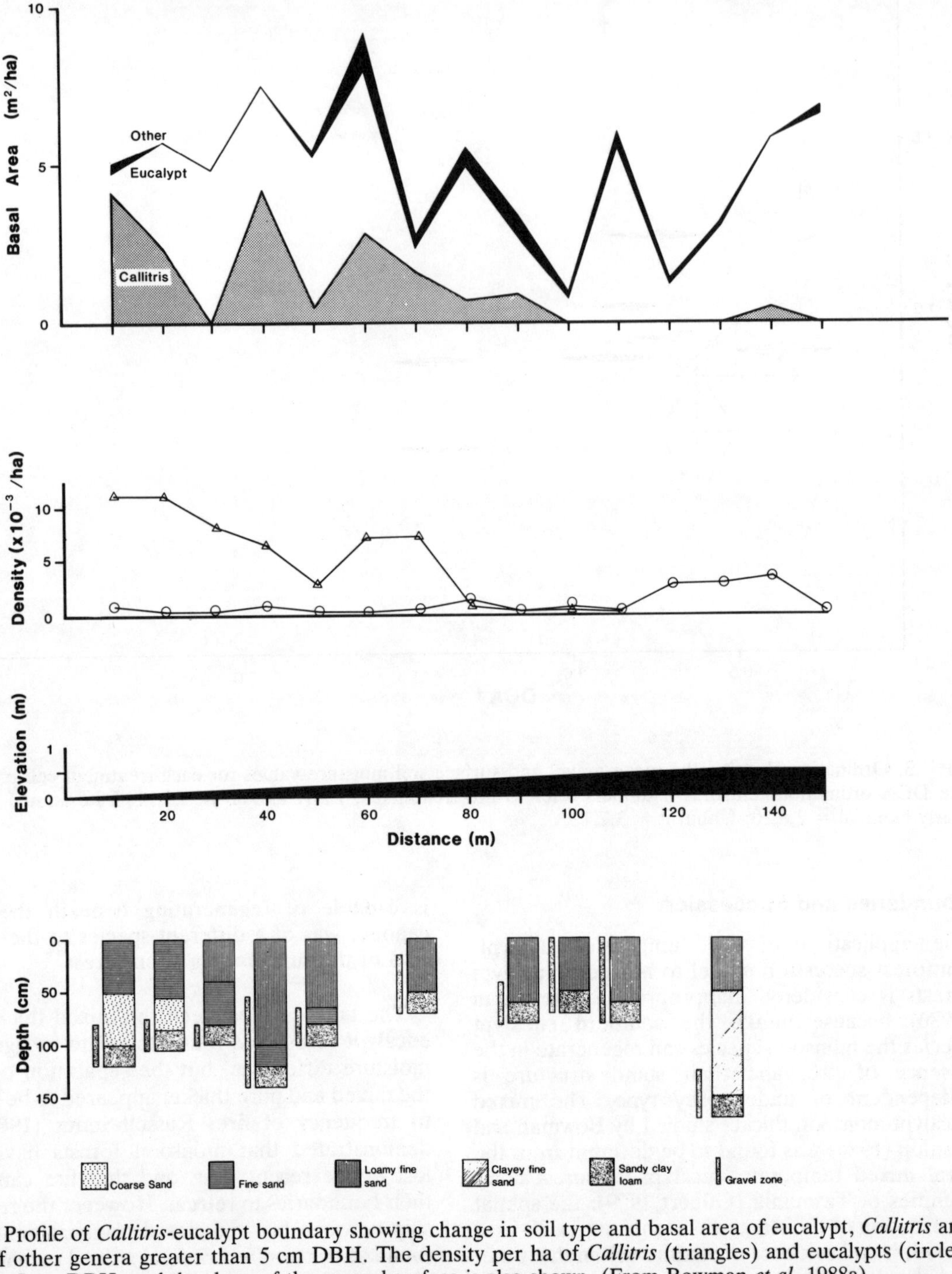

FIG. 4. Profile of *Callitris*-eucalypt boundary showing change in soil type and basal area of eucalypt, *Callitris* and stems of other genera greater than 5 cm DBH. The density per ha of *Callitris* (triangles) and eucalypts (circles) less than 5 cm DBH, and the slope of the ground surface is also shown. (From Bowman *et al.* 1988a).

fire sensitive. Unlike the monsoon species *Callitris* is extremely drought tolerant, often occurring on sandstone escarpments. Eighteen years of fire management at Murgenella has enabled *Callitris* to expand from small clumps into the surrounding eucalypt forest (Fig. 4) (Bowman *et al.* 1988a). The distribution of *Callitris* trees that pre-date fire protection corresponds with deep sandy soils. The invasion of *Callitris* is a consequence of reducing the occurrence of fire. However, even with this protection lightning strikes have started fires which have burnt a significant proportion of the management area (Fig. 5). Thus under natural conditions *Callitris* could never take over eucalypt sites because of the frequency of fire. There is some evidence that eucalypts out-compete *Callitris* on the more fertile laterite soils as evidenced by the overtopping of *Callitris* plantation by regeneration of cleared eucalypt forests.

Wilson & Bowman (1987) and Bowman & Minchin (1987) suggest that vegetation is in equilibrium with fire through a complex interaction with soils. Bowman & Wilson (in press) have found some evidence to support this hypothesis. They found that inflammability and energy content of fuels is inversely related to fire frequency whereas moisture content and biomass of grass is positively related to fire frequency. Thus monsoon forests provide a barrier to fire because they have a moist litter layer and few grasses in contrast to the grassy dry eucalypt forests that are frequently burnt. The *Callitris* community they studied was found to have drier litter than the monsoon forest but it was more moist and the stand had fewer grasses than the eucalypt forest. Litter moisture contents in all communities sampled increased twofold after the summer rains. Late and early dry season sampling did not change the order of communities when ranked by moisture content.

The equilibrium hypothesis is not supported by the destruction of *Callitris* and monsoon forests.

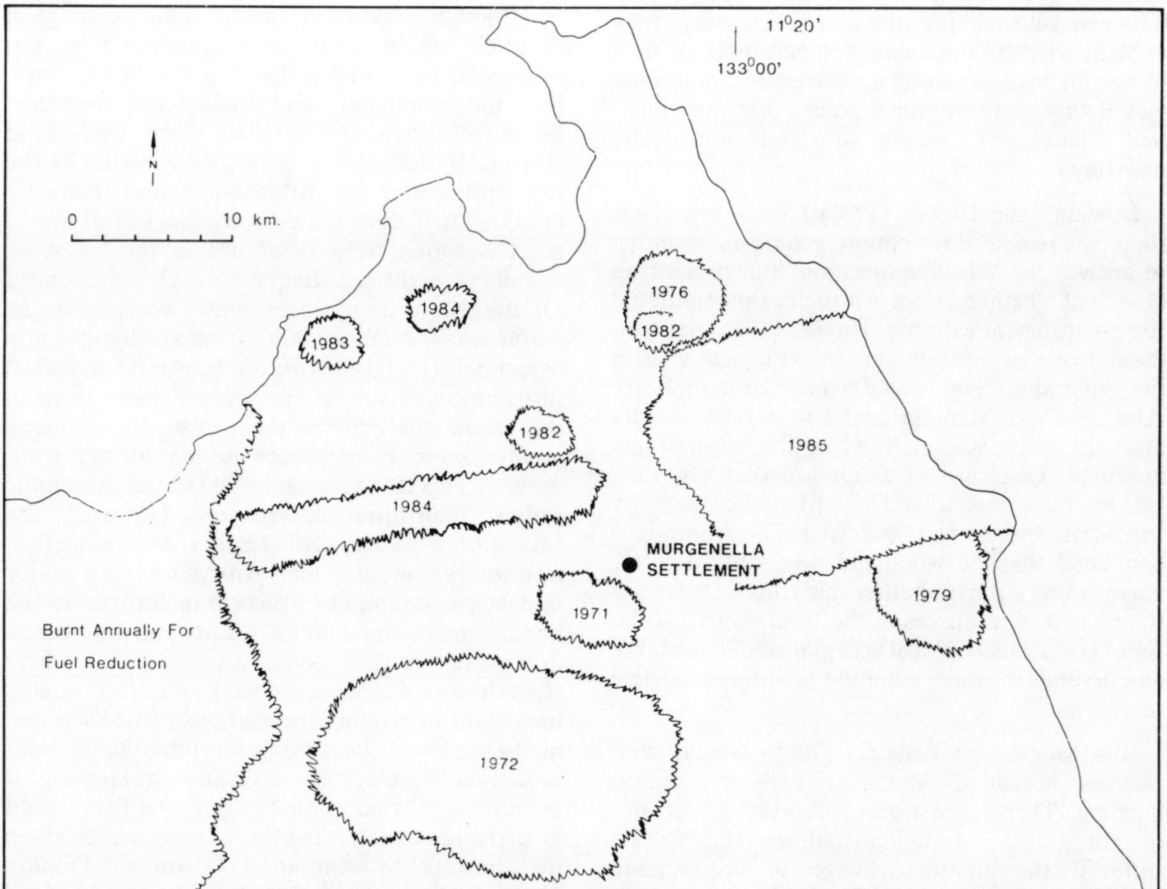

FIG. 5. Map of the Murgenella area depicting the approximate area burnt by lightning-caused fires during the period of fire protection. All the fires were fought and subsequently brought under control.

The damage has in part been attributed to a change in the timing of fires (Russell-Smith 1986; Bowman & Wightman 1985) although there is evidence that feral animal damage interacts with fire (Russell-Smith 1986; Braithwaite *et al.* 1984).

Aboriginal Use of Fire

Braithwaite & Estbergs (1985) suggest that the time of year can characterize the fire regimes that occurred before, during and after Aboriginal custodianship of the forests and woodlands of Kakadu National Park. The major difficulty with this approach is that fire regimes vary according to the time elapsed since the summer rains, which have enormous variability in their timing and intensity (Taylor & Tulloch 1985).

Thus calendar months are a poor guide to fire regimes. Aborigines recognized seasons according to climatic conditions, not by the Gregorian calendar (Haynes 1985). The poor relationship between calendar months and fire regimes may explain why the maximum temperatures of fires in woodlands and forests measured by Braithwaite and Estbergs (1985) which occurred in the "mid" and "late" dry season are not statistically different.

Bowman and Brown (1986) have argued that there is a real need for enthno-ecological research to answer the "burning question" put by Horton (1981) of whether or not Aborigines manipulated the environment with fire. Haynes (1985) has conducted one of the few enthno-ecological studies in Australia. He noted that contemporary Aborigines at Maningrida burnt forests shortly after the wet season rains when fires were of low intensity. One case of a hot fire that was mistakenly lit in a woodland was treated casually by the Aborigines he was observing. The Aborigines explained that the woodland was "only desert". Haynes was unsure whether this attitude reflected the dearth of resources in the community (at the time?) or a philosophical acceptance of a mistake, an etic-emic dilemma common to anthropological research.

Braithwaite and Estbergs (1985) concur with Haynes' notion of Aboriginal early dry season burning. They suggest that following the demise of traditional Aboriginal culture the forests suffered the greatest change in fire regime with forests being burnt later in the dry season. The problem with this argument is that all fires that occur before the wet season build-up, when lightning storms are common, are lit by people. If there is a real difference in the timing of fires in forests and woodlands within any year then the difference either reflects a current cultural preference or indicates that woodlands dry out earlier than the forests. Twelve years of experience associated with the early season burns at Munmarlary has shown that woodlands are more difficult to burn than forests possibly because the woodlands are less well drained and have a greater shrub biomass than the open forests. Some woodlands sampled by Braithwaite and Estbergs (1985) may have been situated on ridges, and would dry out earlier than the forests or low-lying woodlands, and thus be capable of carrying early season fires.

Braithwate and Estbergs (1985) suggest that traditional management was focused on forests because the woodlands were thought to be poor in resources and were "held in low esteem".

Although there is much ethno-ecological research to be done this conclusion is not supported by Taylor & Dunlop (1985) who note that the woodlands and forests are the most floristically rich communities they studied at Kakadu National Park. Nor is it supported by the data presented by Braithwaite and Estbergs (1985). Their third table shows that out of the 20 most common trees or shrubs in the forest or woodland eight are unique to each community. Of the 12 co-occurrences only two species, *E. tetrodonta* and *Planchonia careya* are significantly associated ($P<0.05$) with the forest. The vertical distribution of shrubs and trees is more even in the woodland than in the forest, suggesting a greater biomass of shrubs in the former community. This distinction may be related to edaphic rather than fire differences. However the Munmarlary experiment clearly shows that a fire-free interval greater than two years has a major impact on community structure in contrast to the annual and biennial regimes. Late and mid season fires are hotter than early season fires (Braithwaite & Estbergs 1985) and are probably important in eroding the margins of *Callitris* and monsoon forests because at this time the gradient in fuel moisture content is shallowest. However it is unlikely that occasional late season fires would be capable of destroying these communities given their powers of regeneration (Bowman & Dunlop 1986). More research is required to understand the relationship between fire and monsoon forest ecotones (*cf.* Unwin *et al.* 1985).

Faunistic Responses

Braithwaite and Estbergs (1985) note that structure and floristics are likely to be important in controlling faunal abundance and diversity. There have been few studies of animal-plant community relations in the Northern Territory coastal forests.

Kerle (1985) found that the northern brushtail possum *(Trichosurus arnhemensis)* favoured shrubby *E. miniata* forests that developed on sites with water tables close to the surface. She found that these sites had a greater fire-free interval than floristically distinct grassy forests.

Woinarski *et al.* (in press) studied the relationship between bird species and plant communities at Howards Peninsula, near Darwin. They found that the shrubby eucalypt forests had twice as many bird species as the grassy eucalypt forests. Neither of these studies is able to isolate the effect of vegetation structure from floristic composition. Kerle demonstrates the interrelated nature of these variables by showing that a classification based on presence of plant species also segregates stands according to their structure. Similarly the plant communities studied by Woinarski *et al.* were differentiated by Wilson & Bowman (1987) according to structure and floristics. The Munmarlary experiment provides the opportunity to break the nexus between floristic composition and vegetation structure and to explore the relative importance of these components on associated fauna. Anderson (pers. comm.) found that with the exception of ant species there was no systematic change in ground foraging invertebrates. Ant species abundance showed a difference between the unburnt and late annual and early biennial fire treatments with hot climate specialists favouring burnt sites and cryptic and sub-cryptic species favouring unburnt sites. However, site factors also appear to explain some of the variation in these groupings of ants. To overcome the problem of site factors confounding treatment effects, continuous sampling will be required to ascertain if animal species react to structural changes induced by various fire-free intervals.

Conclusion

The eucalypt forests that occur in the coastal regions of the Northern Territory vary floristically in response to edaphic conditions. The edaphically determined communities also interact with fire which is a pervasive factor, originating early in the winter drought from people and, before the summer rains, from lightning. Fire intensity increases with the time elapsed since the summer rains. Experiment has shown that annual fires at these different times has little effect upon the structure or floristics of eucalypt forest understoreys, which are controlled by dry season moisture supply and overwood competition. One, however, cannot be sanguine about the impact of intense late season fires on monsoon forests and *Callitris* communities. Fire-free intervals greater than biennial have significant effect on eucalypt forest structure but not on floristic composition, and upon the abundance of cryptic and hot climate ant species. It is unclear if fire-induced structural changes impacts upon vertebrates, although the floristically and structurally distinct eucalypt forest understoreys that are associated with elevated water tables are faunistically richer.

Although some monsoon forest species are drought tolerant and like *Callitris* capable of expansion into unburnt eucalypt forests it is unlikely that they could replace the eucalypts given the high natural incidence of fire and the predilection of people to start dry season fires. Similarly the frequent burning is unlikely to eliminate the eucalypt forests given that the fires consume grass fuels and burn quickly at low temperature. The eucalypts, unlike the monsoon forest species and *Callitris,* are able to survive these fires. Why grasslands occur within forests (e.g. Yapilika, Melville Island) is a mystery, the resolution of which will help in understanding of eucalypt forest ecology. Similarly further work on the eucalypt-monsoon forest ecotone and the competitive relations of eucalypts with the broadleaf shrubs, which are important food plants for people and animals, will help in the solution of the enigma of geographically stable forest communities in a temporally tumultuous environment.

Acknowledgements

Bill Freeland has encouraged my study of northern eucalypt forests. I wish to thank John Woinarski, Dick Braithwaite, Glen Wightman, Clyde Dunlop, John Taylor and Bruce Wilson for stimulating argument.

References

Beard J. S. (1976) The monsoon forests of the Admiralty Gulf, Western Australia. *Vegetatio* **31**, 177-92.

Beard J. S. Clayton-Greene K. A. & Kenneally K. F. (1984) Notes on the vegetation of the Bougainville Peninsula, Osborn and Institut Islands, North Kimberly District, Western Australia. *Vegetatio* **57,** 3-13.

Bell A. (1981) Fire damages top-end forests. *Ecos* **30,** 18-20.

Bowman D. M. J. S. (1986) Stand characteristics, understorey associates and environmental correlates of *Eucalyptus tetrodonta* F. Muell. forests on Gunn Point, northern Australia. *Vegetatio* **65,** 105-13.

Bowman D. M. J. S. & Brown, M. J. (1986) Bushfires in Tasmania: A botanical approach to anthropological questions. *Archaeology in Oceania.* **21,** 166-71.

Bowman D. M. J. S. & Dunlop C. R. (1986) Vegetation patterns and environmental correlates in coastal forests of the Australian monsoon tropics. *Vegetatio* **65,** 99-104.

Bowman D. M. J. S. & Minchin P. R. (1987) Environmental relationships of woody vegetation patterns in the Australian monsoon tropics. *Aust. J. Bot.* **35,** 151-69.

Bowman D. M. J. S. & Wightman G. M. (1985) Small scale vegetation pattern associated with a deeply incised gully, Gunn Point, Northern Australia. *Proc. R. Soc. Qd.* **96,** 63-73.

Bowman D. M. J. S., Wilson B. A. & Davis G. W. (1988a) Response of *Callitris intratropica* R. T. Baher & H. G. Smith to fire protection, Murgenella, northern Australia. *Aust. J. Ecol.* **13,** 147-59.

Bowman D. M. J. S., Wilson B. A. & Hooper R. J. (1988b) Response of *Eucalyptus* forest and woodland to four fire regimes, Munmarlary, Northern Territory, Australia. *J. Ecol.***76,** 215-32.

Bowman D. M. J. S. & Wilson B. A. (in press) Fuel characteristics of coastal forests, Northern Territory, Australia. *J. Biogeogr.* **15.**

Braithwaite R. W. & Estbergs J. A. (1985) Fire patterns and woody vegetation trends in the Alligator Rivers Region of northern Australia. In: Ecology and management of the worlds savannas (eds J. C. Tothill & J. J. Mott) pp. 359-64. Australian Academy of Science, Canberra.

Braithwaite R. W., Dudzinski M. L., Ridpath M. G. & Parker B. S. (1984) The impact of water buffalo on the monsoon forest ecosystems in Kakadu National Park. *Aust. J. Ecol.* **9,** 309-22.

Carnahan J. A. (1976) Natural Vegetation. In: *Atlas of Australian Resources,* Division of National Mapping, Canberra.

Christian C. S. & Stewart G. A. (1953) In: *General report on survey of Katherine-Darwin region, 1946.* Land Research Series No. 1. CSIRO, Australia.

Clayton-Greene K. A. & Beard J. S. (1985) The fire factor in vine thicket and woodland vegetation of Admiralty Gulf region, north-west Kimberley, Western Australia. *Proc. ecol. Soc. Aust.* **13,** 225-30.

Fox R. E. & Clark N. B. (1972) The incidence of termites in eucalypts of the Darwin area. *Aust. For. Res.* **5,** 29-36.

Gilbert J. M. (1959) Forest succession in the Florentine Valley, Tasmania. *Pap. Proc. Roy. Soc. Tas.* **93,** 129-51.

Haynes (1985) The pattern and ecology of *munwag*: traditional Aboriginal fire regimes in north-central Arnhemland. *Proc. ecol. Soc. Aust.* **13,** 203-14.

Horton D. R. (1982) The burning question: Aborigines, fire and Australian ecosystems. *Mankind* **13,** 237-51.

Kerle J. A. (1985) Habitat preference and diet of the northern brushtail possum *Trichosurus arnhemensis* in the Alligator Rivers region, N.T. *Proc. ecol. Soc. Aust.* **13,** 161-76.

Kikkawa J., Webb L. J., Dale M. B., Monteith G. B., Tracey J. G. & Williams W. T. (1981) Gradients and boundaries of monsoon forests in Australia. *Proc. ecol. Soc. Aust.* **11,** 39-52.

Kirkpatrick J. B., Bowman D. M. J. S., Wilson B. A. & Dickinson K. J. M. (1987) A transect study of the *Eucalyptus* forests and woodlands of a dissected sandstone and laterite plateau near Darwin, Northern Territory. *Aust. J. Ecol.* **12,** 339-59.

Langkamp P. J., Ashton D. H. & Dalling M. J. (1981) Ecological gradients in forest communities on Groote Eylandt, Northern Territory, Australia. *Vegetatio* **48,** 27-46.

Pedley L. & Isbell R. F. (1971) Plant communities of Cape York Peninsula. *Proc. R. Soc. Qd.* **82,** 51-74.

Russell-Smith J. (1985) A record of change: studies of Holocene vegetation history in the South Alligator Rivers region, Northern Territory. *Proc. ecol. Soc. Aust.* **13,** 191-202.

Russell-Smith J. (1986) *The forest in motion: exploratory studies in Western Arnhem Land, Northern Australia.* Unpublished PhD. Thesis, Australian National University.

Russell-Smith J. & Dunlop, C. R. (1987) The status of monsoon vine forest in the Northern Territory: a perspective. In: *The rainforest legacy. Australian National Rainforests Study volume 1 — The nature, distribution and status of rainforest types.* Pp. 227-88. Australian Government Publishing Service, Canberra.

Specht R. L., Salt R. B. & Reynolds S. T. (1977) Vegetation in the vicinity of Weipa, North Queensland. *Proc. R. Soc. Qd.* **88,** 17-38.

Stocker G. C. (1966) Effects of fires on vegetation in the Northern Territory. *Aust. For.* **30,** 223-30.

Stocker G. C. (1971) The age of charcoal from old jungle fowl nests and vegetation change on Melville Island. *Search* **2,** 28-30.

Stocker G. C. (1976) *Report on cyclone damage to natural vegetation in the Darwin area after cyclone Tracey 25 December 1974.* Forestry and Timber Bureau Leaflet No. 127.

Story R. (1976) Vegetation of the Alligator Rivers Area. In: *Lands of the Alligator Rivers Area, Northern Territory* (eds R. Story, R. W. Galloway, J. R. McAlpine, J. M. Aldrick & M. A. J. Williams). Land Research Series No. 38, CSIRO, Australia.

Story R., Galloway R. W., McAlpine J. R., Aldrick J. M. & Williams M. A. J. (1976) *Lands of the Alligator Rivers Area, Northern Territory*. Land Research Series No. 38, CSIRO, Australia.

Story R., Williams M. A. J., Hooper A. D. L., O'Ferall R. E. & McAlpine R. J. (1969) *Lands of the Adelaide-Alligator Area, Northern Territory*. Land Research Series No. 25, CSIRO, Australia.

Taylor J. A. & Dunlop C. R. (1985) Plant communities of the wet-dry tropics of Australia: the Alligator Rivers region, Northern Territory. *Proc. ecol. Soc. Aust.* **13**, 83-127.

Taylor J. A. & Tulloch D. (1985) Rainfall in the wet-dry tropics: Extreme events at Darwin and similarities between years during the period 1870-1983. *Aust. J. Ecol.* **10**, 281-95.

Unwin G. L., Stocker G. C. & Sanderson K. D. (1985) Fire and the forest ecotone in the Herberton highland, north Queensland. *Proc. ecol. Soc. Aust.* **13**, 215-24.

Webb L. J. (1968) Environmental relationships of the structural types of Australian rain forest vegetation. *Ecology* **49**, 296-311.

Webb L. J. & Tracey J. G. (1981) Australian rainforests: Patterns and change. In: *Ecological biogeography of Australia* (ed. A. Keast) pp. 605-94. Dr W. Junk, The Hague.

Wilson B. A. & Bowman D. M. J. S. (1987) Fire, storm, flood and drought: The vegetation ecology of Howards Peninsula, Northern Territory, Australia. *Aust. J. Ecol.* **12**, 165-74.

Woinarski J. C. Z., Tidemann S. C. & Kerin S. H. (In press) Birds in a tropical mosaic: the distribution of bird species in relation to vegetation patterns at Howards Peninsula, Northern Territory. *Aust. Wildl. Res.*

Proc. Ecol. Soc. Aust. 15:161-165

An aerial photographic study of the long-term effect of wildfires on Magnetic Island

C. S. SANDERCOE

Queensland National Parks and Wildlife Service, P.O. Box 42, Kenmore, Qld. 4069

Abstract

A method was devised to quantify patterns of change in canopy cover over time, in relation to the known fire history of Magnetic Island (19°08′S, 148°50′E). A time series of aerial photographs was examined and five plots representing a range of fire histories were selected. The plots were then assessed for "relative openness" of the canopy, that is percentage absence of tree and shrub cover. Aerial photographic interpretation showed that the areas which had been subjected to the most frequent wildfires also showed the greatest increase in "relative openness". Ground confirmation supported the technique by showing extensive tree and shrub death in areas most affected by fire.

Introduction

In recent years considerable interest has developed in the use of fire history and the effect of fire on vegetation, to help provide a background on which to base fire management practices. Without precise long-term data, this task usually proves difficult. In some areas in Australia, especially close to the coastline, aerial photography is available over a number of decades, and it is possible to measure gross structural change of the vegetation from these photographs. This paper addresses the problem of measuring these changes in areas of known fire histories, using data available from Magnetic Island.

Materials and Methods

Study area

Magnetic Island is a continental granitic island 5 km off the coast from Townsville, Queensland, Australia. Details of climate, geology, landform and vegetation are given in Sandercoe (in prep.). A permanent population of approximately 1 000 people is situated in the lowland bay areas from where most of the fires originate. Some fires spread from there into the hilly areas covered by Magnetic Island National Park.

Aerial photographs and fire records

Sets of aerial photographs taken during the years 1941, 1952, 1961, 1974 and 1980 were available. Records of wildfires taken by national park staff since 1964 were also available. These records usually included area burnt, duration, intensity and cause of fire. Some were more comprehensive than others, but it was possible to build a composite map of all major wildfires to 1980. These are summarized in Fig. 1. Fires and cyclones are the only major causes of vegetation damage known on the island since settlement.

Choice of plots

Several areas of similar mixed eucalypt woodland that appeared to be intact in the 1941 aerial photographs, but which had experienced varying fire histories subsequently, were selected as the basis for this study. Altogether five triangular-shaped plots were accurately located (Fig. 1), using reliable features such as boulders to pinpoint angle boundaries. A triangular plot shape was chosen because it was considered the most efficient and replicable area over rough terrain. Three plots were located in the area behind Horseshoe Bay where the greatest number of fires have been recorded; one plot was adjacent to

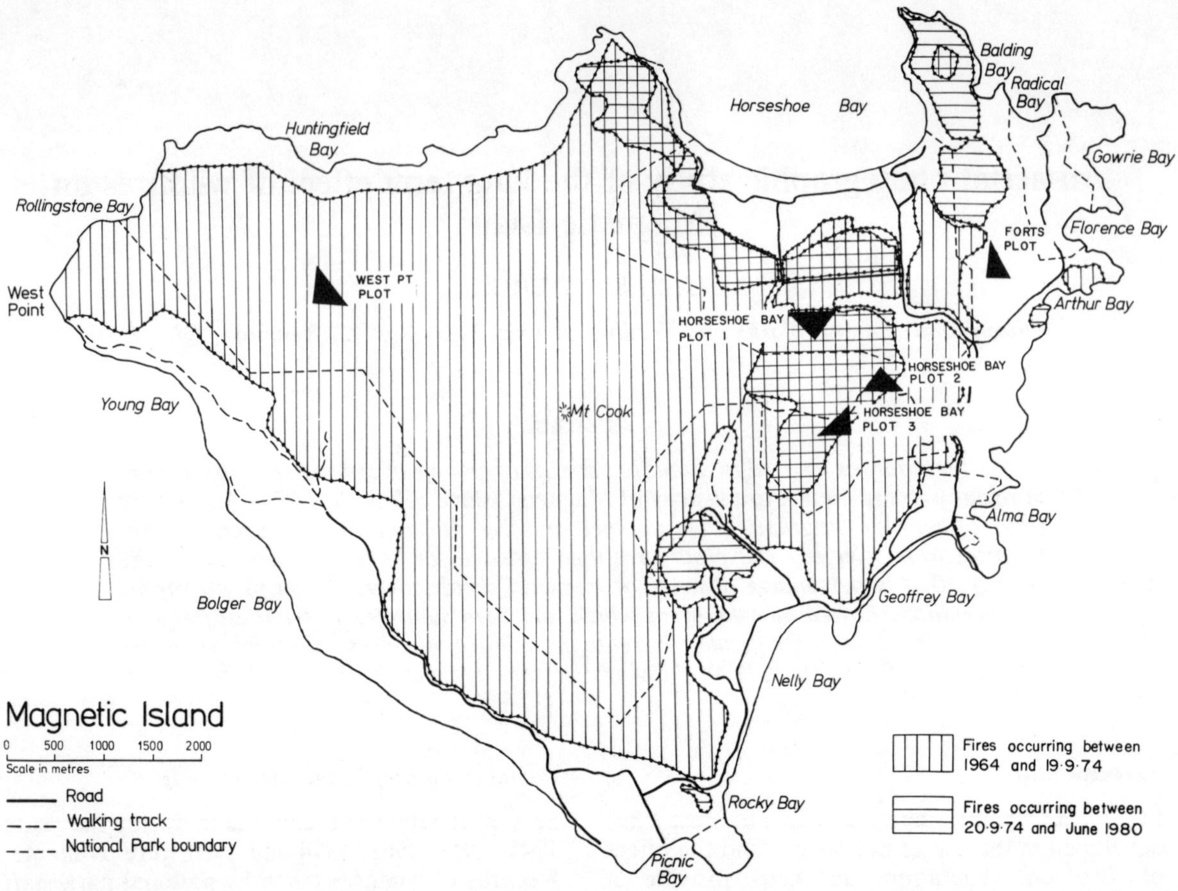

FIG. 1. Magnetic Island fire records 1964-74 and 1974-80. Location of plots assessed for "relative openness" are shown.

West Point with one recorded fire; and a control plot was selected in the Forts area where there have been no recorded fires in the last 40 years.

Technique

These plots were then assessed in the five available photographs for the degree of "openness", that is absence of tree and shrub cover. Trees and shrubs were represented by dark images whereas grassland was represented by light images. As original photographs ranged in scale from approximately 1:12 000 to 1:30 000, a zoom stereomicroscope was used to magnify each plot to the same level. A 10×10 eyepiece micrometer was used to divide each plot into 500 to 1 000 squares. Greater than 50% cover of a light image per square provided a register for "openness". As a test to detect possible measuring bias, the measurements were conducted by two people counting each plot twice. This resulted in a total of 100 measurements (five photographs, five triangles, two persons counting twice).

Analyses

A significant "between-count" difference for each person occurred in six out of the 50 cases (Chi-square test with Yates' correction, d.f.=1 and $p<0.05$). A comparison between people, using pooled counts for each person, showed a significant difference in three of the 25 cases. Two of these three also had shown a "between-count" difference for either one or both persons. A test of "individual-bias" by regression analysis showed there was no evidence of a consistent bias between the two people carrying out the counts.

Since the above tests showed only minor inconsistencies in the data, the means and standard deviations were calculated using pooled data and

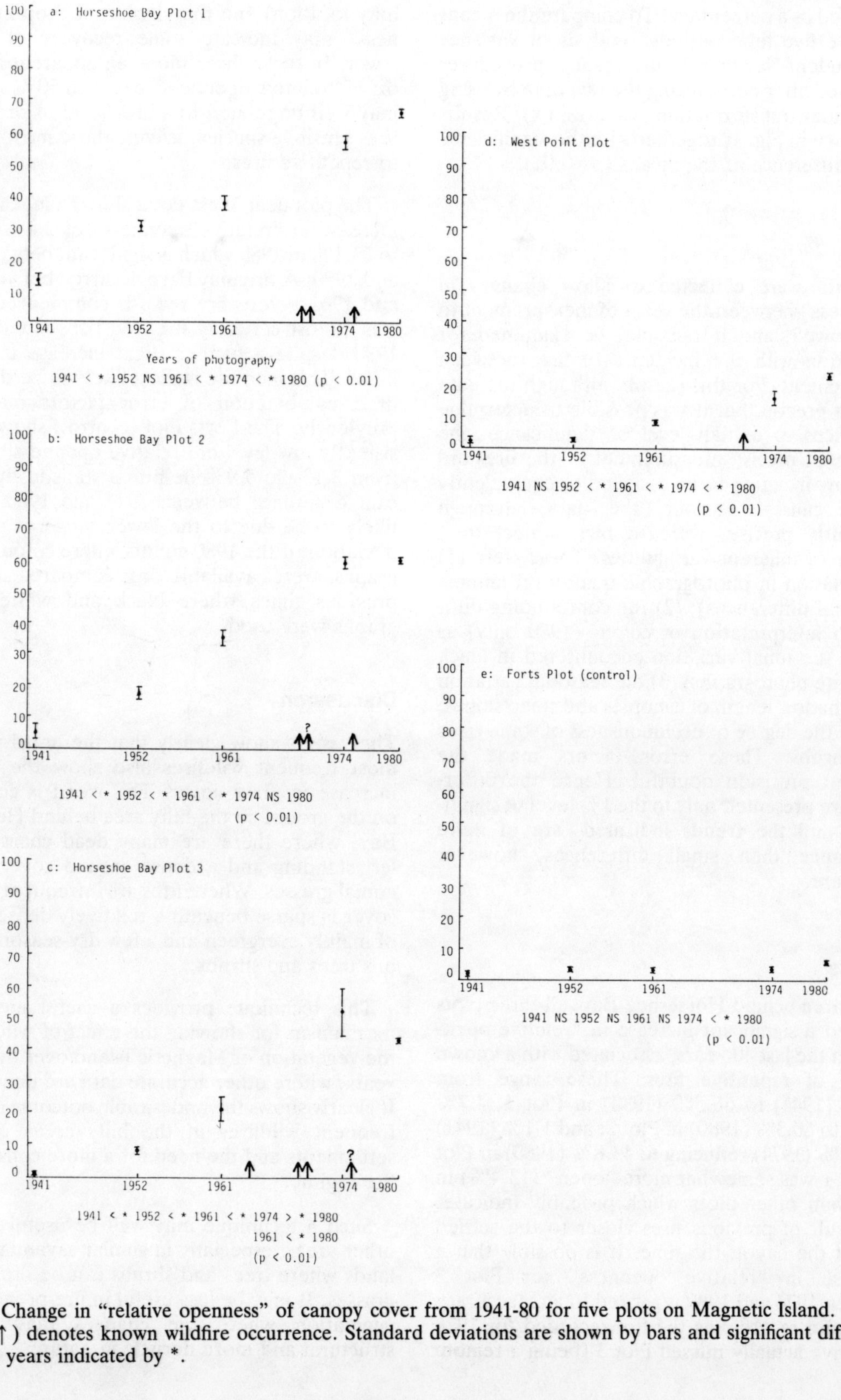

FIG. 2. Change in "relative openness" of canopy cover from 1941-80 for five plots on Magnetic Island. Vertical arrow (↑) denotes known wildfire occurrence. Standard deviations are shown by bars and significant differences between years indicated by *.

expressed as a percentage. To compare the means over the five time periods, analysis of variance and Student-Newman-Keuls testing procedures were used after normalizing the raw data by using the angular transformation ($x' = \arcsin x$). Results are shown in Fig. 2, together with the significance of the difference of the means ($p<0.01$).

Error factors

The data were collected to allow changes in "openness" between the years of measurement to be followed, and if possible, be examined for association with the incidence of fires between measurement. For this reason, although the data were so precise that it was possible to determine differences to a high level of significance, the absence of precise measurement of the fires did not allow meaning to be assigned to equivalently precise causes. Further, the data, although apparently precise, were in fact subject to a number of inherent variabilities. These were (1) the variation in photographic quality (graininess and tonal differences); (2) the confounding difficulty in interpretation of colour (1980 only) as well as the tonal variation encountered in black and white photographs; (3) the seasonal variation in the shadow length of canopies and steep slopes; and (4) the degree of deciduousness of some trees and shrubs. These error factors made the apparent precision doubtful. Hence the differences are presented only to the 1% level of significance, and the trends indicated, are of more importance than small differences, however significant.

Results

In the area behind Horseshoe Bay, all three plots revealed a significant increase in "relative openness" in the last 40 years, associated with a known history of repetitive fires. These range from 13.4% (1941) to 66.5% (1980) in Plot 1, 4.7% (1941) to 60.3% (1980) in Plot 2, and 1.1% (1941) to 52.4% (1974) reducing to 43.8% (1980) in Plot 3. Plot 1 was somewhat more "open" (13.4%) in 1941 than other plots which probably indicates the result of previous fires closer to the settled area of the bay at the time. It is possible that a decrease in "relative openness" for Plot 3 between 1974 and 1980 is related to an inaccuracy in the fire records, as the fire recorded for 1976 may have actually missed Plot 3 (being a remote hilly location) and this significant drop in "openness" may indicate some recovery of canopy cover. In these three plots, an apparent levelling off of "relative openness" between 60% and 70% may well be related to a high level of removal of fire- sensitive species, leaving those most resilient to repetitive fires.

The plot near West Point shows a much smaller increase in "relative openness" of 2.6% in 1941 to 23.1% in 1980 which is significant between 1952 and 1980. A fire may have occurred between 1952 and 1961 before fire records commenced, but it was almost certainly fire-free between 1974 and 1980 despite a further slight increase in "openness". This increase is most likely to be the result of a combination of error factors mentioned previously. The Forts Plot (control) shows a consistently low level of "relative openness" ranging from 2.1% to 5.9% despite a statistically significant difference between 1974 and 1980. This is likely to be due to the lower range of standard deviation of the 1980 counts where colour photographs were available, as compared with all previous times where black and white photographs were used.

Discussion

The results show clearly that the areas with the most frequent wildfires also show the greatest increase in grass cover. This trend is confirmed on the ground in the hilly area behind Horseshoe Bay, where there are many dead charred trees left standing and a dense ground cover of perennial grasses. Where fires are infrequent, ground cover is sparse beneath a relatively dense canopy of mainly evergreen and a few dry-season deciduous trees and shrubs.

This technique provides a useful monitoring mechanism for showing the effect of wildfires on the vegetation of Magnetic Island over the last 40 years, where other accurate data are unavailable. It clearly shows the undesirable outcome of recent frequent wildfires in the hilly areas closer to settlements and the need for a more conservative fire regime.

Such a technique may well be useful in many other areas, especially in similar savannah woodlands where trees and shrubs can be replaced by grasses. It may be less useful in fire-prone heathy vegetation where the changes may be less structural and more floristic in nature.

Acknowledgements

I gratefully acknowledge the assistance of Ms J. Curnow and Dr W. Drake for undertaking measurements of "openness", and Ms C. Nance for analysing the data. Mr G. Blackman provided useful suggestions in devising this technique.

Reference

Sandercoe C. S. (in prep.) Vegetation of Magnetic Island. *Queensland National Parks and Wildlife Service. Technical Bulletin Series.*

Proc. Ecol. Soc. Aust. 15:167-175

Comparisons of the extent of fire in different land management systems in the Top End of the Northern Territory

A. J. PRESS

Australian National Parks and Wildlife Service, Kakadu National Park Headquarters, P.O. Box 71, Jabiru, N.T. 0886

Abstract

Landsat imagery was used to analyse the extent of fires in lowland forests in the Top End of the Northern Territory. Comparisons were made between Arnhem Land, Kakadu National Park (Stages I and II) and pastoral leases to the west of Kakadu, for the years 1980 to 1985 inclusive. The extent of early dry season and late dry season fires was analysed.

Analyses of variance showed significant interactions between the season of fire (early dry and late dry) and the land management system, and between season and years. There was great variability in the amount of fires in different land management systems in different years.

A significant negative correlation between the amount of early dry season burning and the amount of late dry season burning was found.

Each year over 50% of all lowland forests in all land management systems carried fire.

Introduction

Fires occur annually across large areas of the monsoonal tropics of the Top End. These fires are chiefly restricted to the dry season and early wet season (Braithwaite & Estbergs 1985; Day 1985). Before settlement of the Top End by Aboriginal people (*c.* 40 000 BP) fires probably were restricted to the months of greatest lightning activity (October to December, Braithwaite & Estbergs 1985). It is generally accepted that Aboriginal people dramatically changed the fire regime in Northern Australia. Jones (1969), Haynes (1985) and Stevenson (1985) have shown that fire was and still is used extensively by Aboriginal people of northern Australia for hunting and gathering, movement and ritualistic reasons. Recent ethnographic and ecological studies have pointed to changes in fire patterns in the Top End over the last 100 years. Braithwaite & Estbergs (1985) have produced a summary of pre-Aboriginal, Aboriginal and contemporary fire regimes. They found that tall open forests were subjected to later (=hotter) dry season fires than under traditional Aboriginal regimes but that contemporary and Aboriginal fire regimes in the woodlands are similar.

This paper analyses the extent of fires in different land management systems — pastoral, national park and Arnhem Land — for the years 1980 to 1985 inclusive. The study concentrates on the lowland forests — predominantly woodlands and tall open forests. Sandstone rocklands and floodplain habitats are less prone to fire than the lowland forests (Stocker 1966; Braithwaite & Estbergs 1985; Day 1985). Data are derived from interpretation of Landsat imagery.

Methods

Study area

The study area extended from near the King River in the east and north (approximately 133°30′E, 11′30′S) to Cape Hortham in the west and north (approximately 131°15′E, 12°00′S) and south to approximately 13°20′S. The area falls within the northern tallgrass zone of Tothill &

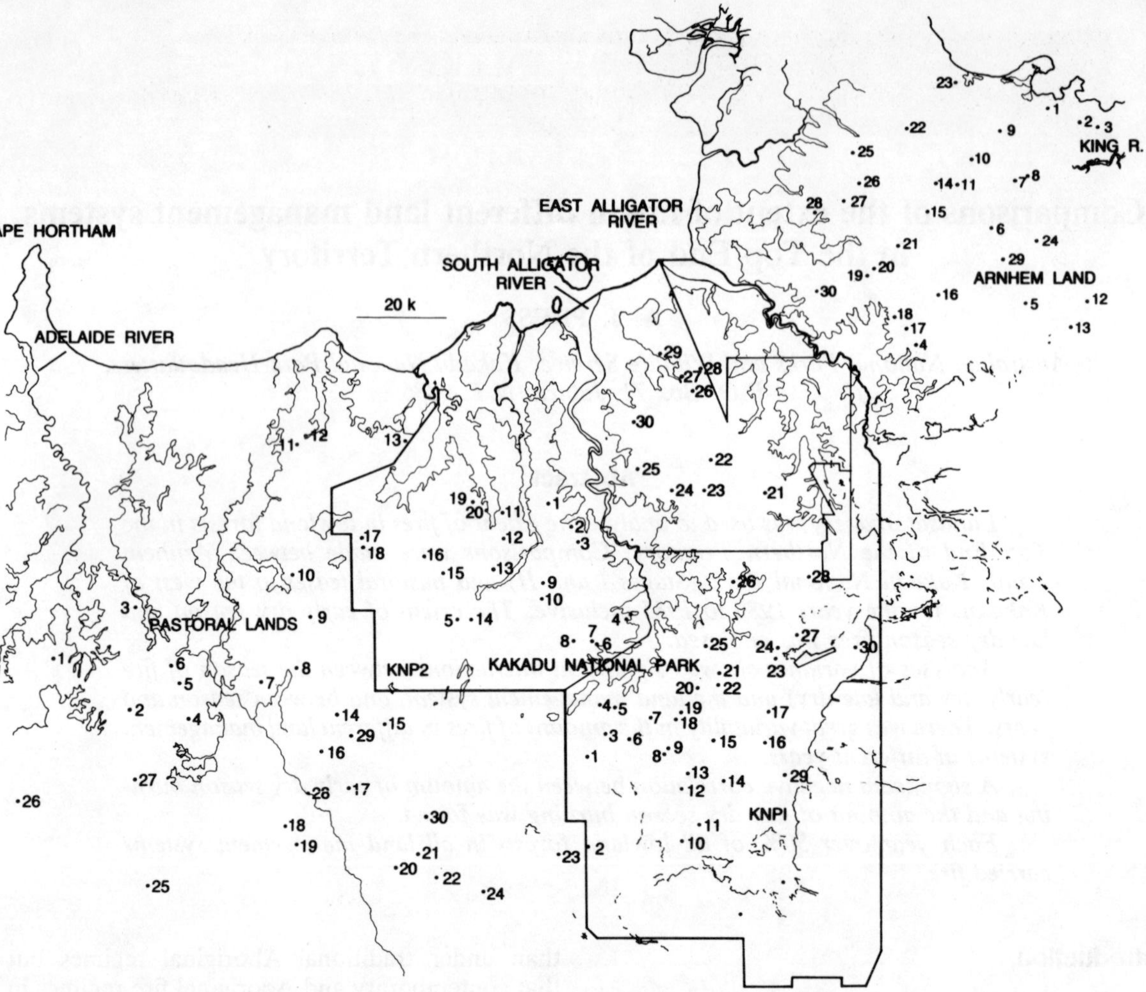

FIG. 1. The study area. The Sampling points in each land management system are numbered 1-30.

Mott (1983) and Mott & Tothill (1984). The seasonal changes from wet to dry are dramatic but very variable (Taylor & Tulloch 1985). Mean annual rainfall is around 1 500 mm but decreases with distance from the coast to about 1 300 mm in the southern part of the study area. The bulk of the rain falls from December to March with virtually no rain falling from June to August (but note the variability in rainfall shown by Taylor & Tulloch 1985).

The study area was divided into four land management systems: Arnhem Land Aboriginal Land; Kakadu National Park (Stage I), declared in 1979; Kakadu National Park (Stage II), declared in 1984 and prior to 1984 included the Munmarlary and Mudginberry Pastoral Leases and Commonwealth Land; and Pastoral Land. These areas are shown in Figure 1.

Analysis of Landsat Imagery

Landsat imagery covering all or part of the study area was obtained for the years 1980 to 1985. The scale of the imagery obtained was either 1:500 000 or 1:250 000. The imagery was either single band (Band 7) or colour composite (Bands 4, 5 and 7).

Day (1985) discusses the problems associated with the interpretation of fire patterns using this imagery. In summary some low intensity fires or fires of a small size are difficult to recognize or are undetectable. Precise location of points from image to image was difficult due to fluctations in satellite orbit (900 ± 30 km). Difficulty was also found in interpreting Landsat imagery along water courses and in some rocky hill areas. Fire

FIG. 2. Landsat Image of part of the study area taken on 2nd August 1984. The image is ALS 105-69B (Band 7). A, areas recently burnt by early dry season fires. B, a fire still burning at time of image. C, areas burnt by fires at least one month prior to image.

patterns were also difficult to interpret in areas that may have carried fire early in the dry season and a subsequent low intensity fire later in the dry season. Despite these difficulties, fire could be easily detected in most instances without difficulty (Fig. 2), especially with the aid of sequential images for the area (Day 1985).

Data collection

In order to analyse the fire patterns in the lowland forests in the four land management systems, 30 randomly chosen points were placed in each area. Floodplains, the sandstone rocklands and outliers, and southern hills were excluded (see Fig. 1). The points were chosen by dividing a composite 1:250 000 map of the area drawn onto translucent graph paper, into a grid of 1 kilometre squares numbered along two borders of the map. The points were chosen from a random numbers table. Each point (numbered 1-30 in each area — Fig. 1) represented a square of 1 kilometre square. This map was overlaid on the 1:250 000 Landsat images. Through each point along two axes north-south and east-west, in the shape of a +, nine squares, including the central point, were sampled for presence or absence of fire. If a fire had burned 50% or more of a square it was given a score of one; if not a score of zero. For each of the sampling points a maximum score of nine was obtained. These points were transferred to a composite map of a scale of 1:500 000 and the same procedure adopted for sampling Landsat imagery of that scale.

TABLE 1. Dates and coverage of Landsat imagery used for sampling of fires. In 1984 and 1985 separate images were needed to cover all of the study area, hence two dates given in the table.

Year	KNP1	KNP2	West	Arnhem
1980	10.viii 3.x	10.viii 3.x	10.viii 3.x	Incomplete coverage
1981	5.viii 16.x	5.viii 16.x	5.viii 16.x	Incomplete coverage
1982	9.viii 21.x	9.viii 21.x	Incomplete coverage	Incomplete coverage
1983	8.viii 11.x	8.viii 11.x	8.viii 11.x	Incomplete coverage
1984	2.viii/26.vii 21.x/30.x	2.viii 21.x	2.viii 21.x	26.vii 30.x
1985	5.viii/14.viii 8.x/17.x	5.viii 8.x	5.viii 8.x	14.viii 17.x

The fire season was divided into the early dry season (before the end of the first week in August) and the late dry season (to the end of October). Data was collected as described above using the Landsat imagery available nearest these two times. Cloud cover late in October often made interpretation impossible so that an earlier image (e.g., mid-October) had to be used. The dates of the early and late dry season images used are given in Table 1.

Data analysis

Data from each of the 30 points in each land management system were analysed by analyses of variance. Three factors were analysed: Season (early dry season and late dry season), Land Management System, Kakadu National Park (Stage I) — KNP1; Kakadu National Park (Stage II) — KNP2; Western Pastoral Leases — West; and Arnhem Land — Arnhem, for the years (1980 to 1985). Because of incomplete coverage in all years of all land management systems (Table 1), the analyses were done in two parts. The first analysis of variance (Table 2) analyses Kakadu National Park (Stage I), Kakadu National Park (Stage II) and the Western Pastoral Leases for the the years 1980, 1981 and 1983. The second analysis of variance (Table 3) analyses all four land management systems for the years 1984 and 1985. The Landsat imagery for 1982 covered only KNP1 and part of KNP2 and was not included in the analyses of variance.

In order to meet the assumption of independence in the analyses, 10 of the 30 points in each land management system were randomly chosen for each season in each year. As all factors were fixed, this allowed each factor and interaction to be tested against residual mean squares with 162 (Table 2) and 144 (Table 3) degrees of freedom. Where significant interactions were found, the interaction means were tested using the protected Least Significant Difference (Snedecor & Cochran 1980).

TABLE 2. Analysis of variance, extent of fires at the sampling points in KNP1, KNP2 and West in the early dry season and late dry season for the years 1980, 1981 and 1983. Probabilities: *P<0.05; **P<0.01; ***P<0.001; NS,P>0.05.

Source of Variation	Sums of Squares	Degrees of freedom	Mean Squares	F.ratio	
Season	6.81	1	6.81	<1	NS
Land	11.48	2	5.74	<1	NS
Years	45.65	2	22.83	1.92	NS
Season×Land	176.94	2	88.47	7.45	**
Season×Years	212.04	2	106.02	8.92	**
Land×years	40.32	4	10.08	<1	NS
Land×Season×Years	48.46	4	12.12	1.02	NS
Residual	1 925.30	162	11.88		
TOTAL	2 467.00	179			

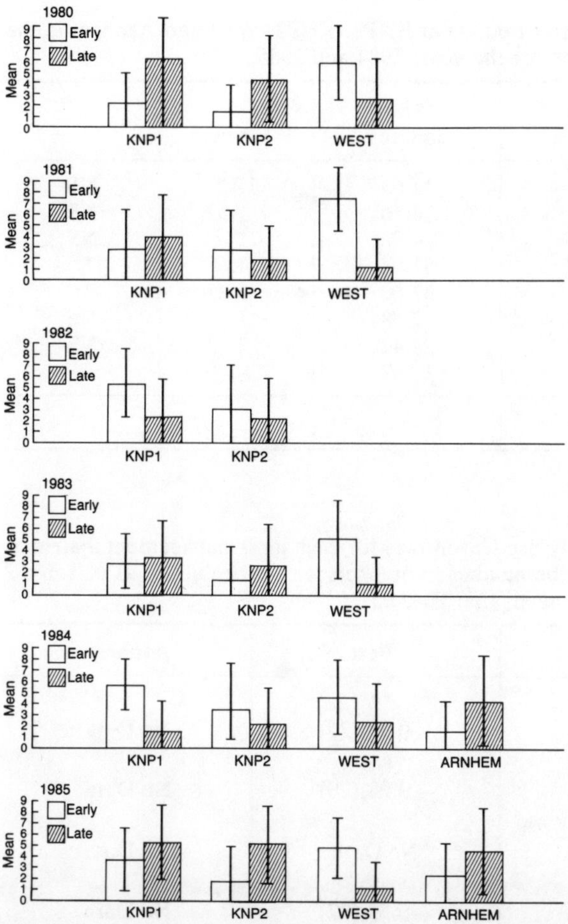

FIG. 3. Histograms showing extent of early and late dry season fires. Open bars, early dry season fires; hatched bars, late dry season fires. Standard deviations are shown for each histogram. KNP1, Kakadu National Park (Stage I), KNP2, Kakadu National Park (Stage II); West, the western pastoral leases; Arnhem, Arnhem Land. Note that a value of 9 is equivalent to 100% of all sampling points having recorded fire.

Data for each point in each land management system for each season in each year were pooled and averaged to produce the histograms in Figure 3. The standard deviations were calculated from the scores of each of the 30 points. In some cases not all 30 points were included due to incomplete coverage of the image or cloud cover. The number of points actually sampled is given in Table 4. Incomplete sampling of all 30 points did not affect the analyses of variance as only 10 points for each season were sampled for the analyses and these were allocated randomly.

Correlation analyses between early dry season scores and late dry season scores at each point were undertaken in each land management system in each year (Table 4), to test for any relationship between the amount of early dry season burning and the subsequent amount of late dry season fires. In these analyses x=the amount of early dry season burning at one point (from 0 to 9) and y=the amount of late dry season burning at the same point. On the basis of a significant relationship in most cases (Table 5) and overall correlation was undertaken by randomly selecting 160 points from all points sampled in all areas in all years. These data were pooled to produce the overall correlation. Figure 4 shows the overall relationship between early dry season burning (x) and late dry season burning (y) in this analysis.

Results

Both the analyses of variance (Tables 2 and 3) showed that there were significant interactions between Seasons and Land, and between Seasons and Years. No other factors were significant. These results indicated that the amounts of early and late dry season fire varied between land management systems and between years. The results of the protected LSD tests for the first analysis (KNP1, KNP2 and West for 1980, 1981 and 1983) are given in Tables 5 and 6. The results show that there were no significant differences between the three land management units in the amounts of late dry season fires averaged over the three years, but that significantly more early dry season burning occurred in the Western Pastoral Leases than in the two parts of Kakadu National Park (Table 5). Table 6 shows that there were significantly more late dry season fires in 1980 (averaged over all land management systems) than in 1981 and 1983. This result corresponds with significantly more early dry season fires in 1980 than in 1981, although the amount of early dry season burning was not significantly different between 1981 and 1983, and between 1980 and 1983 (Table 6).

The results of the protected LSD tests for the second analysis of variance (KNP1, KNP2, West and Arnhem for 1984 and 1985) are given in Tables 7 and 8. The results show that there were significantly less late dry season fires averaged over years in the Western Pastoral Leases than in Kakadu National Park (Stage II) and Arnhem Land. There was no significant difference in the amount of late dry season fires between the Western Pastoral Leases and Kakadu National Park (Stage I), but KNP1 was not significantly different from KNP2 or Arnhem (Table 7).

TABLE 3. Analysis of variance, extent of fires at the sampling points in KNP1, KNP2, West and Arnhem in the early dry season and late dry season for the years 1984 and 1985.

Source of variation	Sums of squares	Degrees of freedom	Mean squares	F.ratio	
Season	15.63	1	15.63	1.42	NS
Land	43.73	3	14.56	1.32	NS
Years	4.90	1	4.90	<1	NS
Season×Land	153.13	3	51.04	4.65	**
Season×Years	57.60	1	57.60	5.25	**
Land×Years	22.15	3	7.38	<1	NS
Land×Season×Years	73.25	3	24.42	2.23	NS
Residual	1 579.60	144	10.97		
TOTAL	1 949.96	159			

TABLE 4. Correlations of late dry season fires against early dry season fires for each land management system in each year. The first number is the correlation co-efficient, the number in brackets is n. Probabilities as in Table 2 except that no asterisk means that P>0.05.

	KNP1	KNP2	West	Arnhem
1980	*** —0.73,(29)	—0.22,(26)	*** —0.95,(25)	No Data
1981	*** —0.68,(30)	* —0.43,(26)	*** —0.98,(20)	No Data
1982	** —0.54,(30)	* —0.43,(22)	No Data	No Data
1983	—0.35,(30)	—0.29,(30)	* —0.53,(22)	No Data
1984	* —0.42,(30)	** —0.56,(30)	*** —0.68,(30)	—0.30,(30)
1985	*** —0.84,(29)	*** —0.62,(30)	* —0.45,(30)	* —0.64,(30)

TABLE 5. Protected LSD tests of the mean frequency of fire for 1980, 1981 and 1983 (data pooled over years). (L), late dry season; (E), early dry season. The underlines indicate means which are not significantly different.

West(L)	KNP1(E)	KNP2(E)	KNP2(L)	KNP1(L)	West(E)
2.23	2.50	3.10	3.80	3.87	5.40

TABLE 6. Protected LSD tests of the mean frequency of fire for 1980, 1981 and 1983 (data pooled across land management units). Conventions as in Table 5.

1983(L)	1981(L)	1980(E)	1983(E)	1981(E)	1980(L)
2.17	2.30	2.90	3.60	4.67	5.43

TABLE 7. Protected LSD tests of the mean frequency of fire for 1984 and 1985 (data pooled over years). Conventions as in Table 5.

West(L)	Arnhem(E)	KNP2(E)	KNP1(L)	KNP2(L)	Arnhem(L)	West(E)	KNP1(E)
1.75	2.23	3.15	3.35	3.95	3.95	4.90	5.40

TABLE 8. Protected LSD tests of the mean frequency of fire for 1984 and 1985 (data pooled over land management units). Conventions as in Table 5.

1984(L)	1985(E)	1985(L)	1984(E)
2.54	3.25	4.08	4.25

The amount of early dry season fires was greatest in Kakadu National Park (Stage I) and lowest in Arnhem Land although the LSD test showed much overlap (Table 7). There were significantly more late dry season fires in 1985 (averaged over all land systems) than in 1984 but no significant differences in the amounts of early dry season fires (Table 8).

Figure 4 shows the mean amounts of early and late dry season fires in the different land management systems and different years. In general the amount of late dry season fires varied in all land systems over different years. The absolute amount of late dry season fire in Kakadu National Park (Stage I) generally fell between 1980 and 1984 but rose sharply in 1985. The pattern was similar in Kakadu National Park (Stage II). The amounts of early dry season fires varied between land management systems but usually were greatest in the Western Pastoral Leases and the extent of late dry season fires were least in these areas. The pattern of early and late dry season fires in Arnhem Land was similar in 1984 and 1985.

Table 4 shows that there is a negative relationship between the amount of late dry season fire and the amount of early dry season fire at any point. This relationship is significant in most combinations of Years and Land Management System (Table 4) and is highly significant overall ($r=-0.55$, $P<0.001$, $n=160$).

Discussion

The results of this study show that fires are widespread annually in all land management systems in the Top End of the Northern Territory. Each year more than 50% of the lowland forests were subjected to fires during the dry season. There was great variability in all land management systems (Arnhem Land, national park and pastoral land) in the relative amounts of early and late dry season fires each year, although the western pastoral leases generally had greater amounts of early dry season fires and less amounts of late dry season fires than the other land management systems.

Pastoral practices in the Top End rely on widespread burning to create green pick for cattle and buffalo, and for the removal of old rank forage (Andrew 1986). Hazard reduction burning to reduce risk of wildfire is also undertaken. Pastoralists burn extensively early in the dry season either on the ground along tracks and roads or with the assistance of the Northern Territory

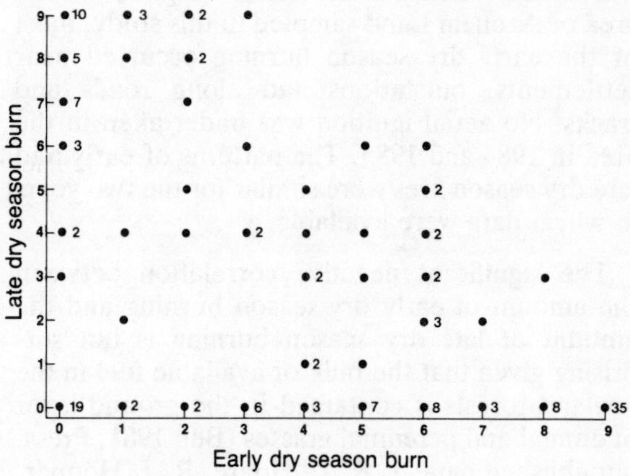

FIG. 4. The relationship between amount of early dry season fires and late dry season fires. The data are from 160 samples randomly selected from all areas in all years. The number next to the points on the figure is the number of times that value was recorded.

Bushfires Council using aerial ignition (M. Brennan, pers. comm.). Since the declaration of Kakadu National Park (Stage I) in 1979 and Kakadu National Park (Stage II) in 1984, an active programme of early dry season burning from roads and tracks, combined with aerial ignition, has been implemented (usually from May to July). The purpose of this programme is to re-establish as far as possible the patterns of traditional Aboriginal burning; to reduce the frequency and extent of late dry season wildfire and to create diversity of habitat (ANPWS 1980; HRSCEC 1984 p. 40; Day 1985). The amount of area burnt during the early dry season has varied from year to year and in general has been less than that carried out in the western pastoral leases (Fig. 3). The years in which the early dry season burning was most extensive in KNP1 were also the years in which late dry season wildfires were the least (1982 and 1984, Fig. 3). The patterns of burning in KNP1 found in this study during the years 1980 to 1982 were similar to those found by Day (1985). Differences in absolute amounts of burning between the two studies is most likely due to differences in sampling methods (e.g., sampling technique and timing of early and late dry season). Prior to the declaration of Kakadu National Park (Stage II) much of that area was used for buffalo harvesting and fire was used for both pastoral reasons (as above) and for ease of access for hunting and harvesting. Aerial ignition was undertaken by the Northern Territory Bushfires Council prior to the declaration of Stage II. The pattern of early and late dry season fires was similar in KNP1 and KNP2 (Fig. 3). In the area of Arnhem Land sampled in this study, most of the early dry season burning occurred near settlements, outstations and along roads and tracks. No aerial ignition was undertaken in the area in 1984 and 1985. The patterns of early and late dry season fires were similar for the two years in which data were available.

The significant negative correlation between the amount of early dry season burning and the amount of late dry season burning is not surprising given that the bulk of available fuel in the lowland forests is contained in the ground layer of annual and perennial grasses (Bell 1981; Press, unpublished data; J. R. L. Hoare, R. J. Hooper, N. P. Cheney & L. S. Jacobsen, unpublished report). The strong relationship, though, and inspection of Fig. 3, suggests that if the sole purpose of any fire management policy in the Top End was reduction in late dry season fires, one method of achieving this would be widespread and intensive early dry season burning.

The ecological consequences of widespread annual fires in the Top End have been the subject of much study and debate in recent years (Bell 1981; Taylor & Dunlop 1985; Braithwaite & Estbergs 1985; Bowman 1986; Bowman & Dunlop 1986; Bowman and Wightman 1986). Although the extent to which fire determines vegetation composition is debated among ecologists (e.g., Hoare *et al.* unpublished, and various publications by Bowman), much evidence exists that plant community structure is altered by fire (Braithwaite & Estbergs 1985). Fire intensity, and hence resultant fire effects, are largely determined by the time of the year at which a fire occurs (Bell 1981; Braithwaite & Estbergs 1985). Fires occurring late in the dry season are more intense (as measured by scorch height, patchiness, size and temperature) than those occurring early in the dry season or early in the wet season (Braithwaite and Estbergs 1985). Hoare *et al.* (unpublished) have shown that late dry season fires had the most marked effects on habitat structure.

The extent of annual fires shown in this study concurs with the findings of Braithwaite & Estbergs (1985) that annual fires were most common in open forest. They concluded that most fires in the woodland occurred early in the dry season (June) and late in the dry season (September), in open forest. The present study indicates that many of late dry season fires affect the open forests.

This study shows that early dry season fires can limit the extent of late dry season fires. Hoare *et al.* (unpublished) have shown that annual early dry season fires had less effects on habitat structure than annual late dry season fires. Braithwaite and Estbergs (1985) argued that the apparent rarity in the mid-storey of open forests of some species such as *Planchonia careya* and *Persoonia falcata* may have been the result of contemporary fire regimes favouring late dry season wildfires. In the absence of logistic and financial ability to fight major late dry season wildfires over such large and sparsely populated areas, early dry season burning may be used to the same effect.

The objectives of any fire management programme necessarily must meet the needs of the land manager. Thus in the pastoral leases the widespread use of early dry season burning achieves the aims of creating pasture and reducing late dry season wildfire. In Kakadu National Park fire management aims to create a diversity of fire history, reduce the extent of late dry season wildfire and re-establish the patterns of traditional Aboriginal burning. Braithwaite (unpublished report) recommends a diversity of fire regimes in Kakadu National Park from annual to pentennial and no fires to enhance diversity and to create conditions suitable for different floral and faunal groups. In Arnhem Land fire management is carried out by Aboriginal people around settlements and outstations and along roads and tracks (Haynes 1985, this study). The scale of fire management in Arnhem Land is smallest and most intense close to settlements and outstations and remote areas may be subjected to large wildfires late in the dry season. Aerial ignition could be used in these areas to reduce the extent of late dry season wildfire.

Acknowledgements

This study was undertaken as part of the research and monitoring programmes conducted by the Australian National Parks and Wildlife Service in Kakadu National Park. I wish to thank Professor J. D. Ovington, the Director of A.N.P.W.S. and the staff of Kakadu National Park. John Day began the process of acquiring Landsat imagery and his work in this area has been invaluable. Discussions with many naturalists and ecologists have contributed to the production of this paper. I wish to thank especially Dr R. W. Braithwaite, Dr P. Werner, Dr M. H. Andrew, Mr P. Wellings, Mr G. Ryan, Mr D. Lindner and Mr B. Bailey for many such discussions. I wish to thank Mr M. Brennan of the Northern Territory Bushfires Council for the information he gave me. I wish to thank Adrienne MacCartie for typing the manuscript and the Office of the Supervising Scientist for access to word processing and computer facilities. The assistance of Mr G. McEntyre of the Australian Survey Office in the production of maps and figures is appreciated.

References

Andrew M. H. (1986) Use of fire for spelling monsoon tallgrass pasture grazed by cattle. *Trop. Grassl.* **20,** 69-78.

A.N.P.W.S. (1980) *Kakadu National Park Plan of Management*, Commonwealth of Australia, Canberra.

Bell A. (1981) Fire damages top-end forests. *Ecos.* **30,** 18-20.

Bowman D. M. J. S. (1986) Stand characteristics, understorey associates and environmental correlates of *Eucalyptus tetrodonta* F. Muell. forests on Gunn Point, Northern Australia. *Vegetatio.* **65,** 105-13.

Bowman D. M. J. S. & Dunlop C. R. (1986) Vegetation pattern and environmental correlates in coastal forests of the Australian monsoon tropics. *Vegetatio.* **65,** 99-104.

Bowman D. M. J. S. & Wightman G. M. (1986) Small scale vegetation pattern associated with a deeply incised gully, Gunn Point, Northern Australia. *Proc. Roy. Soc. Qd.* **96,** 163-73.

Braithwaite R. W. & Estbergs J. A. (1985) Fire patterns and woody vegetation trends in the Alligator Rivers Region of Northern Australia. In: *Ecology and Management of the World's Savannas* (eds J. C. Tothill & J. J. Mott) pp. 359-64. Australian Academy of Science, Canberra.

Day J. C. (1985) Applications of satellite imagery for fire management in Kakadu National Park. In: *Towards an Expert System for Fire Management in Kakadu National Park . Technical Memorandum 85/2* (eds J. Walker, J. R. Davis & A. M. Gill) CSIRO, Division of Water and Land Resources, Canberra.

Haynes C. P. (1985) The pattern and ecology of *munwag*: traditional Aboriginal fire regimes in north central Arnhem Land. *Proc. ecol. Soc. Aust.* **13,** 203-14.

H.R.S.C.E.C. (1984) *Bushfires and the Australian Environment*. House of Representatives Standing Committee on Environment and the Australian Environment. Australian Government Publishing Service, Canberra.

Jones R. (1969) Fire-stick farming. *Aust. Nat. History* **16,** 224-28.

Mott J. J. & Tothill J. C. (1984) Tropical and subtropical woodlands. In: *Management of Australian Rangelands* (eds G. N. Harrington, A. P. Wilson & M. Young) pp. 255-69. CSIRO, Melbourne.

Snedecor G. W. & Cochran W. G. (1980) *Statistical Methods*, 7th ed. Iowa State University Press, Ames.

Stevenson P. M. (1985) Traditional Aboriginal resource management in the wet-dry tropics: Tiwi case study. *Proc. ecol. Soc. Aust.* **13,** 309-15.

Stocker G. C. (1966) Effects of fire on vegetation in the Northern Territory. *Aust. For.* **30,** 223-30.

Taylor J. A. & Dunlop C. R. (1985) Plant communities of the wet-dry tropics of Australia: the Alligator Rivers Region, Northern Territory. *Proc. ecol. Soc. Aust.* **13,** 83-127.

Taylor J. A. & Tulloch P. (1985) Rainfall in the wet-dry tropics: Extreme events at Darwin and similarities between years during the period 1870-1983. *Aust. J. Ecol.* **10,** 281-95.

Tothill J. C. & Mott J. J. (1985) Australian savannas and their stability under grazing. *Proc. Ecol. Soc. Aust.* **13,** 317-22.

Proc. Ecol. Soc. Aust. 15:177-189

The distribution, characteristics and conservation status of lakes in the wet tropics and subtropics of Queensland

A. H. ARTHINGTON

Division of Australian Environmental Studies, Griffith University, Nathan, Qld. 4111

Abstract

This paper presents an overview of the classes of lakes found in the wet tropics and subtropics of Queensland, i.e., floodplain lakes, dune lakes, lakes in terminal drainage basins, lakes associated with lava flows, and crater lakes and maars. The limnological characteristics of each class of lake are described and their distributions outlined. From this basis the conservation status of Queensland lakes is analysed, with particular reference to the representation of lakes in National Parks. Floodplain lakes, billabongs in river meanders, waterholes in main river channels, lakes in terminal drainage basins and lakes associated with lava flows are poorly represented in Queensland's system of National Parks. The analysis highlights the need for enhanced limnological research and inventory work in Queensland and a greater emphasis on the conservation of wetlands in any state conservation strategy.

Introduction

In 1986 World Wildlife Fund Australia joined an international effort, the "Wetlands Campaign", to conserve the world's diminishing wetlands. As part of this campaign, World Wildlife Fund Australia sponsored a national Wetlands Workshop in February 1986. The workshop brought together experts from around Australia to present and discuss review papers on the distribution and conservation status of athalassic and tidal wetlands in each state and the Northern Territory.

The Queensland review by Arthington and Hegerl (1988), entitled "The Distribution, Conservation Status and Management Problems of Queensland's Athalassic and Tidal Wetlands", gives an account of the major concentrations of wetlands in each of twelve mainland biogeographic regions (Fig. 1). The biogeographic regions are those recognized by Stanton & Morgan (unpublished report, 1977) in the RAKES Report ("The Rapid Selection and Appraisal of Key and Endangered Sites: The Queensland Case Study"). The review drew a major part of its material from Stanton's (1975) assessment of wetlands in Queensland, although many additional sources of published and unpublished information, and maps, were consulted. Stanton (1975) related high altitude aerial photographs (7 500 m) at a scale of 1:84 000 to information taken from the 1:1 000 000 map series, and in this way recorded 142 major aggregations of wetlands, representing, by his estimation, about 90% of total wetland surface area in the State (excluding the main channels of rivers and artificial wetlands).

In addition to providing a general description of the major wetland resources of the State, the Queensland review assessed the number of reserves (i.e., National Parks, Environmental Parks etc.) which fully or partially protect wetlands within each biogeographic region. The review did not discuss in any detail the classification and biophysical characteristics of wetlands, although much of the available literature on these topics was briefly mentioned and a fairly comprehensive bibliography forms part of the paper.

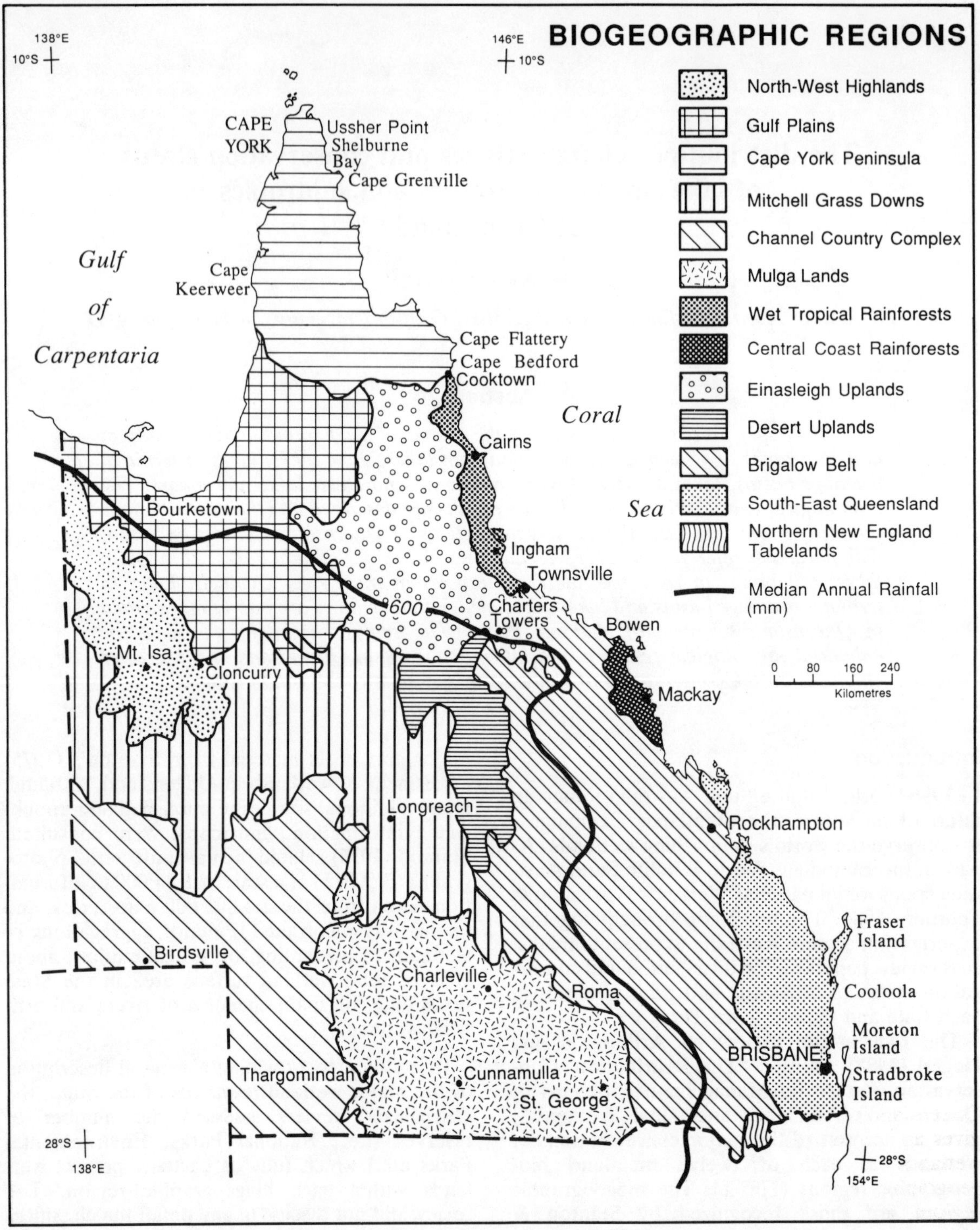

FIG. 1. Terrestrial biogeographic regions of Queensland, redrawn from Stanton & Morgan (unpublished report) and associated map produced by the Queensland National Parks and Wildlife Service.

However, two schemes for the classification of wetlands were presented in tabular form (Tables 1 and 2, Arthington & Hegerl 1988).

The first classification was that of Stanton (1975), based primarily on the form of the water body (lake, swamp etc.), salinity and frequency of inundation, with some distinctions to accommodate major vegetation formations. The second classification was proposed by Briggs (1981) in her overview of Australian wetland vegetation. This was reworked as a table giving the structural and floristic characteristics of athalassic wetland vegetation and major areas of occurrence of each plant formation in Queensland.

Reference was also made to the hierarchical classification of wetlands developed for Australia by Paijmans *et al.* (1985, Table 1). Arthington and Hegerl (1988) suggested that this classification be adopted and refined to include information from the state-by-state inventories presented at the Workshop, so that an Australia-wide inventory of wetlands can be compiled, national conservation issues addressed and management solutions developed for common problems.

This paper takes up that suggestion, in part, by presenting an overview of information on the distribution and conservation status of one major category of athalassic wetlands, in Queensland, that is, lakes as defined by Paijmans *et al.* (1985). It is confined largely to that part of Queensland

TABLE 1. Classification of Wetlands. (modified from Paijmans *et al.* 1985; full details of classification given only for lakes).

I LAKES: areas of open water generally over 1 m deep with little or no persistent emergent vegetation.

1. Permanent and near permanent lakes
   - (a) Permanent floodplain lakes including billabongs and waterholes in channels.
   - (b) Permanent lakes of coastal dunes and beach ridge plains.
   - (c) Permanent lakes in terminal drainage basins.
   - (d) Permanent lakes associated with lava flows.
   - (e) Permanent crater lakes and maars.
   - (f) Permanent karst lakes.
   - (g) Permanent glacial lakes.
   - (h) Permanent man-made lakes.
2. Seasonal lakes: alternately wet and dry every year according to season.
   - (a) Seasonal floodplain lakes.
   - (b) Seasonal lakes in terminal drainage basins.
   - (c) Seasonal lakes associated with lava flows.
3. Intermittent lakes: alternately wet and dry but less frequently and regularly than seasonal lakes.
   - (a) Intermittent floodplain lakes.
   - (b) Intermittent coastal dune lakes.
   - (c) Intermittent lakes in terminal drainage depressions.
   - (d) Intermittent man-made lakes.
4. Episodic lakes: dry most of the time with rare and very irregular wet phases.
   - (a) Episodic lakes in terminal drainage depressions.
   - (b) Episodic lakes on present or former floodplains.
   - (Other sub-classes rare or non-existent).

II SWAMPS: dominantly vegetated; where present water generally less than 1 m deep; persistent emergent vegetation.

1. Permanent swamps: wet most of the time.
2. Seasonal swamps: seasonally wet and dry each year.
3. Intermittent swamps: alternately, but irregularly, wet and dry.
4. Episodic swamps: very rarely contain water and lacking swamp vegetation.

III LAND SUBJECT TO INUNDATION: water not present long enough for typical wetland vegetation to develop, but may be important waterbird habitat.

1. Seasonally inundated.
2. Intermittently inundated.

IV RIVER AND CREEK CHANNELS

V TIDAL FLATS: subject to tidal inundation at least once per year.

VI COASTAL WATER BODIES: estuaries, lagoons and inlets with access to the sea.

receiving a median annual rainfall of more than 600 mm (Fig. 1), in keeping with the theme of this Symposium, and therefore includes the wet tropics, the wet-dry tropics and the wetter parts of subtropical Queensland. Approximately 65% of the 142 major aggregations of wetlands identified by Stanton (1975) occur in these wetter parts of Queensland (see Fig. 2a,b). However, for ease of comparison between the analyses of this paper and those of Arthington and Hegerl (1988), this treatment goes beyond the 600 mm rainfall isohyet by mentioning lakes occurring within some of the drier biogeographic regions.

Types and Distribution of Queensland Lakes

Lakes are defined by Paijmans *et al.* (1985) as areas of open water generally over 1 m deep with little or no persistent emergent vegetation (Table 1). Permanent lakes have water at all times, but water levels may fluctuate and the development of emergent, floating-leaved and submerged vegetation depends upon exposure, basin shape, shoreline development, substrate and the variability of water levels, as well as water chemistry and clarity. Permanent freshwater lakes are of various types, and most are found in high rainfall areas of Queensland (Figs. 2a and c).

Lakes may also be seasonal, intermittent or episodic, although the differences between these classes are not always clear-cut (Paijmans *et al.* 1985). The distributions of all permanent and near permanent wetlands and of episodic lakes in Queensland are shown in Figs. 2b and d. Although this paper emphasizes permanent lakes, mention is made of the less permanent classes of lakes where information is available.

Permanent floodplain lakes. These lakes occur on riverine floodplains that are to some extent still actively forming (Paijmans *et al.* 1985). They persist only where there is sufficient water from overbank flow, rainfall and groundwater to fill

TABLE 2. Number and category of reserves covering all or part of a major wetland aggregation identified by Stanton (1975) (from Arthington & Hegerl 1988).

KEY TO RESERVES:

Terrestrial reserves — EP = Environment Park, FR = Fauna Reserve, NP = National Park, SF = State Forest

Marine reserves — FHR = Fisheries Habitat Reserve, WR = Wetlands Reserve, FS = Fish Sanctuary, MP = Marine Park

Biogeographic Region	Number Wetland Aggregations	Number of Wetland Reserves								Total No. Wetland Reserves
		NP	EP	FR	SF	FHR	WR	FS	MP	
1. North-western Highlands	0	0	—	—	—	—	—	—	—	0
2. Gulf Plains	7	1	—	—	—	—	—	—	—	1
3. Cape York Peninsula	21	4	—	—	—	4	—	1	—	9
4. Mitchell Grass Downs	1	—	—	—	—	—	—	—	—	0
5. Channel Country Complex	15	—	—	—	—	—	—	—	—	0
6. Mulga Lands	10	—	—	—	—	—	—	—	—	0
7. Wet Tropical Rainforests	18	20	6	—	—	7	4	—	—	37
8. Central Coast Rainforests	5	2	—	—	—	1	—	—	—	3
9. Einasleigh Uplands	18	—	—	—	—	—	—	—	—	0
10. Desert Uplands	6	—	—	—	—	—	—	—	—	0
11. Brigalow Belt	16	1	—	—	—	—	—	—	—	1
12. South-eastern Queensland	25	10	3	1	2	13	7	3	1	40
Total Number	142	38	9	1	2	25	11	4	1	91

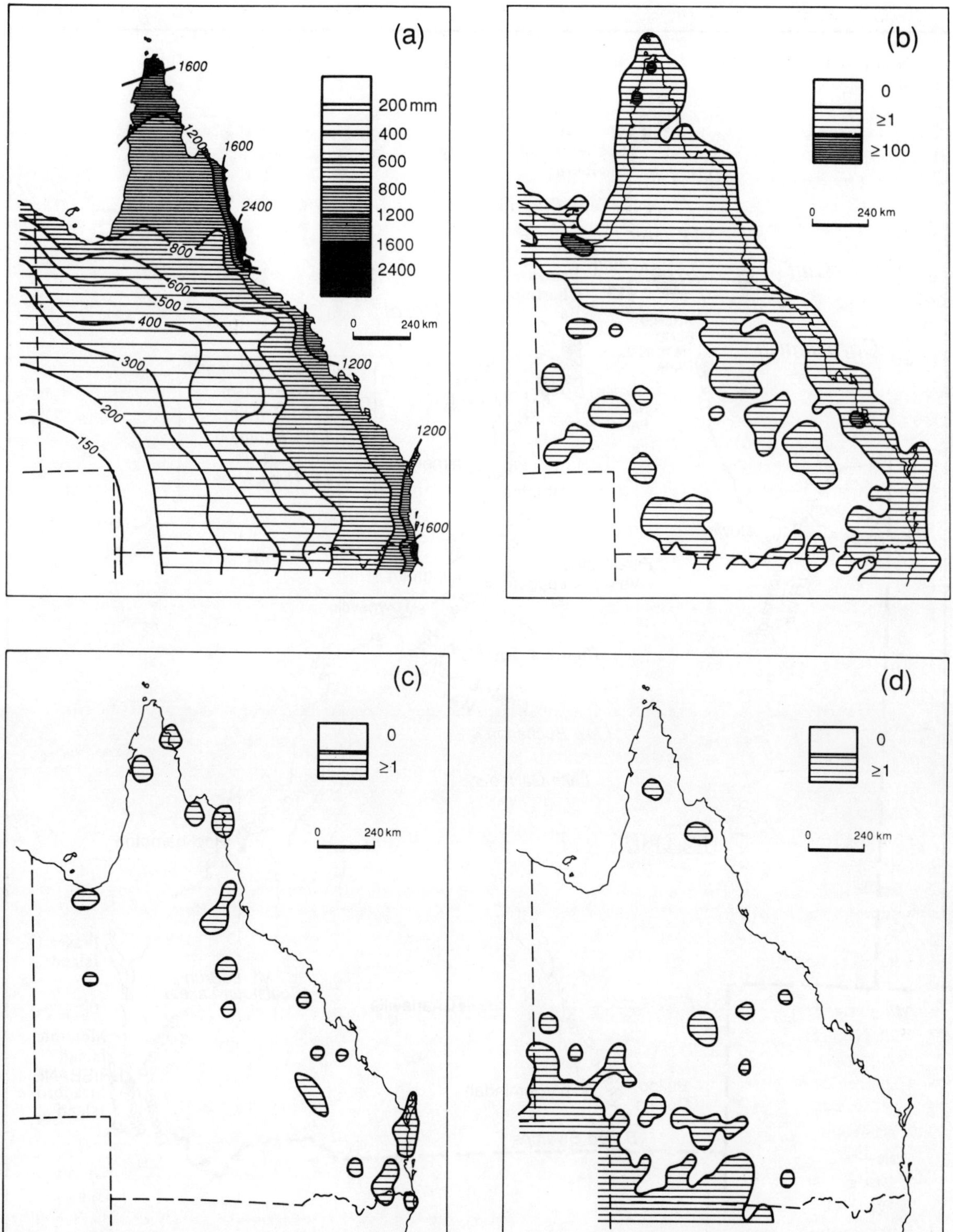

FIG. 2. Distribution of median annual rainfall in Queensland (a) and the distribution of all permanent and near-permanent wetlands (b), permanent lakes (c) and episodic lakes (d) (adapted from Paijmans *et al.* (1985)).

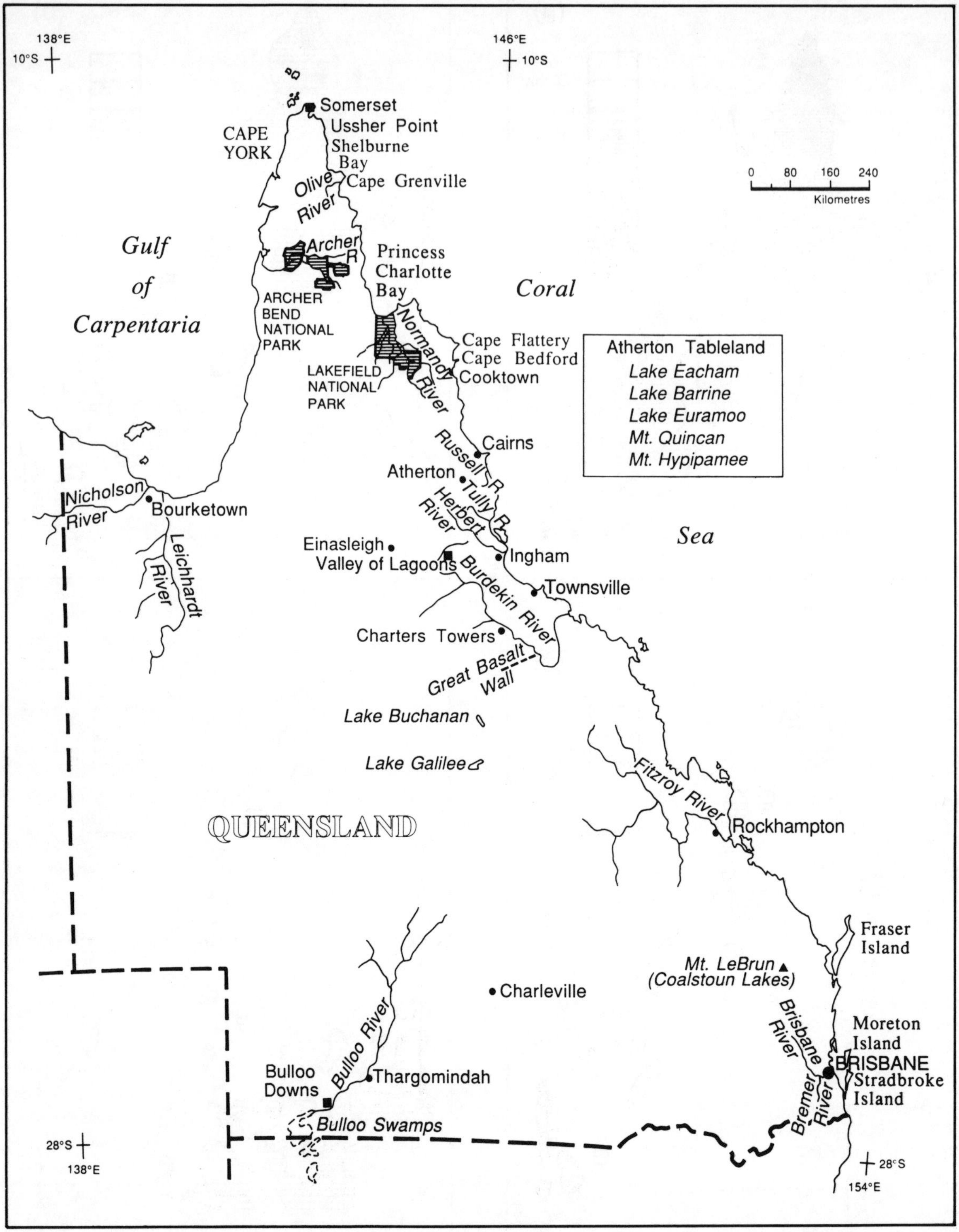

FIG. 3. Locations of some major lakes and rivers in Queensland.

depressions in floodplain sediments, or where there is a channel connecting the depression to the main stream. This class of lake includes persistent "waterholes" in the main channels of rivers, and "billabongs" or "oxbow lakes" formed in river meanders, although strictly speaking, waterholes in main channels belong in the category of wetlands termed "River and Creek Channels" (Table 1).

According to Stanton (1975) the most extensive occurrences of floodplain lakes and billabongs in Queensland are along streams draining into the southern part of the Gulf of Carpentaria, e.g., along the Lower Nicholson River and the Lower Leichhardt River (Fig. 3). Many of these lakes are seasonal, but about 10% are permanent. Permanent floodplain lakes also occur along Saltwater Creek on Cape York Peninsula where they have been formed by blockages in the southern tributaries of the creek by river-borne sand (Paijmans *et al.* 1985).

Other examples of permanent floodplain lakes in tropical Queensland are found along the floodplain of the upper Burdekin River (Fig. 3). Wairuna Lake is a permanently filled depression between levee banks and adjoining low hills, and the Pelican Lakes to the east of the Valley of Lagoons are backplain lakes and swamps of the Burdekin River which appear to be associated also with lava flows. The Pelican Lakes and Mt. St. John Lagoon in the floodplain of the Bohle River provide important habitat for waterbirds, particularly Black Swan, and Mt. St. John Lagoon has been quite intensively studied in relation to the seasonal dynamics of vegetation and waterbirds (Lavery & Blackman 1971). However, there have been no other published studies of floodplain lakes in Queensland.

Permanent and seasonal floodplain lakes, waterholes and billabongs occur along the lower reaches of many streams and rivers draining to the east coast of Queensland south of Princess Charlotte Bay, e.g., on the floodplains of the Tully, Herbert, Burdekin, Fitzroy and Russell Rivers (Fig. 3). Intermittent and episodic floodplain lakes are found mainly in the semi-arid and arid parts of Queensland (Figs. 2a and d).

Permanent lakes of coastal dunes and beach ridge plains. These lakes develop in wind-formed depressions amongst sand dunes where local and regional groundwater tables accumulate past rainfall and surface runoff. Water is lost by evaporation, and by seepage through the podzolized B horizon which forms a relatively impermeable, cemented sand ("coffee-rock") underlying dune lakes perched above the regional groundwater table. Perched lakes may therefore fluctuate in water level by more than 4 m, as Brown Lake has done in its recent past (Lee-Manwar *et al.* 1980). Because they lie in wind-deflated hollows, perched lakes are usually shallow (<8 m maximum depth) whereas watertable window lakes may be as deep as 13 m (Arthington *et al.* 1986). Typical examples of perched dune lakes are Brown Lake on North Stradbroke Island (Bensink and Burton 1975) and most of the Fraser Island lakes (Bayly 1964; Bayly *et al.* 1975; Arthington *et al.* 1986).

Watertable window lakes usually occur at lower altitudes and closer to the coast than perched lakes, and represent surface manifestations of regional groundwater tables. Some have an outflowing creek opening directly to the sea (e.g., Lake Freshwater at Cooloola, Reeve *et al.* 1985) or into coastal freshwater swamps (e.g., Blue Lake on North Stradbroke Island, Bensink and Burton 1975). Water may also seep through a permeable sand aquifer to the coast, causing fluctuations in the water level which parallel those occurring in perched lakes. Seepage, and perhaps regional variations in groundwater, may explain the falling water level in Lake Freshwater over the past five years (Miller, Arthington & Outridge unpublished report).

Dune lakes achieve their greatest concentrations on the large sand islands off the coast of south-eastern Queensland, but similar lakes have formed amongst dunes between Somerset and Ussher Point on Cape York Peninsula, at Shelburne Bay and near Cape Grenville, and in the Cape Bedford-Cape Flattery areas north of Cooktown (Fig. 3). Although situated within coastal dunes, the Cape York lakes are associated with ancient laterite and Mesozoic sandstones and differ chemically from the southern dune lakes because of this (Timms 1986a).

A typology of dune lakes has been published by Timms (1982) and the limnology of such lakes has received thorough treatment in three recent publications (Timms 1986a, b; Arthington *et al.* 1986). The palaeolimnology of a perched lake on Fraser Island is reviewed by Longmore (1986). Additional information on the dune lakes at Shelburne Bay is presented in the Shelburne Silica Sand Joint Venture, Draft Environmental Impact Statement (Macdonald Wagner Pty Ltd 1986), and in an unpublished commentary on this statement (Arthington 1986).

The water of permanent dune lakes in Queensland is always of very low salinity, and nutrient levels are also low (e.g., phosphorus is <10 ug/l, Arthington *et al.* in press). Major ions are derived by atmospheric supply from the sea coupled with insignificant land weathering. The water of perched lakes is often stained a deep brown colour by dissolved humic acids leached from swamps, peat bogs and the podzolized B horizon (Bayly 1964; Bayly *et al.* 1975; Timms 1986b).

These coloured humic acids give the perched lakes their characteristic low pH (<5) and low transparency (<1-3 m). The waters of window dune lakes are generally much less coloured because dissolved humic acids are greatly diluted in the regional watertable. However, in some dune lakes, for example, Lake Freshwater at Cooloola, there appears to be a mixture of "black" water from swamps and "white" water from the regional water mound (Reeve *et al.* 1985).

The emergent vegetation of permanent dune lakes in Queensland includes various sedges *(Lepironia, Cyperus, Baumea, Eleocharis, Gahnia, Lepidosperma, Scirpus)* and restiads *(Empodisma minus = Calorophus minor, Restio tetraphyllus)* and *Sphagnum* is often present in swampy areas and bogs associated with the lakes. Submerged aquatic plants are uncommon, but the shallows of perched lakes and lakes with high transparency may support beds of Cyperaceae, *Nitella* and *Utricularia*. The bottom substrates of Blue Lake on North Stradbroke Island have quite extensive beds of Cyperaceae even though this lake has a maximum depth of 11.2 m (Bensink and Burton 1975). Introduced *Nymphaea* has been recorded in several dune lakes on Cape York Peninsula (Timms 1986a) and in Lake Freshwater at Cooloola (Arthington unpublished data).

The invertebrate faunas of Queensland dune lakes are distinctive and, like dune lakes of New South Wales and Victoria, include a number of rare and unique species, such as representatives of the Copepoda (Bayly 1964), Cladocera (Smirnov & Timms 1983), Oligochaeta (Jamieson 1978) Odonata (Watson & Arthington 1978; Arthington & Watson 1982), Trichoptera (Neboiss 1978) and Chironomidae (Brundin 1983; Arthington *et al.* 1986).

Most dune lakes have only one or two species of fish, of which the commonest are *Hypseleotris galii* and *Rhadinocentrus ornatus*. There are evidently no fish species unique to the Queensland dune lakes, but some species occurring in dune lakes have disjunct distributions (e.g., *Nannoperca oxleyanna*) or restricted mainland distributions (e.g., *Rhadinocentrus ornatus*) (Arthington 1984). Deep watertable window lakes of high transparency have a more diverse fish fauna: for example, 11 species occur in Lake Wabby on Fraser Island (Arthington *et al.* 1986).

Although scientifically interesting for their mode of origin and the rare and unique invertebrates, fish and frogs (Ingram & Corben 1975) they support, dune lakes in Queensland and elsewhere are mostly very unproductive (Miller *et al.* unpublished; Arthington *et al.* in press). As in dilute, acidic waters elsewhere, phytoplankton biomass is low and the community is dominated by desmids (Bayly *et al.* 1975; Outridge *et al.* in press). Perhaps because primary productivity is so low, the dominant food chain in perched lakes appears to be dissolved organic matter — heterotrophic bacteria — zooplankton (Bayly & Williams 1973; Lee-Manwar *et al.* 1979). However, several species of fish, and the turtle, *Emydura krefftii,* maintain surprisingly large populations in unproductive perched lakes by feeding on wind-blown terrestrial insects, and the seeds and pollen of terrestrial vegetation (Bayly *et al.* 1975; Georges 1982). In the more transparent, window lakes with deeper euphotic zones and higher productivity fish seem to rely almost entirely on autochthonous food resources such as benthic invertebrates and algae (Arthington *et al.* 1986). Nevertheless, dune lakes are not an important resource for waterbirds because of their overall low productivity (Vernon & Barry 1972).

Intermittent dune lakes are uncommon, occurring along the east coast of Cape York Peninsula (see Timms 1986a) whereas Tortoise Lagoon on North Stradbroke Island is an episodic dune lake. Before 1975, it was a shallow sedge swamp lying in an interdune valley, but during the 1975 Brisbane floods, after a period of extraordinarily high and extended rainfall, Tortoise Lagoon was filled to depths of over 7 m (Lee-Manwar *et al.* 1980). The "Depression" on Fraser Island is, likewise, an episodic perched lake (Longmore 1986).

Lakes in terminal drainage basins. Endorheic basins are common in Australia and often have extensive drainage areas (De Deckker 1983). The lakes found in such situations, often called playa lakes, are usually not permanent, but a few are deep enough, or fill often enough, to be regarded as permanent despite fluctuating water levels (Paijmans *et al.* 1985). Lakes Buchanan and Galilee in central Queensland (Fig. 3) are saline playas formed by tectonic warping of the earth's crust (Bayly & Williams 1973). They are the largest lakes in a complex of shallow clear saline lakes, swamps and springs which are seasonally or intermittently flooded, thus varying in salinity and importance as waterbird habitat (Stanton 1975; Gasteen unpublished). Salinity has been shown to affect the diversity of the benthic fauna in these lakes (Bayly & Williams 1973; Timms 1979).

Other examples of seasonal and intermittent playa lakes occur south-west of Thargomindah, in the Bulloo Swamps area, where there is a large expanse of terminal swamps and lakes set amongst the anastomosing channels of the Bulloo River (Stanton 1975; Fig. 3). Playa lakes also occur in ancient drainage systems and such lakes usually have a characteristic elongated shape (De Deckker 1983).

Lakes associated with lava flows. Such lakes develop where basalt flows have blocked river and stream channels in geologically recent times (Paijmans *et al.* 1985). Basalt flows are concentrated in relatively high rainfall areas of Queensland, and the lakes associated with them are therefore frequently permanent. Many lakes of this type occur in the Valley of Lagoons area on the upper Burdekin River (Fig. 3), where Anabranch Lake and Saltern Lagoon are freshwater examples (Paijmans 1978; Paijmans *et al.* 1985), whereas Lakes Pelican and Agnes, which also appear to have formed by lava damming (Twidale 1966), are saline (Paijmans *et al.* 1985). Other examples are associated with the Great Basalt Wall to the north-west of Charters Towers (Fig. 3), where lava supports deep crater-like lakes with clear water that are thought to be spring-fed (Stanton 1975). Springs in the basalt feed Lolworth and Fletcher Creeks, and overflow from these creeks supplies several seasonal lakes and swamps.

Apart from observations on their vegetation (Paijmans 1978), very little is known of the limnology of lakes associated with lava flows even though some of them provide important habitat for waterbirds (Stanton 1975). Timms (1979) studied the benthos of shallow, silty lagoons near Lolworth Creek which probably originated by sediments from the main stream blocking side valleys and by lava-damming. He found these seasonal lagoons to be depauperate compared to other lakes in north-eastern Queensland, perhaps because they dry completely at times. Three larger lagoons lying in a former distributary path of the Burdekin River near Ayr were slightly richer in benthic species.

Crater lakes and maars. Permanent crater lakes are found near the summit of distinctly raised, extinct volcanoes with orifices that have been enlarged during the final stages of eruption. Such lakes occur in recent volcanic areas on the Atherton Tableland (Fig. 3). Mt Quincan is a young scoria cone with a crater about 0.8 km wide containing a single small lake. Other examples are the Coalstoun Lakes, two small, almost perfectly circular lakes in the two craters of Mt Le Brun near Gayndah (Bayly & Williams 1973) (Fig. 3).

Maars are craters formed by a violent eruption caused when hot lava comes in contact with ground water, or by degassing of magma. Their morphology is usually subcircular and the lakes lying within them are often very deep (Bayly & Williams 1973). Three good examples of maars occur on the Atherton Tableland (Fig. 3). Two of these, Lakes Eacham and Barrine, are both more than 100 m deep, and Lake Barrine, with a diameter of 1.9 km, is exceptionally wide for this class of lake (see Timms 1976).

Lakes Eacham and Barrine are very similar in many physiographic and physicochemical features (Bayly & Williams 1973; Timms 1979); both are extremely fresh (salinity <100 mg/l), but their benthic communities are very different, with only 9 shared species among the 40 species occurring in the two lakes (Timms 1979). Interestingly, the species of *Conochironomus* (Chironomidae) found in Lake Eacham occurs also in two Queensland dune lakes with very fresh, clear waters, but not elsewhere in Australia so far as is known (Arthington *et al.* 1986; J. Martin, personal communication).

Mt. Hypipamee on the Atherton Tableland contains a crater of the type known as a gas maar on diatreme, with sheer walls and a diameter of about 60 m and depth of 140 m. The lake within this crater is about 70 m deep (Bayly & Williams 1973).

Permanent karst lakes. Karst lakes develop in solution depressions where limestone occurs at the surface or in shallow depths (Paijmans *et al.* 1985). They require high rainfall to maintain the water balance and are usually fresh. Queensland examples are unknown.

Permanent glacial lakes do not occur in Queensland and *man-made lakes* are not a subject of this review, although it should be remarked that a good deal of the wetland surface area of Queensland is now man-made, mostly within the last 25 years.

Conservation Status of Queensland Wetlands

The conservation of all wetlands, both tidal and athalassic, in Queensland, is summarized in Table 2 in relation to the twelve biogeographic regions delineated in the RAKES Report (Stanton & Morgan unpublished). Full details of the locations of these reserves, and of the types of wetland they protect, can be found in Appendix 1, Arthington & Hegerl (1988).

The outstanding feature of Table 2 is that 85% of the reserves which currently protect wetlands in Queensland are concentrated in two biogeographic regions, the Wet Tropical Rainforests and the South-eastern Queensland region (see Fig. 1). In these two regions there are 30 National Parks, 9 Environmental Parks and 20 Fisheries Habitat Reserves. The remaining biogeographic regions are extremely poorly endowed with wetland reserves and several regions afford no protection at all to wetlands. To summarize, there are no wetland reserves on the Mitchell Grass Downs, in the Mulga Lands, on the Einasleigh Uplands, the Desert Uplands or in the interior areas of the Brigalow Belt and the extensive wetland areas of the Channel Country, both in Queensland and South Australia (Arthington & Hegerl 1988).

Conservation Status of Queensland Lakes

Lakes of the types described in this paper are variously protected by the current system of reserves in Queensland. The following section discusses the conservation status of each class of permanent lake and outlines some of the threats to these wetland systems. Areas recommended for conservation by Stanton and Morgan (unpublished) in the RAKES Report are also mentioned.

Floodplain Lakes. These lakes are very poorly represented in Queensland National Parks and other terrestrial reserves, although there are probably a few examples in the inland extremity of Lakefield National Park, bordering Princess Charlotte Bay, and in Archer Bend National Park, Cape York. None of the major areas of floodplain lakes in the Gulf of Carpentaria or along the eastern coastal plains of Queensland are of National Park status. This situation largely reflects a lack of inventory work and research in Queensland river systems and the waterbodies of their floodplains, presenting a dismal contrast to the major studies of floodplain billabongs and wetlands in the Alligator Rivers region of the Northern Territory (Bishop 1980, 1983; Bishop *et al.* 1986).

Conservation of the wetlands bordering the southern part of the Gulf of Carpentaria is a high priority because of their importance to waterbirds, particularly migratory waders designated for protection under the Australia-Japan Treaty (see Braithwaite 1980; Garnett 1983; Williams 1983).

Coastal dune lakes. The dune lakes of south-eastern Queensland are moderately well protected by the National Parks currently gazetted on North Stradbroke, Moreton and Fraser Islands and at Cooloola. A deep, clear and beautiful window lake is protected in the Blue Lake National Park on North Stradbroke Island and the management plan for Moreton Island prohibits the use of power boats on Blue Lagoon. Several perched dune lakes and the watertable window, Lake Freshwater, lie within the Cooloola National Park.

On Fraser Island the Great Sandy National Park includes all of the northern dune lakes (see Fig. 3, Arthington *et al.* 1986), and many lakes in the southern part of the island are situated in State Forests, thereby coming under the management of the Queensland Department of Forestry. All fauna within State Forests is automatically protected by their Sanctuary status (Fauna Conservation Act). Although camping and recreation are permitted at these lakes, there are restrictions on various activities within a boundary zone around each lake, and sites for the disposal of sewage are carefully located.

Lake Wabby, the deepest and most biologically diverse dune lake on Fraser Island, is slowly being invaded by a mobile dune or blowout which has reduced the volume of the lake, smothered benthic vegetation and reduced benthic invertebrate diversity in recent years (Arthington *et al.* 1986). Although the movement of this dune is a natural process, human trampling of the shoreline has undoubtedly contributed to the collapse of sand into the lake (see Figs. 1, 2 and 4 Arthington *et al.* 1986). Recent analysis shows that the rate of movement of the blowout into Lake Wabby has slowed considerably in the last 10 years, but, despite this, the long-term future of Lake Wabby is not assured. The Queensland Department of Forestry has prevented vehicular access to Lake Wabby and camping is prohibited, but many people walk across the dunes from the coast to swim in this clear deep lake. Nevertheless, there are at present no indications of water quality problems due to human actions. This is in sharp contrast to Lake Freshwater in the Cooloola National Park, where there has been an unusual increase in phosphorus levels in the past ten years, possibly due to recreational use, with the result that the lake shows classic symptoms of eutrophication (Outridge *et al.* in press; Arthington *et al.* in press). In these oligotrophic lakes the conflict between recreation and conservation is a difficult issue which cannot be resolved without a better understanding of nutrient dynamics and loadings.

Dune lakes at Shelburne Bay are situated close to leases under consideration for silica sand mining. These lakes have an unusually diverse fish fauna, including some tropical species shared with Papua New Guinea. Although various protective measures are proposed in relation to mining activities, the future of these dune lakes seems dubious. Other dune lakes situated in the Olive River Dunefield to the south may be included in a new National Park in that region. Dune lakes in the Cape Flattery-Cape Bedford area are unprotected, even though Stanton and Morgan (unpublished) identified the entire complex of landforms and wetlands from the capes to Mt Stucky as a Regional Key Area for conservation.

Lakes in terminal drainage basins. These lakes occur mainly in the drier areas of the State where there are no reserves gazetted to protect wetlands and their catchments (Table 2). There are very considerable problems in the determination of boundaries to reserves in this region, where wetlands may be extensive during periods of flooding, and relatively dry at other times. These systems must be studied so that wetland values can be determined and conservation as well as catchment management strategies devised to protect them, particularly from grazing pressure.

Lake Galilee and the small, spring-fed Lake Dunn to the south are the only areas within the Desert Uplands (see Fig. 1) given Regional Key Area status by Stanton & Morgan (unpublished). This area holds coal reserves and the wetland vegetation has been damaged by over-burning and over-stocking. Lake Buchanan was nominated as a Regional Important Area of lesser significance for conservation than Lake Galilee because of its higher salinity, lower faunal diversity and intermittent importance for waterbirds.

Permanent lakes associated with lava flows. These lakes occur in the Einasleigh Uplands where there are no reserves of any type protecting wetlands (Table 2). Remote and relatively inaccessible, these lakes provide major habitat for large populations of waterbirds in the region, and they are of such interest in terms of mode of origin and geological formations that the conservation of representative types would seem to be a priority for Queensland. Stanton & Morgan (unpublished) selected the Valley of Lagoons, the Undara Lava Tunnels and associated vegetation formations as a Regional Key Area for conservation in this biogeographic region.

Permanent crater and other volcanic lakes. The more spectacular crater lakes and their surrounds have been gazetted as National Parks in two areas of Queensland. Lakes Eacham and Barrine on the Atherton Tableland are both situated within National Parks but the species of rainbowfish unique to Lake Eacham *(Melanotaenia eachiamensis)* may be endangered (Pollard & Burchmore 1986), evidently by introduced fishes. The Coalstoun Lakes National Park at Mt Le Brun is centred on the two crater lakes from which its name is derived.

It is worth noting that Lakes Eacham and Barrine, Lake Euramoo and The Crater were the only Queensland waterbodies listed for conservation by Project Aqua in 1971 (Williams 1983), reflecting the lack of limnological activity in Queensland at that time, rather than that any lesser degree of priority should be attached to the conservation of waterbodies in Queensland *vis-a-vis* those of other states.

Conclusions

This review of permanent and other lakes in the wetter areas of Queensland highlights the poor limnological knowledge we presently have of all classes of lake except the dune lakes of south-eastern Queensland and further north. It shows that, with the exception again of these dune lakes, and lakes associated with volcanic activity, very few lakes are conserved within National Parks and other types of reserve. Several major classes of lake are poorly represented in reserves, *viz.*, floodplain lakes, billabongs in river meanders, waterholes in main river channels, lakes in terminal drainage basins and lakes associated with lava flows, despite their obvious importance as habitat for endemic aquatic biota and migratory waterbirds, and as a source of water for terrestrial fauna in drier regions. Furthermore, many lakes, including some that lie within National Parks, are presently under stress or endangered in various ways.

Recommendations of the Wetlands Workshop held in February 1986 will eventually be communicated to the various state authorities responsible for the conservation and management of wetlands around Australia. The Workshop review of Queensland's wetlands, and this paper, highlight the need to place greater emphasis on wetlands in the state's conservation strategy. It is hoped that new initiatives in wetland conservation may result from these studies. In Queensland, there remains the urgent and enormous task of gathering site-specific information on wetlands so that sensible priorities can be developed for wetland conservation. It is not sufficient to consider only the representativeness of wetland types as the major criterion for conservation, although this is an important aspect of selection.

Finally, Queensland lags far behind other states in wetland research, and limnological research in its broadest sense, an imbalance which must be redressed, particularly in tropical areas and the inland, where water resources are coming increasingly under pressure from multiple and often conflicting uses.

Acknowledgements

I would like to express my appreciation to Mr Aubrey Chandica of Griffith University for the preparation of maps and Ms Diane Conrick for library and mapping research. Mr Ross Blick and Dr Paul Sattler of the Queensland National Parks and Wildlife Service assisted me frequently with information, comments and the loan of source material and maps. Griffith University supported the study.

References

Arthington A. H. (1984) Freshwater fish of North Stradbroke, Moreton and Fraser Islands. In: *Focus on Stradbroke* (eds R. J. Coleman, J. Covacevich & P. Davie) pp. 279-82. Boolarong Press, Brisbane.

Arthington A. H. & Hegerl E. J. (1988) The distribution, conservation status and management problems of Queensland's athalassic and tidal wetlands. In: *The Conservation of Australian Wetlands* (eds H. J. McComb & P. S. Lake) pp. 59-101. Surrey Beatty & Sons, Sydney.

Arthington A. H., Burton H. B., Williams R. J. & Outridge P. M. (1986) The ecology of humic and non-humic dune lakes, Fraser island, with emphasis on sand infilling in Lake Wabby. *Aust. J. mar. freshwat. Res.* **37**, 241-62.

Arthington A. H., Miller G. & Outridge P. M. (in press) Water quality, phosphorus budgets and management of Queensland dune lakes used for recreation. In: *Water Quality and Management for Recreation and Tourism.* Water Science and Technology Series. Pergamon Press, London.

Arthington A. H. & Watson J. A. L. (1982) Dragonflies (Odonata) of coastal sand dune fresh waters of south-eastern Queensland and northern New South Wales. *Aust. J. mar freshwat. Res.* **33**, 517-28.

Bayly I. A. E. (1964) Chemical and biological studies on some acidic lakes of east Australian sandy coastal lowlands. *Aust. J. mar. freshwat. Res.* **15**, 56-72.

Bayly I. A. E., Ebsworth E. P. & Wan H. F. (1975) Studies on the lakes of Fraser Island, Queensland. *Aust. J. mar. freshwat. Res.* **26**, 1-13.

Bayly I. A. E. & Williams W. D. (1973) *Inland Waters and their Ecology*. Longman, Victoria.

Bensink A. H. A. & Burton H. (1975) North Stradbroke Island — a place for freshwater invertebrates. *Proc. R. Soc. Qld.* **86**, 29-45.

Bishop K. (1980) Fish kills in relation to physical and chemical changes in Magela Creek (East Alligator River System, Northern Territory) at the beginning of the tropical wet season. *Aust. Zool.* **20**, 485-500.

Bishop K. (1983) Development of a model to predict freshwater fish community structure. In: *Environmental Protection in the Alligator Rivers Region.* Scientific Workshop, Vol. 2, paper 47, 1-21, Office of the Supervising Scientist for the Alligator Rivers Region.

Bishop K. A., Allen S. A., Pollard D. A. & Cook M. G. (1986) *Ecological Studies on the Freshwater Fishes of the Alligator Rivers Region, Northern Territory.* Research Report **4**, Vol. 1, 1-55. Aust. Govt. Publ. Service: Canberra.

Braithwaite L. W. (1980) Waterfowl resources and their management. In: *An Ecological Basis for Water Resource Management.* (ed. W. D. Williams) pp. 287-94. ANU Press, Canberra.

Briggs S. V. (1981) Freshwater wetlands. In: *Australian Vegetation* (ed. R. H. Groves) pp. 335-60. Cambridge University Press, Cambridge.

Brundin L. (1983) Two new aphrotenian larval types from Chile and Queensland, including *Anaphrotenia lacustris* n. gen., n. sp. (Diptera: Chironomidae). *Entomol. Scand.* **14**, 415-33.

De Deckker P. (1983) Australian salt lakes: their history, chemistry and biota — a review. *Hydrobiologia.* **105**, 231-44.

Garnett S. T. (1983) Report of the fifth aerial survey of migrating wading birds between Weipa and Millingimbi, 9-13 February, 1983. *Stilt* **4**, 15-17.

Georges A. (1982) Diet of the Australian freshwater turtle *Emydura krefftii* (Chelonia: Chelidae), in an unproductive benthic environment. *Copeia* **1982,** 331-36.

Ingram G. J. & Corben C. J. (1975) The frog fauna of North Stradbroke island with comments on the "acid" frogs of the wallum. *Proc. R. Soc. Qld.* **86,** 49-54.

Jamieson B. G. M. (1978) *Rhyacodrilus arthingtonae.* A new species of freshwater Oligochaete (Tubificidae) from North Stradbroke Island, Queensland. *Proc. R. Soc. Qld.* **89,** 39-43.

Lavery H. J. & Blackman J. G. (1971) Studies of waterfowl (Anatidae) in north Queensland. 7. Long-term changes in habitat. *Qd. J. Agric. Anim. Sci.* **28,** 275-80.

Lee-Manwar G., Arthington A. H. & Skerman V. B. D. (1979) Microbiological studies on Blue Lake, Brown Lake and Tortoise Lagoon, Queensland. *Aust. Soc. Limnol. Spec. Publ.* **3,** 45-48.

Lee-Manwar G., Arthington A. H. & Timms B. V. (1980) Comparative studies of Brown Lake, Tortoise Lagoon and Blue Lake, North Stradbroke Island, Queensland. 1. Morphology and origin of lakes. *Proc. R. Soc. Qld.* **91,** 53-60.

Longmore M. E. (1986) Modern and ancient sediments — data base for management of aquatic ecosystems and their catchments. In: *Limnology in Australia* (eds P. de Deckker & W. D. Williams) pp. 509-22. CSIRO, Melbourne and Dr W. Junk, Dordrecht.

Neboiss A. (1978) A review of caddisflies from three coastal islands of south-eastern Queensland (Insecta: Trichoptera). *Aust. J. mar. freshwat. Res.* **29,** 825-43.

Outridge P. M., Arthington A. H. & Miller G. J. (in press) Limnology of naturally acidic, oligotrophic dune lakes in subtropical Australia, including chlorophyll-phosphorus relationships. *Hydrobiologia.*

Paijmans K. (1978) A reconnaissance of four wetland pilot study areas. *CSIRO Div. Land Use Res. Technical Memorandum* **78/3.** pp. 37. CSIRO, Canberra.

Paijmans K., Galloway R. W., Faith D. P., Fleming P. M., Haantjens P. C., Heyligers P. C., Kalma J. D. & Loffler E. (1985) Aspects of Australian wetlands. *CSIRO Div. Water and Land. Res. Technical Paper.* **44.** pp. 77. CSIRO, Melbourne.

Pollard D. A. & Burchmore J. J. (1986) A possible scenario for the future of Australia's freshwater fish fauna. In: *Limnology in Australia* (eds P. De Deckker & W. D. Williams) pp. 615-36. CSIRO, Melbourne and Dr W. Junk, Dordrecht.

Reeve R., Fergus I. F. & Thompson C. H. (1985) Studies in landscape dynamics in the Cooloola-Noosa River area, Queensland. 4. Hydrology and water chemistry. *CSIRO Div. Soils Div. Rep.* **77.** pp. 45. CSIRO, Brisbane.

Smirnov N. N. & Timms B. V. (1983) A revision of the Australian Cladocera (Crustacea). *Rec. Aust. Museum. Suppl.* **1.**

Stanton J. P. (1975) A preliminary assessment of wetlands in Queensland. *CSIRO Div. Land Use Res. Tech. Mem.* **75/10.** pp. 52, CSIRO, Canberra.

Timms B. V. (1976) Morphometry of Lakes Barrine, Eacham and Euramoo, Atherton Tableland, North Queensland. *Proc. R. Soc. Qld.* **87,** 81-84.

Timms B. V. (1979) The benthos of some lakes in north-eastern Queensland. *Proc. R. Soc. Qld.* **90,** 57-64.

Timms B. V. (1982) Coastal dune waterbodies of north-eastern New South Wales. *Aust. J. mar freshwat. Res.* **33,** 203-22.

Timms B. V. (1986a) Reconnaissance limnology of lakes on Cape York Peninsula. *Aust. J. mar. freshwat. Res.* **37,** 167-76.

Timms B. V. (1986b) The coastal dune lakes of eastern Australia. In: *Limnology in Australia* (eds P. De Deckker & W. D. Williams)pp. 421-32. CSIRO, Melbourne and Dr W. Junk, Dordrecht.

Twidale C. R. (1966) Geomorphology of the Leichardt-Gilbert area of north-west Queensland. *CSIRO Div. Land Use Res. Series.* **16,** pp. 56. CSIRO, Canberra.

Vernon D. P. & Barry D. H. (1972) Birds of Fraser Island and adjacent waters. *Mem. Qd. Mus.* **16,** 223-32.

Watson J. A. L. & Arthington A. H. (1978) A new species of *Orthetrum* Newman from dune lakes in eastern Australia (Odonata: Libellulidae). *J. Aust. ent. Soc.* **17,** 151-57.

Williams W. D. (1983) *Life in Inland Waters.* Blackwell Scientific Publications, Oxford.

Proc. Ecol. Soc. Aust. 15:191-202

Soil nitrogen mineralization under closed and open forests on Fraser Island

B. N. RICHARDS, S. P. GOOSEM and J. J. O'NEILL

Department of Ecosystem Management, University of New England, Armidale, N.S.W. 2351

Abstract

On Fraser Island myrtaceous rainforest species dominate on the highest quality sites and are progressively replaced by a pattern of Symphyomyrtus, Monocalyptus *and* Corymbia *eucalypt subgenera associations as site quality declines. Soils from each of these associations were examined for their nitrogen mineralization potentials. Ammonification proceeded freely at all sites and NH_4-N was the dominant inorganic form of nitrogen present. Nitrate pools and fluxes were extremely small, and no real differences were evident between sites. Additions of ammonium as a potentially limiting substrate or phosphorus as a potentially limiting nutrient in factorial experiments had no effect on nitrate production. Nitrification was induced by liming the soils (1% w/w, pH 7.2), but only after a lag period of about eight weeks and only following net mineralization. Very little nitrification occurred in soils from the poorer quality sites even after liming.*

It is inferred from these results that low pH effectively controls the populations of autotrophic nitrifiers in these giant podzols. Furthermore, nitrifiers apparently cannot compete successfully with heterotrophic micro-organisms for available ammonium until such time as the energy sources for the heterotrophs are depleted, even when conditions are otherwise favourable for nitrification. Implications for forest management are briefly discussed.

Introduction

Fraser Island is a large (140 000 ha) sand mass situated close to the mainland, just south of the Tropic of Capricorn. The island's forests grow on high transgressive dunes of siliceous sand and are inherently infertile being particularly deficient in phosphorus and nitrogen. The surface 15 cm contains only 1.7me% total metallic cations, 30 $\mu g\ g^{-1}$ total phosphorus (half of which is organic), and 0.02% total nitrogen (Richards 1976). The levels of phosphorus and nitrogen decline dramatically with depth. The soils are giant podzols similar to those described by Thompson & Moore (1984) from the nearby Cooloola sand mass on the mainland. Mean annual rainfall is about 1 500 mm and markedly seasonal, much of it falling in late summer as a result of tropical cyclone activity. Under this rainfall regime, the soils are highly susceptible to leaching.

Despite its infertile substrate, the island supports 45 000 ha of commercially exploitable forest, of which 10 400 ha is subtropical rainforest (satinay-box scrub), and 17 000 ha high quality wet sclerophyll forest (old-growth eucalypt forest) (Richards 1976).

The major forest types on the island occupy a range of sites dominated by rainforest species or different subgenera of the genus *Eucalyptus*. This supports the view of Florence (1981), who pointed out that when east coast forest communities are aligned along a gradient of soil fertility, the resulting pattern reflects the subgeneric groupings of *Eucalyptus* recognized by Pryor & Johnson (1981).

TABLE 1. Site Descriptions.

SITE Site Quality (SQ)	1 SQI+ HIGH	2 SQI	3 SQII	4 SQIII	5 SQ << III LOW
Structural Form (*sensu* Specht 1970)	Tall closed forest	Closed forest	Tall open forest	Open forest	Open woodland
Type Description	Rainforest generally > 30 m with 25% of canopy dominated by sclerophyll emergents	Rainforest < 30 m with scattered sclerophyll emergents	Canopy predominantly *E. pilularis* > 30 m	Mixed eucalypt forest 18-24 m	*Banksia* dominated woodland < 18 m with scattered eucalypts
Dominant Tree Species*	*Syncarpia hillii* *Tristania conferta*	*S. hillii* *T. conferta* *Eucalyptus microcorys* (S) *E. resinifera* (S)	*E. pilularis* (M) *E. microcorys* (S) *S. hillii*	*E. pilularis* (M) *E. signata* (M) *E. resinifera* (S) *E. gummifera* (C) *E. intermedia* (C)	*Banksia serratifolia* *E. signata* (M) *E. gummifera* (C) *E. intermedia* (C) *Angophora costata* (A)
Subgenus Category	(Rainforest)	(Rainforest) — *Symphyomyrtus* transition	*Symphyomyrtus*-*Monocalyptus* transition	*Monocalyptus*-*Corymbia* transition	*Corymbia-Angophora* dominated

*Letters in brackets after *Eucalyptus* spp. indicate subgenera: S. *Symphyomyrtus,* M. *Monocalyptus,* C. *Corymbia,* A. *Angophora* (*sensu* Pryor & Johnson 1971).

The subgenus *Symphyomyrtus* is generally associated with soils of higher nutrient status and the subgenera *Monocalyptus* and *Corymbia* with soils of lower nutrient status. On Fraser Island myrtaceous rainforest species dominate on the highest quality sites and are progressively replaced by a pattern of mixed rainforest/*Symphyomyrtus*, *Symphyomyrtus*/*Monocalyptus*, *Monocalyptus*/*Corymbia* and *Corymbia*/*Angophora* associations as site quality declines (the genus *Angophora* is considered by Pryor & Johnson to be taxonomically equivalent to a eucalypt subgenus).

If the subgeneric groupings were the result of adaptation to environmental pressures, then it is reasonable to assume that some distinguishing attributes might still persist, permitting an insight into present species distributions, patterns of association and reaction to disturbance. For instance, in response to this evolutionary differentiation of the genus, different physiological mechanisms may have evolved, resulting in differential responses of the several subgenera to environmental factors such as microbial activity and nutrient cycling and availability. This may have important implications for the management of these forests not only for timber production but also for the purposes of biological conservation.

Site selection

Five study areas were selected, representative of five site quality/subgenus categories based on the dominant associations of the subgenera (Table 1). In conjunction with this floristic delineation of sites, soil profile descriptions were undertaken by C. H. Thompson (CSIRO Division of Soils) to provide a soil morphology base to assist in the interpretation of the soil/floristic relationships and to confirm the basic assumption of site selection, that is, that the sites represented a fertility gradient.

This fertility gradient is clearly reflected in soil profile development (Fig. 1). The two major factors of interest are the thickness of the combined A_1 horizons and the depth to the B horizon. All sites have three A_1 sub-horizons (a characteristic of Australian dune podzols — Thompson & Hubble 1980), made up of nutrient impoverished quartz sand grains and small quantities of organic matter. Due to the paucity of elements which might normally stabilize the products of organic matter decomposition (e.g., calcium, aluminium or iron), the degree of development of the A_1 horizons has been found to be related to the quantity of standing biomass (Thompson & Bowman 1984). It can be seen from Figure 1 that the thickness of the combined A_1 horizons decreases progressively from Site 1 to Site 5. The degree of nutrient leaching can be gauged from the depth to the B horizon. The *Monocalyptus* and *Corymbia* dominated sites (Sites 4 and 5) have B horizons which have been leached to much greater depths than in the three higher quality sites which reflects of the older age of the dune systems of the lower quality sites (C. H. Thompson pers. comm.).

Nitrogen mineralization and disturbance

Many of the characteristics of forests of the humid tropical lowlands are exaggerated on Fraser Island. Forests 40 metres tall, with a biomass of about 1 500 t ha^{-1} (Applegate 1982), are supported on the deep infertile sands whose continued productivity must depend to a large degree on nutrient mineralization processes at the soil-litter interface. The soil-litter subsystem (*sensu* Charley & Richards 1983) is the gate through which virtually all the minerals that cycle in forests must pass at some stage, and is therefore a focal point in any study of ecosystem dynamics. It is often the rate at which energy and nutrients flow through this subsystem that governs the productivity of the whole system. In ecosystems which operate on a very low nutrient capital such as those of Fraser Island, this regulatory role should be particularly important. Disturbance of these forests by logging and burning will stress the nutrient cycle and any resultant strains should be detectable first at critical points in that cycle, for example, in the mineralization processes of the soil-litter subsystem.

The nitrogen cycle is a good entry point for the study of an ecosystem with respect to possible effects of disturbance. Nitrate losses have increased more consistently and to a greater extent than those of other ions in most disturbed systems (Vitousek *et al.* 1979). Before disturbance, the rates of nitrogen mineralization and nitrification might differ widely among forests as a consequence of differences in biotic and abiotic controlling factors, however in undisturbed systems there is usually only a small loss of mineral nitrogen since it is rapidly recycled by the vegetation. Following destructive disturbance (either natural or man-made), conditions often favour the acceleration of mineralization processes, and factors previously controlling nitrification may be removed. Increased nitrification, in combination with the reduction or elimination of

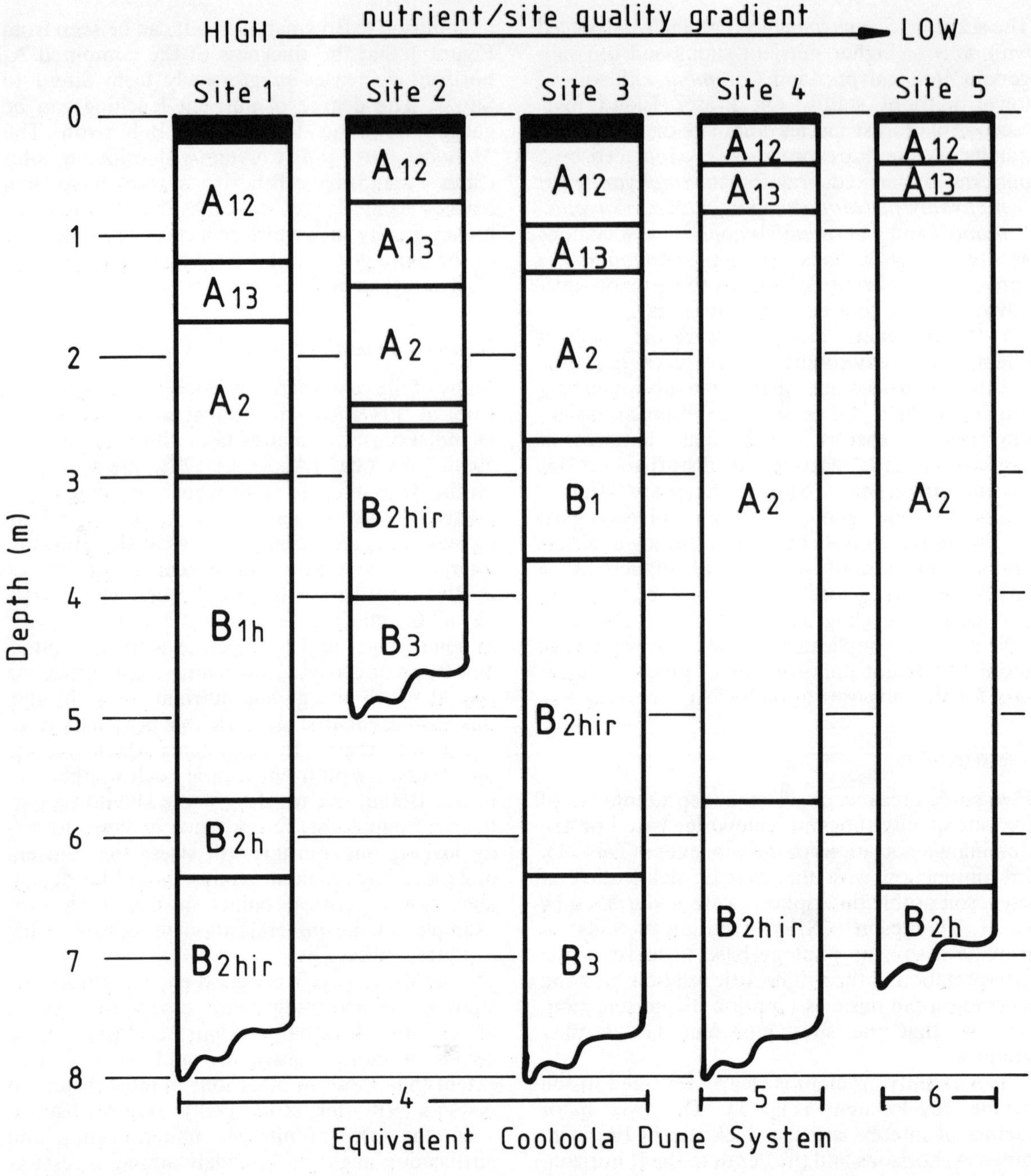

FIG. 1. Soil profile development. Site-soil classification: first three sites — *Giant podzol* (Uc2.22), fourth site — *Giant podzol* (Uc2.21), fifth site — *Giant humus podzol* (Uc2.20). A_{11} horizons represented by the shaded area at top of profiles. The narrow unlabelled soil horizon in site 2 is a B_1 horizon. Descriptions by C. H. Thompson.

nitrogen uptake by the vegetation during the recovery phase may result in large amounts of nitrate being leached, at least until vegetative cover is re-established.

Factors which control nitrification in forest ecosystems are of more than theoretical interest because they are central to the problem of predicting nutrient losses from forested catchments should these be cleared for agriculture or harvested for wood production. Practical interest in the latter has heightened in recent years with the wider use of more intensive logging techniques, notably clearfelling. Nitrate losses are particularly significant in this context, not only due to leaching of the nitrate ion itself, but because the release of hydrogen ions in the formation of nitrate, in conjunction with the high mobility of the nitrate ion, promotes the mobilization and subsequent loss of soil cations. Any factors that inhibit nitrification in forest ecosystems should therefore assist in preventing or delaying nutrient losses. Such inhibitory mechanisms will confer resistance in a nutrient depauperate ecosystem, while the speed of vegetation recovery will confer resilience (*sensu* Vitousek *et al.* 1979). Resistance (or persistence) in this sense is the relative magnitude of the response to disturbance, while resilience is the relative recovery rate following disturbance. It would seem reasonable to expect that a fertile site would revegetate more rapidly than a low quality site, thereby magnifying differences between high and low site quality areas following repeated disturbances. Although there is evidence that nitrate losses following disturbance are greater on high quality sites (Vitousek *et al.* 1979), nutrient losses should be proportionately more significant and recovery time greater on a nutrient poor site.

There have been several possible mechanisms put forward by various researchers to explain the regulation of nitrification. The hypotheses and conclusions reached by different investigators are often conflicting, reflecting both differences among the ecosystems studied and in the methodologies used. Three main biotic mechanisms have been proposed to account for low rates of nitrification. First, nitrification could be allelopathically inhibited by secondary plant compounds (Rice 1974). Second, nitrification could be prevented by a deficiency of available nutrients, especially ammonium or phosphorus (Purchase 1974a,b). And third, autotrophs may be weak competitors for ammonium compared to heterotrophs and so nitrification (a process dependent on ammonium as an energy source) is regulated by heterotrophic demand for nitrogen (Jansson 1958). A corollary to the latter hypothesis is that nitrification will not occur at substantial rates unless there is net mineralization. The main abiotic controlling mechanism proposed has been soil pH. Many forest soils are moderately to strongly acidic, and consequently the activity of autotrophic nitrifying bacteria in these soils is likely to be inhibited (Keeney 1980).

Methods and Results

Effect of soil moisture status

Soil incubation tests showed that ammonification proceeded freely at all sites although SQIII (Site 4) produced significantly less NH_4-N than the other sites (Table 2); this was true at all levels of soil moisture tested. Nitrification rates on the other hand were minimal at all water contents, although marginally enhanced rates were obtained for water contents of 20 to 30% (w/w). Values in Table 2 are corrected for initial levels which were extremely low for nitrate. These tests were performed on bulked soils which had been collected during a protracted period of drought. Soils collected subsequently generally had a greater capacity to ammonify but over a four week incubation period their nitrification rates were similar to the results shown in this table even when incubated at their optimal water contents. The optimal moisture contents for nitrification were equivalent to 60% to 70% field capacity, calculated on the <2 mm soil fraction; these are the values given in Table 2.

TABLE 2. Soil characteristics.

Parameter	Site 1	2	3	4	5
NO_3-N (t_0)	0.12	0.24	0.10	0.11	0.12
NH_4-N (t_0)	2.15	3.63	2.18	1.55	5.20
NO_3-N produced (t_4)	0.00	0.00	0.04	0.00	0.91
NO_3-N produced (t_{32})	2.22	3.42	4.70	0.82	6.45
net NH_4-N produced (t_4)	18.80	18.03	17.83	10.04	14.24
% loss on ignition	0.79	1.09	0.62	4.61	4.92
% total N (Kjeldahl)	0.06	0.14	0.04	0.09	0.12
pH	5.7	5.3	5.1	4.6	4.5

t_0 = before incubation; t_4 = increase after 4 weeks incubation; t_{32} = increase after 32 weeks incubation.
NO_3-N and NH_4-N concentration units = $\mu g\ g^{-1}$.

Effect of adding ammonium and phosphorus

Since the soils are very infertile low nitrification rates may reflect limited substrate availability (i.e. NH_4^+) or a deficiency of some essential nutrient, notably phosphorus. To test this hypothesis, high (0 — 400 μg g^{-1}) and low (0 — 32 μg g^{-1}) levels of NH_4-N (as $(NH_4)_2SO_4$) and PO_4- (as NaH_2PO_4) were added to the soils from the five sites in two 6 × 6 factorial incubation experiments. Neither addition had any significant effect on nitrification rates, nor was there any interaction between the two (results not presented). It is inferred that short-term (4 weeks') inhibition of nitrification was either not regulated by NH_4-N or P deficiency, or that other factors were more strongly limiting.

Effect of lime and period of incubation

Nitrate pools and fluxes were extremely small and there was very little real difference in field levels of nitrate between sites (Table 2, t_o). An outstanding feature of these soils is their negligible capacity to nitrify. Soils of low pH often have low nitrifier activity and this has been shown to be a major factor governing patterns of nitrification in many soils (Morril & Dawson 1967). Autotrophic nitrifying bacteria have a narrow optimum pH range, usually between pH 6.0 and 8.0 (Alexander 1977). The soils in the present study had an average pH of 5.0 (see Table 2 for details). Alexander (1977) reported that nitrification usually decreases below pH 6.0 and becomes almost negligible at pH 5.0. Scatter-diagrams of field pH versus nitrification rates for each site and for all sites combined revealed no apparent relationship between the two, but this may simply be a reflection of sub-optimum field pH values. To test whether soil pH was limiting for nitrification, a series of experiments was conducted involving extended incubation periods in the presence or absence of lime (1% w/w). This amount of lime raised the pH of the soil samples from each site to a common pH of 7.2, all soils being substantially buffered at this pH. Soil samples (20 g) were brought to and maintained at 60% of their respective water holding capacities and incubated at 28°C in glass bottles sealed with polyethylene film.

Figure 2 shows the effect of incubation period and lime on nitrification rates. In the absence of lime, a long lag period (up to 32 weeks for SQIII) was evident before any substantial nitrification occurred, but the overall rates were still very low.

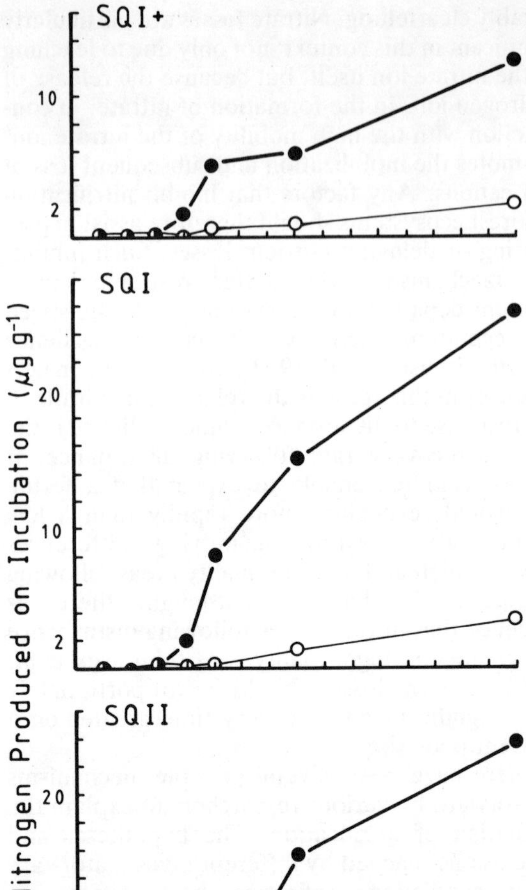

FIG. 2. Rates of nitrification during incubation of disturbed samples at 28°C and 60% water holding capacity; means of 3 replicates. Control treatment (minus lime) represented by open circles, lime treatment by closed circles.

TABLE 3. Interaction of site and lime on nitrification rate[1].

Site	Control	Lime	Main effect of site
1	0.53 **	3.60	2.07
2	0.71 ***	7.25	3.98
3	1.29 ***	6.61	3.95
4	0.41 ns	0.58	0.49
5	2.73 ns	3.58	3.15
Main effect of lime	1.13 ***	4.32	
LSD (main effects)	p<0.05	1.32	2.09
	p<0.01	1.79	2.83
LSD (interaction)	p<0.05	2.96	
	p<0.01	4.01	

[1] means of 14 observations for sites.
means of 35 observations for lime.
means of 7 obervations for interaction.
ns not significant.
** significant at 95% level.
*** significant at 99% level.

TABLE 4. Interaction of time and lime on nitrification rate[1].

Time	Control	Lime	Main effect of site
2	0.04 ns	0.02	0.03
4	0.16 ns	0.13	0.15
6	0.49 ns	0.24	0.36
8	0.80 ns	1.79	1.29
10	0.97 **	4.75	2.86
16	1.94 ***	8.99	5.46
32	3.52 ***	14.34	8.93
LSD (main effects)	p<0.05		2.47
	p<0.01		3.35
LSD (interaction)	p<0.05	3.50	
	p<0.01	4.74	

[1] means of 10 observations for incubation time.
means of 35 observations for lime.
means of 5 obervations for interaction.
ns not significant.
** significant at 95% level.
*** significant at 99% level.

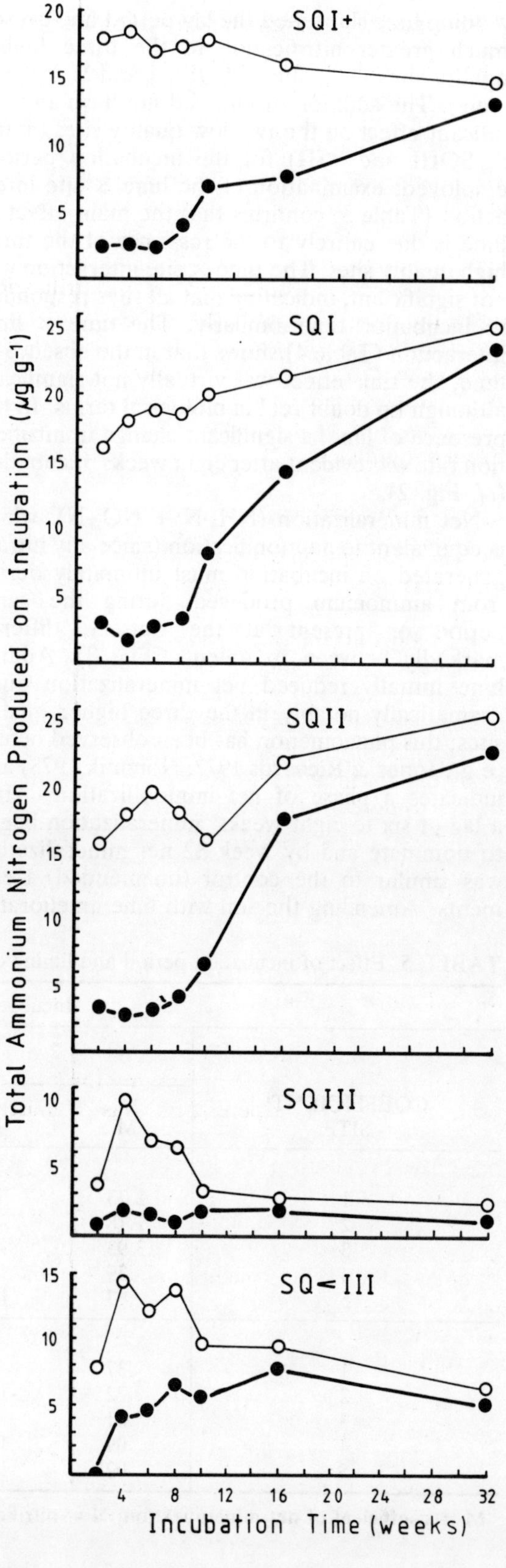

FIG. 3. Rates of net mineralization of organic N during incubation of disturbed samples at 28°C and 60% water holding capacity; means of 3 replicates. Control treatment (minus lime) represented by open circles, lime treatment by closed circles.

Adding lime shortened the lag period and caused much greater nitrification in the three higher quality sites (1, 2 and 3, SQI$^+$, I and II respectively). The addition of lime did not have any significant effect on the two low quality sites (4 and 5, SQIII and <III) for the incubation periods employed: examination of the lime × site interaction (Table 3) confirms that the main effect of lime is due entirely to the response of the three high quality sites. The time × site interaction was not significiant, indicating that all sites responded to incubation time similarly. The time × lime interaction (Table 4) shows that in the absence of lime, the time effect was virtually not significant although no doubt real in biological terms. In the presence of lime, a significant change in nitrification rate was evident after eight weeks' incubation (*cf.* Fig. 2).

Net mineralization (NH_4-N + NO_3-N), which is equivalent to ammonification (since any nitrate generated on incubation must ultimately derive from ammonium produced during the same period or present at the outset), differed markedly between treatments (Fig. 3). Adding lime initially reduced net mineralization quite dramatically notably in the three higher quality sites; this phenomenon has been observed before (e.g., Jones & Richards 1977; Nommik 1978) and indicates a phase of net immobilization. After a lag of six to eight weeks, mineralization began to dominate and by week 32 net mineralization was similar to the control (unamended) treatments. Amending the soil with lime ameliorated the acidic environment, thereby favouring the development of both heterotrophic and autotrophic micro-organisms. The lag in net mineralation was a response to these favourable conditions since the population increase of heterotrophs would result in rapid immobilization of ammonium by assimilation into new microbial protoplasm. The ammonium that accumulated following this initial phase of net immobilization represents the quantity of nitrogen in excess of heterotrophic requirement, and served as a substrate for the autotrophic nitrifiers. Thus nitrification in lime-amended high site quality soils occurred only during the period of net mineralization.

Further interpretation is possible by examining the coefficients of nitrification and net mineralization (Table 5). The nitrification coefficient represents the percentage of available substrate (ammonium initially present before incubation plus the total ammonium produced on incubation) which is nitrified over the incubation period. Adding lime caused a dramatic increase in the nitrification coefficient after eight weeks' incubation in the three higher site qualities (Sites 1, 2, 3) from which it is inferred that nitrifying populations in the more acidic, unamended soils were too small to use the available substrate fully. The slower rise in nitrification coefficients in the lower site qualities (Sites 4, 5) suggest that factors other than low soil pH were inhibiting the development of nitrifying bacteria in these soils. The possibility that alleopathic agents were involved should not be dismissed, and will be discussed later.

TABLE 5. Effect of incubation period and liming on the coefficients of net mineralization and nitrification (%).

	Incubation Time (weeks)					
	2		8		32	
COEFFICIENT[1] SITE	M	N	M	N	M	N
			Control Treatment			
1	2.73	.15	2.63	.76	2.23	13.16
2	1.18	-.64	1.36	1.07	1.86	11.88
3	3.93	-.22	4.52	3.76	6.32	17.14
4	.39	.12	.69	5.30	.27	20.39
5	.67	1.79	1.17	12.81	.55	55.51
			Lime Treatment			
1	.37	.66	.62	26.84	1.99	81.48
2	.22	-1.35	.24	27.99	1.73	93.99
3	.81	.74	1.04	23.53	5.66	95.04
4	.09	1.71	.09	12.89	.13	67.67
5	.02	1.29	.57	29.84	.44	80.35

[1] M = coefficient of net mineralization; N = nitrification coefficient. For explanation, see text.

Coefficients of net mineralization (the ratio of N mineralized to total Kjeldahl N expressed as percentages) (Table 5) indicate that a much lower proportion of total soil nitrogen is mineralized in the two lower quality sites. This is probably a reflection of poor substrate quality as indicated by the much higher C:N ratio of these soils. The C:N ratios calculated on the less than 2 mm fraction were 11.9, 8.0, 15.6, 49.8 and 42.5 for Sites 1 to 5 respectively. Using total soil N as a measure of potentially available substrate when calculating net mineralization coefficients may be misleading however, especially in forest soils with typically high C:N ratios, and containing organic matter at various stages of decomposition. Although nearly all of the soil-N exists in organic form, the importance of the total-N soil pool to an ecosystem is not necessarily reflected by its absolute size, only a small proportion of the total participating in the mineralization-immobilization cycle at any one time (Jansson 1958). In an attempt to overcome this difficulty, the amount of total-N which was extracted by autoclaving at 120°C and 103 kPa for one hour was deemed as the readily available total-N fraction. This autoclave-soluble fraction was 5.16, 3.31, 8.26, 5.94 and 7.65% of total (Kjeldahl)-N for Sites 1 to 5 respectively. Fig. 4 gives the results of the percentage of so-called readily available total-N which was mineralized. The vastly different behaviour of the two lower quality sites is immediately evident. In the absence of lime, net mineralization of the available substrate was very rapid in the three higher quality sites, with up to half being mineralized during the first two weeks of incubation. The lag period observed in the lime amended soils accords well with the hypothesis that lime results initially in net immobilization due to the activities of the heterotrophic soil microflora, followed by a period of net mineralization when the population of heterotrophs stabilizes. The pattern evident in the low quality sites, and their lack of response to liming is assumed to be a consequence of the poor quality (inherent decomposability) of the organic substrates available for mineralization. The alternative hypothesis, *viz.* that of alleopathic inhibition, does not seem tenable in relation to mineralization given the cosmopolitan nature of the ammonifying microflora (Richards *et al.* 1985). Allelopathic inhibition of nitrification seems much more plausible, since we are dealing here with certain specific physiological groups of autotrophic bacteria. Rice and Pancholy (1974) have reported that various phenolic and tannin compounds inhibit the process of nitrification in soils when present at concentrations in the soil solution as low as 10 nM. However, the release and activity of these toxins depends on their solubility (Lodhi 1978), and increasing the soil pH, by practices such as liming, markedly increases the solubility and release of phenolic compounds from the soil, plant roots and litter (Whitehead *et al.* 1981). If such compounds were inhibiting nitrification in these giant podzols then a decrease in nitrification following liming would be expected, not an increase as is shown here. So far as inhibition of nitrification is concerned therefore, it is doubtful whether even this can be attributed to allelopathy.

Finally, the five sites examined fall into two distinct groups on the basis of their mineralization potentials, the three better quality sites (1, 2 and 3) being clearly differentiated from the two poorer quality sites (4 and 5). Thus soils from the rainforest/*Symphyomyrtus*-dominated communities respond to liming more readily (Table 3, Figs. 2 and 3), and are able to utilize a much greater proportion of the substrates available for both ammonification and nitrification (Table 5, Fig. 4), than soils from the *Monocalyptus/Corymbia/Angophora*-dominated communities. This conforms with Florence's (1981) hypothesis that the eucalypt subgenera are aligned along a soil fertility gradient.

Implications for Forest Management

The sites which support the commercial timber stands on the island — satinay *(Syncarpia hillii)*, brush box *(Tristania conferta)*, tallowwood *(E. microcorys)* and high quality blackbutt *(E. pilularis)* (see Table 1) — are all on the more fertile side of the site quality/nutrient status gradient. These sites are therefore more likely to be subjected to destructive disturbance. They are also the sites most likely to nitrify following the types of disturbance envisioned in forestry practice, especially when it is associated with the burning of logging debris or slash to induce regeneration. Ash increases soil pH (O'Connell *et al.* 1981), and increased rates of nitrification in eucalypt forest soils following repeated burning are well documented (e.g., Jones & Richards 1977, 1978; O'Connell *et al.* 1981).

Whether a particular disturbance will destabilize mineral cycling and lead to a decline in soil fertility depends on several factors such as those outlined by Kimmins (1972) *viz.* the degree of soil disturbance; whether or not slash burning

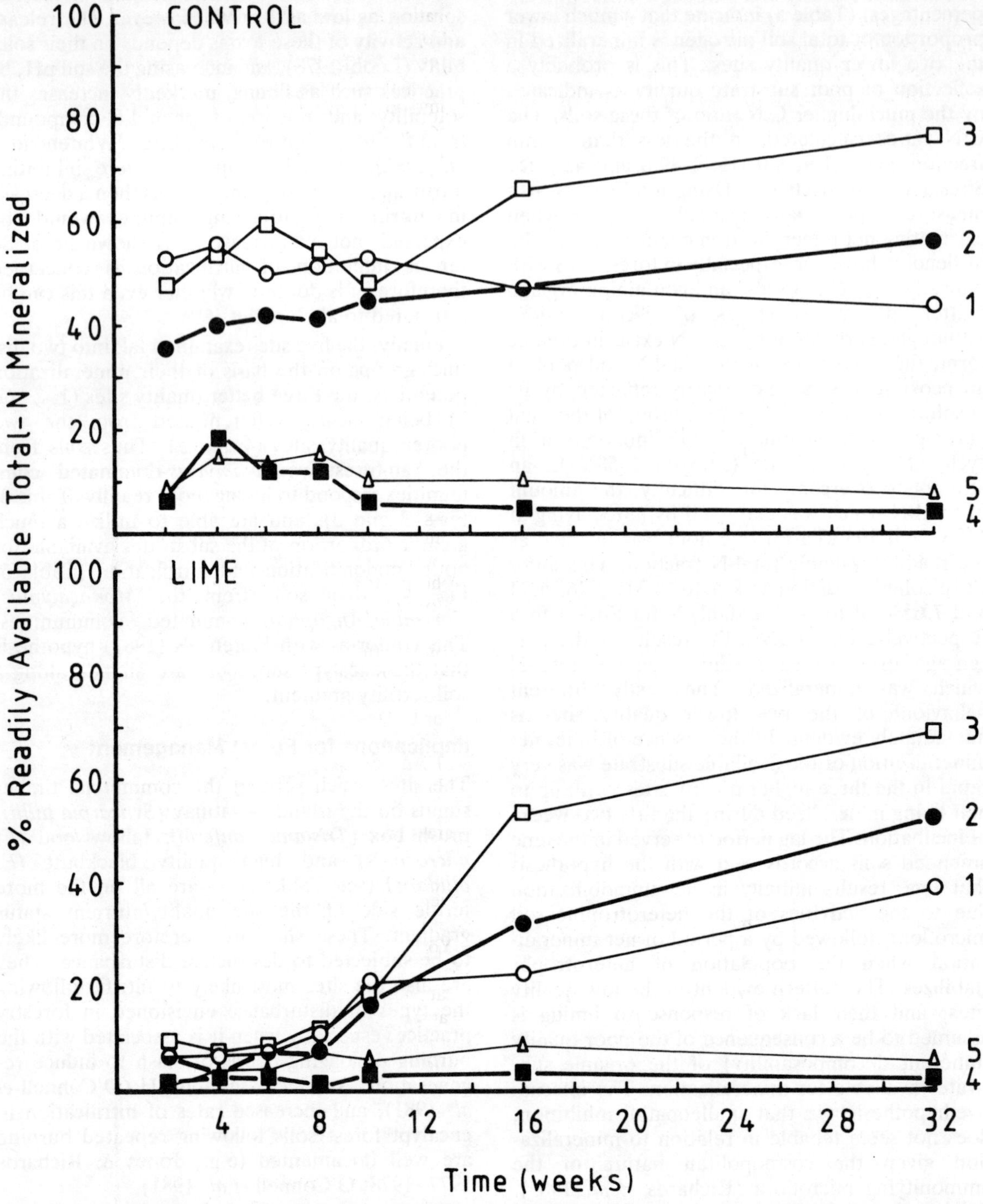

FIG. 4. Effect of incubation period and liming on the percentage of readily available (autoclave soluble) total N mineralized. Open circles, Site 1; closed circles, Site 2; open squares, Site 3; closed squares, Site 4; open triangles, Site 5.

is done; the speed of successional change; the area disturbed at any one time; and the frequency of disturbance. The interaction of these factors will determine the magnitude of nutrient depletion during the reorganization phase (*sensu* Bormann & Likens 1979) following disturbances. Successful management should aim to manipulate these factors in such a way as to hasten the onset of the aggradation phase of mineral nutrient accretion in biomass.

The observed lag phases in both net mineralization and nitrification may also be important factors. The acidic conditions which have prevailed in these soils for a very long time have apparently prevented the development of substantial populations of nitrifiers *in situ,* requiring a substantial interval of time before a marked population increase occurs following the onset of conditions conducive to nitrification. Depletion of total N reserves in soil as a consequence of repeated burning has also been demonstrated with loss being greatest in the more labile (HC1 hydrolysable) fraction (Jones & Richards 1977). If this takes place in the giant podzols of Fraser Island, the quality of the substrates (as indicated by the ratio of C to readily available N) would decline, thus favouring net immobilization.

The critical factor involved in the response of these forest ecosystems to cope with disturbance may be time. The rate of site revegetation (a measure of resilience) may be governed by the balance between the length of time in which the nutrient conserving lags (resistance or persistence) are operative and the resumption of nutrient uptake by the regenerating vegetation.

References

Applegate G. B. (1982) *Biomass of blackbutt* (Eucalyptus pilularis) *forests on Fraser Island.* M. Nat. Res. Thesis, Univer. New England, Armidale.

Alexander M. (1977) *Introduction to Soil Microbiology.* Wiley, New York.

Bormann F. H. & Likens G. E. (1979) *Pattern and Process in a Forested Ecosystem.* Springer-Verlag, New York.

Charley J. L. & Richards B. N. (1983) Nutrient allocation in plants: mineral cycling in terrestrial ecosystems. In: *Encyclopedia of Plant Physiology (New Series): Physiological Plant Ecology IV* (eds. O. L. Lange, C. B. Osmond, P. S. Noble & H. Ziegler) pp. 5-45. Springer-Verlag, Berlin.

Florence R. G. (1981) The biology of the eucalypt forest. In: *Biology of Native Australian Plants* (eds. J. Pate & J. McComb) pp 147-80. Univ. Western Australia Press, Perth.

Jansson S. L. (1958) Tracer studies on nitrogen transformations in soil with special attention to mineralization-immobilization relationships. *K. Lanthr. Hogsk. Annlr.* **24,** 101-361.

Jones J. M. & Richards B. N. (1977) Effect of reforestation on turnover of ^{15}N-labelled nitrate and ammonium in relation to changes in soil microflora. *Soil Biol. Biochem.* **9,** 383-92.

Jones J. M. & Richards B. N. (1978) Fungal development and the transformation of ^{15}N-labelled amino and ammonium nitrogen in forest soils under several management regimes. *Soil Biol. Biochem.* **10,** 161-68.

Keeney D. R. (1980) Prediction of nitrogen availability in forest ecosystems: A literature review. *Forest Sci.* **26,** 159-71.

Kimmins J. P. (1972) The ecology of forestry: the ecological role of man, the forester, in forest ecosystems. *For. Chron.* **48,** 301-07.

Lodhi M. A. K. (1978) Allelopathic effects of decaying litter of dominant trees and their associated soil in a lowland forest community. *Amer. J. Bot.* **65,** 340-44.

Morril L. G. & Dawson J. E. (1967) Patterns observed for the oxidation of ammonium to nitrate by soil organisms. *Soil Sci. Soc. Amer. Proc.* **21,** 385-88.

Nommik H. (1978) Mineralization of carbon and nitrogen in forest humus as influenced by additions of phosphate and lime. *Acta Agric. Scand.* **28,** 221-30.

O'Connell A. M., Grove T. S. & Lamb D. (1981) The influence of fire on the nutrition of Australian forests. In: *Proc. Australian Forest Nutrition Workshop: Productivity in Perpetuity* pp. 277-289. CSIRO, Melbourne.

Pryor L. D. & Johnson L. A. S. (1981) *Eucalyptus,* the universal Australian. In: *Biogeography and Ecology of Australia* (ed. A. Keast) pp 499-536. Junk, The Hague.

Purchase B. S. (1974a) Evaluation of the claim that grass root exudates inhibit nitrification. *Plant Soil.* **41,** 527-39.

Purchase B. S. (1974b) The influence of phosphate deficiency on nitrification. *Plant Soil* **41,** 541-47.

Rice E. L. (1974) *Allelopathy.* Academic Press, New York.

Rice E. L. & Pancholy S. K. (1974) Inhibition of nitrification by climax ecosystems III. Inhibitors other than tannins. *Amer. J. Bot.* **61,** 1095-103.

Richards B. N. (1976) The Utilization, Management and Conservation of the Forest Resources of Fraser Island. School of Natural Resources Project Rep. No. 6, Univ. of New England, Armidale, 176 pp.

Richards B. N., Smith J. E. N., White G. J. & Charley J. L. (1985) Mineralization of soil nitrogen in three forest communities from the New England region of New South Wales. *Aust. J. Ecol.* **10,** 429-41.

Thompson C. H. & Bowman G. M. (1984) Subaerial denudation and weathering of vegetated coastal dunes in eastern Australia. In: *Coastal Geomorphology in Australia.* (ed. B. G. Thom) pp. 263-90. Academic Press, Sydney.

Thompson C. H. & Hubble G. D. (1980) Subtropical podzols (spodosols and related soils) of coastal eastern Australia. In: *Classification and Management of Tropical Soils,* (ed. K. T. Joseph) pp. 203-13. Malaysian Soc. Soil Sci., Kuala Lumpur.

Thompson C. H. & Moore A. W. (1984) Studies in Landscape Dynamics in the Cooloola-Noosa River Area Queensland. *CSIRO Division of Soils, Divisional Rep. No. 73.*

Vitousek P. M., Gosz J. R., Grier C. C., Melillo J. M., Reiners W. A. & Todd R. L. (1979) Nitrate losses from disturbed ecosystems. *Science (N.Y.)* **204,** 469-74.

Whitehead D. C., Dibb H. & Hartley R. D. (1981) Extractant pH and the release of phenolic compounds from soils, plant roots and leaf litter. *Soil Biol. Biochem.* **13**, 343-48.

Proc. Ecol. Soc. Aust. 15:203-211

The influence of seed predators on structure and succession in tropical tidal forests

T. J. SMITH III

*Australian Institute of Marine Science, P.M.B. No. 3, Townsville M.C., Qld. 4810**

Abstract

Effects of seed predators within mangrove forests were studied. Controlled experiments were used to test for differences in predation among five species of mangroves, examine effects of seed and predator density on seed consumption rates, and to examine how differential amounts of predation on seeds may be related to species distribution patterns and patterns of forest succession.

Averaged over all species and forests, 57.7±34.4% ($\bar{X}$±1SD) of the seeds were consumed by predators (small crabs belonging to the family Grapsidae). Significant differences existed in the amounts of predation among species. Avicennia marina *was most heavily preyed upon, followed in decreasing order by* Bruguiera gymnorrhiza, Ceriops tagal, Rhizophora apiculata *and* R. stylosa. *Significant differences also existed among five forest types, which differed in species dominance. Cumulative amounts of predation were significantly higher in a* Bruguiera*-dominated forest for all species. For* A. marina, C. tagal *and both* Rhizophora's, *amounts of predation tended to be highest in forests where conspecifics were rare or absent and lowest where conspecifics were dominant.*

Predation on Bruguiera *was significantly less in a* Rhizophora *forest that had no* Bruguiera *in the canopy, but sapling sized* Bruguiera *in the understorey, than in a* Rhizophora *forest with no* Bruguiera *saplings. This suggests that seed predators may influence observed replacements of* Rhizophora *by* Bruguiera.

Predation rates were significantly greater in a forest with high predator densities compared to a forest with low densities. More importantly, however, is the finding that amounts of predation did not depend on propagule density. Single propagules were just as likely to be consumed as were propagules in a large pile, regardless of predator density.

Introduction

Janzen (1970) and Connell (1971) independently hypothesized that the activity of species specific seed predators were responsible for the high tree species diversity observed in tropical forests. Numerous studies have since shown that seed predators and pathogens have important effects on tree community structure and population dynamics (See Clark & Clark 1984 for a review).

Current hypotheses concerning mangrove zonation patterns do not consider biological factors such as seed predation. Rather, they invoke physiological adaptation by each species to different portions of a gradient in salinity and frequency of tidal inundation which exists across the intertidal zone (Watson 1928; Davis 1940; Macnae 1968, 1969) or tidal sorting of propagules based on differences in their size (Rabinowitz 1978).

Recently Smith (1987a,b) showed that seed predators influenced tree species distributions across the intertidal in low diversity tropical, tidal forests (mangroves). For example, *Avicennia marina*, which is often the dominant species in low and high intertidal regions (Bunt & Williams

*Present address: Florida Co-operative Fish and Wildlife Unit, School of Forest Research and Conservation, University of Florida, Gainesville, Florida 32611 USA.

1981; Elsol & Saenger 1983, respectively), is usually absent from mid-intertidal forests (Johnstone 1983). Smith (1987b) used caging experiments to show that *Avicennia* seedlings could grow successfully in the mid-intertidal. Further experiments revealed that 100% of *Avicennia* propagules were consumed in the mid-intertidal by grapsid crabs (primarily by *Sesarma messa*). In low and high-intertidal forests consumption of *Avicennia* propagules was significantly less (mainly *Neosarmatium smithi* and *N. meinerti* in high-intertidal forests).

The research reported here addresses further questions concerning the role of seed predators in mangrove forests:

(1) Does the amount of predation vary between forest types for a given species?
(2) Do the predators select propagules of certain species?
(3) Is the amount of predation mediated by variation in predator and/or propagule density?
(4) Is the pattern of seed predation between forests related to patterns of tree species replacements (succession) which have been reported in the literature?

Methods

Study sites

Experiments were conducted in the extensive (70 km^2) mangrove forests at Missionary Bay, Hinchinbrook Island (18°15′S, 146°12′E), 160 km north of Townsville, Queensland, and at a site near Lucinda at the mouth of the Herbert River (18°31′S, 146°19′E) 100 km north of Townsville (Fig. 1). Approximately 27 species of mangroves are found at Missionary Bay with four dominants in the canopy: *B. gymnorrhiza, C. tagal, R. apiculata* and *R. stylosa. A. marina* is a conspicuous member of the low and high-intertidal forest communities but never reaches the abundance of the other species in these forests. Missionary Bay has been intensively studied and described in detail by Bunt & Williams (1980, 1981) and Bunt *et al.* (1982). Sixteen species are found at the Lucinda site (high-intertidal) with *A. marina* being dominant in the canopy. [Nomenclature follows Bunt *et al.* (1982) for mangroves and Davie (1982) for crabs.] Predator/prey density experiments were performed in the mangrove forests near the Australian Institute of Marine Science on Bowling Green Bay which have been previously described (Smith, 1987b).

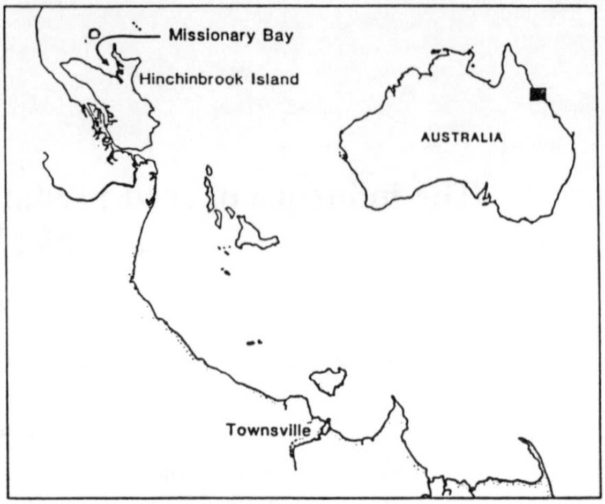

FIG. 1. Location of study sites along the north Queensland coast.

Experimental design

Predation differences between forest types. Experiments were conducted during the 1986 wet season (Jan-March); the season when propagules mature and drop from the parent (Williams *et al.* 1981). Propagules were tethered on 1 m lengths of nylon twine to prevent their floating away on the tides. The twine served as a "tracer" for recovering the propagule, for example, from crab burrows. Although tethering restricted movement of propagules by tidal action, it allowed study of predation in the post-dispersal stage. Propagules of five species (*Avicennia marina*, *Bruguiera gymnorrhiza*, *Ceriops tagal*, *Rhizophora apiculata* and *R. stylosa*) were tied to a single stake. Twenty stakes were used per replicate plot with stakes 2.5 m apart and arrayed in either a 4×5 or 2×10 grid pattern. The plots and grid spacing were large enough to ensure that adjacent propagules did not become entangled and that the addition of propagules did not increase the density at which propagules normally occur in the forest (Williams *et al.* 1981). The background density of propagules was determined by counting all naturally occurring propagules in five 1.0 m^2 quadrats in each replicate plot for each species and forest combination.

Five plots were located randomly in each of five forests, one forest being dominated by each of the five species under study (Table 1). The *A. marina* forest was at Lucinda. All other forests were at Missionary Bay. The rationale for

TABLE 1. (a) Characteristics of the five forests studied. Values given are the means and standard deviations (n=5) for DENsity of individuals (number per hectare) and for Basal Area (sq. metres per hectare). AM=*A. marina*, BG=*B. gymnorrhiza*, CT=*C. tagal*, RA=*R. apiculata*, RS=*R. stylosa*. BA measured by the method of angle counts (Cintron & Novelli 1984). (b) Density (x̄±ISD; #/.01 ha) of understorey (saplings) individuals in study plots. Forest type labels as in Table 1a.

	FOREST TYPE				
Species present	AM	BG	CT	RA	RS
A. marina					
BA	21.6±2.6	0.2±0.4	2.6±1.1	0.2±0.4	0.4±0.5
DEN	2 338±792	0.6±1.3	50±26	3±8	11±18
B. exaristata					
BA	0.8±0.8	—	—	—	—
DEN	72±100	—	—	—	—
B. gymnorrhiza					
BA	—	30.4±4.9	—	0.2±0.4	—
DEN	—	1 087±321	—	45±101	—
B. parviflora					
BA	—	—	—	0.2±0.4	—
DEN	—	—	—	8±18	—
C. tagal					
BA	0.2±0.4	3.8±7.4	36.2±9.4	—	0.4±0.9
DEN	16±36	385±833	12 357±8 585	—	61±137
Heritiera littoralis					
BA	—	1.6±2.6	—	—	—
DEN	—	23±1	—	—	—
Lumnitzera racemosa					
BA	—	—	2.4±2.4	—	—
DEN	—	—	175±175	—	—
R. apiculata					
BA	—	—	—	19.8±7.8	4.4±7.2
DEN	—	—	—	1 804±727	267±349
R. lamarckii					
BA	—	—	—	1.4±1.3	1.0±2.2
DEN	—	—	—	59±56	41±93
R. stylosa					
BA	—	0.8±1.1	—	10.4±6.5	34.0±13.3
DEN	—	56±134	—	538±322	1 404±527
Xylocarpus granatum					
BA	—	9.6±3.0	—	—	—
DEN	—	344±141	—	—	—
Melaleuca spp.					
BA	—	5.8±2.2	—	—	—
DEN	—	17±6	—	—	—
A. marina	0.8±0.8	—	1.4±0.5	—	—
B. exaristata	9.4±6.8	—	—	—	—
B. gymnorrhiza	1.0±0.7	—	—	1.4±1.1	5.6±5.4
B. parviflora	—	—	—	—	2.0±2.9
C. tagal	3.6±1.9	1.6±1.7	7.8±4.7	—	0.8±0.8
H. littoralis	—	1.4±3.1	—	—	—
L. racemosa	—	—	—	—	—
R. apiculata	—	—	—	—	4.6±1.3
R. stylosa	—	—	—	—	1.6±1.5
X. granatum	—	15.0±6.8	—	—	—

different sites is based on the relative abundances of the species among areas, for example, *B. gymnorrhiza* is often dominant at Missionary Bay while *A. marina* is never dominant there but is abundant at Lucinda. Plots were checked at 1, 2, 3, 4, 6, 8, 10 and 18 days after the start of the experiments. After 18 days most propagules had either been eaten or had started to establish themselves in the substrate with pronounced root and leaf development. Propagules were scored as viable (capable of growth) or non-viable (incapable of growth). A propagule was scored as non-viable if: (1) 50% or more of its mass had been consumed by predators, or (2) it was pulled entirely down a crab burrow, or (3) the plumule and cotyledonary buds had been completely removed along with the proximal portion of the propagule (often 10% or less of the mass).

Effects of predator and propagule density on consumption. The abundance of crabs varies across the intertidal and between forest types (Sasekumar 1974; Macintosh 1984). The influence of this variation on consumption of propagules was examined in two high-intertidal forests which had similar tree species composition but which differed in relative abundances of crabs. Abundance was measured by trapping crabs for 10 nights using pit traps. The "high" density forest had approximately seven times the numbers of crabs (primarily *N. mienerti*) as the "low" density forest. Three hundred propagules of each of two species (*A. marina* and *B. sexangula*) were placed in each forest. Propagules were placed in three density treatments: (1) singles, individual propagules with no other propagules within 1 m; (2) piles of 10, 10 piles with 10 propagules per pile; (3) pile of 100, 100 propagules in one "superpile". Propagules were not tethered. Experiments were conducted during the period between spring high tides to ensure that none floated away.

Statistical analyses

Comparisons of mean amounts of predation were performed using a two-factor repeated measure Analysis of Covariance (Winer 1981). Species and treatment (forest type) were the two factors. Time (1, 2, 3, 4, 6, 8, 10 and 18 days) was the within factors repeated measure. Density of naturally occurring propagules was entered as a covariate to adjust for effects of the background seed density on amounts of predation (Wilson & Janzen 1972).

For the predator/propagule density experiments survivorship curves were calculated and analysed using Breslow's (1970, 1974) modification of the Kruskal-Wallis test.

Results

Predation differences among species and forests

Averaged over all forests and species, 57.7±34.4% of the propagules were consumed by crabs. Significant differences were noted among species and forest types accounting for ⩾65% of the overall variance in predation (Table 2 and Fig. 2). An *A. posteriori* one-way contrast showed that significant differences existed int the amount of predation among species in the order: *Avicennia* ⩾ *Bruguiera* = *Ceriops* ⩾ R. apiculata=*R. stylosa* ($F_{1\ 100}$=660.1, P⩽.001; Fig. 2a). Differences also existed in the amount of predation among forests. Predation was highest in the *Bruguiera* dominated forest, least in the *Avicennia* forest and intermediate in the remaining three forests (*A. posteriori* one-way contrast, $F_{1\ 000}$=836.5, P⩽.001; Fig. 2b).

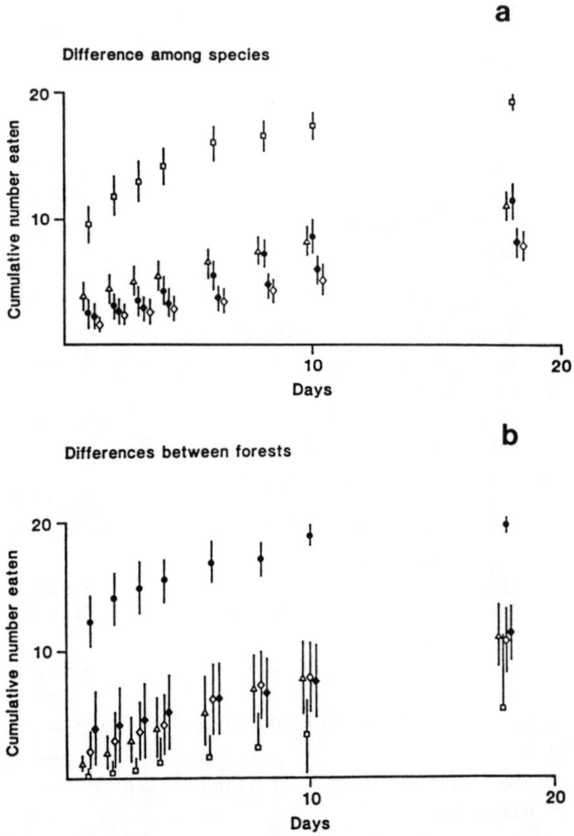

FIG. 2. (a) Differences among species for *A. marina* (□); *B. gymnorrhiza* (●); *C. tagal* (△); *R. apiculata* (◆) and *R. stylosa* (◇). X̄±1SD (n=25) for each species averaged over the five forests. (b) Differences between forests. Dominance by *Avicennia* (□); *Bruguiera* (●); *Ceriops* (△); *R. apiculata* (◆) *R. stylosa* (◇). X̄±1SD (n=25) for each forest averaged over the five species.

TABLE 2. Analysis of covariance results for overall and individual species comparisons. SS, sum of squares; DF, degrees of freedom; %VAR, percent of the total variance explained by a particular effect; F, F value; P, probability; ns, not significant; SP, effect due to species; FOR, effect due to forest type; REG, effect due to covariate; T, time, the within factors repeated measure.

Effect	SS	DF	%VAR	F	P
OVERALL COMPARISONS					
SP	13 243	4	27.0	137.8	≤.001
FOR	19 473	4	39.7	202.6	≤.001
SP×FOR	4 029	16	8.2	10.5	≤.001
REG	7	1	—	0.3	ns
RESIDUAL	2 379	99			
T	5 464	7	11.1	350.6	≤.001
T×SP	436	28	0.9	7.0	≤.001
T×FOR	396	28	0.8	6.4	≤.001
T×FOR×SP	2 119	112	4.3	8.5	≤.001
RESIDUAL	1 558	700			
A. marina					
FOR	1 465	4	26.4	14.4	≤.001
REG	5	1	—	0.2	ns
RESIDUAL	482	19			
T	1 770	7	31.9	80.4	≤.001
T×FOR	1 381	28	24.9	15.7	≤.001
RESIDUAL	440	140			
B. gymnorrhiza					
FOR	3 351	4	55.2	73.0	≤.001
REG	9	1	—	0.7	ns
RESIDUAL	231	19			
T	1 724	7	28.4	112.3	≤.001
T×FOR	447	28	7.4	7.3	≤.001
RESIDUAL	307	140			
C. tagal					
FOR	6 390	4	78.2	87.8	≤.001
REG	2	1	—	0.1	ns
RESIDUAL	346	19			
T	994	7	12.2	69.5	≤.001
T×FOR	155	28	1.4	2.7	≤.001
RESIDUAL	286	140			
R. apiculata					
FOR	4 227	4	64.8	18.3	≤.001
REG	8	1	—	0.1	ns
RESIDUAL	1 099	19			
T	713	7	10.9	54.4	≤.001
T×FOR	209	28	3.2	4.0	≤.001
RESIDUAL	262	140			
R. stylosa					
FOR	4 706	4	75.9	114.9	≤.001
REG	1	1	—	0.0	ns
RESIDUAL	205	19			
T	700	7	11.3	53.2	≤.001
T×FOR	323	28	5.2	6.1	≤.001
RESIDUAL	263	140	—	—	—

Amounts of predation on a given species varied between forests as shown by the significant interaction (Table 2). The within factors repeated measure and interaction terms were also significant indicating differences in the time course of predation among both species end forests (Table 2). Background density of propagules (the covariate) was found to be insignificant.

To understand more fully the effect of predators on each species, an ANCOVA was performed on each species separately (Table 2). Differences among forests were highly significant, being the most important effect for all species except *A. marina*. For each species the ANCOVA interaction terms (time within treatment) were significantly different among the five forests (Table 2), indicating differences in the time course of predation on propagules among the various forests. No significant covariate effect was found for any species, hence the background density of propagules was not an influence on the number of propagules consumed.

Although these analyses show differences in the time course of predation among species and forests, an ecologically important question concerns differences in the total amount of predation (that observed by day 18) among forests for the various species. *A. posteriori* contrasts showed significant differences among species in each forest except the *Bruguiera* forest (Fig. 3a-e). In this forest seed predators consumed ≥98% of all propagules.

Effects due to predator/prey density variation

Propagules of both *A. marina* and *B. sexangula* were consumed fastest in the forest having the greatest density of predators as would be expected (Fig. 4a,b). An important result is that only marginal differences in the consumption of propagules based on propagule density were found. Single propagules were consumed slightly more quickly than propagules in piles of 10 or 100 in both forests for both species. The cumulative result was that most propagules were consumed regardless of their initial density, at both high and low crab densities.

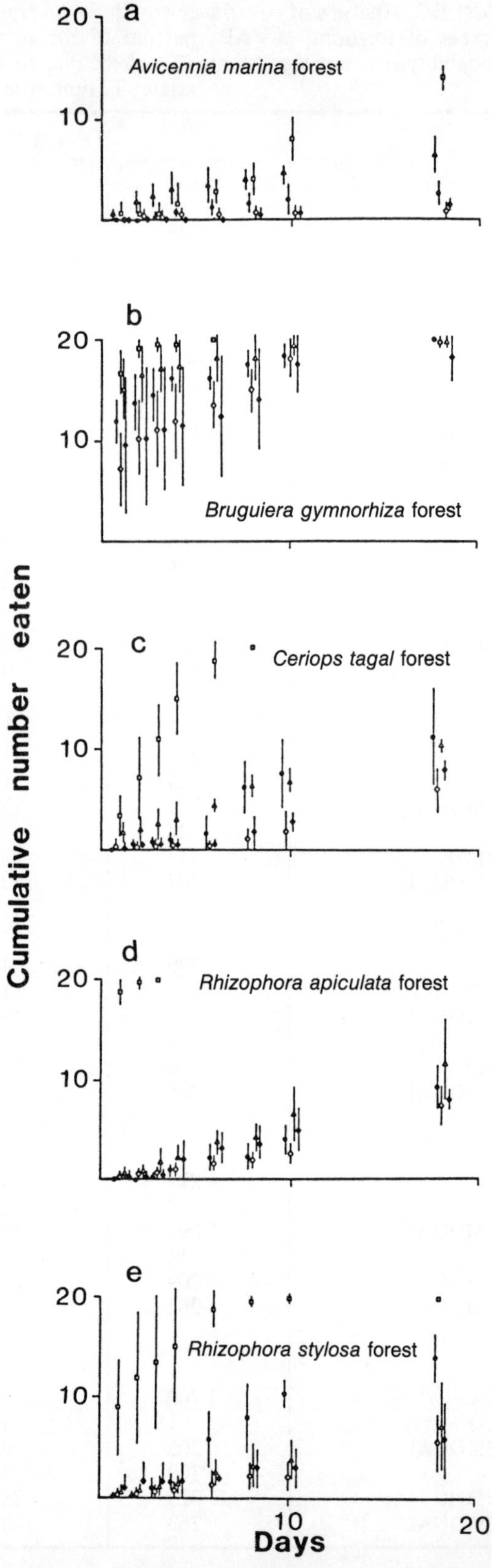

FIG. 3. Cumulative amount of seed predation over the study period. Symbols as in Fig. 1a. Shown are the $\bar{X} \pm 1SD$ for n=5. *A. posteriori* contrasts for differences between species within each forest at day 18: (a) *Avicennia* forest, $F_{4,20}=130.2$, $P \leq .001$; (b) *Bruguiera* forest, $F_{4,20}=2.2$, not significant; (c) *Ceriops* forest, $F_{4,20}=15.7$, $P \leq .001$; (d) *R. apiculata* forest, $F_{4,20}=17.3$, $P \leq .001$; (e) *R. stylosa* forest, $F_{4,20}=19.5$, $P \leq .001$.

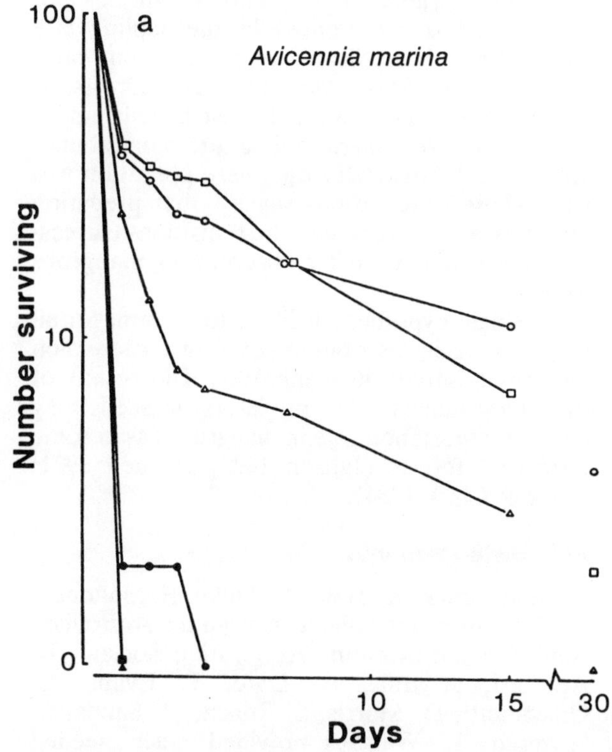

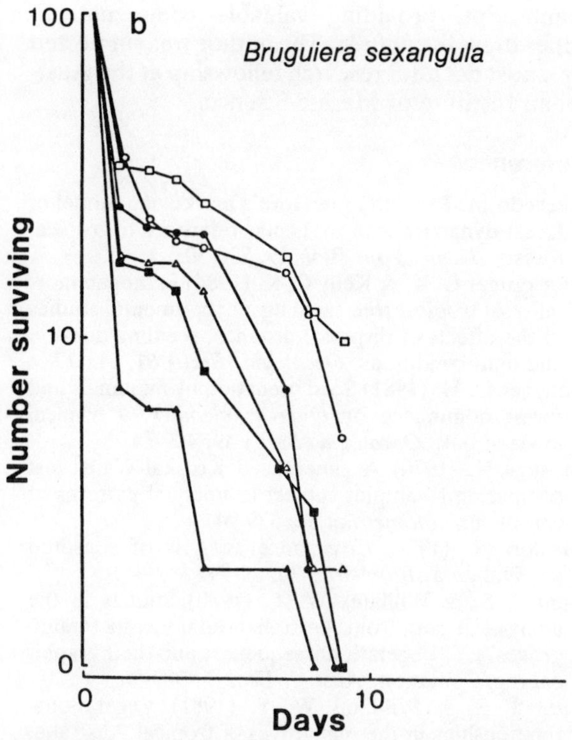

Discussion

Amounts of predation measured during this study are comparable to those of Chan *et al.* (1982) who found that 44-83% of *R. apiculata* and *R. mucronata* propagules were consumed by crabs, monkeys and insects in Malaysia, and Smith (in press, a) whose previous studies (1984-85) showed that 76±28% of the propagules of five mangrove species were eaten in north Queensland forests. Similar levels of predation have been noted for tropical terrestrial forests (for example, Janzen 1975; Boucher 1982; Clark & Clark 1984; DeSteven & Putz 1984; Howe *et al.* 1985).

Avicennia was preyed upon more heavily than any other species, suffering 100% losses in four forests and 96% or more in its "home" forest. This is evidence that the predators are selecting *Avicennia* preferentially over the remaining species. The basis for selectivity is probably related to the chemical composition of the propagules. *Avicennia* propagules have higher concentrations of mono-, di- and tri-saccarides and lower levels of structural components such as lignins and cellulose than due *Bruguiera, Ceriops* and *Rhizophora* (Smith, in press a).

Between forest differences in the amount of predation are most likely due to differences in predator density among forests, changes in the predator community across the intertidal, or a combination of these. Wilson and Janzen (1972) showed that seeds of the rainforest palm *Scheelea rostrata* suffered less mortality from predators when found singly rather than in clumps. Augspruger & Kelly (1984) showed similar density effects for pathogen induced mortality on seedlings of the tropical tree *Platypodium elegans*. Seedlings escaped from pathogens when they occurred at low densities at some distance removed from the parent. Boucher (1981) found the opposite for the tropical, lowland oak *Quercus oleoides*. Single seeds of this species were invariably consumed, while at high seed densities some escaped predation, but only where

FIG. 4. Numbers of propagules of (a) *Avicennia marina* and (b) *Bruguiera sexangula* surviving at different predator and propagule densities (log scale). Predator density: high (solid symbols); low (open symbols). Propagule density: singles (▲ △); piles of 10 (■ □); pile of 100 (● ○). For *A. marina* Breslow's B=110.9 (P⩽.001) for differences between predator density; B=6.2 (P⩽.10) for differences in propagule density. For *B. sexangula*: B=352.0 (P⩽.001) and B=4.6 (P⩽.10) for predator and propagule effects, respectively.

predators were in low densities. In the present study propagules were consumed significantly faster where crabs were most abundant, regardless of propagule density. Propagule density effects were less apparent, although single propagules disappeared most quickly at both predator densities for both species. It is interesting that in forests where the parent tree is abundant seeds suffer less predator induced mortality only at high densities, in contrast to the opposite condition in high diversity, low dominance terrestrial forests where adults are rare and seeds escape predation by rarity as well.

A more likely explanation for between forests differences is changes in species composition of the crab community across the intertidal. Some predators (e.g., *Sesarma messa*) are found throughout the intertidal, while others appear to be confined to a particular region or forest type. For example, *Neosarmatium meinerti* is abundant in certain high-intertidal forests while *N. smithi* is usually found only in association with *Ceriops* thickets. Unfortunately the ecology of the crab community in mangrove forests is poorly understood.

Predation and patterns of forest succession

Several authors have discussed successional patterns in mangrove forest systems. Watson (1928), Macnae (1968, 1969) and later Johnstone (1983) described a typical sequence as an *Avicennia* stage followed by *Rhizophora* which, in turn, is replaced by *Burguiera gymnorrhiza*. Interestingly, Watson (1928) observed that when it occurred in pure stands, *B. gymnorrhiza* failed to regenerate and this species did not replace itself in the canopy. Thus the forest returned to an earlier stage. Watson gave no reasons to account for this observation.

Results from this study can be placed within the context of forest succession if: (1) predation affects recruitment and hence the composition of the sapling community found beneath the forest canopy, and (2) the density and composition of the sapling community is an indication of the future composition of the canopy. Both assumptions have been supported by investigations in a variety of forest ecosystems (Waggoner & Stephens 1970; Horn 1976; Enright and Ogden 1979; Acevedo 1981). Within the *Avicennia*-dominated forest, *Avicennia* was itself the most heavily preyed upon of the five species (Fig. 3a). Of the saplings in this forest, *Avicennia* is the rarest; *Ceriops* and *B. gymnorrhiza* are more abundant (Table 1b). Neither species of *Rhizophora* is represented in the sapling size class, although seedlings are present, but rare. Within the two *Rhizophora*-dominated forests, *B. gymnorrhiza* was consumed most heavily under *R. apiculata*. *Bruguiera* saplings are more abundant in the *R. stylosa* forest where predation was less. These observations suggest that predation may play a role in some of the transitions (successions) which have been observed in mangrove forests.

No single hypothesis is likely to account for all of the species distribution or forest succession patterns observed in mangroves. The results of this study indicate that predators on seeds may have as important a role in mangroves as in some terrestrial forests (Janzen 1970; Connell 1971; Clark & Clark 1984).

Acknowledgements

A. Aulsebruck, K. Boto, N. Duke, F. Mahoney, A. Robertson, and volunteers from the Australian-New Zealand Scientific Exploration Society (B. Anastasi, J. Bills, A. Date, T. Evans, C. Giacometti, D. Morris, S. Roach, M. Sands, E. Seymour, P. Watson) provided much needed assistance in the field. D. Alongi, K. Boto, B. Clough, and A. Robertson critically reviewed the manuscript, providing valuable comments. F. Jones drew the figures. The author was supported by a post doctoral research fellowship at the Australian Institute of Marine Science.

References

Acevedo M. F. (1981) On Horn's markovian model of forest dynamics with particular reference to tropical forests. *Theor. Pop. Biol.* **19,** 230-50.

Augspruger C. K. & Kelly C. K. (1984) Pathogen mortality of tropical tree seedlings: experimental studies of the effects of dispersal distance, seedling density, and light conditions. *Oecologia (Berl.)* **61,** 211-17.

Boucher D. H. (1981) Seed predation by mammals and forest dominance by *Quercus oleoides*, a tropical lowland oak. *Oecologia (Berl.)* **49,** 409-14.

Breslow N. (1970) A generalized Kruskal-Wallis test comparing k samples subject to unequal patterns of censorship. *Biometrika* **57,** 579-94.

Breslow N. (1974) Covariance analysis of censored survival data. *Biometrics* **30,** 89-99.

Bunt J. S. & Williams W. T. (1980) Studies in the analysis of data from Australian tidal forests ("mangroves"). I. Vegetational sequences and their graphical representation. *Aust. J. Ecol.* **5,** 385-90.

Bunt J. S. & Williams W. T. (1981) Vegetational relationships in the mangroves of tropical Australia. *Mar. Ecol. Prog. Ser.* **4,** 349-59.

Bunt J. S., Williams W. T. & Duke N. C. (1982) Mangrove distributions in north-east Australia. *J. Biogeogr.* **9,** 111-20.

Chan H. T., Ujang R. & Putz F. E. (1984) A preliminary study on planting of *Rhizophora* species in an *Avicennia* forest at the Matang mangroves. In: *Prosiding Seminar II. Ekosistem Mangrove,* (ed. S. Soemodihardjo) pp. 340-45. Lembagu Ilmu Pengetahuan, Jakarta.

Cintron G. & Novelli Y. S. (1984) Methods for studying mangrove structure. In: *The Mangrove Ecosystem: Research Methods. Monographs on oceanographic methodology No. 8* (eds S. C. Snedaker & J. G. Snedaker) pp. 91-113. UNESCO, Paris.

Clark D. A. & Clark D. B. (1984) Spacing dynamics of a tropical rainforest tree: evaluation of the Janzen-Connell model. *Amer. Nat.* **124,** 769-88.

Connell J. H. (1971) On the role of natural enemies in preventing competitive exclusion in some marine animals and in rainforests. In: *Dynamics of populations* (eds P. J. den Boer & G. R. Gradwell) pp. 298-312. Centre for Agricultural Publishing and Documentation, Wageningen.

Davie P. (1982) A preliminary checklist of the Brachyura (Crustacea: Decopoda) associated with Australian mangrove forests. *Operculum* **5,** 204-07.

Davis J. H. Jr. (1940) *The Ecology and Geologic Role of Mangroves in Florida.* Publication No. 517. Carnegie Institute, Washington, D.C., U.S.A.

De Steven D. & Putz F. E. (1984) Impact of mammals on early recruitment of a tropical tree, *Dipteryx panamensis*, in Panama. *Oikos* **43,** 207-16.

Elsol J. A. & Saenger P. (1983) A general account of the mangroves of Princess Charlotte Bay with particular reference to zonation of the open shoreline. In: *Biology and ecology of mangroves. Tasks for vegetation science,* Vol. 8 (ed. H. J. Teas) pp. 37-48. Dr W. Junk, The Hague.

Enright N. & Ogden J. (1979) Applications of transition matrix models in forest dynamics: *Araucaria* in Papua New Guinea and *Nothofagus* in New Zealand. *Aust. J. Ecol.* **4,** 3-23.

Horn H. S. (1975) Markovian properties of forest succession. In: *Ecology and evolution of communities* (eds M. L. Cody & J. M. Diamond) pp. 196-211. Harvard Press, Cambridge, Mass.

Howe H. F., Schupp E. W. & Westley L. C. (1985) Early consequences of seed dispersal for a neotropical tree (*Viola surinamensis*). *Ecology* **63,** 781-91.

Janzen D. H. (1970) Herbivores and the number of tree species in tropical forests. *Am. Nat.* **104,** 501-28.

Janzen D. H. (1975) Intra- and interhabitat variations in *Guazuma ulmifolia* (Sterculiaceae) seed predation by *Amblycerus cistelinus* (Bruchidae) in Costa Rica. *Ecology* **56,** 1009-13.

Johnstone I. M. (1983) Succession in zoned mangrove communities: where is the climax? In: *Biology and ecology of mangroves. Tasks for Vegetation Science,* Vol. 8 (ed. H. J. Teas) pp. 131-39. Dr W. Junk, The Hague.

Macintosh D. J. (1984) Ecology and productivity of Malaysian mangrove crab populations (Decapoda: Brachyura). In: *Proc. Asian Symp. Mangrove Environ. Res. and Manag.* (eds E. Soepadmo, A. N. Rao & D. J. Macintosh) pp. 354-77.

Macnae W. (1968) A general account of the fauna and flora of the mangrove swamps and forests in the Indo-Pacific region. *Adv. Mar. Biol.* **6,** 73-270.

Macnae W. (1969) Zonation within mangroves associated with estuaries in north Queensland. In: *Estuaries,* (ed. G. E. Lauff) pp. 432-41. AAAS, Washington, D.C.

Rabinowitz, D. (1978) Early growth of mangrove seedlings in Panama, and an hypothesis concerning the relationship of dispersal and zonation. *J. Biogeogr.* **5,** 113-33.

Sasekumar A. (1974) Distribution of macrofauna on a Malayan mangrove shore. *J. Anim. Ecol.* **43,** 51-69.

Smith T. J. III (1987a) Seed predation in relation to tree dominance and distribution in mangrove forests. *Ecology* **68,** 266-73.

Smith T. J. III (1987b) Effects of seed predators and light level on the distribution of *Avicennia marina* (Forsk.) Vierh. in tropical, tidal forests. *Est. Coast. & Shelf Sci.* **25,** 43-51.

Waggoner P. E. & Stephens G. R. (1970) Transition probabilities for a forest. *Nature, Lond.* **225,** 1160-61.

Watson J. G. (1928) Mangrove forests of the Malay Peninsula. *Malay. For. Rec.* **6,** 1-275.

Williams W. T., Bunt J. S. & Duke N. C. (1981) Mangrove litterfall in north-eastern Australia. II. Periodicity. *Aust. J. Bot.* **29,** 555-63.

Wilson D. E. & Janzen D. H. (1972) Predation on *Scheela* palm seeds by bruchid beetles: seed density and distance from the conspecific tree. *Ecology* **53,** 954-59.

Winer B. J. (1971) *Statistical Principles in Experimental Design, Second edition.* McGraw-Hill, New York.

Proc. Ecol. Soc. Aust. 15:213-219

Differences in shoot architecture in mature communities of the tropical mangrove *Avicennia marina* (Forsk.) Vierh.

J. D. S. DAVIE

*Department of Botany, University of Queensland, St Lucia, Qld. 4067**

Abstract

The architecture of growing shoots of mature mangrove communities appears to be strongly dependent on the environment in which the plants are found. Sylleptic shoot growth is common among mangroves and has been recorded for a number of tropical mangrove species in the genera Avicennia *and* Rhizophora. *In Moreton Bay it can also be seen in other members of the Rhizophoraceae, in* Aegiceras corniculatum *and* Lumnitzera racemosa.

Studies of shoot production over three years has demonstrated that where available water is never limiting Avicennia marina *grows sylleptically with up to seven active lateral meristems at each node expanding simultaneously with the apical meristem. Under increasing water stress, as is encountered at intertidal levels above the mean high water neap tide mark (MHWN), the lateral meristems are increasingly suppressed to create apically dominant proleptic shoots. In these environments a lateral is activated only after the death of the initial shoot. The extent of apical dominance varied from year to year depending on rainfall.*

Canopy height declines in a graded series across the intertidal zone from about the MHWN tide level, while growth habits become increasingly stunted and tortuous, eventually becoming widely spaced shrubs. These structural changes, and the evenness and density of the canopy, are reflections of the growth pattern of individual shoots. Internodal length, number of lateral shoots and survival of apical shoots are all determining factors.

Increase in apical dominance is associated with reduced vigour in the plants as a direct result of water stress. Suppression of lateral shoots in the plants above MHWN results in a significant reduction in leaf area index (LAI). Litter production falls from 6.4 t ha^{-1} y^{-1} to 1.9 t ha^{-1} y^{-1}. There is a quasi-linear decline in overstorey foliage projective cover (FPC) from 75-80% to about 60% followed by a step down to an open shrubland of 20-30% at about the point where interstitial osmotic potentials fall below -5 to -6 MPa, and water becomes essentially unavailable to the plants for some period of the year. Reduction in overstorey FPC is associated with a compensatory increase in FPC of the understorey as known for wholly terrestrial plant communities.

Introduction

Avicennia marina is the dominant species in a range of mangrove plant formations from closed-forest to low open-shrubland in Moreton Bay (Davie 1984). This diversity of community structure reflects a high degree of phenological plasticity in growth habit which is particularly evident in species of *Avicennia* but can be seen also in a number of other mangrove plants throughout north Australia (Davie 1985). Comparison of individuals demonstrating different growth habits shows a consistent difference in the

*Present address: Queensland National Parks and Wildlife Service, P.O. Box 1395, Rockhampton, Queensland 4700.

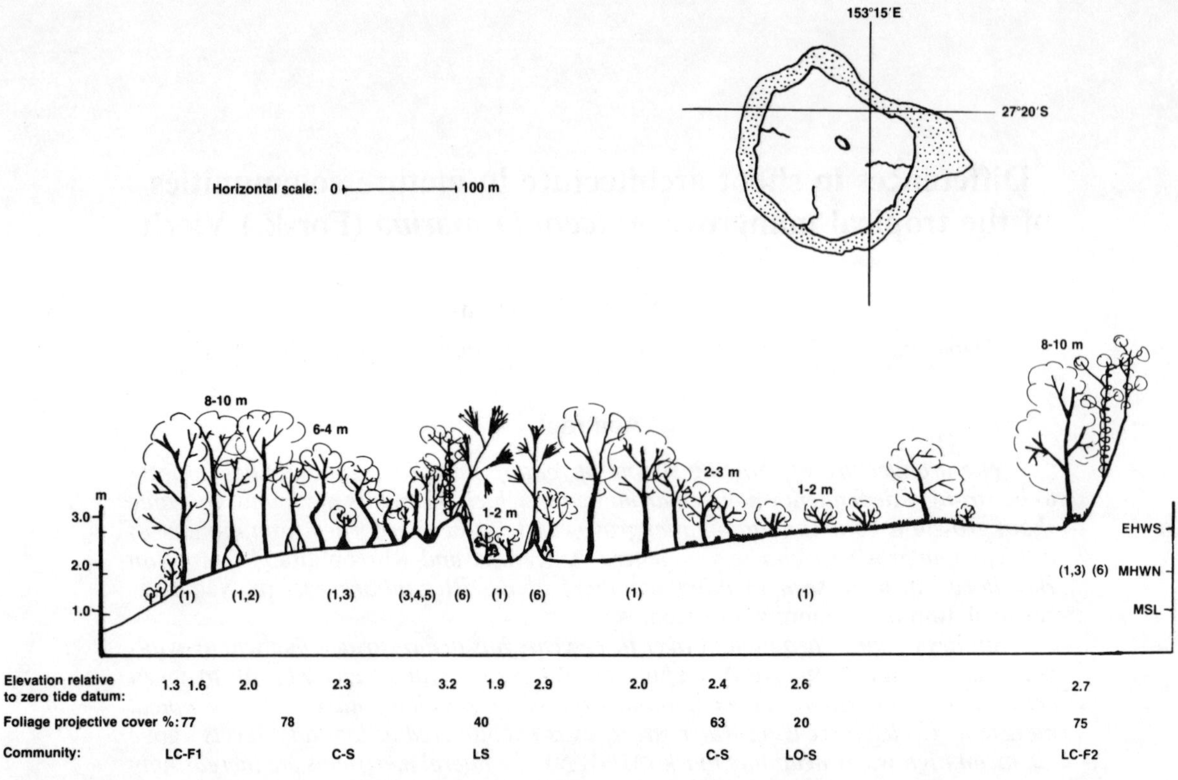

Fig. 1. Profile of mangrove and associated coastal vegetation on the western side of Mud Island. The numbers in brackets refer to the canopy species: (1) *Avicennia marina*; (2) *Rhizophora stylosa*; (3) *Ceriops tagal*; (4) *Excoecaria agallocha*; (5) closed-vine scrub with *Cupaniopsis anarcardioides*; (6) *Pandanus pedunculatus*. Height of the upper stratum is shown on the diagram above the canopy. MSL is mean sea level; MHWN is mean high water neap tide and EHWS is extreme high water spring tide.

pattern of shoot growth, now commonly referred to as shoot architecture (Halle *et al.* 1978).

A classification of shoot growth patterns in tropical trees by Tomlinson and Gill (1973) recognizes a sylleptic pattern where lateral shoots develop simultaneously from a node with the apical shoot; and a proleptic pattern where lateral buds are suppressed by growth of the apical shoot. Halle *et al.* (1978) have pointed out that many mangrove species show a sylleptic habit such as that described for *Rhizophora mangle* in Florida by Gill and Tomlinson (1971), and support the suggestion by Tomlinson and Gill (1973) that this may be associated with vigorous growth in a physically benign environment.

Evidence for this suggestion is provided by Neville (1969) who showed experimentally that syllepsis can be induced in seedlings of *Gleditsea*, which would otherwise have become proleptic, and by a limited number of observations on mature plants in native environments (e.g., Gill and Tomlinson 1971).

The fact that *Avicennia marina* can be found across an intertidal gradient along which habitats can be identified which are spatially very close but physically very different, allows examination of several hypotheses which may help explain differences in plant growth habits and community structure. This paper is concerned with the influence of environmental conditions on shoot growth pattern; and especially on the extent to which syllepsis is associated with vigorous growth. Consideration is also given to the implications of the architecture of shoot growth to the growth habits of mature plants.

Study Area

Mud Island is situated in the centre of Moreton Bay, approximately six kilometres from the mouth of the Brisbane River (27°20′S 153°15′E). The island is a recent, but no longer active coral reef and the presently stable sedimentary intertidal landform slopes continuously inland, except

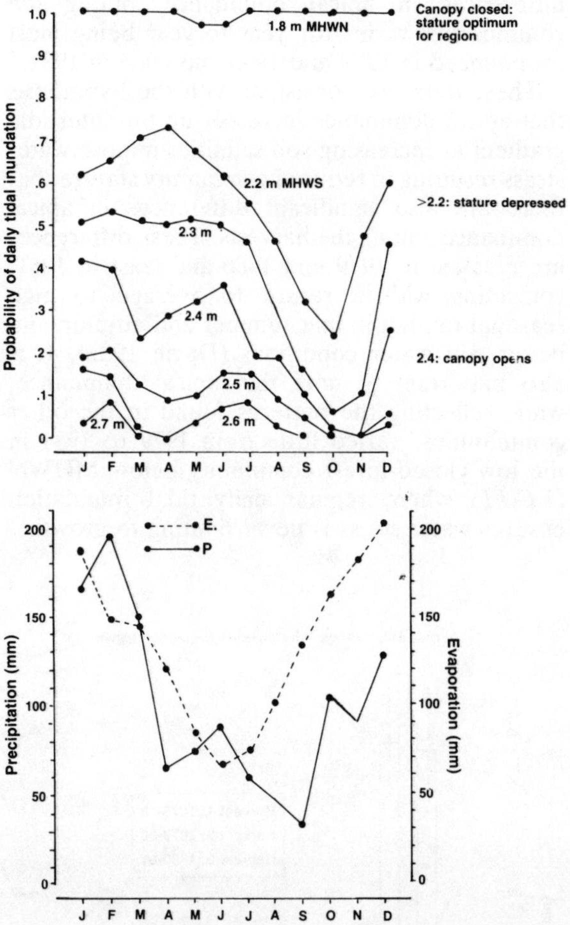

Fig. 2. Predicted seasonal patterns of tidal inundation frequency compared to seasonal patterns of average rainfall and evaporation.

for several residual concentric cheniers, to a central terrestrial knoll 1.5 km from the shore. The intertidal area comprises some 85 ha of mangrove and associated samphire vegetation.

Five mangrove plant communities dominated by *Avicennia marina* have been recognized and described by Davie (1982) and their distribution and general features are shown in Fig. 1.

Although there is some variation in the type of substrate it is only in the special case of the low shrubland community (*LS*) that soil type is a primary, environmental determinant. As previously described by Clarke and Hannon (1970) the most important environmental variable is tidal elevation. As elevation exceeds the mean high water neap tide level (MHWN) there is a marked reduction in the frequency of tidal inundation and increasing dependence on rainfall as the primary source of available water as shown in Fig. 2. Low tides and low rainfall in spring result in very high interstitial salinities at elevations above mean high water spring tide level (MHWS) and available water can be severely limiting (Davie 1982). The variable nature of the annual rainfall pattern in the region (Dick 1958) results in wide differences from year to year in the extent of the water stress experienced by the mangroves and other coastal plants during the critical spring period.

Methods

The analyses presented here are based on three consecutive annual harvests from each of the five *Avicennia marina* communities on Mud Island. Because of clear differences in the colour and texture of the stem of the current year's shoot growth, harvests at the end of the spring/autumn growing season in late May can be confidently used as an index of growth in the absence of tags. Tagged shoots were also used as a check and in the investigation of individual shoot phenology (Davie 1982).

At each harvest 10 shoot systems, defined as all shoots subsequent to the node from which the current season's growth began, were sampled from each of five randomly selected plants in each community. Care was taken in each case to

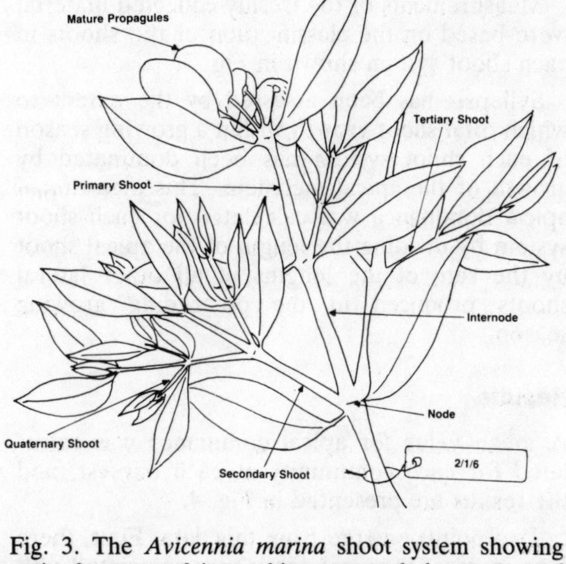

Fig. 3. The *Avicennia marina* shoot system showing main features of its architecture and the system of classification used in this investigation.

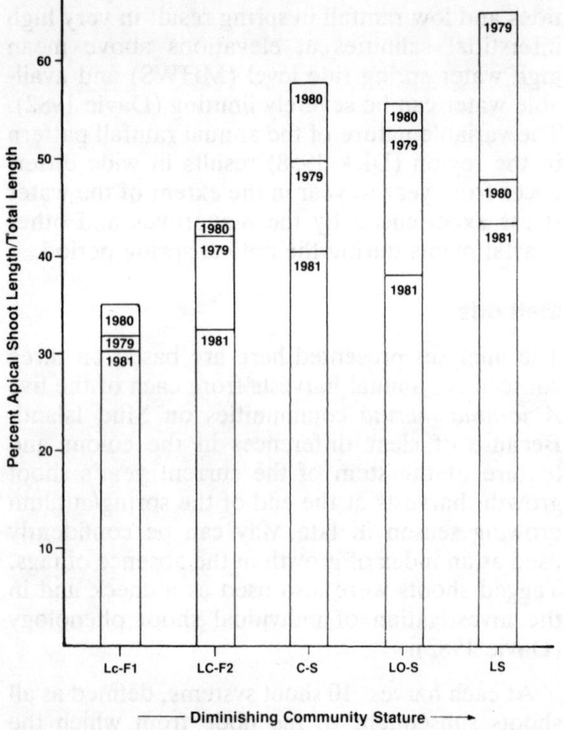

Fig. 4. Relationship between shoot apical dominance and *Avicennia marina* community structure on Mud Island.

harvest from the top centre of the canopy to standardize exposure and canopy height.

Measurements of the freshly collected material were based on the classification of the shoots in each shoot system shown in Fig. 3.

Syllepsis has been assessed by the extent to which total shoot growth within a growing season of each shoot system has been dominated by growth of the apical meristem. This measure of apical dominance was calculated for each shoot system by dividing the length of the apical shoot by the sum of the lengths of all other lateral shoots produced in the preceeding growing season.

Results

A mean value for apical dominance was calculated for each community at each harvest, and the results are presented in Fig. 4.

Two points emerge from this data. First, there is an increase in apical dominance associated with a reduction in community stature. Secondly the differences in apical dominance among the communities varies for year to year being most pronounced in 1979 and 1980 and least in 1981.

These data are consistent with the hypothesis that apical dominance increases up the intertidal gradient as increasing soil salinities impose water stress resulting in reduced community stature, but there are also significant differences in apical dominance among the harvests. These differences are greatest in 1979 and 1980 and least in 1981, coinciding with a return to average to high seasonal rainfall in late summer and autumn and better soil water conditions (Davie 1982). It is also important to note that apical dominance, while reflecting the patterns found in the other communities, varied little from 1979 to 1981 in the low closed-forest community below MHWN (*LC-F1*) where regular daily tidal inundation ensures water stress is never limiting to growth.

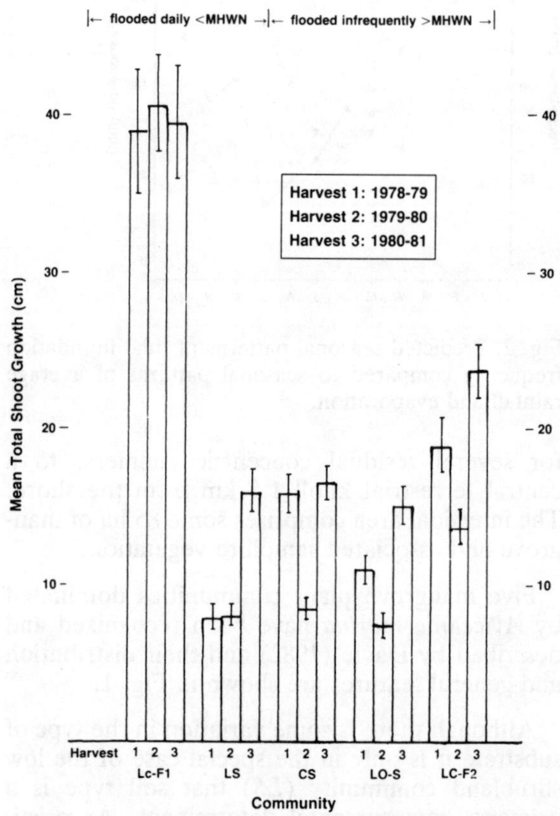

Fig. 5. Total shoot extension growth achieved by *Avicennia marina* at the end of the growing season for three successive years in each community on Mud Island. Vertical lines indicate the 95% confidence limits.

TABLE 1. Mean number of lateral shoots on terminal shoot systems in *Avicennia marina* communities on Mud Island. Figures in parentheses refer to the number of shoot systems sampled on each occasion. Variation is expressed as S.E. of the mean.

Community	1979	1980	1981
LC-F1	4.45±0.35(50)	6.44±0.54(50)	6.94±0.66(48)
C-S	1.08±0.17(56)	1.92±0.23(50)	3.38±0.30(48)
LS	1.14±0.13(50)	2.24±0.22(50)	3.41±0.35(48)
LO-S	1.38±0.17(51)	1.98±0.22(50)	3.85±0.28(48)
LC-F2	2.56±0.20(40)	3.44±0.36(50)	4.74±0.44(48)

The importance of water in limiting and controlling the patterns of growth in mangrove communities is demonstrated further in Fig. 5 which shows differences in annual total shoot growth. The low closed-forest below MHWN sustained twice the shoot growth of the inland low closed-forest (*LC-F2*) averaged over the three harvests, while it was between three and four times as productive as the lower scrub and shrubland communities at the higher tidal levels, where variation in rainfull produced significantly lower growth during the lower rainfall years.

Apical dominance is also regulated by variation in the number of laterals produced, and by survival of the apical shoot. As shown in Table 1 the number of lateral shoots in the low closed-forest below MHWN was consistently greater than in any other community. Lateral shoots where produced at almost all nodes examined. The greatest number of lateral shoots found at one node was six and examples of three to four shoots were common. This contrasted strongly with that found in the other four communities. The proliferation of lateral shoots in each of the communities above MHWN tide level in the 1981 harvest should be noted in Table 1 and compared with the consistent performance of the low closed-forest below the MHWN level.

Details of variation in the internodal lengths of shoots were not documented. General observations suggest that internodal lengths become progressively shorter across the intertidal zone but increase substantially in the low closed-forest on the terrestrial margin. In the low closed-forest below MHWN internodal lengths were in the order of three to six centimetres, whereas in the low open-shrubland (*LO-S*) growing in the drier, highly saline habitats reached by high spring tides only, the intertidal lengths rarely exceeded three centimetres.

In addition to being longer, shoots in the low closed-forest communities were more robust. In consistently watered microhabitats at the terrestrial margin and particularly along the seashore, lateral shoots developed up to three nodes in a growing season. This was rarely the case in the stunted or open communities at higher tidal levels. As for the number of lateral shoots, the number of internodes produced in a growing season remained the same in the communities below MHWN but varied from year to year at higher tidal levels with the distribution and quantity of rainfall.

The growth habits of the plants and the evenness and density of the canopies of each of the *Avicennia marina* communities are reflections of the pattern of growth of individual shoots. Internodal length of the apical and lateral shoots, the numbers of lateral shoots, and the survival of the apical shoot are all determining factors. In the low closed-forests, the development of numerous, robust lateral shoots produces a dense, even and deep canopy with few, if any gaps. Such a canopy typology is one of the most characteristic features of the physiognomy of seaward fringing mangrove formations.

Canopy evenness and density are less evident in the closed-scrub (*C-S*) community between MHWN and MHWS tide levels, and the low open-shrubland formations near the upper margins of the intertidal zone. In the closed-scrub, suppression of lateral shoots results in spindly branches, with many gaps in the shallow canopy. At higher elevations, but still within this formation, the plants tend towards increasingly erratic growth habits. While they retain a single main stem, this and the terminal branches become strongly twisted: a phenomenon which appears to be related to periodic dieback of the dominant shoot at the beginning of the growing season, and its subsequent replacement by a shoot from a lateral meristem. Which of the many potential lateral buds is activated following shoot dieback will determine the subsequent orientation of that shoot system and consequently the axes of growth of the mature plant. There seems little doubt that frequent dieback of the leader shoots because of periodic environmental stress is a major factor controlling the growth habits assumed by *Avicennia marina* in each of these formations. The shrub habit found in the low open-shrubland and the low shrubland formations, where branches derive from lateral meristems on a thick, prostrate, main stem, represents the extreme result of a process which can be observed beginning in the closed-scrub.

Discussion

The patterns revealed from the data obtained in this investigation are consistent with the hypothesis that apical dominance in *Avicennia marina* is sensitive to variation in edaphic conditions during the growing season, and is inversely related to vigour as suggested by Tomlinson and Gill (1973). In the long term a specific habitat effect is expressed in the habit of the mature plants. There is a steady increase in apical dominance coinciding with the decline in community stature, and with the exception of the low shrubland, this increase is associated with stress caused by decreasing availability of water from either tidal inundation or rainfall or fresh-water seepage. Conditions of high salinity are most severe in the low open-shrubland community but decline rapidly in the low closed-forest community on the terrestrial fringe (Davie 1982). However, the data on annual shoot growth within each community for the years 1979-80 compared to 1981, and the associated apical dominance data, emphasize that alternative shoot growth patterns can be expressed by the same plants at different times as environmental conditions change.

Evidence for the importance of water stress to lateral shoot growth and apical dominance comes from a number of other studies, although the exact mechanism operating in *Avicennia marina* has yet to be demonstrated experimentally. Reduction in the length of the internodes could result directly from the loss of turgor in the expanding cells under high water stress; or it may be influenced by hormonal growth inhibition. Abscisic acid has been implicated in the inhibition of shoot elongation, although there is little correlation between endogenous ABA concentration and elongation in those plants in which this has been measured (Aspinall 1980).

Water stress has been shown experimentally to inhibit lateral meristems in seedlings of *Phaseolus vulgaris* and *Pisum sativum* (McIntyre 1971, 1973). There is no similar experimental information available for *Avicennia marina* or other mangrove species. McIntyre considered that the importance of water stress to the mechanism of apical dominance may be greater in broad-leaved plants than in small-leaved xerophytic plants because of the potential of the latter to reduce water loss through the stomates.

This hypothesis is consistant with the data reviewed by Aspinall (1980) on the role of abscisic acid in adaptation to water stress. Abscisic acid accumulates rapidly in plants subjected to water stress. High endogenous concentrations in inhibited axillary buds suggest a direct role for this hormone in apical dominance. Because of the insights into the plasticity of growth habit and survival of plant species in marginal habitats that such information may offer, further experimentation into the phenomenon described for *Avicennia marina* in Moreton Bay would be warranted.

Recognition of the differences in shoot architecture that can occur in mangroves is also useful in understanding differences in the structure of mangrove plant formations.

In Moreton Bay canopy height declines progressively in a graded series of *Avicennia marina* dominated communities across the intertidal zone from about MHWN tide level (Davie 1984). This is a consequence of the reduced number of new shoots initiated in a growing season; a very high rate of shoot mortality; reduced shoot internode elongation, and a drastically reduced growing season due to severe seasonal water stress during periods of potential growth when temperatures are optimal (Davie 1982). It is also a consequence of a modified shoot architecture as described above.

Suppression of lateral shoots in the plants above MHWN results in a significant reduction in the leaf area index (LAI). This clearly distinguishes the seaward fringing closed-forests below MHWN tide level from each of the other communities. LAI is a measure of the vertical spread of foliage in the canopy and in *Avicennia marina* communities this is determined by the degree to which the shoots are functionally sylleptic.

The greater proliferation of shoots arising from each node as shoot growth quickens with the onset of warmer spring temperatures allows the seaward fringing community to operate at a higher energetic level than those further up the intertidal gradient. This is expressed in growth rate and in accumulated biomass and also in litter production which falls from 6.4 t ha^{-1} y^{-1} in the low closed-forests to 1.9 t ha^{-1} y^{-1} in the low open shrubland (Davie 1984).

The increase in water stress associated with the intertidal gradient also produces changes in the Foliage Projective Cover of the canopy as the community adjusts to maintain a sustainable water balance. Two stages appear to be involved. The first is a quasi-linear decline in overstorey FPC from 75-80% to about 60% represented by the spot measurements shown in Fig. 1. The second is a step down to about 20-30% when the overstorey opens and the community assumes an open shrubland structure. While the exact position of the boundary may fluctuate slightly from year to year depending on rainfall, its position appears to coincide with the tidal level where interstitial osmotic potentials fall below about —5 to —6 MPa and water becomes essentially unavailable to the plants for some period of the year (Davie 1982).

Over much of the intertidal gradient the total FPC (overstorey plus understorey) which measures the horizontal spread of foliage remains approximately the same, as it is set by the evaporative potential of the atmosphere. Rather, as overstorey FPC declines there is a compensating increase in the FPC of the understorey. In Moreton Bay the reallocation involves a shift in FPC from an *Avicennia marina* overstorey with a shrub layer of *Ceriops tagal*, to an herbaceous understorey of *Suaeda australis*/*Sarcocornia quinquiflora*. These observations on changes in the overstorey and understorey FPC of mangroves are in agreement with those reported by Specht (1983) in an analysis of 56 terrestrial plant communities throughout Australia.

Conclusions

The different patterns of shoot growth discussed in this paper could be readily observed and interpreted because of the particular nature of the environment. Similiar gradients occur in terrestrial communities in the wet tropics although the spatial scale may be greater and the boundaries between communities more diffuse. Further observation of shoot growth patterns in terrestrial communities should seek to test the generality of the patterns which are clearly recognizable in mangroves.

Acknowledgements

This work was carried out as part of a Ph.D Thesis in the Botany Department of the University of Queensland. Financial assistance was provided by a Commonwealth Post-graduate Award.

References

Aspinall D. (1980) Role of abscisic acid and other hormones in adaptation to water stress. In: *Adaptation of Plants to Water and High Temperature Stress* (eds N. C. Turner & P. J. Kramer) pp. 155-72. New York, John Wiley and Sons.

Clarke L. J. & Hannon N. J. (1970) The mangrove swamp and salt marsh communities of the Sydney District. *J. Ecol.* **58,** 351-69.

Davie J. D. S. (1982) *Pattern and Process in the Mangrove Ecosystems of Moreton Bay, Southeastern Queensland.* Ph.D Thesis, University of Queensland.

Davie J. D. S. (1984) Structural variation, litter production and nutrient status of mangrove vegetation in Moreton Bay. In: *Focus on Stradbroke* (eds R. J. Coleman, J. Covacevich & P. Davie) pp 208-23. Brisbane, Boolarong Press.

Davie J. D. S. (1985) Mangrove ecosystems in Australia. In: *Research for Development: the Mangrove Ecosystem* (ed. C. D. Field). In Press.

Dick R. S. (1958) Variability in rainfall in Queensland. *J. Trop. Geogr.* **11,** 32-42.

Gill A. M. & Tomlinson P. B. (1971) Studies on the growth of the red mangrove *Rhizophora mangle* 3. Phenology of the shoot. *Biotropica* **3,** 109-24.

Halle F., Oldemann R. A. A. & Tomlinson P. B. (1978) *Tropical Trees and Forests.* Berlin, Springer-Verlag.

McIntyre G. I. (1971) Water stress and apical dominance in *Pisum sativum. Nature (London), New Biol.* **230,** 87-88.

McIntyre G. I. (1973) Environmental control of apical dominance in *Phaseolis vulgaris. Can. J. Bot.* **51,** 293-99.

Neville P. (1969) Morphogenese chez *Gleditzia tricanthos* L. III. Étude histologique et experimentale de la senescence des bourgeons. *Ann. Sci. Nat., Botan. Biol. Veg.* **10,** 302-24.

Specht R. L. (1983) Foliage projective covers of overstorey and understorey strata of mature vegetation in Australia. *Aust. J. Ecol.* **8,** 433-39.

Tomlinson P. B. & Gill A. M. (1973) Growth habits of tropical trees: some guiding principles. In: *Tropical Forest Ecosystems in Africa and South America: a Comparative Review* (eds B. J. Meggars, E. S. Ayensu & W. D. Duckworth) pp. 129-43. Washington, Smithsonian Institute Press.

Proc. Ecol. Soc. Aust. 15:221-226

Seasonal patterns of shoot growth in *Avicennia eucalyptifolia* Zipp ex Miq along a natural salinity gradient in the Northern Territory

M. J. J. DIETERS[1] and M. U. SLEE[2]

[1]*Queensland Department of Forestry, M.S. 483, Fraser Road, Gympie, Qld. 4570*

[2]*Department of Forestry, Australian National University, G.P.O. Box 4, Canberra, A.C.T. 2601*

Abstract

Shoot and leaf growth of Avicennia eucalyptifolia *Zipp ex. Miq. were monitored along a natural salinity gradient in the Northern Territory, Australia. Measurements began at the end of the wet season and continued until the middle of the dry season in 1984. Shoot and leaf growth was not correlated with salinity and it is concluded that either temperature, humidity or day length is the major factor(s) controlling the growth of* Avicennia eucalyptifolia *during the dry season and not salinity.*

Introduction

Avicennia eucalyptifolia Zipp. ex Miq. is a mangrove species of the Northern Territory which is widely distributed within the intertidal region with respect to salinity. *A. eucalyptifolia* is found growing under seasonal saline conditions which can range from close to fresh water through to three times that of sea water.

This study examined the shoot growth of *A. eucalyptifolia* along a natural salinity gradient during a period of progressive salination following the cessation of monsoonal rains and the onset of the dry season. Shoot growth was chosen as an indicator of growth as it has been demonstrated to be sensitive to salinity change for a range of species (Boyer 1970; Matsuda & Riazi 1981; Papp *et al.* 1983; Clough 1984; Rawson & Munns 1984).

Materials and Methods

Location and climate

The study site was located adjacent to Micket's Creek (Lat. 12°20′, Long. 130°58′) 16 km north-east of Casuarina Post Office, Darwin (See Fig. 1). The climate is monsoonal with a pronounced wet/dry season — 87% of the annual precipitation falls in the five months November to March.

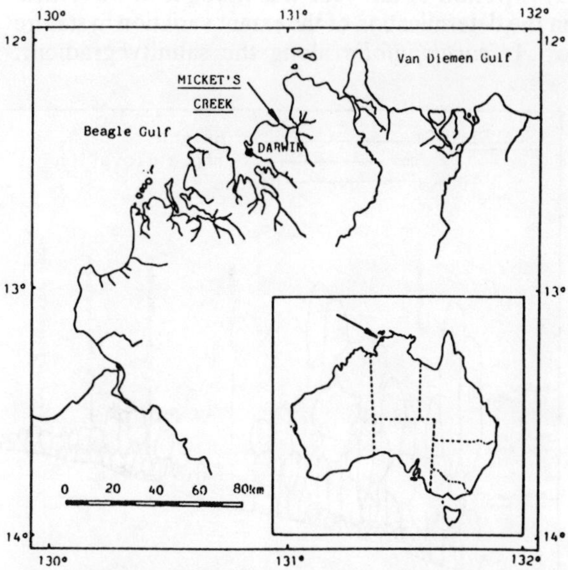

FIG. 1. Location of Micket's Creek Study Area.

The salinity gradient

A single transect through the mangrove from the seaward edge of vegetation to the upper limit of tidal penetration was selected for detailed study of shoot growth. This transect followed the assumed pattern of salinity increase from the seaward edge to land.

Along this transect *A. eucalyptifolia* was the dominant mangrove species in all plant communities from the level of mean high neap tides to land. Below this level, *A. eucalyptifolia* though present, was either co-dominant with or subdominant to trees of the genera *Rhizophora* or *Sonneratia* (Refer Fig. 2). The stature of *A. eucalyptifolia* declined along the transect in a manner directly related to the presupposed salinity gradient.

Six plant communities were subjectively selected along this intertidal salinity gradient as being representative of the ecological types associated with the gradient. The locations of these six plant communitites are marked on Fig. 2, while their characteristics are summarized in Table 1. All measurements of plant growth and edaphic conditions were conducted in these six communities.

Timing of the study

The study was timed so as to commence in the latter part of the wet season (March 1984) continuing into the middle of the dry season (August). This period of the year was thought to be critical in the determination of the extant variation in stature of *A. eucalyptifolia* along the salinity gradient; that is, it was hypothesized that following the end of the monsoonal rains[1] and their leaching effect, growth in the seaward community would be relatively unaffected, while growth in the other plant communities would decline progressively with elevation, decreased frequency of tidal inundation and higher salinity.

Shoot measurements

Four neighbouring trees on which to monitor shoot growth were selected randomly in the centre of each community. On each tree five simple, terminal shoots comprising only the current seasons growth were tagged. The length of each internode and of each leaf was measured at approximately monthly intervals.

The measured change in total shoot and leaf length for each community was expressed as a relative growth rate (RGR) viz.,

$$RGR = \frac{2}{W_n + W_{n+1}} \times \frac{W_{n-1} - W_n}{t_{n+1} - t_n}$$

where W is the total shoot or leaf length for a community
t is the number of days from t_1, and
n indicates the measurement at the start of period "n".

The use of RGR is a proven tool in the quantitative analysis of growth data (Briggs *et al.* 1920; Hughes & Freeman 1967; Radford 1967; Ledig

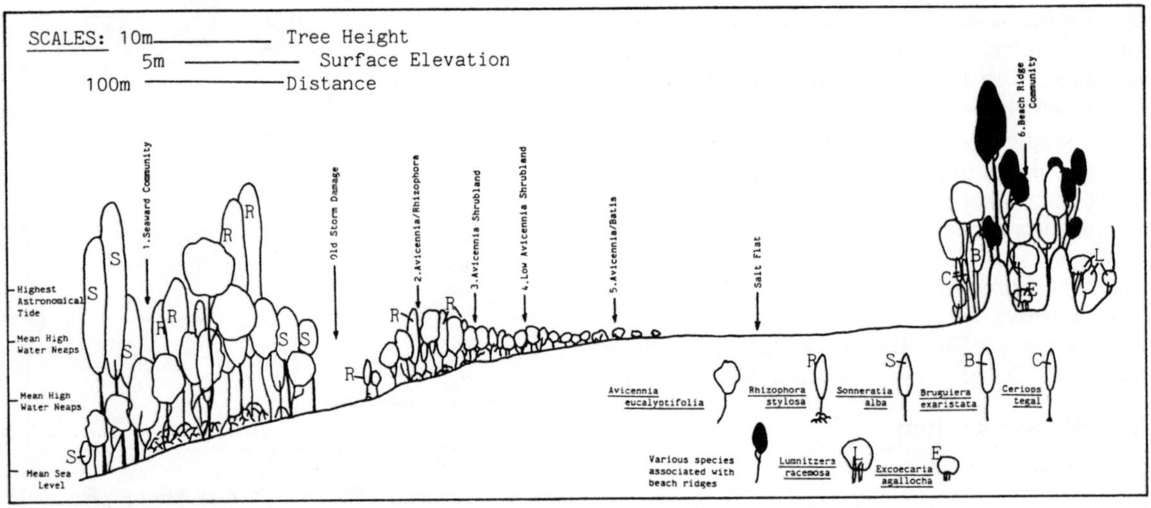

FIG. 2. Profile diagram of Vegetation along the Salinity Gradient at Micket's Creek.

[1]During 1984 the monsoonal rains ceased in the second week of April, with 1 515 mm having been recorded at Darwin airport in the period January to April (long term average – 1 169 mm). No further rainfalls were recorded throughout the remainder of the study.

TABLE 1. Characteristics of six plant communities along the salinity gradient at Micket's Creek, Darwin.

Community	Species	P.F.C.[1] (%)	Total Height (m)	Basal Diameter (cm)	D.B.H.O.B. (cm)	Crown Diameter (m)	Stocking (Stems/ha)
1. Seaward	Avicennia	0.52	5.8	18.2	17.4	—	43
	Rhizophora		7.4	13.1	—	—	128
	Sonneratia		9.4	29.6	34.0	—	22
2. *Avicennia/Rhizophora*	Avicennia	0.55	2.7	9.3	7.3	1.8	788
Shrubland	Rhizophora		2.6	4.7	3.2	0.95	716
3. *Avicennia* Shrubland	Avicennia	0.65	1.8	11.4	—	1.6	1 680
4. Low *Avicennia*	Avicennia	0.53	0.85	7.6	—	1.1	3 640
Shrubland	Batis	0.10					
5. *Avicennia Batis* Mix	Avicennia	0.22	0.65	4.5	—	0.9	3 530
	Batis	0.34					
6. Beach Ridge	Avicennia	0.61	5.3	15.8	17.9	3.0	416

[1]P.F.C. =Projected Foliage Cover is for all species unless otherwise specified.

1974) and has been applied to mangroves in a similar manner by Christensen and Wium-Andersen (1977) and Wium-Andersen (1981).

Soil salinity

Soil Salinity was determined from oven dried surface soil (0.20 m) samples collected twice monthly throughout the study. In each month samples were collected once at mid-neap tides and once at mid-spring tides, usually in consecutive weeks.

The oven dried sample was ground, passed through a 2 mm sieve, and a 5 g subsample shaken three times with distilled water to extract the salts. The filtered extracts were made up to 100 ml and conductivity measured with a "Radiometre CDM3" metre. Conductivities were con-

TABLE 2. Summary of Results from the Analysis of Varience of Shoot Measurements[1].

Source of Variation		Average Size/Shoot/Tree: Shoot Length	Average Size/Shoot/Tree: Total Leaf Length	Relative Growth Rate: Shoot Length	Relative Growth Rate: Total Leaf Length
Periods/Intervals[2]	Mean Square	16.9	75.3	24.9	56.4
	Degrees of Freedom	4	4	3	3
	Significance[3] of Difference	**	n.s	***	***
Communities	Mean Square	31.0	703	5.12	0.616
	Degrees of Freedom	4	4	4	4
	Significance[3] of Difference	***	***	n.s	n.s
Interaction	Mean Square	0.175	15.6		
	Degrees of Freedom	16	16		
	Significance[3] of Difference	n.s	n.s		
Residual	Mean Square	3.73	54.0	2.84	0.330
	Degrees of Freedom	50	50	12	12

[1]The Seaward Community has been ommitted from the anaylsis.
[2]Intervals applies to the time between measurement periods for which relative growth rates were calculated.
[3]n.s indicates not significant at p = 0.05; *, ** and *** indicate significance at p = 0.05, 0.01 and 0.001 respectively.

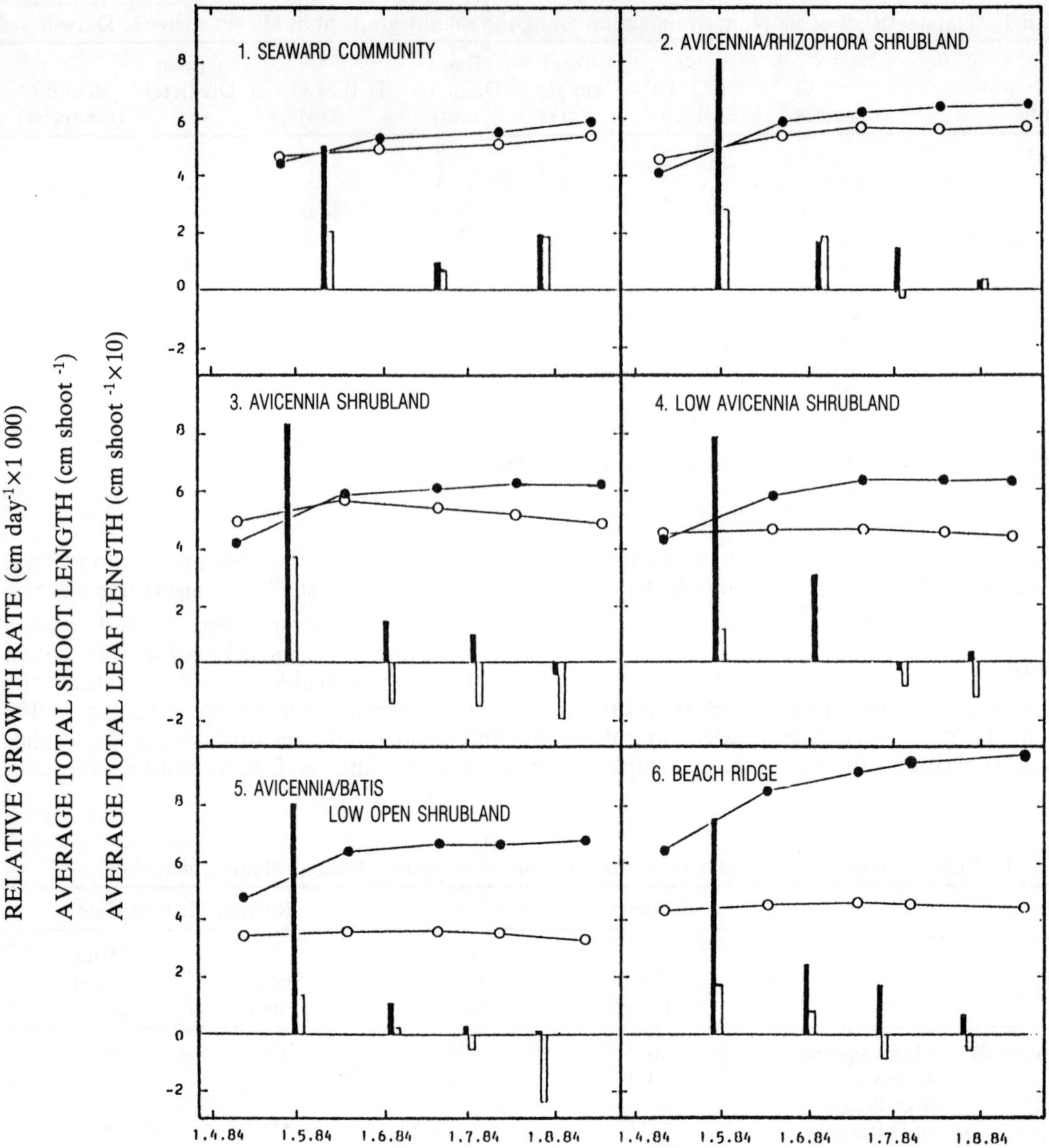

FIG. 3. Average Total Shoot and Leaf Length and Relative Growth Rates of shoots and leaves measured in the six *Avicennia* communities at Micket's Creek. (Shoot size = ○, Leaf size = ●, Relative Growth Rate of the shoot = Open histograms, and Relative Growth of leaves = Closed Histograms).

verted to field equivalents using the previously determined water content of each sample.

Statistical analyses

All data collected were analysed using standard analysis of variance techniques, analyses being carried out using GLIM3 (Baker & Nelder 1978). Row and column means were compared using Duncan's Multiple Range Test (Steel & Torrie 1960) where the interaction term was not significant, but where significant interaction was detected least significant difference (Steel & Torrie 1960) was used.

Results

The results of the shoot and salinity measurements made in each community along the gradient are presented in Figs. 3 and 4 respectively and a summary of the statistical analyses is presented in Table 2. The shoot measurements were expressed as an average per shoot since as a result of mortality and a few breakages there was not a constant number of shoots measured throughout the study in each community.

Shoot growth was rapid between April and May 1984 (the only period in which a significant change in the shoot length was recorded) but during the remainder of the study growth continued only slowly. Leaf growth was similarly rapid during the early part of the study (April/May) but thereafter the amount of leaf material present declined due to abscission and some herbivore damage.

The calculated relative growth rates (RGR) did not differ significantly between communities (Table 2)[2]. In all communities there was a period of high RGR recorded at the end of the wet season (April) and this declined to very low levels during the dry season. In the Seaward Community (No. 1) however there was a slightly different growth pattern observed: only in this community was positive shoot and leaf growth maintained throughout the study.

Further it is important to note that although the net effect of growth and mortality may have been negative at no time did growth actually cease nor did the plants at any time appear to be under severe water stress.

Soil salinity in all communities, except the Beach Ridge Community (No. 6) increased rapidly in the period April to June, and then tended to level off in the latter part of the study (Fig. 4). This levelling off generally occurred at progressively higher elevations above mean sea level. Soil salinity in the Beach Ridge Community rose continuously throughout the study. This reflected the progressive drying of the soil in the Beach Ridge Community which resulted from the total absence of any tidal inputs to this community after May 1984 while all other communities continued to receive regular tidal inundation throughout the study.

Discussion

Two distinct phases of salinity increase can be observed in Fig. 4. The first being one of rapid salination and the second one of levelling off about an upper limit. The first phase coincides with the end of the wet season and reflects the change over from an environment dominated by fresh-water inputs to one dominated by saline inputs. The second phase of levelling off corres-

[2]Changes in the actual shoot of leaf size can not be compared across communities due to variations in the initial amount of material sampled in each community.

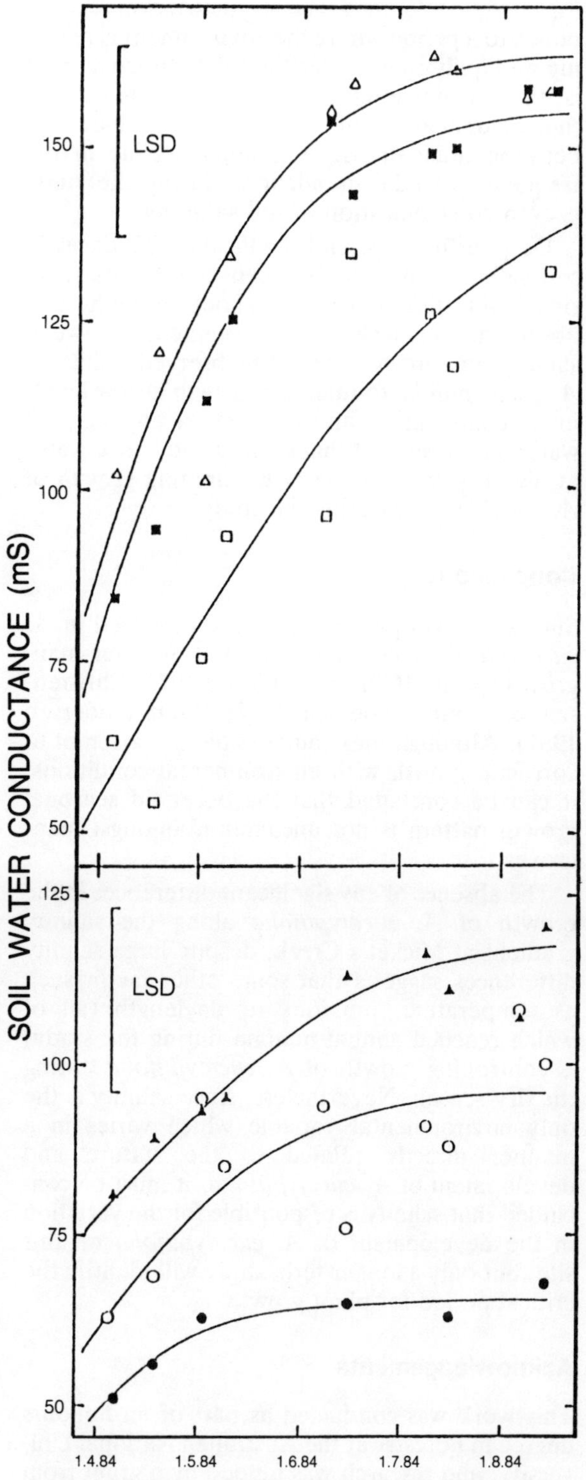

FIG. 4. Soil Water Salinity (Expressed as conductance) 1. Seaward Community (●) 2. Avicennia/Rhizophora Shrubland (○) 3. Avicennia Shrubland (▲) 4. Low Avicennia Shrubland (△) 5. Avicennia/Batis Low Open Shrubland (■) 6. Beach Ridge Community (□).

ponds to a period where the environment is reaching an equilibrium under the tidal influence, that is, the leaching effects of the tidal inputs balance the accumulation of salts in the surface soil between inundations. A gradual decline in the frequency of tidal inundations during the study prevented stabilization of soil salinities.

Despite the large and significant differences in soil salinity observed between communities, no significant differences in shoot growth was observed. This lack of correspondence between salinity and growth, plus the observed ability of *A. eucalyptifolia* to maintain growth at low levels in all communities and the lack of any signs of water stress, suggest that salinity and hence water availability was not the factor limiting growth of *A. eucalyptifolia* during the study.

Conclusions

Similar growth patterns to those observed in *A. eucalyptifolia* have been reported in other mangroves species (Gill & Tomlinson 1971; Christensen & Wium-Andersen 1977; Wium-Andersen 1981). Although these authors did not attempt to correlate growth with environmental conditions, it can be concluded that the observed seasonal growth pattern is not uncommon amongst mangroves.

The absence of any significant differences in the growth of *A. eucalyptifolia* along the salinity gradient at Micket's Creek, despite large salinity differences suggests that some other factor such as temperature, humidity or daylength (all of which reached annual minima during the study) is controlling growth of *A. eucalyptifolia* during the dry season. Nevertheless, since salinity is the only environmental variable which varies in a manner directly related to the stature and development of *A. eucalyptifolia,* it must be concluded that salinity is responsible for the variation in the development of *A. eucalyptifolia* on this site, but only a longer term study will identify the critical period for plant growth.

Acknowledgements

This work was conducted as part of an honours degree in Forestry at the Australian National University, and research was funded by a grant from the North Australian Research Unit. Thanks are also due to the Queensland Department of Forestry for making the time available to allow the presentation of this paper.

References

Baker R. J. & Nelder J. A. (1978) *The Glim system release 3, Generalized Linear Interactive Modelling.* Numerical Algorithm Group, Oxford.

Boyer J. S. (1970) Leaf enlargement and metabolic rates in corn, soybean and sunflower at various leaf water potentials. *Plant Physiol.* **46,** 233-35.

Briggs G. E., Kidd F. & West C. (1920) A quantitative analysis of plant growth. *Ann. appl. Biol.* **7,** 103-23.

Christensen B. & Wium-Andersen S. (1977) Seasonal growth of mangrove trees in Southern Thailand. 1. The phenology of *Rhizophora apiculata* B1. *Aquat. Bot.* **3,** 281-86.

Clough R. I. (1984) Growth and salt balance of the mangroves *Avicennia marina* (Forsk.) Vierh. and *Rhizophora stylosa* Griff. in relation to salinity. *Aust. J. Plant. Phisiol.* **11,** 417-30.

Gill A. M. & Tomlinson P. B. (1971) Studies on the growth of Red Mangrove (*Rhizophora mangle* L.) 3. Phenology of the shoot. *Biotropica* **3,** 109-24.

Hughes A. P. & Freeman P. R. (1967) Growth analysis using frequent small harvests. *J. appl. Ecol.* **4:** 553-60.

Ledig F. T. (1974) Concepts of growth analysis. In: *Proc. of the Third North American Forest Biology Workshop* (eds. C. P. O. Reid & G. H. Fechner) pp. 168-82. Colorado State University, Fort Collins, Colorado.

Matsuda K. & Riazi A. (1981) Stress induced osmotic adjustment in growing regions of barley leaves. *Plant Physiol.* **68,** 571-76.

Papp J. C., Ball M. C. & Terry N. (1983) A comparative study of the effects of NaCl salinity on respiration, photosynthesis, and leaf extension growth in *Beta vulgaris* L. (sugar beet). *Plant, Cell and Environment* **6,** 675-77.

Radford P. J. (1967) Growth analysis formulae – their use and abuse *Crop Sci.* **7,** 171-75.

Rawson H. M. & Munns R. (1984) Leaf expansion in sunflower as influenced by salinity and short-term changes in carbon fixation. *Plant, Cell and Environment* **7,** 207-13.

Steel R. G. D. & Torrie J. H. (1960) *Principles and Procedures of Statistics.* McGraw-Hill Book Company Inc., New York, Toronto, London.

Wium-Andersen S. (1981) Seasonal growth of mangrove trees in Southern Thailand. 3. Phenology of *Rhizophoria mucronata* Lamk. and *Scyphiphora hydrophyllacea* Gaertn. *Aquat. Bot.* **10,** 371-76.

Proc. Ecol. Soc. Aust 15:227-228

Management of a subtropical coastal salt-marsh for mosquito control

K. HULSMAN and P. E. R. DALE

School of Australian Environmental Studies, Griffith University, Nathan, Qld. 4111

Introduction

The long term aim of our study is to develop integrated programmes of effective mosquito management while minimizing both costs and biological impacts on the salt-marsh ecosystem.

In this paper we discuss the development of an integrated programme of mosquito control. In this context we present the results of our monitoring the effectiveness of the programme in reducing the numbers of mosquitoes breeding successfully on the marsh as well as the short-term impacts it has had on the salt-marsh ecosystem.

The study area is on Coomera Island in south-east Queensland where large numbers of mosquitoes are produced during the summer months. The main pest species is *Aedes vigilax* (Skuse). Not only is this species a nuisance but also a health hazard because it transmits arbovirus to people and heartworm to dogs (Kay *et al.* 1981). Large areas of marsh close to centres of human populations make it necessary to keep mosquito numbers below at least nuisance level.

Habitat Modification

Methods of control may focus on some critical characteristic of the organism and/or its environment. The larval stage is perhaps the most susceptible to habitat modification. Water is a critical characteristic of the larvae's environment, thus, modifying frequency of flooding and patterns of water movement are likely to affect their survival.

Our modifications have increased the flushing of the marsh thereby (1) increasing the likelihood of larvae being transported into the more hazardous estuarine environment; (2) increasing access of fish to larvae; and (3) decreasing the amount of time that water remains in pools to less than that needed for larvae to complete their development.

A traditional method of habitat modification has been ditching. The modification that we are testing differs from ditching in that the channels ("runnels") are shallow (<30 cm deep) and follow natural drainage lines wherever possible.

Evaluation

We are measuring the number of mosquito larvae and non-target organisms, the height of the water table, its salinity, substrate moisture and salinity, vegetation height and density and to determine the effectiveness of the method and its effect on the marsh ecosystem.

Impacts of the programme on the marsh ecosystem have so far been evaluated over a nine-month period, but we propose to continue the monitoring for several years.

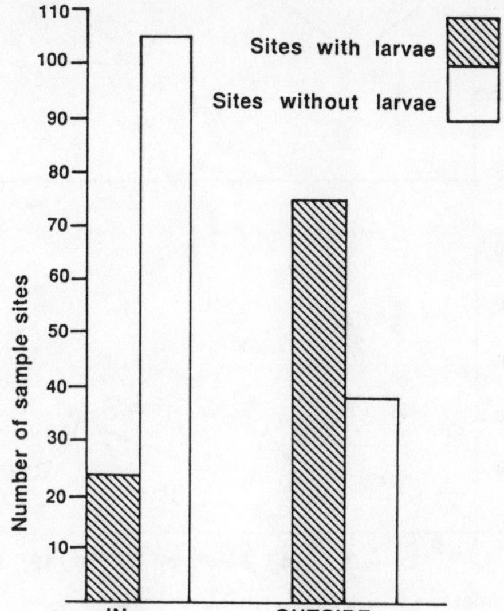

FIG. 1. Comparison of the number of sites with larvae and without larvae inside the modified area and outside it.

Short-term changes in fauna that we have measured are similar to those reported by Resh and Balling (1983). For example, there has been a significant decrease in the number of larvae in the modified area (X^2=45.0 df 1, P<0.001. Fig. 1) because larvae are carried off the marsh by the ebbing water and there is increased predation on them. Also the number and diversity of fish utilizing the marsh and adjacent mangrove ecosystem have increased (R. Morton, pers. comm.). However, in contrast to Balling and Resh's findings (1983), there have been few changes in water table, substrate or vegetation. Close to runnels these have so far not changed in any way which is significantly different from other areas (see Fig. 2).

Data are not yet available on long-term changes in the saltmarsh ecosystem. However, there is one potential change that would be undesirable. The increased rate at which nutrients are released from the marsh may cause a decrease in the store of nutrients on it and eventually lead to a decrease in the number and diversity of fish utilizing it as well as decrease the suitability of mangroves as habitat for juvenile fish and crustaceans of commercial and/or ecological importance.

References

Balling S. S. & Resh V. H. (1983) The influence of mosquito control recirculation ditches on plant biomass, production and composition on two San Francisco Bay salt marshes. *Est. Coast. Shelf Sci.* **16**, 151-61.

Kay B. H., Sinclair P. & Marks E. N. (1981) Mosquitoes: their interrelationships with man. In: *"The Ecology of Pests"* (eds. R. L. Kitching & R. E. Jones) pp. 157-74. C.S.I.R.O., Melbourne.

Resh V. H. & Balling S. S. (1983) Tidal circulation alteration for salt marsh mosquito control. *Env. Manag.* 7, 79-84.

FIG. 2

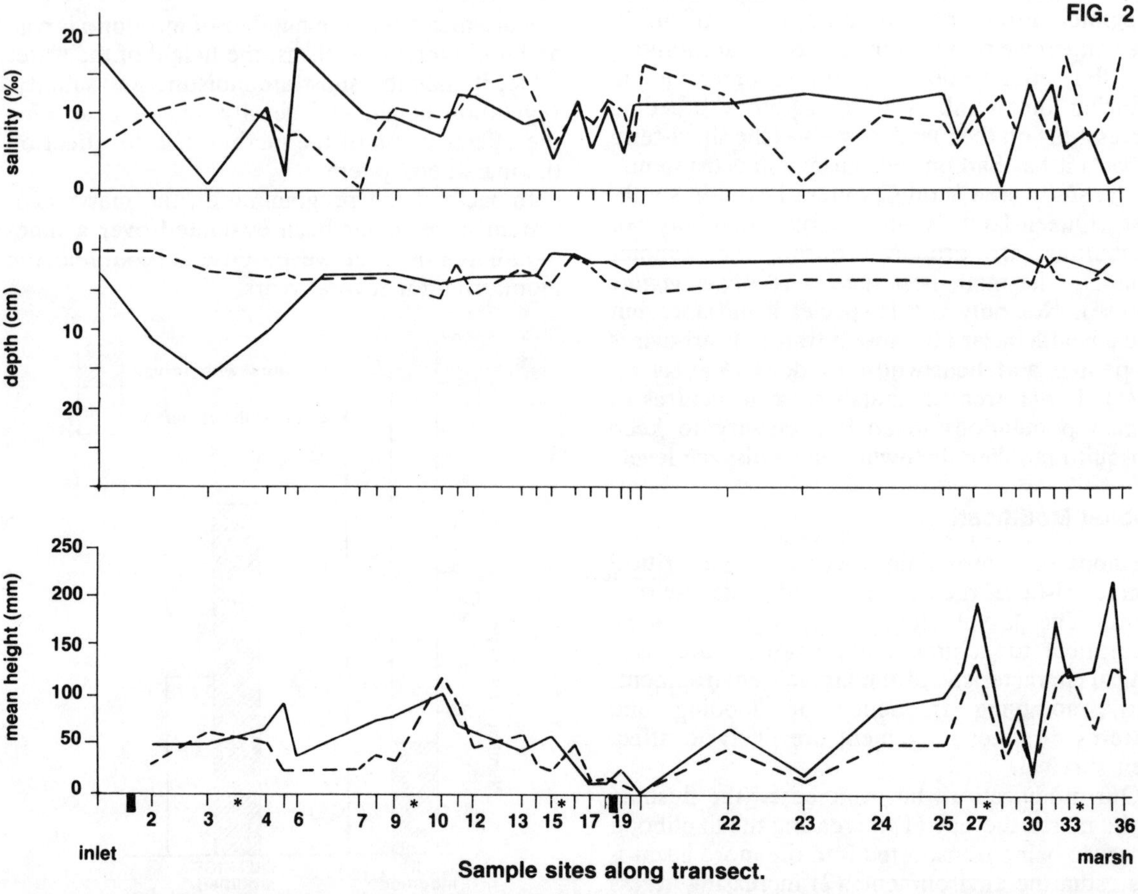

FIG. 2. Relative variation in soil moisture salinity (‰), depth water table is from the surface and mean height of *Sporobolus virginicus* along a transect from an inlet onto the marsh during the pre-runnel stage (broken line; and sic months after the runnels were dug (solid line). Salinity and water table measurements were each standardized such that the minimum reading was zero. ▮ indicates a runnel and * indicates a pool.

Proc. Ecol. Soc. Aust. 15:229-237

Seasonal and longer term patterns in the macrobenthos versus benthic stability in a subtropical estuary

P. SAENGER[1], J. MOVERLEY[2], and W. STEPHENSON[3]

[1]*Centre for Coastal Management, Northern Rivers College of Advanced Education, P.O. Box 157, Lismore, N.S.W. 2480*
[2]*Queensland Electricity Commission, P.O. Box 10, Brisbane, Qld. 4001*
[3]*Zoology Department, University of Queensland, St. Lucia, Qld. 4067*

Abstract

Quantitative sampling of the macrobenthos of the Calliope River and Auckland Creek estuary was carried out at approximately three monthly intervals at 55 sites over a nine-year period beginning in November 1974. At each site-time, 0.1 m^2 samples were taken with a Van Veen grab and sieved using a 1 mm mesh. A total of 595 species were recorded during this survey and a separate data listing is available upon request.

Severe flooding (and flood-induced scouring) in December 1973/January 1974 had virtually eliminated the soft-bottom benthos from this estuary before this study began. Consequently, the opportunity existed to study the recolonization of the estuary, and to investigate the changes in the benthos during and after this recolonization period.

Several studies have examined the recolonization of benthos after flooding or such other natural disasters as storm surges, deoxygentation and red-tide induced defaunation, and others have studied the recolonization of benthos after man-induced changes such as dredging, dredge spoil dumping or the laying of submarine pipelines. In most of these studies, the density of the benthos rapidly returned to normal levels, generally by rapid increases in the density of opportunists which subsequently declined when equilibrium species returned and competition reduced the numbers of these opportunists.

A different pattern of recolonization was found in this study. Throughout the study, there were irruptions of some species but the overall trend has been one of increasing numbers of species and of individuals until plateau densities were attained approximately five years after the flooding. However, even after plateau densities were reached, species turnover continued at more or less a constant rate until the completion of the study, indicating that species-stability had not been established. In addition, seasonal cycles were pronounced at all times. This, together with the high species turnover, suggests that the concept of benthic stability with its inherent assumptions of constancy of species composition, has only limited application to this situation.

From studies of past records of rainfall and floods in the area, it appears that a monthly rainfall exceeding 500 mm leads to severe flooding. Since 1875, this has occurred on 15 occasions with an average interval of 74 months, a maximum of 191 and a minimum of one. These data indicate that severe flooding is a phenomenon which recurs with a frequency which precludes the development of a species-stable ("climax") community.

Introduction

After early assumptions of benthic stability (e.g., Petersen 1914), came the realization that considerable seasonal variation existed in benthic populations (e.g., Raymont 1949; Kitamori 1950; Sandison & Hill 1959; Stephenson 1980 Rudnick *et al.* 1985). In addition, other studies indicated that benthic communities had a high temporal instability resulting from either abiotic (Stephenson *et al.* 1977; Probert 1981; Thistle 1981; Shaw *et al.* 1983; Poiner & Kennedy 1984) or biotic (Dayton & Hessler 1972; Russ, 1980; Moran 1980; Stephenson 1981) factors, or from inherent, unexplainable variability (Stephenson *et al.* 1978; Coull 1985).

More recently, there has been considerable debate over whether marine and estuarine communities exhibit succession and develop in a predictable sequence of species towards a climax state or one of several "patch climax" states (Osman 1977; Sutherland & Karlson 1977; Saenger *et al.* 1979; Keen & Neill 1980; Ayling 1981; Hughes 1984). Although there appears to be some evidence suggesting that there may be a succession in trophic groups within the benthos (Grassle & Grassle 1974; Moverley *et al.* 1986), species succession is more dubious.

As part of a monitoring programme on the effects of the thermal discharge from the Gladstone Power Station, a long-term study of the estuarine macrobenthos of the Calliope River and Auckland Creek estuary was undertaken. An initial analysis covering the first two years of the present data set (Saenger *et al.* 1980) showed that the benthic communities of this estuary were recovering following the severe flooding of December 1973 and January 1974. Severe flooding (and flood-induced scouring) had virtually eliminated the soft-bottom benthos from this estuary before this study began (Saenger *et al.* 1980). Consequently, the opportunity existed to study the recolonization of the estuary over an extended period (November 1974-May 1983) and to investigate the temporal changes in the benthos during and after this recolonization period.

Methods

Survey design

Eleven transects were selected on the river and creeks (Fig. 1); each transect extends across the river from about 1 m below L.W.M. on each bank. Two transects were located in Auckland Creek, one in Black Harry Creek, two in the anabranch of the Calliope River and six were distributed along the main course of the Calliope River. Selections of the transects were based on their spread along the rivers and the ease with which they could be accurately relocated. Relocation errors are about ± 2 m except for transects 1 and 11, where they are approximately ± 4 metres.

At each transect, five equally-spaced stations were sampled using a 0.05 m^2 Van Veen grab sampler; two replicate grabs were combined to give a total of 0.1 m^2 sampled at each of 55 sites in all.

Sample treatment

Sampling was carried out at approximately three-monthly intervals as follows: pre-thermal surveys (i.e., before the power station began operation) – November 1974; March 1975; June 1975; October 1975; February 1976; May 1976; August 1976. Transitional surveys (i.e., when the power station started to operate but not at full generating capacity) – November 1976; February 1977; May 1977; October 1977; January 1978; March 1978; July 1978; November 1978; January 1979; April 1979; July 1979; November 1979; March 1980; June 1980; October 1980; February 1981; June 1981; September 1981; December 1981. Post-thermal surveys (i.e., when the power station reached its full generating capacity) – March 1982; August 1982; November 1982; January 1983; May 1983. These times are referred to below as $t_{1\text{-}31}$ respectively. On all occasions all sites were sampled although at some site-times, no organisms were found.

All samples were sieved through an 8 mm (to break up the sediments) and a 1 mm sieve in the field and that fraction retained by the fine sieve was fixed in 5% formalin/seawater and subsequently transferred to glycerol:water:ethanol (5:25:70 v/v).

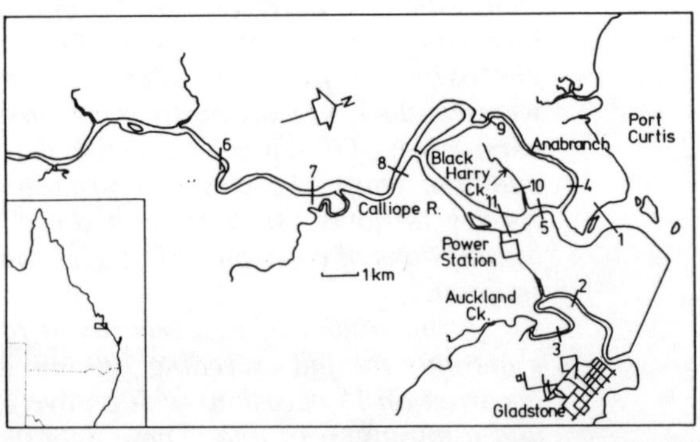

FIG. 1. Map of study area showing benthic sampling transects 1-11.

Numerous taxonomic problems were encountered while working up the collections and the assistance of many specialists was enlisted (see acknowledgements). Once species had been identified, subsequent identifications were generally made using validated reference material, except in a few difficult groups such as some polychaetes, echinoderms and ascidians which were all identified by the taxonomic specialists.

Representative material of all taxa has been deposited in the Queensland Museum (crustaceans, pycnogonids, ascidians, fish), the Australian Museum (molluscs, polychaetes, echinoderms, gobiid fish), the National Museum of Victoria (pycnogonids), the Smithsonian Institution (ostracods), Northern Territory Museum (Natantia), the Queensland Herbarium (algae, seagrasses) and the University of Adelaide Herbarium (algal type specimens).

Analysis of data

Curve fitting. Data on the numbers of organisms (N) and of species (S) summed over the 55 sites, suggested that a sigmoid or logistic model was most suitable. However, fitting such an equation requires the assumption of an asymptote which proved difficult with the seasonal variation in the data. As a result, trend lines were fitted to both the $\log_{10}$-transformed N and S data using the generalized equation:

$$y=t/(a+bt)$$

where y is the log-transformed data for N and S
t is the time in months from July 1974
1/a is the initial slope of the trend line
1/b is the asymptote of the trend line

which is the logarithmic equivalent of a sigmoid or logistic curve. This curve can be linearized as follows:

$$t/y=a+bt$$

and constants a and b can be determined by regression without any assumptions regarding an asymptote.

The cyclical components in the residuals were subsequently analyzed using the generalized equation

$$y=0.5A_o+A\cos(2\pi t/T)+B\sin(2\pi t/T)$$

where y is the log-transformed data for N and S
t is the time in months from July 1974
$0.5A_o$ is the midpoint of the oscillating curve
A is the contribution of the cosine component
B is the contribution of the sine component, and
T is the wavelength of the cyclic component (T=12)

Diversity calculations. Three measures of diversity were used with all data sets. These include Shannon diversity (H), Gleason diversity (G) and Shannon equitibility (J) defined as follows:

$$H=-\Sigma([n/N]\log_2[n/N])$$
$$G=(S-1)/\log_e N$$
$$J=H/\log_2 S$$

where n is the number of individuals of each species
N is the sum of the number of individuals in the sample, and
S is the number of species in the sample.

Results

Biotic composition

A total of 595 faunal taxa were recorded during this survey although some of these taxa possibly require

TABLE 1. Faunistic composition of the Calliope River and Auckland Creek estuary, November 1974-May 1983.

Taxa	Total No. of organisms		Total No. of species	
Porifera	147	(0.1%)	3	(0.5%)
Coelenterata				
Anthozoa	282	(0.3%)	7	(1.2%)
Pennatulida	11	(0.1%)	3	(0.5%)
Alcyonaria	135	(0.1%)	1	(0.2%)
Hydrozoa	3	(0.1%)	2	(0.3%)
Nemertea	925	(0.9%)	2	(0.3%)
Echiura	34	(0.1%)	3	(0.5%)
Platyhelminthes				
Turbellaria	25	(0.1%)	7	(1.2%)
Bryozoa	6	(0.1%)	1	(0.2%)
Annelida				
Polychaeta	39 679	(36.7%)	227	(38.2%)
Oligochaeta	1	(0.1%)	1	(0.2%)
Hirudinea	3	(0.1%)	1	(0.2%)
Mollusca				
Pelecypoda	22 952	(21.2%)	45	(7.6%)
Gastropoda	4 319	(4.0%)	50	(8.4%)
Scaphopoda	229	(0.2%)	1	(0.2%)
Arthropoda				
Crustacea	31 420	(29.0%)	164	(27.6%)
Pycnogonida	573	(0.5%)	10	(1.7%)
Insecta	30	(0.1%)	4	(0.7%)
Sipunculida	2	(0.1%)	1	(0.2%)
Phoronida	39	(0.1%)	1	(0.2%)
Branchiopoda	12	(0.1%)	1	(0.2%)
Echinodermata				
Asteroidea	1	(0.1%)	1	(0.2%)
Ophiuroidea	1 663	(1.5%)	19	(3.2%)
Holothuroidea	4 743	(4.4%)	3	(0.5%)
Chordata				
Ascidiacea	603	(0.6%)	19	(3.2%)
Pisces	380	(0.4%)	18	(3.0%)
Total	108 217		595	

subdivision (e.g., *Cirolana, Squilla, Alpheus*). The distribution of these taxa amongst the higher taxonomic groupings is shown in Table 1, in terms of both the numbers of individuals and species. The faunal taxa showed a pre-ponderance of polychaetes (38.2%), followed by crustaceans (27.6%) and molluscs (16.2%). In terms of the total number of individuals recorded, 36.7% were polychaetes, 29.0% crustaceans and 25.4% molluscs.

Changes in summed biotic data with time

The meaned or summed data (over the 55 sites) are given for the 31 times in Table 2. The data for the number of species (actually $\log_e S$) and organisms (actually $\log_e N$), given in Figs. 2 and 3, show an underlying curvilinear trend with a superimposed annual periodicity. Combining the curvilinear and cyclic components resulted in the following equations, which are superimposed on the data in Figs. 2 and 3:

$$\log_e N = t/(0.1225t + 0.2483) + 0.055 + 0.176\cos(2\pi t/12) + 0.899\sin(2\pi t/12)$$
$$R^2 = 0.76^{***}\ (F = 20.6,\ d.f. = 4, 26)$$

$$\log_e S = t/(0.1935t + 0.3084) + 0.011 + 0.047\cos(2\pi t/12) + 0.338\sin(2\pi t/12)$$
$$R^2 = 0.73^{***}\ (F = 17.6,\ d.f. = 4, 26)$$

where N is the total number of organisms
S is the total number of species
t is the time in months where t=0 in July 1974.

The coefficient of determination (R^2) for both of these curves are significant at less than the 0.005 level. The residuals from these equations were related to rainfall using a rainfall index defined as follows:

$$RI = ([r_1 + 0.5r_2] - [R_1 + 0.5R_2])/(R_1 + 0.5R_2)$$

where RI is the rainfall index
r_1 is the rainfall in the month prior to sampling
r_2 is the rainfall in the second month prior to sampling
R_1 is the long-term mean rainfall for the period of r_1
R_2 is the long-term mean rainfall for the period of r_2

The following relationships were found:

Residuals of $\log N = 0.0224 - 0.2987\ (RI)$ $\quad R^2 = 0.47^{***}$

Residuals of $\log S = 0.0041 - 0.1045\ (RI)$ $\quad R^2 = 0.29^{**}$

When these relationships were incorporated into the overall equations, these became:

$$\log_e N = t/(0.1225t + 0.2483) + 0.0774 + 0.176\cos(2\pi t/12) + 0.899\sin(2\pi t/12) - 0.2987\ (RI)$$
$$R^2 = 0.867^{***}\ (F = 26.1,\ d.f. = 6, 24)$$

$$\log_e S = t/(0.1935 + 0.3084) + 0.0151 + 0.047\cos(2\pi t/12) + 0.338\sin(2\pi t/12) - 0.1045\ (RI)$$
$$R^2 = 0.806^{***}\ (F = 16.6,\ d.f. = 6, 24)$$

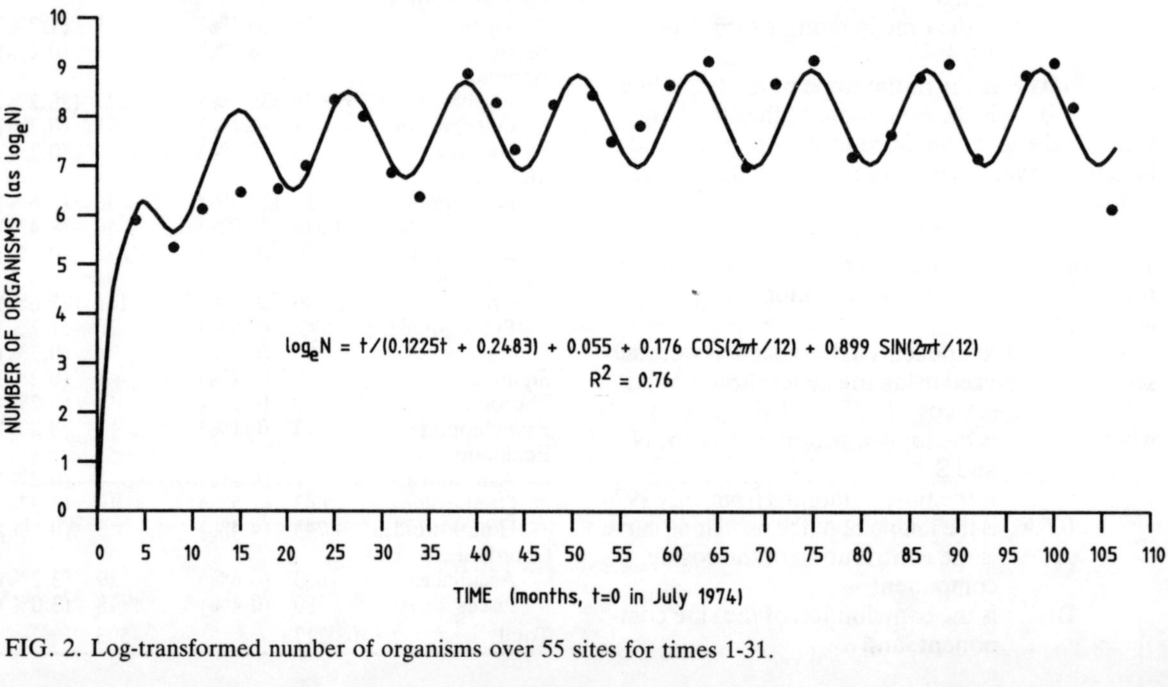

FIG. 2. Log-transformed number of organisms over 55 sites for times 1-31.

In other words, a total of 86.7% and 80.6% of the variance in the summed N and S data respectively can be explained, in order of importance, by (1) the time from the flood, (2) the season and (3) the deviation of the rainfall before sampling when compared with the respective long-term means. The seasonal means of N and S per survey became asymptotic after approximately 25 months. Similar trends are apparent in the diversity data (Fig. 4)

The cumulative number of species recorded during the survey is shown in Fig. 5, and these data show that except for the last survey, previously unrecorded species were encountered at a more or less constant rate throughout.

Changes with upstream distance

The data for the Shannon diversity index for all transects over time are shown in Fig. 6. Significant positive correlations (n=31) were found between all transects, ranging from 0.43 between 3 and 6, to 0.91 for transects 8 and 11. No indications of thermal effects were noted in the thermally exposed sites (transects 5 and 11) when compared with more distant sites Fig. 6.

Discussion

Several studies have examined the recolonization of benthos after flooding or other such natural disasters as storm surges, deoxygenation and red-tide induced defaunation, whereas others have studied the recolonization of benthos after man-induced changes such as dredging, dredge spoil dumping or the laying of submarine pipelines. In most of these studies, the density of the benthos returned rapidly to pre-disturbance levels, generally by rapid increases in the density of opportunist species which subsequently declined when equilibrium species returned.

The data presented here show that the macrobenthos of this estuary recolonized rapidly after the severe flooding of December 1973 to January 1974. All parameters (S, N, H, G, and J) increase more of less logarithmically during the first two and a half to five years (depending on the particular parameter). After this phase, all of the parameters attain an equilibrium level with superimposed seasonal variations as well as variation correlated with rainfall characteristics immediately before each sampling period. These trends (Fig. 6) occur throughout the estuary, from

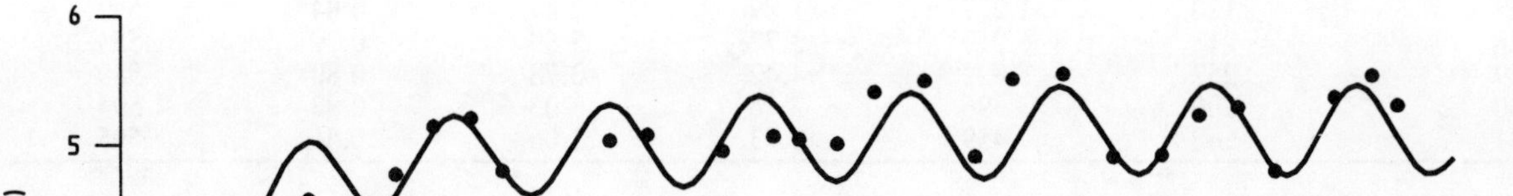

FIG. 3. Log-transformed number of species over 55 sites for times 1-31.

TABLE 2. Summed or averaged parameters of the benthic fauna over the survey period (see text for further explanation).

Survey No.	No. species (S)	No. organisms (N)	Mean G	Mean H	Mean J	Cumulative species
1	57	365	0.85	1.31	0.68	57
2	37	208	0.54	0.80	0.51	73
3	67	454	1.03	1.51	0.68	99
4	97	636	1.30	1.82	0.78	138
5	84	686	1.15	1.75	0.78	170
6	118	1 092	1.58	2.23	0.80	200
7	171	4 194	2.56	2.88	0.77	239
8	182	2 982	2.78	3.24	0.84	278
9	121	951	1.59	2.23	0.85	289
10	83	586	1.01	1.48	0.74	298
11	153	7 099	2.38	2.71	0.71	315
12	160	3 938	2.53	2.93	0.81	334
13	86	1 528	1.29	1.83	0.73	336
14	141	3 796	2.03	2.48	0.76	343
15	158	4 607	2.37	2.81	0.79	355
16	154	1 799	2.24	2.73	0.82	367
17	149	2 471	2.44	2.94	0.85	382
18	223	5 630	3.64	3.57	0.83	414
19	243	9 180	3.94	3.57	0.80	449
20	135	1 080	1.53	1.97	0.72	463
21	246	5 846	4.02	3.70	0.82	499
22	256	9 333	4.24	3.59	0.76	512
23	134	1 315	1.54	2.06	0.74	518
24	136	2 064	1.79	2.22	0.69	525
25	185	6 593	2.83	2.77	0.70	533
26	197	8 711	3.06	2.97	0.72	541
27	120	1 277	1.89	2.44	0.84	547
28	213	6 919	3.72	3.39	0.79	555
29	252	8 822	4.42	3.78	0.80	580
30	200	3 596	2.76	3.05	0.82	594
31	68	459	0.73	1.06	0.54	595

approximately 15 km upstream (transect 6) to the mouth (transect 1).

In view of the above trends, it seems that the recolonization process is following the general pattern of ecosystem succession as envisaged by Odum (1969) and as described by a number of benthic recolonization studies (Dean & Haskin 1964; Rosenberg 1972, 1973, 1976; Wolff *et al.* 1977; Oviatt *et al.* 1984).

On the other hand, the more or less linear rate at which previously unrecorded species were recorded during the course of the study (Fig. 5) suggests that species turnover is continuing and that species-stability has not been attained more than nine years after flooding.

The stability in total numbers of species and density coupled with instability in the species composition may be due to a number of features, including (a) succession with regard to the species composition is not complete; (b) certain trophic groups are gradually being replaced by others as the ecosystem matures in much the same way that these changes were observed in the polychaete fauna (Moverley *et al.* 1986); and (c) minor but on-going disturbances are maintaining species turnover and a "patch climax" community but are not markedly affecting total numbers of species and density.

It is clear that in some respects, the benthic community of this estuary has reached a "climax" phase but without a species-stable ("climax") community. While it is tempting to suggest that given adequate time, such a species-stable community will develop, in reality, such a situation is highly improbable.

From studies of past records of rainfall (99 years) and floods in the area, it appears that a monthly rainfall of 500 mm or more leads to flooding as severe as that which occurred during December 1973-January 1974 (Moverley *et al.* 1986). Since 1875, rainfall events of this magnitude and intensity have occurred in 15 occasions with an average interval of 74 months, a maximum of 191 and a minimum of one month. A species-stable community was not attained although the present survey period is in excess of the mean interval between major floods. Consequently, these data suggest that while a quasi-stable community can develop, severe flooding is a phenomenon which recurs with a frequency which precludes the development of a species-stable ("climax") community.

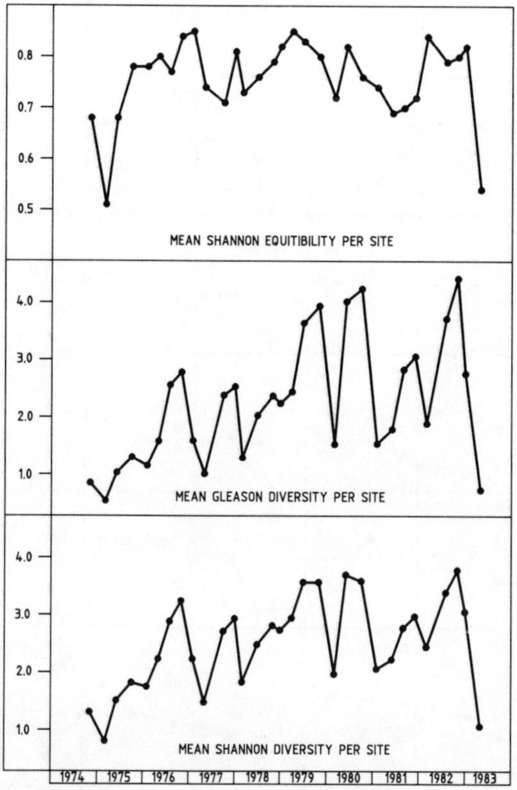

FIG. 4. Mean diversity indices of 55 sites for times 1-31.

Acknowledgements

We thank the following specialists who have identified material for us: Dr P. Alderslade (Octocorallia), Dr A. Bruce (Macrura), Dr N. Bruce (Isopoda), Mr R. Burns (Opisthobranchia), Dr L. Cannon (Platyhelminthes), Mr P. Davie (Brachyura), Dr S. Edmonds (Echiura, Sipunculida), Dr D. Hoese (Gobies), Dr T. Hove (Serpulidae), Dr P. Hutchings (Polychaeta), Mr M. Johnston (Cumacea), Dr L. Kornicker (Ostracoda), Mr R. Mackay (Fish), Dr P. Mather (Ascidiacea), Mr R. Monroe (Amphipoda), Dr J. Ross (Bryozoa), Dr F. Rowe (Echinodermata) and Mr D. Staples (Pyconogonida).

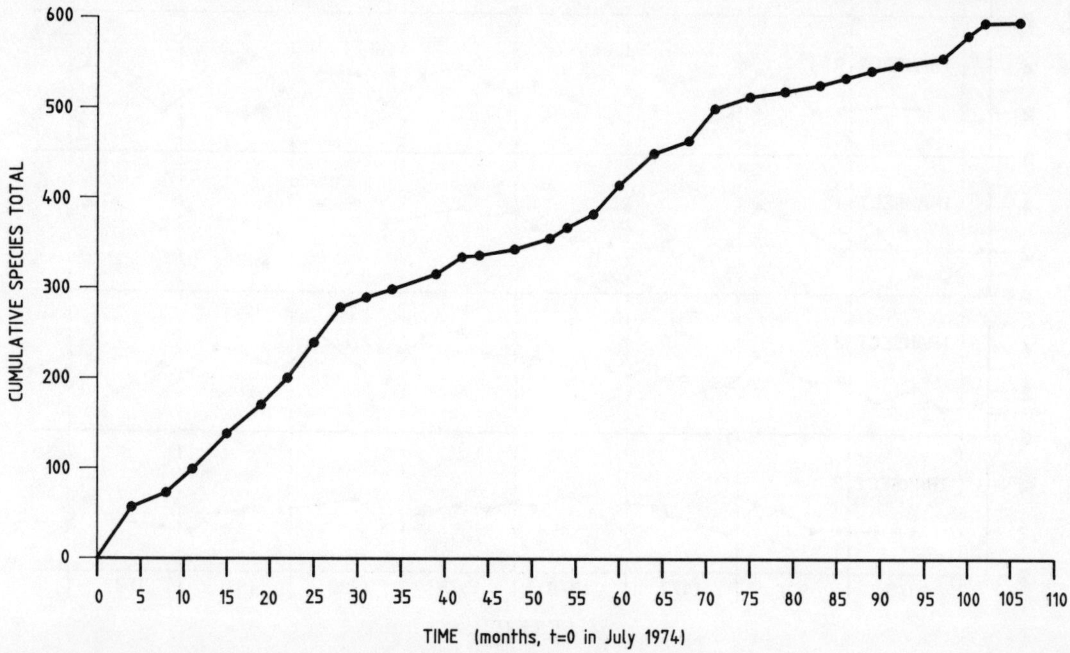

FIG. 5. Cumulative species totals recorded over 55 sites for times 1-31.

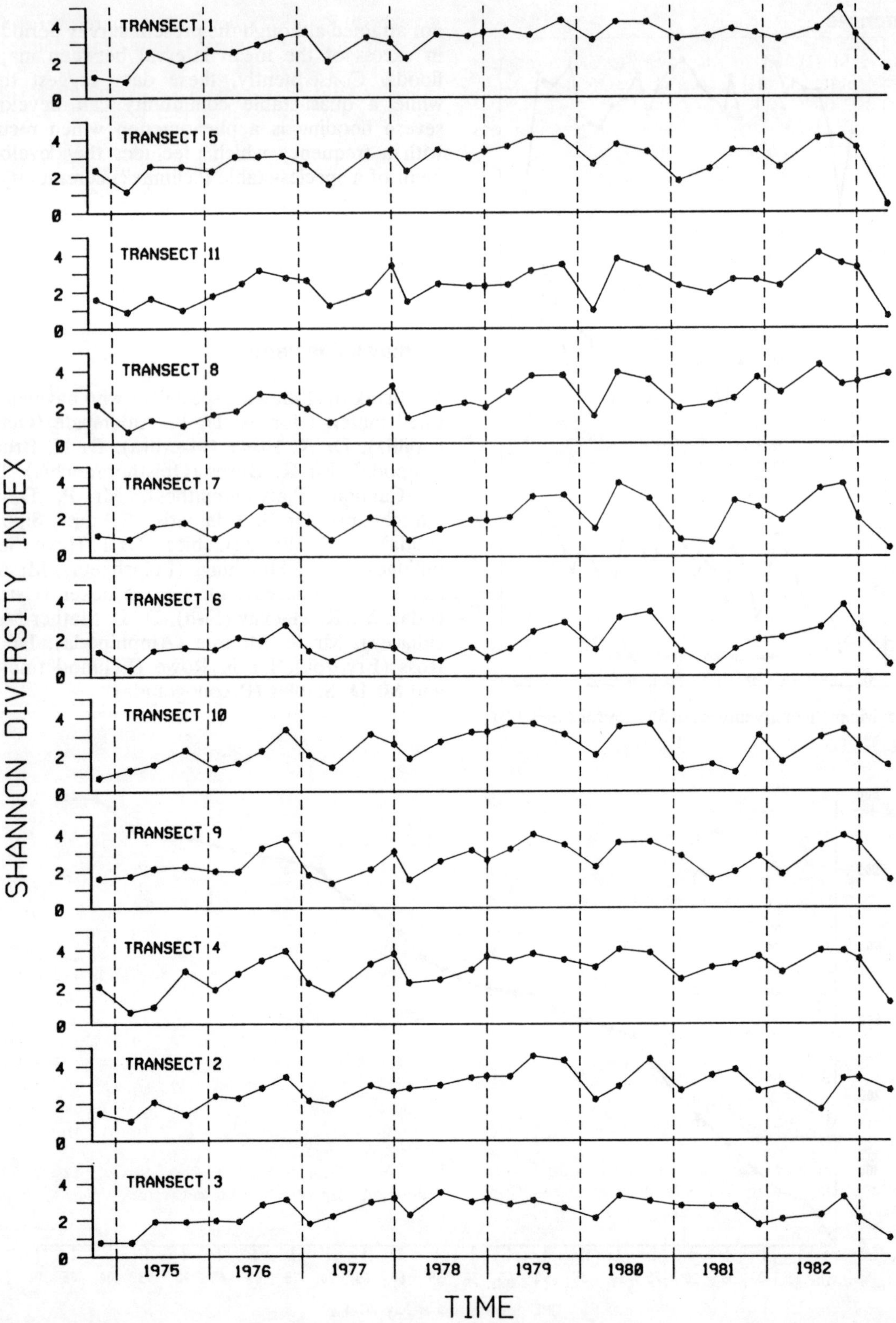

FIG. 6. Variation in Shannon diversity index with time for transects 1-11.

References

Ayling A. M. (1981) The role of biological disturbance in temperate subtidal encrusting communities. *Ecology* **62**, 830-47.

Coull B. C. (1985) Long-term variability of estuarine meiobenthos: an 11 year study. *Mar. Ecol. Prog. Ser.* **24**, 205-18.

Dayton P. K. & Hessler R. R. (1972) Role of biological disturbance in maintaining diversity in the deep sea. Deep Sea Res. **19**, 199-208.

Dean D. & Haskin H. H. (1964) Benthic repopulation of the Raritan River estuary following pollution abatement. *Limnol. Oceanogr.* **9**, 551-63.

Grassle J. F. & Grassle J. P. (1974) Opportunistic life histories and genetic systems in marine benthic polychaetes. *J. Mar. Res.* **32**, 253-84.

Hughes R. G. (1984) A model of the structure and dynamics of benthic marine invertebrate communities. *Mar. Ecol. Prog. Ser.* **15**, 1-11.

Keen S. L. & Neill W. E. (1980) Spatial relationships and some structuring processes in benthic intertidal animal communities. *J. exp. mar. Biol. Ecol.* **45**, 139-55.

Kitamori R. (1950) Studies on the benthos of Tokyo Bay. II. On the distribution and seasonal change of the benthos. *Bull. Jap. Soc. Scient. Fish.* **16**, 265-80.

Moran P. J. (1980) Natural physical disturbance and predation, their importance in structuring a marine sessile community. *Aust. J. Ecol.* **5**, 193-200.

Moverley J. H., Saenger P. & Curtis M. A. (1986) Patterns of polychaete recolonization in Queensland subtropical estuaries following severe flooding. *Hydrobiologia* **143**, 227-36.

Odum E. P. (1969) The strategy of ecosystem development. *Science, N.Y.* **164**, 262-70.

Osman R. W. (1977) The establishment and development of a marine epifaunal community. *Ecol. Monogr.* **47**, 37-63.

Oviatt C. A., Pilson M. E. Q., Nixon S. W., Frithsen J. B., Rudnick D. T., Kelley J. R., Grassle J. F. & Grassle J. P. (1984) Recovery of a polluted estuarine system: a mesocosm experiment. *Mar. Ecol. Prog. Ser.* **16**, 203-17.

Petersen C. G. J. (1914) Valuations of the sea. II. The animal communities of the sea bottom and their importance for marine zoogeography. *Rep. Dan. Biol Sta.* **21**, 1-44.

Poiner I. R. & Kennedy R. (1984) Complex patterns of change in the macrobenthos of a large sandbank following dredging. I. Community analysis. *Mar. Biol.* **78**, 335-52.

Probert P. K. (1981) Changes in the benthic community of china clay waste deposits in Mevagissey Bay following a reduction in discharge. *J. mar. Biol. Ass. U.K.* **61**, 789-804.

Raymont J. E. G. (1949) Further observations on changes in the bottom fauna of a fertilized sea-loch. *J. mar. Biol. Ass. U.K.* **28**, 9-19.

Rosenberg R. (1972) Benthic faunal recovery in a Swedish fjord following the closure of a sulphite pulp mill. *Oikos* **23**, 93-108.

Rosenberg R. (1973) Succession in benthic macrofauna in a Swedish fjord subsequent to the closure of a sulphite pulp mill. *Oikos* **24**, 344-58.

Rosenberg R. (1976) Benthic faunal dynamics during succession following pollution abatement in a Swedish estuary. *Oikos* **27**, 414-27.

Rudnick D. T., Elmgren R. & Frithsen J. B. (1985) Meiofaunal prominence and benthic stability in a coastal marine ecosystem. *Oceologia (Berl.)* **67**, 157-68.

Russ G. R. (1980) Effects of predation by fishes, competition, and structural complexity of the substratum on the establishment of a marine epifaunal community. *J. exp. mar. Biol. Ecol.* **42**, 55-69.

Saenger P., Stephenson W. & Moverley J. (1979) The subtidal fouling organisms of the Calliope River and Auckland Creek, Central Queensland. *Mem. Qd. Mus.* **19**, 399-412.

Saenger P., Stephenson W. & Moverley J. (1980) The estuarine macrobenthos of the Calliope River and Auckland Creek, Queensland *Mem. Qd. Mus.* **20**, 143-61.

Sandison E. E. & Hill M. B. (1959) The annual repopulation of Lagos Harbour by sedentary marine animals. In: *Reprints of the 1st International Oceanographic Congress, 31st August-12th September 1959* (ed M. Sears) pp. 584-85. Amer. Assoc. Adv. Science, Washington.

Shaw K. M., Lambshead P. J. D. & Platt H. M. (1983) Detection of pollution-induced disturbance in marine benthic assemblages with special reference to nematodes. Mar. Ecol. Progr. Ser. **11**, 195-202.

Stephenson W. (1980) Time-patterns of macrobenthic species in Moreton Bay. *Aust. J. Ecol.* **5**, 245-62.

Stephenson W. (1981) Long term cycles caused by patchy predation. *Aust. J. Ecol.* **6**, 357-64.

Stephenson W., Cook S. D. & Raphael Y. I. (1977) The effect of a major flood on the macrobenthos of Bramble Bay, Queensland. *Mem. Qd. Mus.* **18**, 95-119.

Stephenson W., Cook S. D. & Newlands S. J. (1978) The macrobenthos of the middle banks of Moreton Bay. *Mem. Qd. Mus.* **19**, 185-212.

Sutherland J. P. & Karlson R. H. (1977) Development and stability of the fouling community at Beaufort, North Carolina. *Ecol. Monogr.* **47**, 425-46.

Thistle D. (1981) Natural physical disturbances and communities of marine soft bottoms. *Mar. Ecol. Prog. Ser.* **6**, 223-28.

Wolff N. J., Sandee A. J. J. & De Wolf L. (1977) The development of a benthic ecosystem. *Hydrobiologia* **52**, 107-15.

Proc. Ecol. Soc. Aust. 15:239-247

Patterns of differentiation in tropical species of *Melaleuca* L. (Myrtaceae)

B. A. BARLOW

Australian National Herbarium, C.S.I.R.O., G.P.O. Box 1600, Canberra, A.C.T. 2601

Abstract

Melaleuca *is a Gondwanan genus which has apparently originated in seasonally drowned habitats at the margins of tropical rainforests. The* M. leucadendra *species complex appears to represent the ancestral state among extant taxa. The evolution of scleromorphy was an early development associated with the radiation of the genus into edaphically nutrient-deficient, temperate and semi-arid habitats. Late Tertiary and Quaternary climatic cycles have resulted in extensive ecogeographic differentiation both in northern and southern Australia. A number of examples of east-west differentiation in tropical Australia/Papuasia is presented, illustrating the limited significance of present-day coastlines in explaining geographic distribution patterns.*

Introduction

The large genus *Melaleuca* (250 spp.) is widespread through most plant associations in Australia. By far its largest centre of diversity is the south-west of Western Australia, especially in the transition zone between the high rainfall area of the Leeuwin region and the semi-arid interior. In this area, and indeed through most of its range, it is most familiar as a scleromorph, and is commonly found on alluvial, nutrient deficient soils, often with a high water table.

Nevertheless *Melaleuca* falls well within the ambit of this symposium, because the tropical ancestry of the group is clearly evident even though only about 15% of the species occur in tropical Australia. The genus diminishes in species richness southwards in eastern Australia, and there are relatively few species in Victoria and Tasmania. There are *c.* 28 spp. in tropical Queensland, compared with *c.* 18 in temperate Queensland, *c.* 25 in New South Wales, 12 in Victoria and five in Tasmania. The massive radiation in the transitional rainfall zone of temperate Western Australia (cf. Hopper 1979) thus contrasts with the general distribution pattern in eastern Australia. To a large extent, then, *Melaleuca* parallels the eucalypts in distribution patterns, and probably in biogeographic history (see below).

Evolutionary Trends and Ancestral State

There are many attribute states in *Melaleuca* to which polarity can be assigned on *a priori* grounds (Table 1). In general the primitive states tend to be correlated in their occurrence, so that together they provide a strong indication of what the ancestral state in the genus may have been.

With respect to these attributes, the *M. leucadendra* complex of *c.* 15 species is distinctive in exhibiting the entire suite of ancestral states. It may be suggested, therefore, that the hypothetical ancestor of the entire genus was similar to the least specialized members of this group. These include the type of genus, *M. leucadendra* L., as well as species such as *M. dealbata* S. T. Blake, *M. argentea* W. Fitzger., and *M. cajuputi* Powell. The distribution and ecology of these species thus allow some additional speculation on the origin of the genus. The entire species complex is tropical in distribution (Fig. 1); the least specialized members are tall trees, and they grow in seasonally inundated freshwater or brackish situations, notably in gallery forests and in swampy depressions adjacent to rainforests. It may be suggested, then, that the genus has differentiated at the margins of tropical rainforest, in response to selection for tolerance of prolonged seasonal high water table.

TABLE 1. Attribute states and polarity in *Melaleuca*.

Organ/Attribute	Ancestral state	Derived state
Habit	tree	shrub
Leaves		
Shape	flat	terete
Size	macrophyllous	microphyllous
Surface morphology	isobilateral	dorsiventral
Attachment	petiolate	sessile, peltate
Stomate subsiduaries	simple	complex
Wax platelets	absent	present
Surface secretions	absent	present
Inflorescence		
Axial growth	indeterminate	determinate
Uniflorescence	triad	monad
Conflorescence	spike	head or solitary flower
Involucre	absent	present
Flowers		
Sepals	deciduous	persistent
Petals	present	absent
Stamens:		
Uniformity	monomorphic	polymorphic
Claw	short or absent	long
Number per bundle	few	many
Filament insertion	single series terminal on claw	many series superficial on claw
Anthers	versatile	immobile*
Ovules	many	few or solitary
Ovulodes	absent	present
Fruits		
Arrangement	solitary	in cone-like aggregates
Form	simple	adorned
Valves	simple	crested
Position	superficial	immersed
Breeding System		
Sexual expression	hermaphrodite	andromonoecious gynodioecious subdioecious
Pollination	entomophilous	ornithophilous

*Truly basifixed immobile anthers occur in derived, segregate genera *(Regelia, Phymatocarpus, Beaufortia)*.

In this respect, the comparison between *Melaleuca* and the eucalypts is of special interest. The hypothetical ancestors of the eucalypts may also have differentiated to the margins of tropical/subtropical rainforest (Pryor & Johnson 1981; Johnson & Briggs 1981). In tropical Australia today there are situations where *Melaleuca* replaces *Eucalyptus (sens. lat.)* as the dominant forest tree, forming extensive "paperbark" forests. It may be suggested that the paperbarks and the eucalypts have differentiated in parallel, one group as a colonizer of seasonally drowned habitats and the other of nutrient deficient seasonally drier soils. Subsequent evolution and radiation in both groups has led to broad ecological overlap and coexistence, but in the ancestral situation the two groups may have been relatively distinct.

Melaleuca as a Gondwanan Genus

In view of the overall distribution of *Melaleuca,* its Gondwanan (and presumably Australian) origin could hardly be questioned. However, the present geographic distribution is of particular interest with respect to the postulated events of the Miocene collision between the Australian and

Sundaland plates. The distribution of *M. leucadendra* is shown in Fig. 2, in relation to the postulated boundary between the Australian and Sundaland crustal plates (*cf.* Audley-Charles 1981). Whilst *M. leucadendra* occurs in the Moluccas, the Vogelkop and the Aru Islands, it is still effectively confined to putatively Gondwanan land surfaces.

The distribution of *M. cajuputi* (Fig. 3) is even more striking. Of the three subspecies of this species, one shows a remarkable distributional conformity with this boundary line, occurring in Timor, Tanumbar, Kei, Aru and southern Moluccan islands, but not on any of the Sundaland surfaces of the inner Banda Arc. The western and eastern subspecies are geographically disjunct; the western one is entirely distributed on Laurasian land surfaces, and represents the only natural occurrence of *Melaleuca* beyond the Australasian geophysical boundary (see below). It apparently represents a colonization event across Wallace's Line from the east. Because of the wide cultivation of *M. cajuputi* over some centuries for its oil, the natural distribution is difficult to assess, and that shown is the best estimate available from herbarium label data.

The distributions of all members of the *M. leucadendra* complex (Fig. 1) show, firstly, the essentially tropical distribution of the group, and secondly the entirely Gondwanan distribution, apart from the western subspecies of *M. cajuputi*. Thirdly, the relative unimportance of Torres Strait as a distribution barrier is shown, with nine taxa narrowly to widely distributed in New Guinea as well as Australia. Other broadleaved melaleucas which do not belong to the *leucadendra* complex, including *M. symphiocarpa* F. Muell., *M. brassii* Byrnes, *M. angustifolia* Gaertner and *M. acacioides* F. Muell., also occur on both sides of Torres Strait.

It might thus be suggested that *Melaleuca* differentiated in the northern part of the Australian plate, and attained a wide distribution before the Miocene contact with Sundaland. This contact has seemingly had little effect *per se* on the later

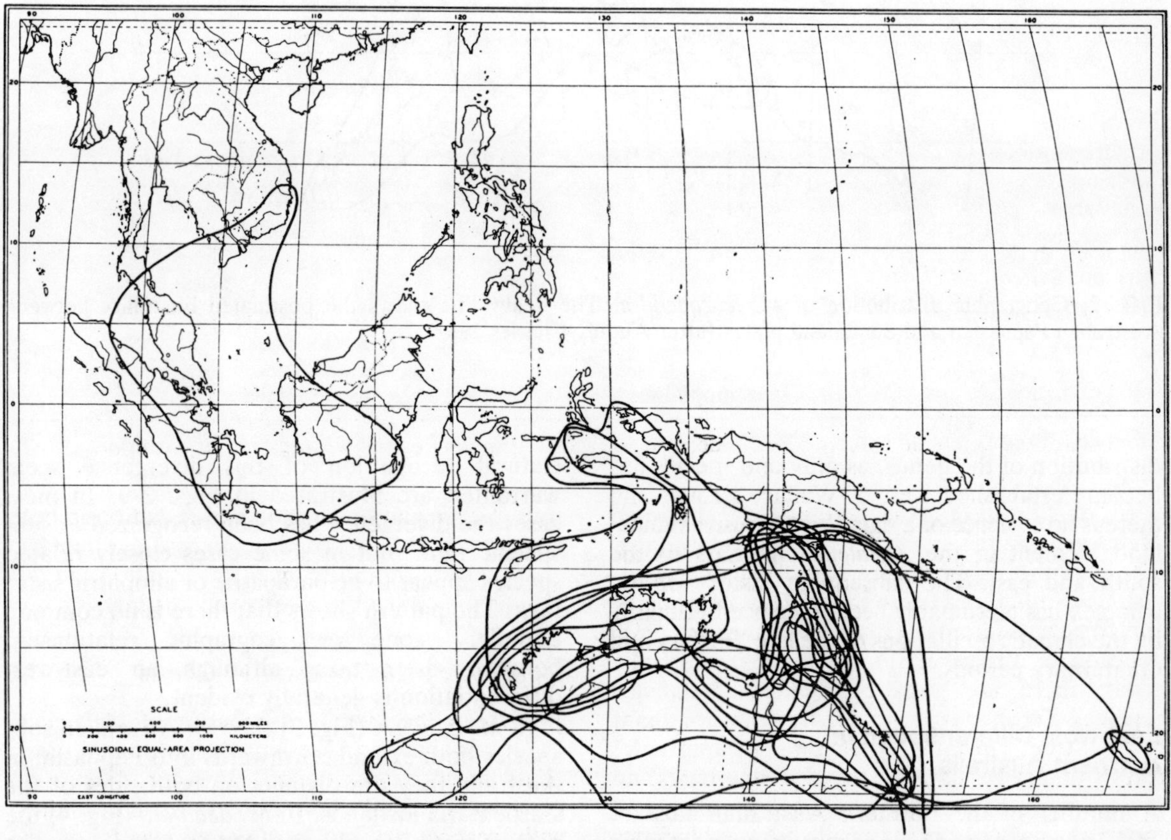

FIG. 1. Geographic distributions of the species of the *M. leucadendra* complex

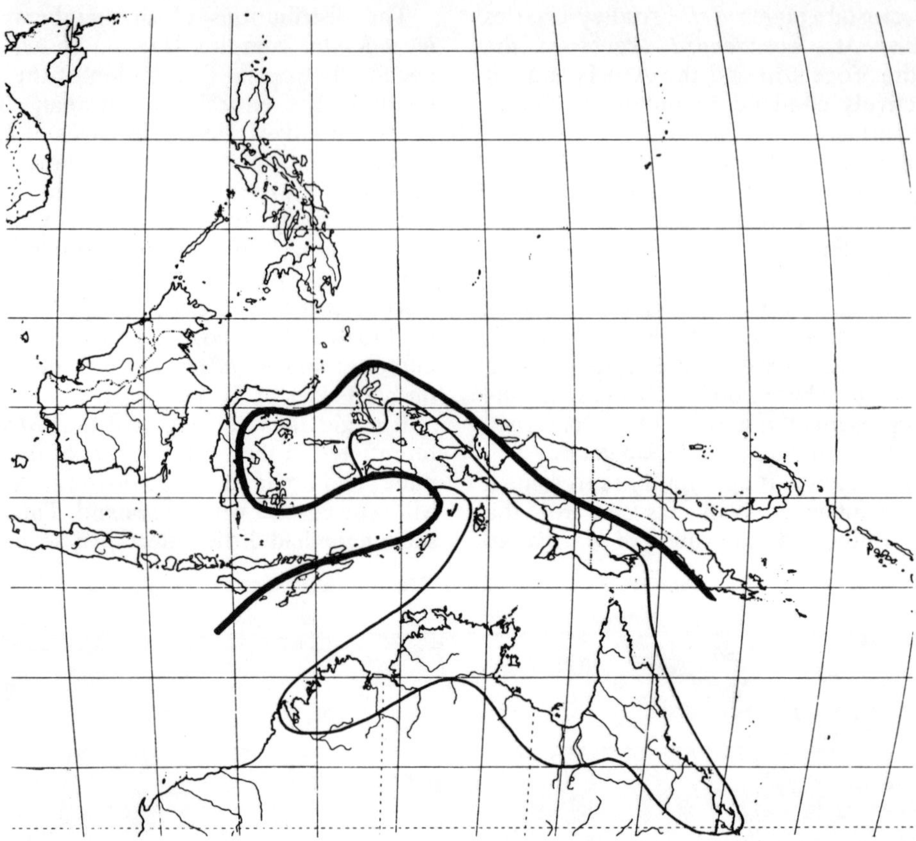

FIG. 2. Geographic distribution of *M. leucadendra*. The heavy line shows the postulated boundary between Australian/Papuasian and Sundaland plates (after Audley-Charles 1981).

distribution of the genus, as only one species has become established west of Wallace's Line, and there is no evidence of a significant intrusive Sundaland element in the *Melaleuca* habitats to the south and east. The subsequent history of the complex has presumably been influenced primarily by the climatic oscillations of the late Tertiary and Quaternary periods.

East-west Differentiation in Northern Australia

A number of the northern Australian taxa of *Melaleuca* show evidence of recent ecogeographic divergence. Several examples in which nomenclatural recognition of this divergence seem warranted are illustrated in Figs. 3-9. In most cases the divergence has been recognized at subspecific rank, but in some cases closely related species appear to be parapatric or allopatric sister taxa. The pattern shows that there is no common boundary zone or geographic relationship between sister taxa, although an east-west differentiation is generally evident.

In *M. cajuputi* (Fig. 3) the two Australian subspecies both extend northwards into Papuasia; in Australia they are disjunct on both sides of the Carpentaria lowlands. In *M. dealbata* (Fig. 4) the pattern is similar, although the eastern subspecies recurs in western Arnhem Land. In *M. acacioides*

(Fig. 5), the subspecies boundary separates the Kimberley region in the west from Arnhem Land and Cape York Peninsula in the east. In *M. minutifolia* F. Muell. (Fig. 6) the disjunction between western (Kimberley) and eastern (Cape York Peninsula) subspecies is much greater. *Melaleuca argentea* and *M.* sp. indesc. *("fluviatilis")* appear to be sister species with closely similar habitat requirements (stream beds and banks) and a narrow contact zone which bisects Cape York Peninsula (Fig. 7); *M. argentea* is further differentiated into a western "Ashburton biotype" which has not been given formal taxonomic status. *Melaleuca stenostachya* S. T. Blake and *M. sericea* Byrnes are disjunct, closely related sister taxa (Fig. 8); both species are in turn parapatric with the closely related *M. lasiandra* F. Muell., which extends southwards and westwards into seasonally arid northern Australia.

In *M. nervosa* (Lindley) Cheel (Fig. 9) an east-west differentiation is less evident, as the "eastern" subspecies also occurs to the north of the "western" one, apparently in less arid situations. The two subspecies have a complex contact zone in which intergradation is relatively common. Finally, in *M. viridiflora* Sol. ex Gaertner there are several regional biotypes which have not been given taxonomic status; these appear to have differentiated along gradients of water table depth and persistence, and the species ranges from gallery forests to arid woodlands.

Similar east-west ecogeographic divergence has been identified in northern Australia in Loranthaceae and Viscaceae (Barlow 1983, 1984), and in *Heliotropium* (L. A. Craven, pers. comm.). It doubtless occurs in many other taxa as well (see, for example, Hnatiuk & Pedley 1985 on *Acacia*).

The patterns suggest that there may have been two types of response to the cycles of aridity which occurred, with increasing frequency and amplitude, through the Quaternary period. One is range disjunction, with survival and differentiation in refugia; such refugia seem to be the topographically diverse Kimberley-Arnhem regions and eastern Cape York Peninsula. The second is

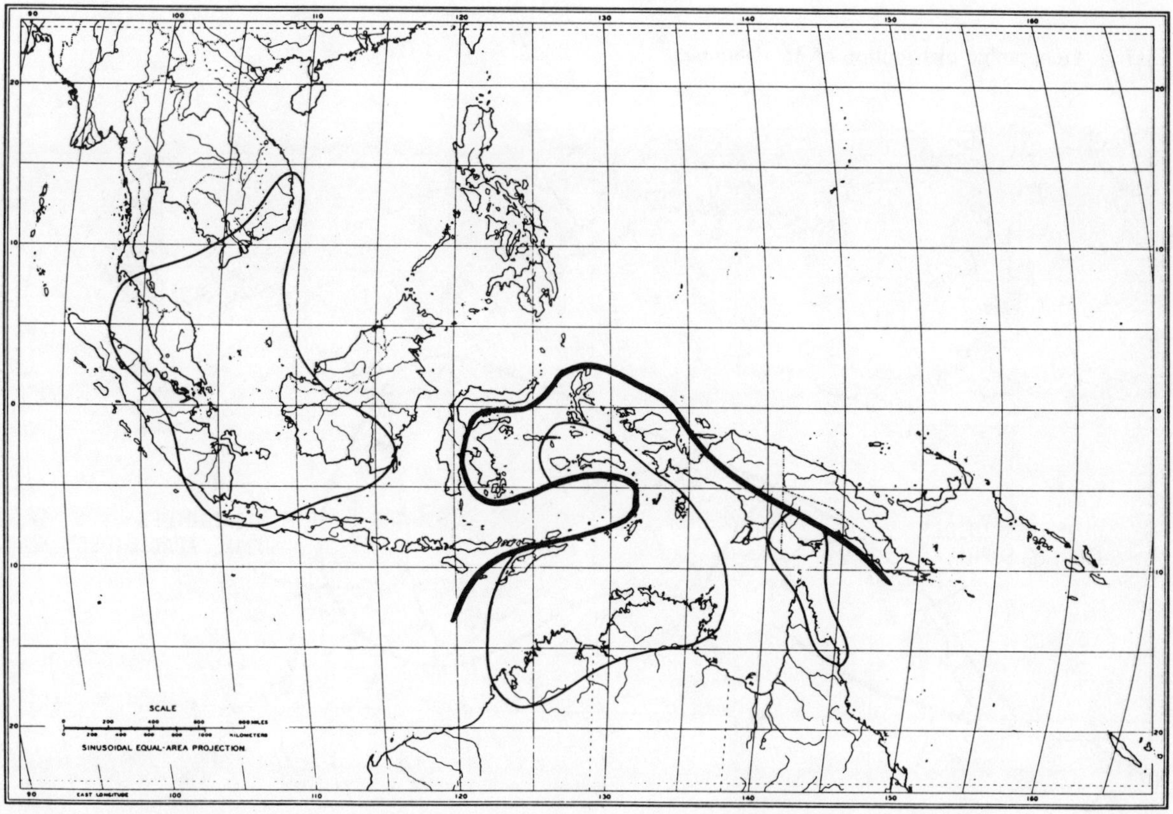

FIG. 3. Geographical distributions of the three subspecies of *M. cajupti.* The heavy line shows the postulated boundary between the Australian/Papuasian and Sundaland plates (after Audley-Charles 1981).

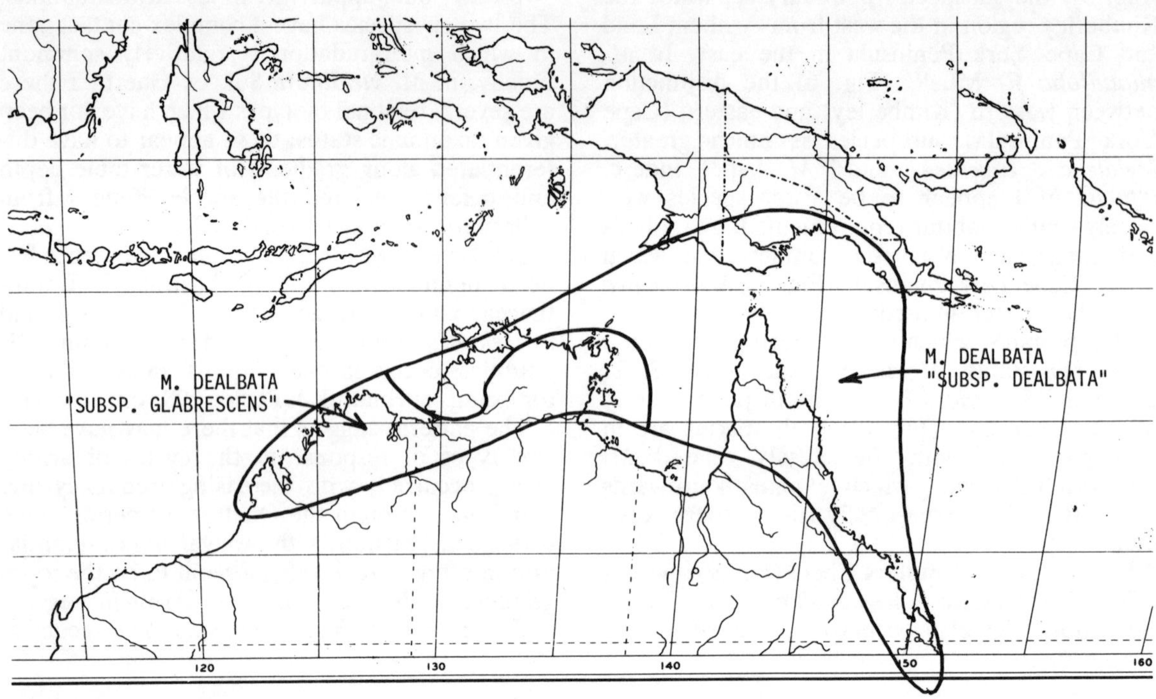

FIG. 4. Geographic distribution of *M. dealbata.*

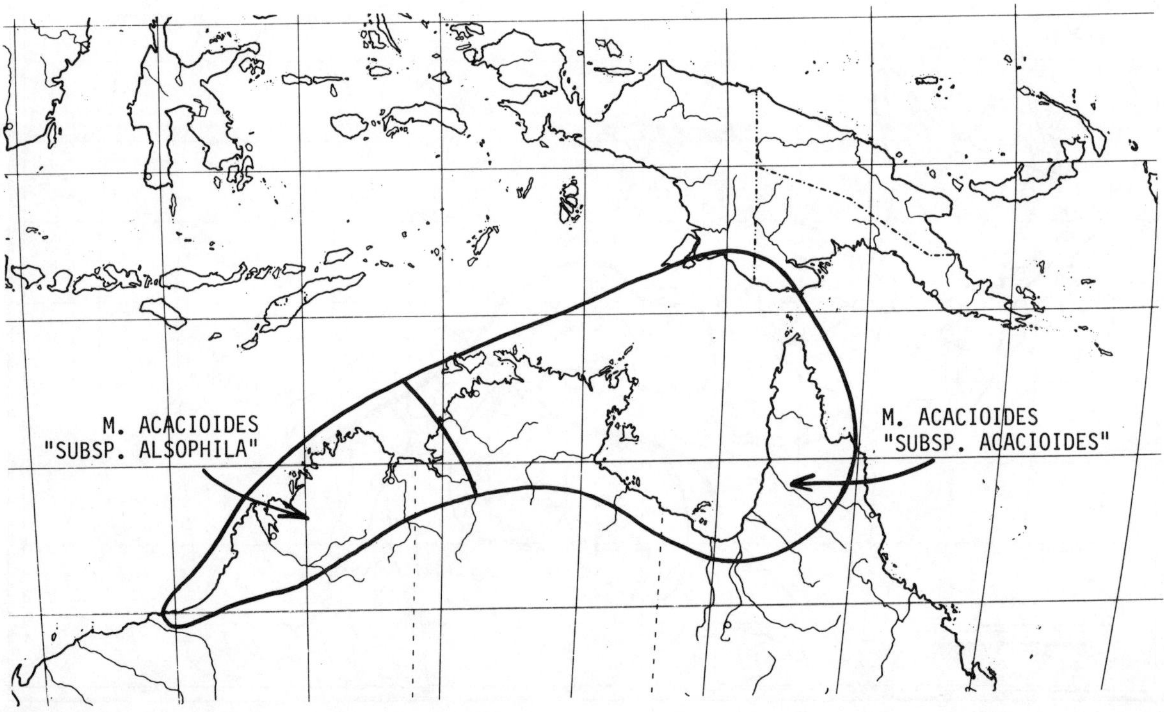

FIG. 5. Geographic distribution of *M. acaciodes.*

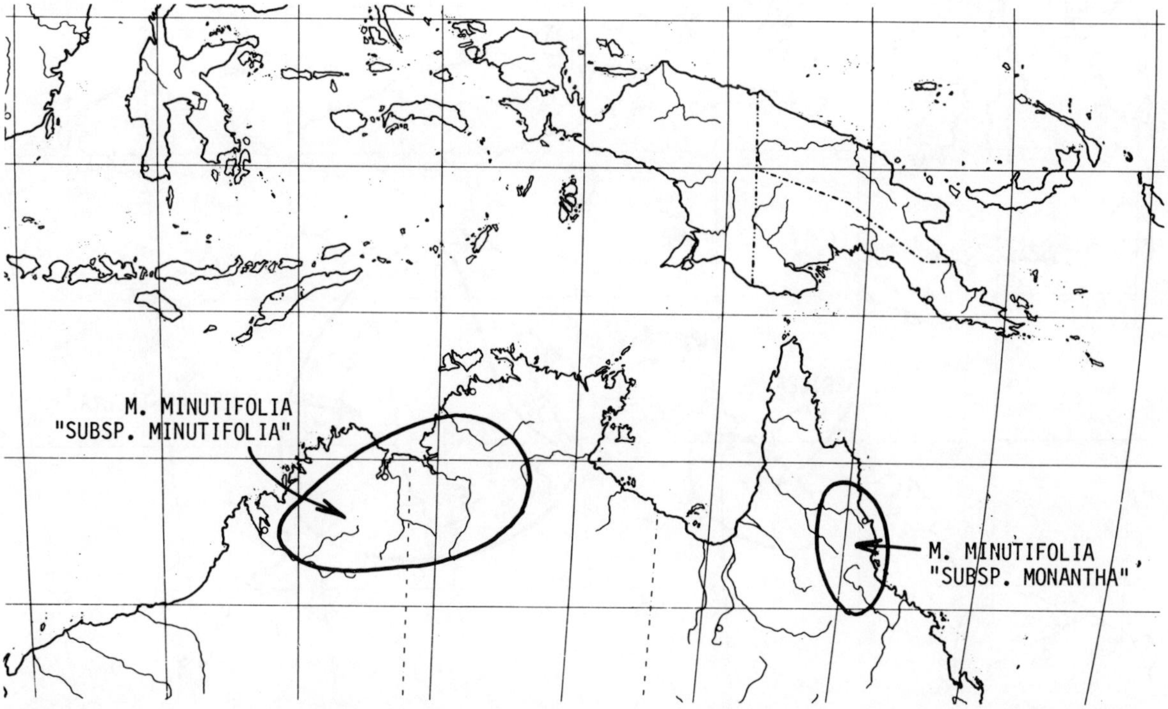

FIG. 6. Geographical distribution of *M. minutifolia*.

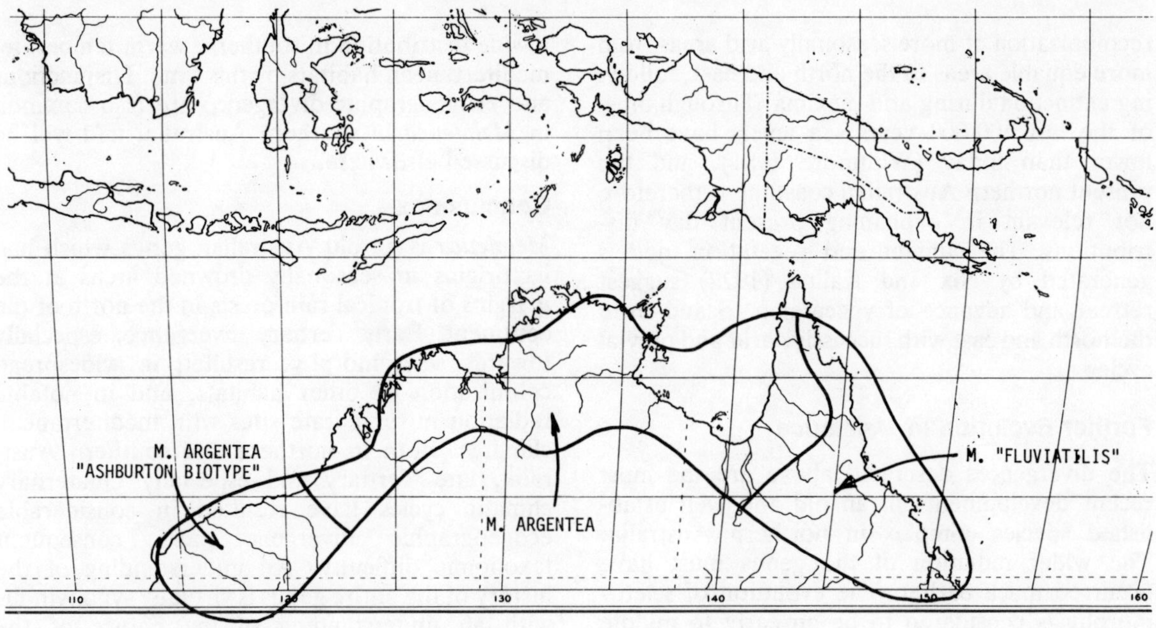

FIG. 7. Geographic distributions of *M. argentea* and *M.* sp. indesc. ("*fluviatilis*").

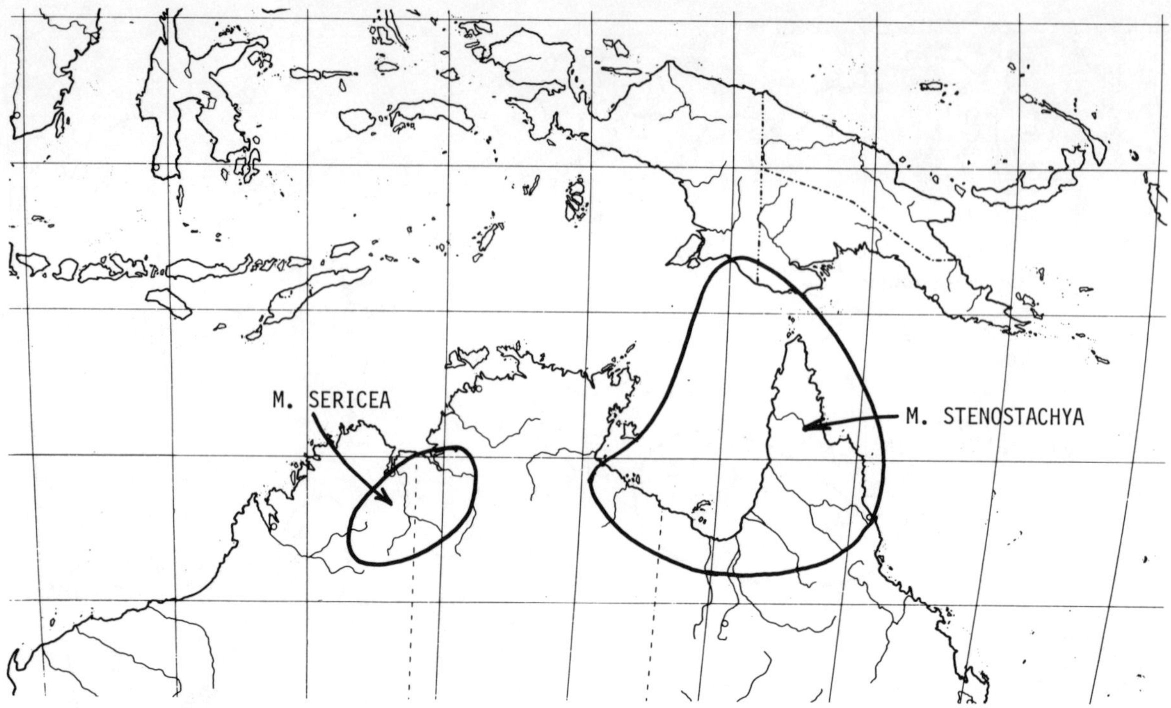

FIG. 8. Geographic distributions of *M. sericea* and *M. stenostachya.*

recolonization of more seasonally arid areas from more equable areas to the north and east, following extinction during arid maxima. Through most of the last 100 000 years sea levels have been lower than today (Whitmore 1981), and the present northern Australian coastline is therefore not relevant in explaining present day distributions. The climatic and vegetation models generated by Nix and Kalma (1972) suggest retreat and advance of vegeataion to and from the north and east with successive arid and pluvial cycles.

Further Evolution in *Melaleuca*

The divergences described above are the most recent developments in an old and well established species complex in northern Australia. The wider radiation of the genus must have occurred much earlier. The evolution of scleromorphy is considered to be an early to middle Tertiary development (Beadle 1981; Specht 1981), and *Melaleuca* had presumably achieved a wide distribution in southern, warm temperate, mediterranean habitats by this time. Disjunctions and ecogeographic divergence are also common in *Melaleuca* in southern Australia, and will be discussed elsewhere.

Conclusions

Melaleuca is an old Australian genus which had its origins in seasonally drowned areas at the margins of tropical rainforests in the north of the continent. Early Tertiary divergence, especially towards scleromorphy, resulted in widespread colonization of other habitats, and in notable radiation in temperate sites with mediterranean climates. Both in northern and southern Australia, late Tertiary and especially Quaternary climatic cycles have resulted in considerable ecogeographic divergence and consequent taxonomic difficulty. An understanding of the history of the entire genus is virtually synonymous with an understanding of the nature of the autochthonous subelement of the Gondwanan element of the Australian flora.

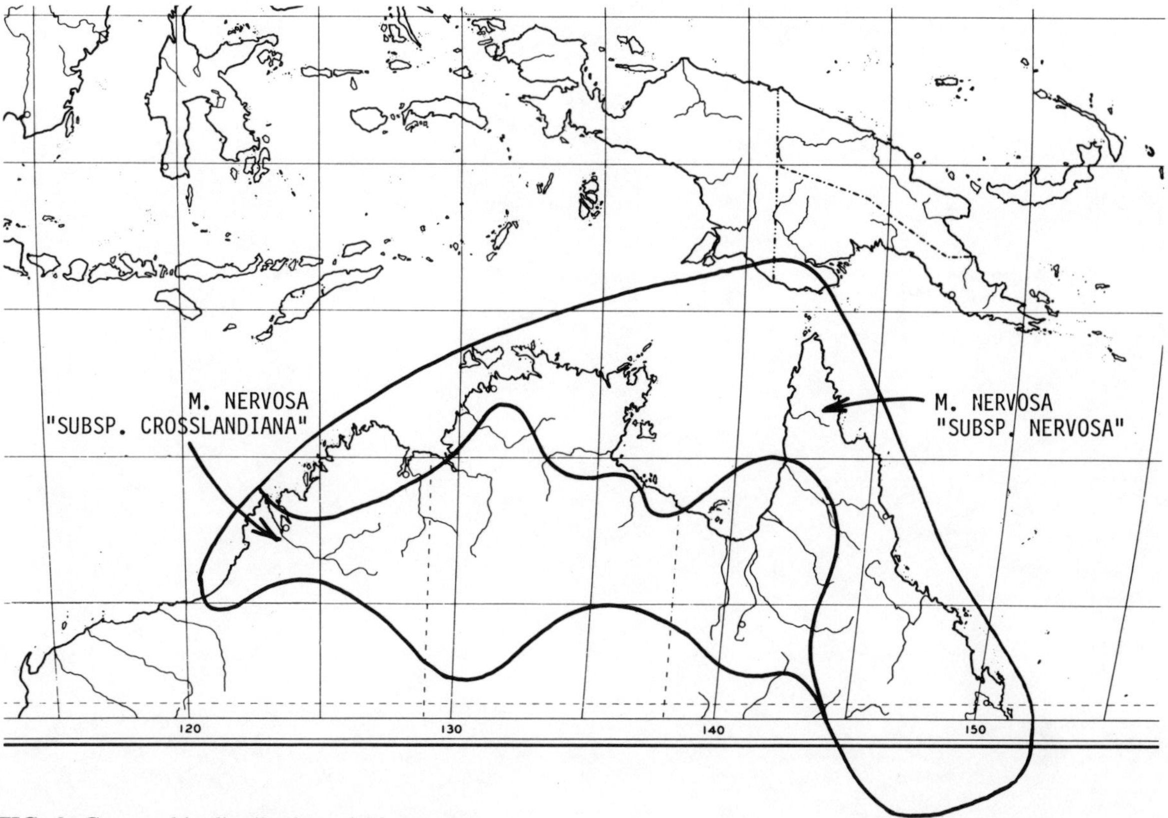

FIG. 9. Geographic distribution of *M. nervosa*.

Acknowledgements

I am grateful to John Clarkson and Helen Hewson for their helpful criticisms.

References

Audley-Charles M. G. (1981) Geological history of the region of Wallace's Line. In: *Wallace's Line and Plate Tectonics* (ed. T. C. Whitmore) pp. 24-35. Clarendon, Oxford.

Barlow B. A. (1983) A revision of the Viscaceae of Australia. *Brunonia* **6,** 25-57.

Barlow B. A. (1984) Loranthaceae. In: *Flora of Australia,* (ed. A. S. George) Vol. 22 pp. 68-131. *Aust. Govt. Publ. Service, Canberra.*

Beadle N. C. W. (1981) Origins of the Australian angiosperm flora. In: *Ecological biogeography of Australia* (ed. A. Keast) pp. 407-26. W. Junk, The Hague.

Hnatiuk R. J. & Pedley L. (1985) Biogeography of *Acacia* in northern Australia. *Proc. Ecol. Soc. Aust.* **13,** 33-42.

Hopper S. D. (1979) Biogeographical aspects of speciation in the south-west Australian flora. *Annual Rev. Ecol. Syst.* **10,** 399-422.

Johnson L. A. S. & Briggs B. G. (1981) Three old southern families – Myrtaceae, Proteaceae and Restionaceae. In: *Ecological biogeography of Australia* (ed. A. Keast) pp. 427-70. W. Junk, The Hague.

Nix H. A. & Kalma J. D. (1972) Climate as a dominant control in the biogeography of northern Australia and New Guinea. In: *Bridge and Barrier* (ed. D. Walker) pp. 61-91. Aust. Nat. Univ., Canberra.

Pryor L. D. & Johnson L. A. S. (1981) *Eucalyptus,* the universal Australian. In: *Ecological biogeography of Australia* (ed. A. Keast) pp. 499-536. W. Junk, The Hague.

Specht R. L. (1981) Ecophysiological principles determining the biogeography of major vegetation formations in Australia. In: *Ecological biogeography of Australia* (ed. A. Keast) pp. 299-333. W. Junk, The Hague.

Whitmore T. C. (1981) Palaeoclimate and vegetation history. In: *Wallace's Line and Plate Tectonics* (ed. T. C. Whitmore) pp. 36-42. Clarendon, Oxford.

Proc. Ecol. Soc. Aust. 15:249-257

Australian Annonaceae in an Asian-Pacific context

L. W. JESSUP

Queensland Herbarium, Meiers Road, Indooropilly, Qld. 4068

Abstract

The Annonaceae is a mostly tropical family of about 132 genera and 2 300 species of trees, shrubs and lianes. In the eastern tropics and subtropics from Asia to the Pacific, there are approximately 54 genera and around 900 species. Only six of these genera and none of these species extend beyond this area. Relatively few genera are restricted to small parts of the region, the majority being widespread. Local species endemism is high.

Australia has 16 genera, two endemic; and 45 species, 34 endemic. Seven genera are represented by one or more non-endemic species, with at least five species transgressing Wallace's Line.

While it is likely that many of the Australian Annonaceae are descendants of post-Miocene immigrants from Sundaland or a Melanesian island arc that collided with the northern margin of New Guinea, it is suggested that at least some of the Annonaceae seen today in eastern Queensland, New Caledonia, Fiji and New Guinea may be surviving descendants of an east Gondwanan stock. An early arrival in east Gondwana is postulated for the pantropical Xylopia, *some* Polyalthia *and possible descendants of early* Polyalthia *– like ancestors namely* Haplostichanthus, Polyaulax *and* Ancana, *and also* Fitzalania, *a likely segregate from ancestors of* Miliusa.

Introduction

The majority of the 132 genera and 2 300 species in the Annonaceae occur in or near the tropics. In America there are around 36 endemic genera, in Africa and Madagascar around 40 endemic genera and in Asia, Malesia and the Pacific including Australia about 50 endemic genera.

A single genus *Xylopia* with over 170 species is known from all three centres of distribution. The genus *Annona* occurs in America and Africa and *Anaxagorea* occurs in America and Asia. There are four genera found in both Africa and the eastern tropics but not elsewhere. They are *Uvaria, Polyalthia, Artabotrys* and *Melodorum*. Australia has representative species of five of the six widespread genera found in Asia.

Suprageneric Taxa

Although the family is generally regarded as a very natural one the suprageneric classifications of the family have long been considered rather artificial. Walker (1971, 1972) proposed an informal classification of the family based on pollen morphology and chromosome numbers as an alternative to the earlier schemes of Bentham & Hooker (1862-1867), Diels (1932), Sinclair (1955), Fries (1959) and Huchinson (1923, 1964). Walker suggested three subfamilies could be recognized, the *Malmea* subfamily with the *Malmea, Uvaria* and *Guatteria* tribes, the *Fusaea* subfamily with no recognized tribes and the *Annona* subfamily with the *Hexalobus, Asimina, Annona* and *Cymbopetalum* tribes.

Although Walker's classification is a significant step and provides a foundation for a future formal classification, recent work has shown that some details need modification. Ultrastructural studies of the pollen of African Annonaceae by Le Thomas (1980, 1981) showed that pollen with similar external morphology may be structurally quite different, and that genera regarded by Walker as having primitive morphology are in fact structurally most advanced. Walker's (1972) suggested phylogeny of the family included an erroneous chromosome count for *Anaxagorea* taken from Ehrendorfer *et al.* (1968). This was corrected by Morawetz (1984). Morawetz (1986) also suggested that the Asian *Disepalum* should be removed from the Cymbopetalum tribe of Walker (1971).

Karyology

Chromosome numbers x=7, 8 and 9 are found in the Annonaceae. Although diploids are most common, hexaploidy and tetraploidy are known in some genera. Walker (1972) considered x=7 to be the base number for the family but Okada & Ueda (1984) suggested x=8 or 9 and Morawetz (1985) suggested x=8.

In the eastern tropics x=7 is rare while x=8 and x=9 are represented in approximately equal numbers of the 24 genera so far studied (Okada & Ueda 1984). A similar ratio between x=8 and x=9 applies to the Australian Annonaceae genera shown by counts made recently by Morawetz (*in litt.*).

Diversity and Endemism

A review of the literature on the Annonaceae of the eastern tropics as part of a taxonomic revision of the family in the Australian region quite early revealed considerable local species endemism particularly in north-eastern Queensland.

In the region from India to the Pacific including Australia there are about 54 genera and around 900 species in Annonaceae, only six genera and no species extending beyond this region. The majority of genera cover a large proportion of the region but as in Australia, local species endemism is high. Australia has 16 genera (two endemic) and about 45 species (34 endemic). Although there are very few recent taxonomic revisions of eastern tropical genera of Annonaceae some useful comparisons may be drawn from the available literature to highlight the areas of diveristy and endemism as shown in Fig. 1.

Attention is drawn particularly to the five genera, *Artabotrys, Melodorum, Polyalthia, Uvaria* and *Xylopia* common to Africa and Malaysia. These are the same five genera shared between Africa and Australia. Four of these genera occur in Madagascar, India and Sri Lanka, and two, *Polyalthia* and *Xylopia,* also occur in New Caledonia and Fiji. Such a distribution attests to the likely great age of these genera.

Taking into account two likely new and undescribed genera in New Guinea (M. Heads, *in lit.*), there are 13 genera confined to one of the Asian-Pacific subregions circled on the map. These include *Cheiniodendron* (China), *Dasoclema* (Thailand), *Guamia* (Guam), *Phoenicanthus* (Sri Lanka), *Tetrapetalum* and *Woodiellantha* (Borneo), *Oreomitra, Petalolophus* and *Schefferomitra* (New Guinea), and

FIG. 1.
Circles show $\frac{\text{ENDEMIC GENERA}}{\text{TOTAL GENERA}}$ for each region. Numbers of SHARED GENERA on lines between regions.

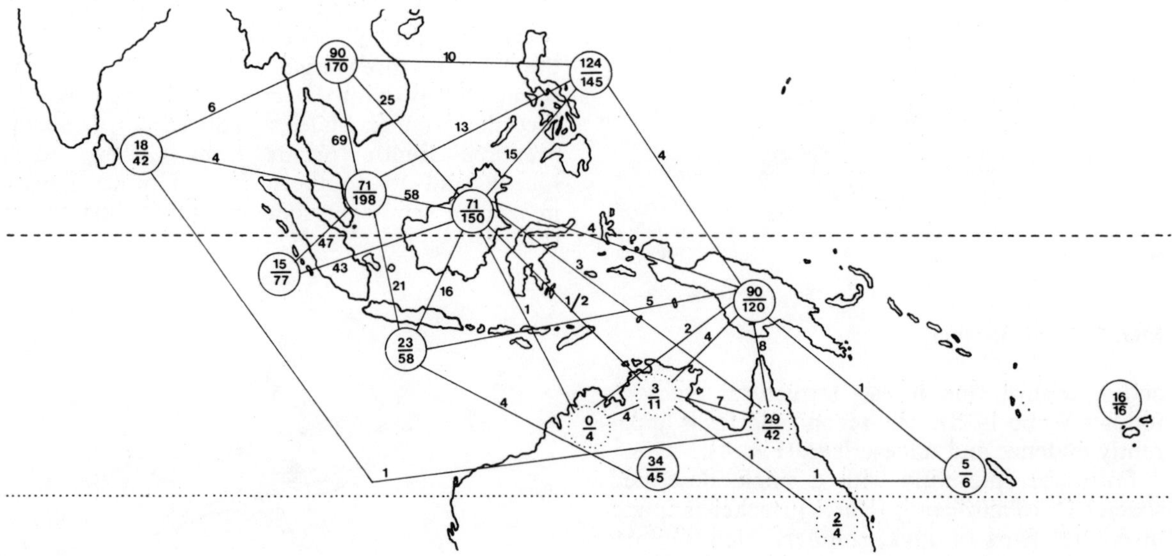

FIG. 2.
Circles show $\frac{\text{ENDEMIC SPECIES}}{\text{TOTAL SPECIES}}$ for each region. Numbers of SHARED SPECIES on lines between regions.

and *Fitzalania* (Australia). Although it could be argued that some of the endemic genera, for example those in New Guinea, are likely to have arisen relatively recently due to the rapid and marked topographical changes that have occurred there since Miocene time it is nevertheless significant that seven out of the 13 subregional endemic genera and over half of the total genera occur on the old Australian Plate.

A further eight genera (*Anomianthus, Disepalum, Enicosanthellum, Dendrokingstonia, Marsypopetalum, Mezzettia, Polyaulax* and *Richella*) are found in two or three of the Asian-Pacific subregions only, leaving over half the genera relatively widespread.

Fig. 2 shows that the number of species shared between regions is low except where Pleistocene land connections have existed. It appears that the majority of species are more or less restricted endemics and have very low vagility but there are some species that are highly vagile. Some species may also be extremely old.

Genera with some Non-endemic Species

The Australian Annonaceae are represented by several genera with one or more non-endemic species, some of which transgress the tectonic boundary between the Australian and Asian plates.

Cananga, has two species, one widespread on mainland south-east Asia only. The other *C. odorata* (Lamk.) J. D. Hook & Thomson which is distributed from Burma to north-eastern Queensland, appears to be a rather vagile species of lowland secondary forest. It is a recent introduction to Fiji and probably also Vanuatu (Smith 1981) (Fig. 3).

Cyathostemma, with about ten species is represented in Australia by two species. *C. micranthum* (A.DC.) J. Sinclair has two centres of distribution, associated presumably with a more of less seasonal climate. In Australia it occurs in evergreen notophyll and semi-deciduous mesophyll

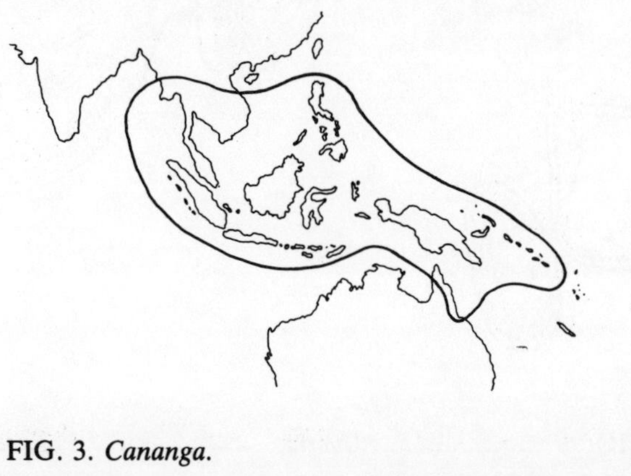

FIG. 3. *Cananga.*

FIG. 4. *Cyathostemma.*

and notophyll vine forests (structural typology follows Webb 1978). The second species is apparently endemic and undescribed (Fig. 4).

Polyaulax, presently with a single described species *P. cilindrocarpa* (Burck) Backer is found in coastal parts of Java, southern New Guinea and in Australia where it occurs in littoral semi-deciduous notophyll vine forests and deciduous vine thickets. There is an undescribed species on the east coast of Cape York Peninsula and a third species, possibly a *Polyaulax* and undescribed, is found in upland notophyll vine forests on NE Queensland. The distribution of this genus strongly suggests a long history in east Gondwana (Fig. 5).

Uvaria, with about 100 species total has about 60 species in the eastern tropics and six in Australia, three still poorly known. *U. rufa* Blume is widespread throughout most of Malesia as well as Thailand and Indochina and occurs in several rainforest types on Cape York Peninsula. *U. holtzei* F. Muell. is a Northern Territory species that probably occurs in New Guinea under the name *U. rosenbergiana* Scheff. and may bear some affinity with *U. cordata* (Dunal) Alston recorded from E. Malesia to Sri Lanka. *U. membranacea* Benth. occurs from Cairns, north Queensland to southern New Guinea mostly in coastal semi-evergreen notophyll vine forest (Fig. 6).

FIG. 5. *Polyaulax.*

Rauwenhoffia, regarded as a distinct genus for many years was reduced to a synoym of *Melodorum* by Ban (1974). This has been commented on recently by Ueda (1986). If Ban's generic concept is accepted the genus *Melodorum* then includes about 12 species of climbing shrubs including three endemics in Australia and one found also in New Guinea. Two are undescribed species. The presence of *Melodorum* in Africa and the bicentric distribution in the eastern tropics

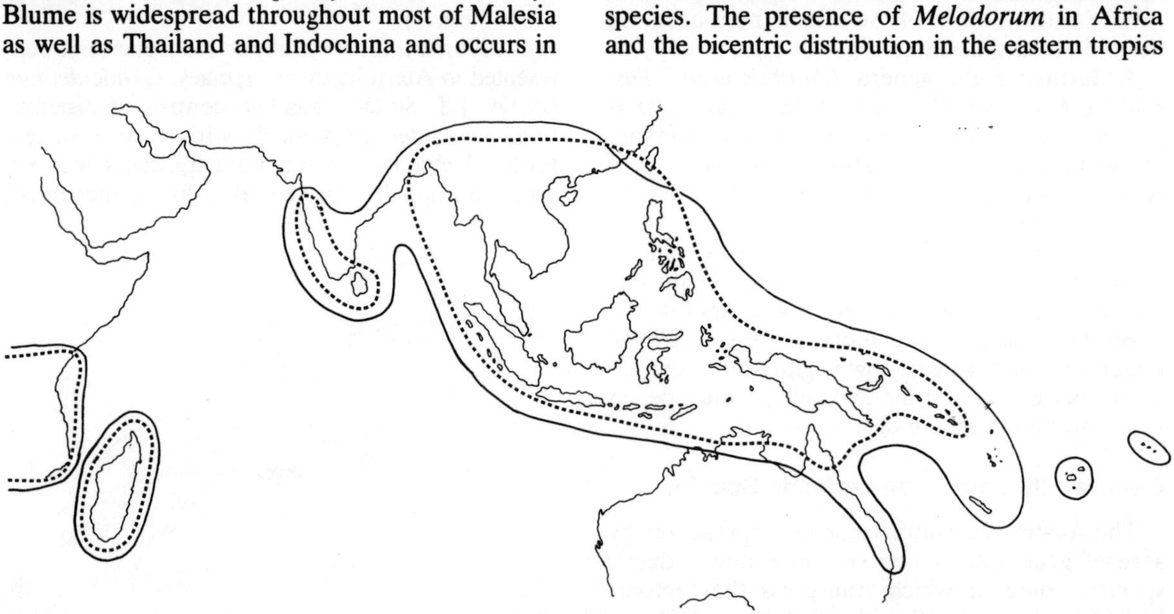

FIG. 6. .... *Uvaria,* ————— *Polyalthia.*

FIG. 7. *Melodorum*.

strongly suggests two early lines of migration from Africa, one to eastern Gondwana the other to eastern Laurasia, probably via India (Fig. 7).

Polyalthia consists of about 100 species of trees mostly in the eastern tropics and subtropics and some in Africa and Madagascar. Traditionally two sections based on ovule number are recognized. Sinclair (1955) regarded *Polyalthia* as ancestral to many present-day Asian genera and pollen ultrastructure studies of Le Thomas (1981) support the view that the genus is an ancient one. Although ovule reduction has probably occurred independantly several times in *Polyalthia* the genus most likely evolved in Africa according to Le Thomas, the multiovulate species since becoming extinct there. Le Thomas suggested that *Polyalthia* probably arrived in Asia via India (Fig. 6).

Polyalthia nitidissima (Dunal) Benth. is common in eastern and northern Australia in several types of notophyll vine forest, particularly on coastal sands. Its range includes New Caledonia, Vanuatu, New Guinea and there is an unconfirmed record from Borneo. *P. australis* (Benth.) Jessup is a northern Australian species of semi-deciduous meso-notophyll vine forest and appears to have some affinity with *P. lateriflora* (Blume) King which occurs from Java to Thailand. *P. michaelii* C. White and an undescribed species are both endemics in north-eastern Queensland in mesophyll vine forest. The presence of the genus in New Caledonia (there appears to be an undescribed endemic species as well) is evidence that the genus has had a long history on both the Australian and

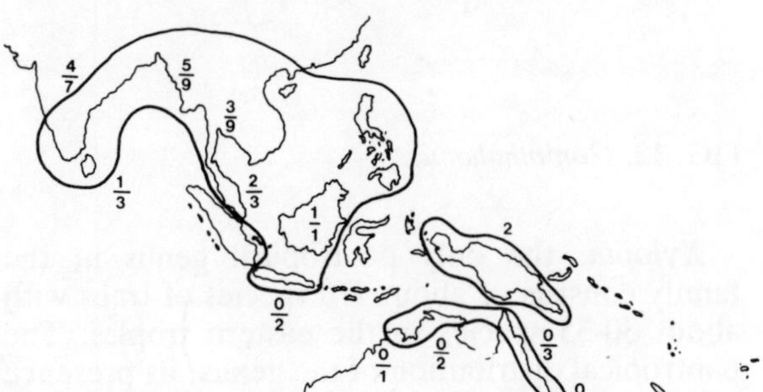

FIG. 8. *Miliusa*.

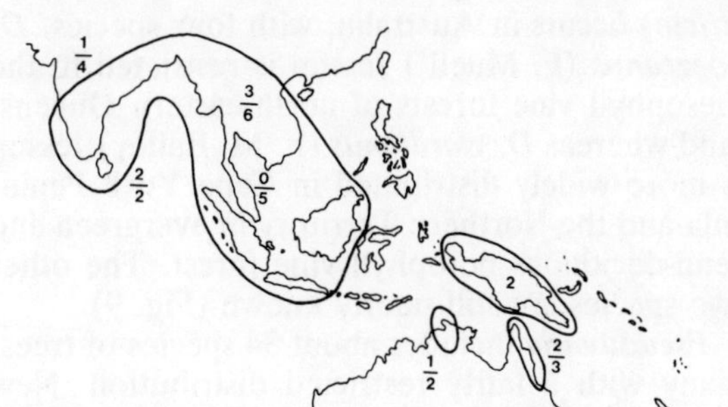

FIG. 9. *Desmos*.

Asian plates with post-collision exchange of species.

Miliusa, with about 28 species of trees is represented in Australia by three species. *M. horsfieldii* (Benn.) Baill. ex Pierre is found in Java and Cape York Peninsula where it occurs in semi-deciduous mesophyll vine forest. *M. brahei* (F. Muell.) Jessup, scattered across northern Australia in semi-evergreen and semi-deciduous forests appears to have some affinity with *M. longipes* King of Java, Thailand and Malaysia or *M. vidalii* J. Sinclair from the Philipines. An apparently undescribed species occurs in the drier deciduous and semi-deciduous vine forests and thickets in Cape York Peninsula and the Northern Territory (Fig. 8).

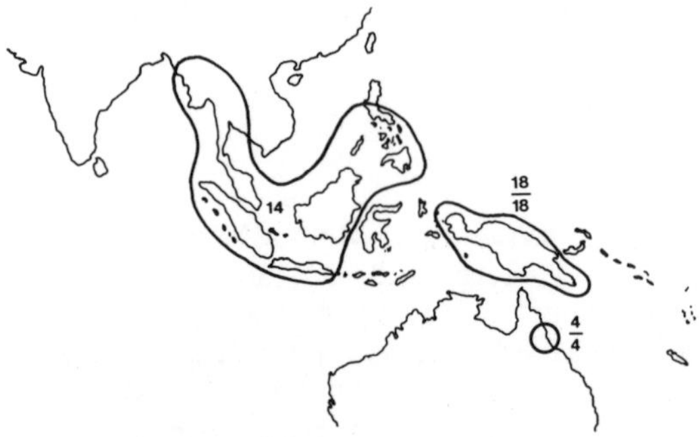

FIG. 10. *Pseuduvaria.*

Genera with Endemic Species Only

Seven genera appear to be represented in Australia only by endemic species. All except *Haplostichanthus* are more or less widespread genera in the eastern tropics.

Desmos, with 18 species of climbing shrubs and the related *Dasymaschalon* with five species, treated as a section of *Desmos* by some authors are widespread genera. Only *Desmos (sensu stricto)* occurs in Australia, with four species. *D. goezeanus* (F. Muell.) Jessup is restricted to the mesophyll vine forests of north-eastern Queensland whereas *D. wardianus* (F. M. Bailey) Jessup is more widely distributed in Cape York Peninsula and the Northern Territory in evergreen and semi-deciduous notophyll vine forest. The other two species are still poorly known (Fig. 9).

Pseuduvaria includes about 34 species of trees, many with a fairly restricted distribution. New Guinea has about 15 described species, several undescribed, and all apparently endemic. There are four species in north-eastern Queensland, all of limited and allopatric or parapatric distribution

FIG. 11. *Mitrephora.*

in mesophyll vine forest, one species found only in the upper catchments of the Mulgrave and Russell Rivers (Fig. 10).

Mitrephora, with about 33 species, found frequently in semi-deciduous forests, is represented by a single species of unknown affinity in Cape York Peninsula. There are three species recorded for New Guinea, all endemic (Fig. 11).

Goniothalamus with about 100 species is represented in Australia by the relatively restricted endemic *G. australis* Jessup in the meso-notophyll vine forests of north-eastern Queensland. The genus exhibits greatest diversity in Malaysia and Borneo but there are at least 10 endemic species in New Guinea. It is probably related to the genus *Richella* which is represented at present by one endemic species in Fiji, one in New Caledonia and one in Borneo (Fig. 12).

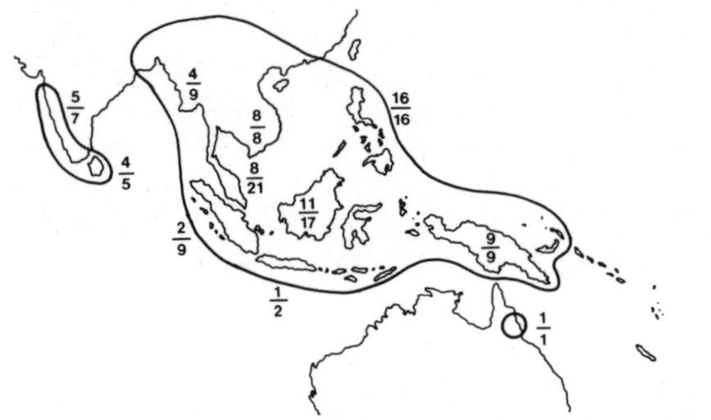

FIG. 12. *Goniothalamus.*

Xylopia, the only pantropical genus in the family consists of about 170 species of trees with about 30-35 species in the eastern tropics. The pantropical distribution of the genus, its presence in New Caledonia (two endemic species), north Queensland (two endemic species), Fiji (three endemic species) and New Guinea (three or four endemic species) and the high species endemism

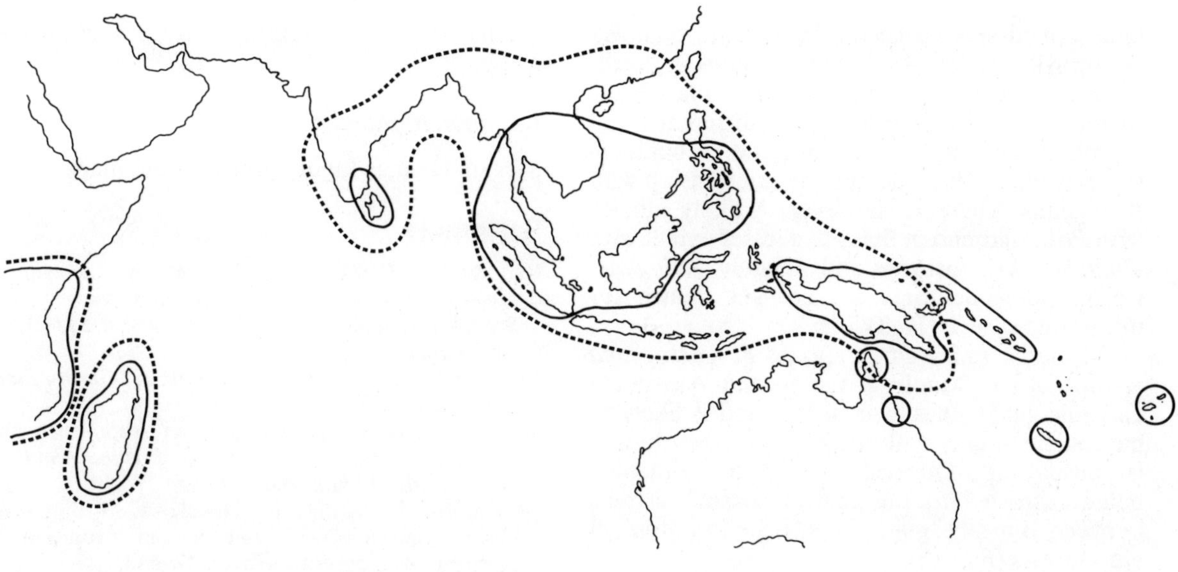

FIG. 13. .... *Artabotrys,* ____________ *Xylopia.*

generally (islands of West Malesia connected during the Pleistocene epoch being the only exceptions) suggest great antiquity and its likely early presence in east Gondwana. *X. maccrae* (F. Muell.) L. S. Smith is restricted to north-eastern Queensland in several rainforest types from the McIlwraith Range to Ingham. The second Australian species occurs in several localities in Cape York Peninsula (Fig. 13).

Artabotrys, with about 100 species of climbers is found in tropical Africa and Madagascar as well as being distributed from India to Australia. In Australia there is a single species of unknown affinity in Cape York Peninsula (Fig. 13).

Haplostichanthus, previously thought to be an endemic genus of understorey shrubs is almost certainly congeneric with *Papualthia* with at least eight species from New Guinea. There are an additional eight species described from the Philippines under *Papualthia* which may or may not be congeneric with the New Guinea and Australian species. There are about five species in north Queensland, four with a distribution of less than 50 km. Some specimens seen from New Caledonia suggest *Haplostichanthus* may be represented there as well. The genus exhibits considerable advancement in floral structure with the six petals fused into a single whorl at the base (Fig. 14).

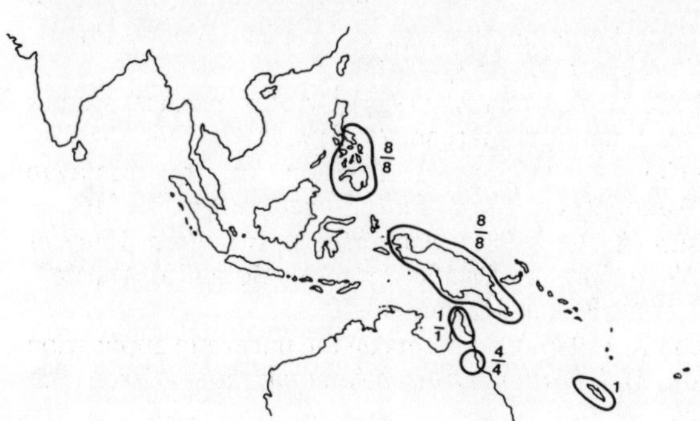

FIG. 14. *Haplostichanthus.*

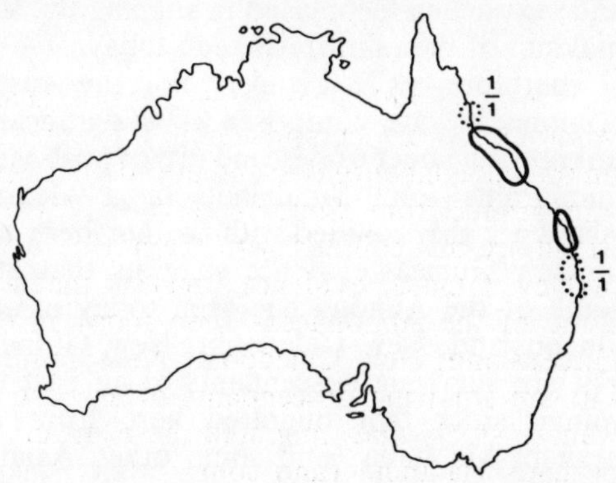

FIG. 15. ____________ *Fitzalania,* .... *Ancana.*

Ancana is an Australian endemic genus of two species of understorey shrubs. *A. stenopetala* F. Muell. occurs in south-eastern Queensland and north-eastern New South Wales in warm subtropical complex notophyll vine forest. An undescribed species occurs in lowland complex mesophyll vine forest in a few localities in north-eastern Queensland. Fries (1959) placed *A. stenopetala* under *Fissistigma* but different floral morphology and chromosome numbers (Morawetz, *in litt.*) rule out any relationship with that genus. There is, however, a likely affinity with *Polyaulax* and perhaps to a lesser extent with *Guamia, Oncodostigma* and possibly *Meiogyne*. Chromosome numbers are not yet known for those four genera (Fig. 15).

Fitzalania, with a single species *F. heteropetala* (F. Muell.) F. Muell. is the second Australian endemic genus. It is a small tree and is likely to have some affinity with *Miliusa*. The distribution is unusual for a Queensland rainforest endemic, being restricted to the central coastal districts between Bundaberg and Townsville in notophyll vine forests (Fig. 15).

Discussion

This preliminary review of the Australian Annonaceae and their suggested affinities to early African as well as later Asian taxa should serve as a caution against overgeneralization when discussing the origin and evolution of the Australian rainforest flora. There is considerable evidence suggesting that the Tertiary Australian rainforests were extensive and this has been emphasized by the important recent discovery of a diverse rainforest-dwelling Miocene marsupial fauna at Riversleigh in north-western Queensland. There must have been considerable habitat and floristic diversity and this surely underlines the importance extinction has played in shaping the floristic makeup of the rainforests seen today.

Therefore, while it is likely that the Australian Annonaceae are comprised of a significant but unknown number of descendants of post-Miocene immigrants from Sundaland or a Melanesian island arc that collided with the northern margin of New Guinea, evidence suggests that at least some of the Annonaceae seen today in eastern Queensland, New Caledonia, New Guinea and Fiji are surviving descendants of an east Gondwanan stock that migrated here from Africa perhaps via India (and some other Asian subcontinent?) which appears to have been in more or less close juxtaposition with western Australia until well into the Cretaceous period (Hamilton 1979). Such an early arrival in east Gondwana is postulated at least for the pantropical *Xylopia*, some *Polyalthia* and possible descendants of early *Polyalthia*-like ancestors such as *Haplostichanthus, Polyaulax* and *Ancana,* as well as *Fitzalania,* a likely segregate from ancestors of *Milusa*.

Acknowledgement

Megan Thomas kindly prepared the maps.

References

Ban N. T. (1974) Critical notes on the genera *Melodorum* Lour., *Mitrella* Miq. and *Rauwenhoffia* Scheff. (Annonaceae Juss.) *Bot. Zhurn.* **59,** 237-45 (in Russian).

Bentham G. & Hooker J. D. (1862-1865) *Genera Plantarum* I. **1,** 20-29.

Diels L. (1932) Die Gliederung der Anonaceen und ihre Phylogenie. *Sitzungsber. K. Preuss. Akad. d. Wiss., Berlin Physik. math. Klasse* 77-85.

Ehrendorfer F., Krendl F., Habeler E. & Sauer W. (1968) Chromosome numbers and evolution in primitive Angiosperms. *Taxon* **17:** 340.

Fries R. E. (1959) Annonaceae in *Die naturlichen Pflanzenfamilien,* ed. 2, *17aII.* Duncker & Humblot, Berlin.

Hamilton W. (1979) Tectonics of the Indonesian Region. *Geological Survey Professional Paper* 1078. United States Government Printing Office, Washington.

Hutchinson J. (1923) Contributions towards a phylogenetic classification of flowering plants II. The Genera of Annonaceae. *Kew Bull.* 241-61.

Hutchinson J. (1964) *The Genera of Flowering Plants Vol. 1.* Oxford University Press, London.

Le Thomas A. (1980) Ultrastructural characters of the pollen grains of African Annonaceae and their significance for the phylogeny of primitive angiosperms. Part 1. *Pollen et Spores* **22,** 267-342.

Le Thomas A. (1981) Ultrastructural characters of the pollen grains of African Annonaceae and their significance for the phylogeny of primitive angiosperms. Part 2. *Pollen et Spores* **23,** 5-36.

Morawetz W. (1984) 6. Karyology. In: *Studies in Annonaceae II.* Anaxagorea, *Part I* (eds P. J. Maas & L. Y. T. Westra). *Bot. Jahrb.* **105,** 89-91.

Morawetz W. (1986) Remarks on Karyological Differentiation Patterns in Tropical Woody Plants. *Pl. Syst. Evol.* **152,** 49-100.

Okada H. & Ueda K. (1984) Cytotaxonomical studies on Asian Annonaceae. *Pl. Syst. Evol.* **144,** 165-77.

Sinclair J. (1955). A revision of the Malayan Annonaceae. *Gard. Bull. Singapore* **14,** 149-516.

Smith A. C. (1981) Annonaceae In: *Flora Vitiensis Nova Vol. 2.* Pacific Tropical Botanical Garden, Lawai, Kauai, Hawaii, 13-40.

Ueda K. (1986) Comments on the uncertain application of *Melodorum* Lour. (Annonaceae). *Taxon* **35,** 347-49.

Walker J. W. (1971) Pollen morphology, phytogeography and phylogeny of the Annonaceae. *Contr. Gray Herbarium, Harvard* **202,** 1-131.

Walker J. W. (1972) Chromosome numbers, phylogeny, phytogeography of the Annonaceae and their bearing on the (original) basic chromosome number of angiosperms. *Taxon* **21,** 57-65.

Webb L. J. (1978) A general classification of Australian rainforests. *Australian Plants* **9,** 349-63.

Note added in proof: Verdcourt in *Kew Bulletin* **41,** 293-94 (1986) has now applied the Asian genus *Sphaerocoryne* to the east African plant previously included in *Melodorum*. *Sphaerocoryne* has not been found in Australia. van Setten (*in litt.*) has suggested one Australian species of *Melodorum* may be an *Ellipeia*. I agree with this.

Proc. Ecol. Soc. Aust. 15:259-266

The floras of Cape York and the Kimberley: a preliminary comparative analysis

J. R. CLARKSON[1] and K. F. KENNEALLY[2]

[1]*Botany Branch, Department of Primary Industries, P.O. Box 1054, Mareeba, Qld. 4880*
[2]*Western Australian Herbarium, Department of Conservation and Land Management, P.O. Box 104, Como, W.A. 6152*

Abstract

Using data compiled from computerized data bases of specimens held in the state herbaria of Queensland and Western Australia the vascular plant flora of Cape York and the Kimberley are analysed by various methods and compared. Comparisons are also made with published data for the floras of Western Australia, Queensland and the combined eastern mainland states.

The flora of Cape York with 216 families, 1 033 genera and 2 533 species is richer in families, genera and species than that of the Kimberley with 149 families, 599 genera and 1 673 species. The average number of species per family, genera per family and species per genus for each area is comparable. The figures for both species per genus and species per family for each area are lower than those reported for either Queensland, Western Australia or the combined eastern mainland states.

Gymnosperms and pteridophytes make only a small contribution to the total number of species recorded for either Cape York or the Kimberley although, in both areas, the pteridophytes assume a more important role at the family level. Naturalized introduced species account for 4.8% of the total species recorded for each area, considerably lower than their representation in the total flora of Queensland or Western Australia.

Poaceae and Fabaceae sens. lat. *are the major families in both the Kimberley and Cape York in terms of both numbers of genera and species. Although only three families recorded from the Kimberley are absent from Cape York, 65 families occurring on Cape York have not been recorded from the Kimberley.*

The floras of Cape York and the Kimberley are shown to have a higher level of similarity at the species level than do the total floras of Queensland and Western Australia. This pattern is reversed at the family level.

Introduction

There have been recent analyses of most state floras (Northern Territory — Chippendale 1972; Dunlop & Bowman 1986; Western Australia — Green 1985; Victoria — Ross 1976; Queensland — Johnson 1983) and an interstate comparison concentrating on the eastern states and Western Australia by Johnson (1983). Other comparative studies have concentrated on large families or genera such as Poaceae (Simon 1981), *Acacia* (Hnatiuk & Pedley 1985) and *Eucalyptus* (Gill *et al.* 1985). This is the first study to concentrate solely on Australia's tropical floras considering Cape York in northern Queensland and the Kimberley in northern Western Australia. The present analysis is a precursor to more detailed studies which will attempt to compare sites of similar structural type and habitat across tropical northern Australia. This habitat matching approach was suggested by Webb & Tracey in reviewing patterns and change in Australian rainforests (Webb & Tracey 1981). It was used by them and Williams in a comprehensive floristic classification of Australian rainforests and monsoon forests (Webb *et al.* 1984) but does not seem to have been applied elsewhere in studies of the Australian flora. It is an attractive concept which may lead to a clearer picture of floristic affinities than comparisons based on political boundaries. For the purposes of this preliminary analysis, however, traditional geographical boundaries have been adopted.

Methods

The data on which this study is based are check lists compiled from specimens held in the state herbaria of Queensland and Western Australia. Only published names were included. Because each herbarium uses different family classifications it was necessary to standardize the lists. The system adopted for the angiosperms was that proposed by Cronquist (1981). Johnson (1983) used this classification so family comparisons made here with that study will be valid. Pteridophyte families follow Pichi-Sermolli (1977).

The Kimberley region of Western Australia covers 302 580 km^2 and lies to the north of latitude 19°S. It encompasses the "Northern Botanical Province" of Beard (1980). Pedley & Isbell's (1971) delineation of Cape York as that part of Queensland north of latitude 16°S is followed here for convenience though there is evidence to suggest that the boundary is unnatural and needs redrawing as was attempted by J. P. Stanton and M. G. Morgan (unpublished). As defined here Cape York covers 130 625 km^2.

Results

The simplest way of comparing the floras of any two areas is to contrast their floristic richness in terms of the numbers of families, genera and species recorded for each (Table 1). At all three taxonomic ranks the flora of Cape York is richer than that of the Kimberley. In terms of native species, the flora of Cape York is richer by a factor of almost ×1.5. At the generic level this factor is in the order of ×1.7 and for families approximately ×1.4. This is similar to the pattern for the state-wide floras of Queensland and Western Australia noted by Johnson (1983) for both families and genera but reverses that recognized by him for species. On this whole state basis the flora of Western Australia is richer in species than that of Queensland by a factor of approximately ×1.1 but, if the floras of the three eastern mainland states are considered as a single eastern Australian unit, the number of families, genera and species in this eastern bloc exceed those for Western Australia in roughly the same proportions as that shown here for this Cape York-Kimberley comparison. In other words, in terms of raw numbers, the comparisons shown here for the floras of Cape York (tropical eastern) versus the Kimberley (tropical western) are of the same order and magnitude as those for a whole east versus west comparison.

From these raw data, estimates of species richness can be determined by calculating the number of species per unit area. Care is needed in the interpretation of these ratios for such comparisons can be misleading where the study areas differ significantly in size. Cape York, as defined for the purpose of this paper, is substantially smaller than the Kimberley. This area difference has been compensated for by employing a log transformation to the data before calculating the ratio. The ratio for Cape York using these transformed data is 0.66 whereas for the Kimberley it is 0.58. For the state-wide floras the ratios are intermediate. For both Queensland and the three eastern mainland states combined the ratio is 0.61 and for Western Australia 0.60.

When the native angiosperm floras of Cape York and the Kimberley are compared on the basis of species per family, genera per family or species per genus a similar pattern is observed for each (Table 2). The high species to genus ratio of 6.9:1 for Western Australia as a whole reported by Johnson (1983) is not reflected in the ratio shown here for the Kimberley. In fact the species per genus and the species per family ratios for both northern tropical areas under consideration here are quite a deal lower than those reported

TABLE 1. Comparison of the vascular flora of Cape York and the Kimberly.

	Cape York			Kimberley		
	Native	Naturalized	Total	Native	Naturalized	Total
Families	212	4	216	149	0	149
Genera	977	56	1 033	568	31	599
Species	2 412	121	2 533	1 592	81	1 673

TABLE 2. Comparison of the species per family, genera per family and species per genus ratios for the native angiosperm floras of Cape York, Kimberley, Queensland, Western Australia and the combined eastern mainland states Queensland, New South Wales and Victoria.

	Cape York	Kimberley	Queensland	Western Australia	Eastern Australia
Species/family	13.0	12.1	28.6	37.7	38.5
Genera/family	5.2	4.2	6.7	5.5	7.4
Species/genus	2.5	2.9	4.2	6.9	5.2

for either Queensland, Western Australia or the combined eastern mainland states indicating a lower species diversification in the tropical areas.

The proportion of pteridophytes, gymnosperms, monocotyledons and dicotyledons in the floras of Cape York and the Kimberley is illustrated in Fig. 1. Gymnosperms make up an insignificant proportion of the flora of either area with only seven species recorded from Cape York and five from the Kimberley. The fern flora is somewhat richer but still relatively unimportant at the species level. Ferns and fern allies account for only 6.1% of the native vascular species from Cape York and 2.7% from the Kimberley. These figures are however above the proportional representation of pteridophytes in the respective state-wide floras where calculations based on figures given by Johnson (1983) show ferns and fern allies comprising 5.9% of the total Queensland native vascular flora and 1.3% of the Western Australian flora. Pteridophytes contribute 5.0% to the native vascular flora of the eastern mainland states and 3.9% to the combined New South Wales and Victorian flora. Although the vascular flora of Cape York was noted earlier to be richer in species than that of the Kimberley by a factor of ×1.5, the factor for ferns and fern allies is much higher at almost ×3.4. The contribution of the pteridophytes to the total flora at the family level is more important. In the northern floras under consideration, pteridophytes account for 15.1% of the families in Cape York and 12.8% of the families in the Kimberley.

Amongst the native angiosperm species the ratio of dicotyledons to monocotyledons is 2.5:1

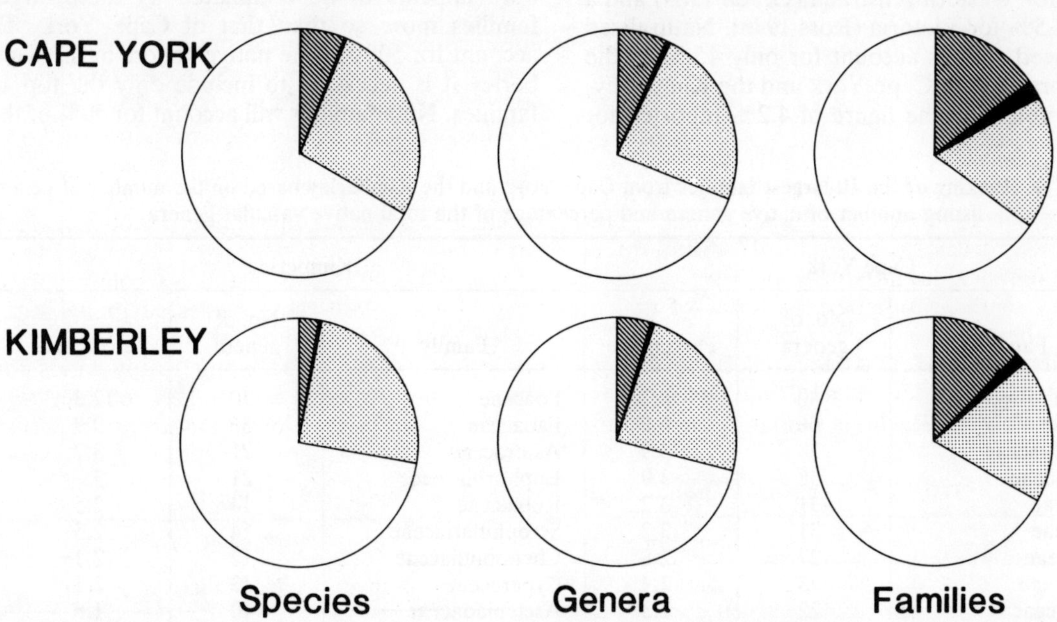

FIG. 1. Proportion of gymnosperms, pteridophytes, monocotyledons and dicotyledons in the native vascular flora of Cape York and the Kimberley. Cross hatching — pteridophytes; solid — gymnosperms; dots — monocotyledons; open — dicotyledons.

TABLE 3. Numbers of naturalized introduced species for Cape York and the Kimberley (figures in parentheses are these numbers expressed as percentages of total naturalized introduced species recorded for each area).

	Cape York	Kimberley
Poaceae	30 (24.7)	22 (27.2)
Fabaceae	26 (21.5)	22 (27.2)
Asteraceae	12 (9.9)	7 (8.6)
Other families	53 (43.8)	30 (37.0)
Total	121	81

for Cape York and 2.9:1 for the Kimberley. These figures are comparable with the ratio for the Victorian flora of 2.5:1 calculated from data presented by Ross (1976) but contrast markedly with the figure of 4.2:1 calculated from Green's (1985) data for the whole of Western Australia. Johnson (1983) did not consider the monocotyledons and dicotyledons separately.

When compared with figures reported for other areas in Australia, introduced taxa which have become naturalized do not feature prominently in the flora of either area under consideration. Recent estimates of the contribution of naturalized introduced species to state floras include 12.8% for Queensland (Johnson 1983), 10.5% for Western Australia (Green 1985) and a high 22.5% for Victoria (Ross 1976). Naturalized introduced species account for only 4.8% of the total flora of both Cape York and the Kimberley, comparable with the figure of 4.2% reported for the Northern Territory by Chippendale (1972). These low figures for Australia's tropical areas reflect sparse settlement and land use patterns which are largely based on grazing of native pastures with minimal clearing. Three plant families, Poaceae, Fabaceae *sens. lat.* and Asteraceae, together account for more than 50% of the naturalized introduced species in the tropical areas under consideration (Table 3).

Of the introduced species which have become naturalized in the Kimberley but as yet have not been recorded from Cape York, none are likely to cause serious problems if introduced to Cape York. The same is not true for a number of naturalized introduced species present on Cape York but as yet not recorded from the Kimberley. The most potentially threatening species are *Themeda quadrivalvis* and *Cryptostegia grandiflora* both of which are aggressive and troublesome weeds in parts of tropical Queensland.

The importance of families in a regional flora can be rated by considering the number of genera or species per family. The top 10 families ranked by these criteria are presented in Tables 4 and 5. These 10 families contribute significantly to the indigenous floras of each area accounting for 38% of the genera and 46% of the species recorded for Cape York and 42% of the genera and 54% of the species from the Kimberley. The Kimberley flora appears to be dominated by these larger families more so than that of Cape York. To account for 50% of the native genera in the Kimberley it is necessary to include only the top 15 families. Nine families will account for 50% of the

TABLE 4. Ranking of the 10 largest families from Cape York and the Kimberley based on the number of genera listing number of native genera and percentage of the total native vascular genera.

Cape York			Kimberley		
Family	No. of genera	Percentage	Family	No. of genera	percentage
Poaceae	80	8.2	Poaceae	70	12.3
Fabaceae	66	6.8	Fabaceae	56	9.9
Euphorbiaceae	38	3.9	Asteraceae	21	3.7
Orchidaceae	38	3.9	Euphorbiaceae	21	3.7
Rubiaceae	31	3.2	Rubiaceae	14	2.5
Myrtaceae	31	3.2	Scrophulariaceae	14	2.5
Asteraceae	27	2.8	Chenopodiaceae	12	2.1
Cyperaceae	23	2.4	Cyperaceae	12	2.1
Sapindaceae	22	2.3	Asclepiadaceae	10	1.8
Proteaceae	19	1.9	Myrtaceae	10	1.8
Total	375	38.4	Total	240	42.3

TABLE 5. Ranking of the 10 largest families from Cape York and the Kimberley based on the number of species listing the number of native species and percentage of the total native vascular species.

Cape York			Kimberley		
Family	No. of species	Percentage	Family	No. of species	Percentage
Poaceae	228	9.5	Fabaceae	234	14.7
Fabaceae	200	8.3	Poaceae	209	13.1
Cyperaceae	152	6.3	Cyperaceae	100	6.3
Myrtaceae	149	6.2	Myrtaceae	76	4.8
Euphorbiaceae	107	4.4	Euphorbiaceae	54	3.4
Orchidaceae	86	3.6	Asteraceae	41	2.6
Rubiaceae	69	2.9	Amaranthaceae	40	2.5
Sapindaceae	56	2.3	Malvaceae	40	2.5
Asteraceae	40	1.7	Rubiaceae	36	2.3
Convolvulaceae	38	1.6	Convolvulaceae	32	2.0
Total	1 125	46.6	Total	862	54.1

native species. To reach a similar percentage for the Cape York flora it is necessary to include the top 19 families to reach the mid-point in the number of genera and 12 families to account for half of the species. When expressed as percentages of the total native families in each area, however, the number of families required to account for 50% of the genera or species in each area under consideration are comparable. In either area, half of the native genera can be accounted for by about 9% of the families and half of the native species by roughly 6% of the families.

Poaceae and Fabaceae *sens. lat.* are clearly the major families in both Cape York and the Kimberley regardless of the measure used. At the generic level, this is also the pattern recognized by Johnson (1983) for the floras of Queensland, Western Australia and the whole of eastern Australia. At the species level, Myrtaceae and Proteaceae feature prominently in the state-wide flora of West Australia relegating the grasses to fourth position in the list of major families. This probably reflects the species rich shrub dominated communities of West Australia's south-west.

A comparison of the families featuring in Tables 4 and 5 gives an indication of one of the fundamental differences in the floras under study. Whilst the majority of the families featuring in each table are similar, when ranked by the number of genera, the families Orchidaceae, Sapindaceae and Proteaceae in the list for Cape York are replaced by Scrophulariaceae, Chenopodiaceae and Asclepiadaceae in the Kimberley list. When species are employed to rank families, Orchidaceae and Sapindaceae appear in the top 10 families for Cape York while Amaranthaceae and Malvaceae replace them in the Kimberley list. Orchidaceae, Sapindaceae and Proteaceae include many plants of closed forest communities. Due to the generally lower annual rainfall in the tropical west, the diversity of closed forest communities which occurs on Cape York is not matched in the Kimberley. The major families in the Kimberley tend to be those better adapted to drier environments.

Genera with 10 or more native species are listed in Table 6 in numerical order. Numbers of genera in each list are comparable, Cape York having 29 genera and the Kimberley, 28. The same genera *Acacia, Cyperus, Eucalyptus* and *Fimbristylis* fill the top four positions though their relative positions vary. In terms of numbers of species, the genus *Acacia* assumes an importance in the Kimberley flora not matched by any other genus in either area. The number of *Acacia* species recorded from the Kimberley is more than twice that for Cape York. The high level of speciation which has apparently occurred in this genus in the Kimberley area was noted by Hnatiuk & Pedley (1985). While over half of the large genera in the flora of each area are common, the xeromorphic nature of the Kimberley flora is highlighted by the presence of genera such as *Cassia, Ptilotus, Goodenia, Triodia* and *Sida* in Table 6 in place of *Syzygium, Phyllanthus, Cryptocarya, Rhynchospora* and *Endiandra* in the list for Cape York.

The common elements in the native angiosperm floras of Cape York and the Kimberley are summarized in Fig. 2 and contrasted with similar

TABLE 6. Genera with 10 or more native species in Cape York or the Kimberley ranked by the number of species.

Cape York		Kimberley	
Genus	No. of Species	Genus	No. of Species
Cyperus	48	*Acacia*	79
Eucalyptus	40	*Fimbristylis*	46
Fimbristylis	36	*Eucalyptus*	44
Acacia	36	*Cyperus*	28
Ficus	27	*Tephrosia*	21
Syzygium	26	*Cassia*	20
Dendrobium	22	*Stylidium*	19
Melaleuca	19	*Ptilotus*	18
Phyllanthus	19	*Utricularia*	17
Ipomoea	17	*Euphorbia*	17
Scleria	15	*Mitrasacme*	16
Aristida	14	*Solanum*	16
Eragrostis	14	*Terminalia*	16
Pandanus	14	*Eriachne*	15
Utricularia	13	*Grevillea*	15
Cryptocarya	12	*Ipomoea*	14
Eriachne	12	*Goodenia*	13
Grevillea	12	*Triodia*	13
Solanum	12	*Aristida*	12
Desmodium	11	*Hibiscus*	12
Eleocharis	11	*Eragrostis*	12
Indigofera	11	*Ficus*	12
Rhynchospora	11	*Melaleuca*	12
Tephrosia	11	*Heliotropium*	11
Terminalia	11	*Panicum*	11
Endiandra	10	*Crotalaria*	11
Eriocaulon	10	*Indigofera*	10
Mitrasacme	10	*Sida*	10
Stylidium	10		

figures derived from Johnson's (1983) study for the flora of the whole of Western Australia compared to eastern Australia.

Only three families of vascular plants, Crassulaceae (one genus/one species), Frankeniaceae (one genus/one species) and Gyrostemonaceae (two genera/three species), found in the Kimberley have not been recorded for Cape York whereas 65 families (Table 7) occurring on Cape York do not extend to the Kimberley. This list includes three gymnosperms, 13 pteridophytes and 49 angiosperms. The majority of the families found on Cape York but not recorded from the Kimberley have only a few genera and species and are found mainly in closed forest or swamp communties. These habitats are not well represented in the north of Western Australia.

A further measure of the relationship of the Cape York and Kimberley floras is given in Fig. 3 in the form of paired east-west comparisons similar to those presented in Fig. 2 but using Czekanowski's coefficient (Clifford & Stephenson 1975) based on data for the native angiosperms as a measure of similarity. These reveal a much higher degree of similarity between the floras of Cape York and the Kimberley at the species level (0.47) than that between Western Australia and eastern Australia (0.28). The comparisons at the generic level are quite similar. The pattern observed at the species level is reversed at the family level where the comparisons show a higher degree of similarity between the flora of the whole of Western Australia and that of the eastern bloc (0.92) than between the Kimberley and Cape York (0.82). This is due to fact that although 98% of the native angiosperm families recorded for Australia occur in the eastern mainland states, and 85% of these are to be found on Cape York, 86% of families recorded for Australia occur in Western Australia, but only 70% of these extend to the Kimberley. Families found on Cape York which are absent from the Kimberley but occur elsewhere in Western Australia include Araliaceae, Brassicaceae, Casuarinaceae, Epacridaceae, Oxalidaceae, Restionaceae and Rosaceae.

Conclusions

Cape York and the Kimberley remain amongst the botanically least known areas in Australia and it is possible that some of the differences observed here may be attributed to this lack of basic knowledge alone. At the family and generic level the absence of so many closed forest and swamp taxa from the Kimberley might be explained by the

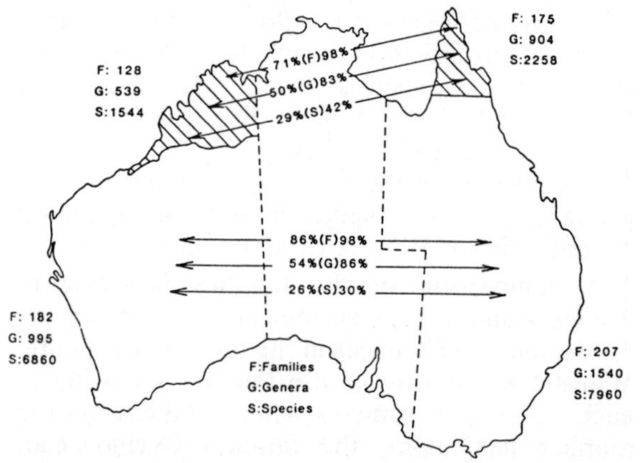

FIG. 2. Families, genera and species of native angiosperms in eastern Australia and Western Australia and percentages held in common.

TABLE 7. Families present on Cape York but absent from the Kimberley listing the number of native genera and species for each.

GYMNOSPERMS					
Araucariaceae	2/2				
Podocarpaceae	1/2				
Zamiaceae	1/1				
PTERIDOPHYTES					
Angiopteridaceae	1/1	Davalliaceae	2/4	Marattiaceae	1/2
Aspidiaceae	2/4	Grammitidaceae	3/7	Osmundiaceae	1/1
Aspleniaceae	1/7	Hymenophyllaceae	8/16	Tmesipteridaceae	1/1
Azollaceae	1/2	Lomariopsidaceae	2/3	Vittariaceae	2/4
Cyatheaceae	1/3				
ANGIOSPERMS					
Agavaceae	2/4	Elaeocarpaceae	3/10	Monimiaceae	8/11
Alangiaceae	1/1	Epacridaceae	1/1	Musaceae	1/1
Araliaceae	4/10	Ericaceae	2/2	Nelumbonaceae	1/1
Austrobaileyaceae	1/1	Eupomatiaceae	1/2	Nepenthaceae	1/1
Balanophoraceae	1/1	Flacourtiaceae	4/7	Oxalidaceae	1/1
Brassicaceae	1/1	Gesneriaceae	1/1	Piperaceae	1/2
Brunoniaceae	1/1	Grossulariaceae	3/4	Restionaceae	2/5
Cardiopteridaceae	1/1	Hamamelidaceae	1/1	Rosaceae	2/2
Casuarinaceae	1/4	Hanguanaceae	1/1	Sphenocleaceae	1/1
Connaraceae	1/1	Himantandraceae	1/1	Stemonaceae	1/1
Corynocarpaceae	1/1	Hippocrateaceae	2/3	Symplocaceae	1/3
Costaceae	1/1	Hydrophyllaceae	1/1	Theaceae	1/1
Cunoniaceae	3/3	Icacinaceae	5/5	Winteraceae	2/3
Cymodoceaceae	3/4	Leeaceae	1/1	Xanthophyllaceae	1/1
Davidsoniaceae	1/1	Linaceae	1/1	Zingiberaceae	5/8
Dichapetalaceae	1/1	Malpighiaceae	1/1	Zosteraceae	1/1
Elaeagnaceae	1/1				

lack of suitable habitats in the tropical west reflecting the generally drier climate experienced there. These taxa may have been present in past geological times and eliminated as the climate became generally drier since the Miocene or the absences could equally suggest fundamental differences in the biogeographic origins of the two floras. The future comparative studies foreshadowed earlier will attempt to test these hypotheses and explore other factors which could explain the patterns demonstrated here.

Acknowledgements

We acknowledge the work of those collectors on whose specimens this study is so much dependent and wish to thank the directors of the Queensland and Western Australian herbaria, Dr Bob Johnson and Dr John Green, for access to computerized data bases held by their institutions. Dr John Green read the manuscript and made a number of useful comments. John Clarkson would like to acknowledge a grant from the Australian Biological Resources Study which provided

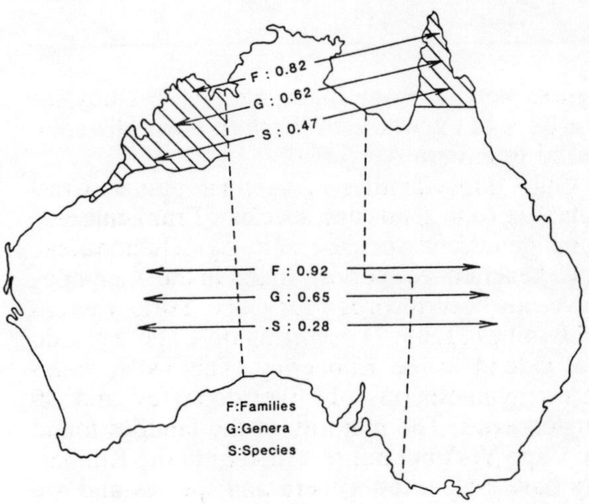

FIG. 3. Coefficients of similarity (Czekanowski) for eastern Australia and Western Australia. 1=total similarity, 0=total dissimilarity.

financial support for technical assistance for floristic studies on Cape York Peninsula from 1981-1985.

References

Beard J. S. (1980) A new phytogeographic map of Western Australia. *W.A. Herb. Res. Notes* **3,** 37-58.

Chippendale G. M. (1972) Check list of Northern Territory Plants. *Proc. Linn. Soc. N.S.W.* **96,** 207-67.

Clifford H. T. & Stephenson W. (1975) *An Introduction to Numerical Classification.* Academic Press, London.

Cronquist A. (1981) *An Integrated System of Classification of the Flowering Plants.* Columbia University Press, New York.

Dunlop C. R. & Bowman D. M. J. S. (1986) *Atlas of the Vascular Plant Genera of the Northern Territory.* Australian Flora and Fauna Series **6**. Australian Government Publishing Service, Canberra.

Gill A. M., Belbin L. & Chippendale G. M. (1985) *Phytogeography of* Eucalyptus *in Australia.* Australian Flora and Fauna Series **3**. Australian Government Publishing Service, Canberra.

Green J. W. (1985) *Census of the Vascular Plants of Western Australia.* 2nd edition. Western Australian Herbarium, Perth.

Hnatiuk R. J. & Pedley L. (1985) Biogeography of *Acacia* in northern Australia. *Proc. ecol. Soc. Aust.* **13,** 33-42.

Johnson R. W. (1983) The Queensland flora and its collection. *Proc. R. Soc. Qld.* **94,** 1-18.

Pedley L. & Isbell R. F. (1971) Plant communities of Cape York Peninsula. *Proc. R. Soc. Qld.* **82,** 51-74.

Pichi-Sermolli R. E. G. (1977) Tentamen pteridophytorum genera in taxonomicum ordinem redigendi. *Webbia* **31,** 313-512.

Ross J. H. (1976) An analysis of the flora of Victoria. *Muelleria* **3,** 169-76.

Simon B. K. (1981) An analysis of the Australian grass flora. *Austrobaileya* **1,** 356-71.

Webb L. J. & Tracey J. G. (1981) Australian rainforests: patterns and change. In: *Ecological Biogeography of Australia* (ed. A. Keast) pp. 607-94. W. Junk, The Hague.

Webb L. J., Tracey, J. G. & Williams W. T. (1984) A floristic framework of Australian rainforests. *Aust. J. Ecol.* **9,** 169-98.

Proc. Ecol. Soc. Aust. 15:267-269

The biogeography of tropical Australian grasses

B. K. SIMON

Queensland Herbarium, Meiers Road, Indooropilly, Qld. 4068

Abstract

An attempt is made to understand the biogeographic relationships of the indigenous grasses of tropical Australia with those of other floristic regions of Australia and of the world. Matrices of similarity coefficients of these regions in terms of major taxonomic groups and genera provide data for a number of analyses, of which the minimum spanning tree, the two-neighbour network and the flexible dendrogram (used with principal co-ordinate and component analyses) are illustrated. At the species level the similarity coefficients of the Australian regions are compared directly with a number of contemporary floras. Different kinds of associations are produced depending on the taxonomic rank used. The present-day distribution of major taxonomic groups can be interpreted as a reflection of past continental movements, modified by climatic adaptations, whereas the distribution of genera and species appears to be more a reflection of the proximity of the floristic regions to each other.

Introduction

Grasses are one of the most successful plant families in an evolutionary sense in that they have a place in almost all ecological niches of the world. Biogeographic studies of grasses have been previously conducted in some detail at the levels of major taxonomic group, genera and species, the findings of which are well summarized by Clayton (1981).

The aim of the present study is to attempt to relate tropical Australia to other areas of Australia and the world in terms of its major taxonomic groups, genera and species. For the first two we can use the coefficients of dissimilarity obtained for another study (Simon 1986), when all floristic regions of the world were compared with each other in terms of their genera and major taxonomic groups. For species the basic data for a comparision of all floristic regions in a similar manner is not available although the chorological analyses undertaken at Kew for all areas of the world except Central and South America have a comparable methodology (Clayton & Cope 1980a, 1980b). In these studies however Australia was not subdivided beyond one floristic kingdom whereas I have regarded it as comprising four floristic regions of which tropical Australia is one. A degree of affinity of tropical Australian species with some other regions can be obtained by comparing the Czekanowski coefficients of this region with the species composition of known floras.

Methods

Matrices of coefficients of dissimilarity (Czekanowski) of the floristic regions in terms of their genera and major taxonomic groups were obtained to run through the Numerical Taxonomy Package (NTP) of Belbin (1984). In the case of genera the matrix was compiled by a FORTRAN programme from the raw data. For the major groups a programme produced an amalgamation of the genera into their seven groups, from which a table of percentage representation was produced for each region and used to generate a matrix, using the programme ASO. Each of the matrices was subjected to an agglomerative (flexible) cluster analysis (FUSE) and output produced in the form of dendrograms (DEND). Ordinations were undertaken (SCAT), the genera being analysed by principal co-ordinate analysis and the major groups by principal component analysis using

GOWC and PCA. Minimum spanning trees (MST) and two-neighbour network analyses from tables of nearest neighbours (NN), bond strengths (BOND) and outgroups (GDF) were produced and used together for the purpose of illustration. For species a table comparing tropical Australia with a number of other regions was drawn up and the Czekanowski coefficients were calculated and compared with coefficients obtained for genera and major taxonomic groups (Table 1).

Results

Genera

A selection of nine clusters was made from an examination of the dendrogram and these were transposed onto the principal co-ordinate analysis. These clusters can be further grouped into four blocks: *viz.*

1. Eurasia and North America
2. Tropical Old World
3. Neotropics
4. Southern temperate areas and associated islands.

The two-neighbour network analysis used in conjunction with the minimum spanning tree procedure produced 10 clusters which grouped into five blocks: *viz.*

1. Eurasia and North America
2. Africa
3. Tropical Asia, tropical Australia and the tropical Pacific
4. Tropical and South America
5. Temperate Australia, New Zealand and southern island groups.

It can be seen that there are five closely knit blocks of floristic regions from their relative bond strengths and the fact that they emerge as adjacent in both analyses. One of these blocks is from tropical Asia to tropical Australia.

Major groups

A selection of eight clusters was made from an examination of the dendrogram and these were transposed onto the principal component analysis. These clusters, together with those produced by the two-neighbour network analysis and minimum spanning tree procedure, can be partitioned into three blocks: *viz.*

1. Areas with polar or cold temperate climates
2. Areas with warm temperate and mediterranean climates
3. Areas with tropical climates

The taxonomic content of the blocks in general terms is as follows.

Polar and cold temperate. Pooids account for 40 to 100% of the genera, with the other groups making up the deficit or not being represented.

Warm temperate and mediterranean. A fairly even representation of the pooids, chloridoids, panicoids, andropogonoids and arundinoids.

Tropical. Panicoids, andropogonoids, chloridoids and, sometimes, bambusoids increase at the expense of the other groups.

The tropical Australian region is grouped together with other Old World savannah regions from Africa, Madagascar and eremaean Australia.

Species

Most Australian native species are endemic. Of the relatively small number of species which extend their range beyond Australia, those from tropical Australia account for the large majority. The Czekanowski coefficients of similarity at species level between tropical Australia and other floristic regions within and outside Australia (Table 1) shows a fairly strong link with the tropical regions directly adjacent, although the connection is somewhat weaker than that exhibited at generic level.

TABLE 1. Czekanowski coefficients of similarity between tropical Australia and selected other floristic regions at three taxonomic ranks.

Floristic Regions	Species	Genus	Major group
Temperate south-east Australia	22.4	45.3	65.0
South-west Australia	5.8	19.1	50.0
Eremaean Australia	49.6	44.8	76.7
New Guinea	31.8	72.2	81.7
Malesia	12.3	66.7	82.1
India	8.7	53.6	84.0
Tropical east Africa	4.2	43.8	86.2
Saharo-Sind	2.8	45.0	70.5
Fiji	3.9	35.4	81.7
New Zealand	3.6	20.9	49.0
East Asia	3.4	35.8	52.0

Conclusions

The analysis of generic distributions in terms of floristic regions reflects geographical location as the regions are grouped, more or less, in adjacent blocks. In contrast the analysis of the major taxonomic group cuts across these geographical boundaries, reflecting more, the areas with similar climates. This is graphically portrayed when the minimum spanning tree is plotted directly onto the geographical positions of the floristic regions. The inference from these different patterns is that the major groups were well dispersed throughout the World before the final break-up of the Gondwanaland super-continent (Clayton 1981, 1983).

The fact that widespread genera, of which there are about 70, and, to a lesser extent, species (e.g., *Heteropogon contortus* and *Themeda triandra)* do exist, suggests that these particular taxa are ancient and possibly Gondwanic. These contrast with what appear to be younger genera which have more restricted distributions, about two-thirds of them being restricted to single continents (Clayton 1975), and most species.

Within the context of tropical Australia, the Czekanowski coefficients of similarity between tropical Australia and other floristic regions are correlated directly with taxonomic rank (Table 1). In other words, and following on from the general pattern for the world, the higher the taxonomic rank the earlier has been the dispersal of these taxa and the greater the distance from tropical Australia. This is exemplified by a consideration of the coefficient between tropical Australia and its nearest neighbour at different ranks. At the level of the major group its nearest neighbour is tropical East Africa; at the generic level, New Guinea; and, at the species level, eremaean Australia. The strong link with New Guinea at both generic and species level reflects the relatively recent formation of the Torres Straits, during the Pleistocene.

A large number of grasses from other tropical regions of the world are naturalized in tropical Australia and those from tropical East Africa have the best representation. This is correlated directly to the high similarity coefficient between the two regions at the level of the major group, both facts being a reflection of climatic similarity.

References

Belbin L., Faith D. P. & Minchin P. R. (1984) Some algorithms contained in the numerical taxonomy package NTP. *CSIRO Division of Land and Water Resources, Technical Memorandum* 84/23.

Clayton W. D. (1975) Chorology of the genera of Gramineae. *Kew Bull.* **30**, 111-32.

Clayton W. D. (1981) Evolution and distribution of grasses. *Ann. Missouri Bot. Gard.* **68**, 2-14.

Clayton W. D. (1983) Tropical grasses. In: *The Genetic Resources of Forage Plants* (eds J. G. McIvor & R. A. Bray) pp. 39-46. CSIRO, East Melbourne.

Clayton W. D. & Cope T. A. (1980a) The chorology of Old World species of Gramineae. *Kew Bull.* **35**, 135-71.

Clayton W. D. & Cope T. A. (1980b) The chorology of North American species of Gramineae. *Kew Bull.* **35**, 567-76.

Simon B. K. (1986) The biogeography of the genera and major groups of grasses. *Int. Symp. Grass Systematics & Evolution, Abstracts* Smithsonian Inst., p. 33.

A Key to the genera of Australian Vitaceae

B. R. JACKES

Botany Department, James Cook University of Northern Queensland, Townsville, Qld. 4810

The family Vitaceae consists of vines, rarely shrubby, characterized by leaf-opposed tendrils. F. von Mueller and G. Bentham considered *Vitis* to be the only Australian genus, but, since 1909 (Gagnepain), five genera have been recognized as occurring naturally in Australia.

Although these genera are easily distinguished when complete specimens are available, those found chiefly in rainforest or monsoonal areas are usually collected (if at all) in a vegetative state thus making conventional identification difficult. However, it is possible using vegetative material to place these specimens in a genus and often even to the level of species. (Numbers in brackets refer to the number of species).

Key using Vegetative and Reproductive Characters

1. Flowers 5-merous 2
1. Flowers 4-merous 4
    2. Leaves pedate .... *Ampelocissus* (3)
    2. Leaves palmate 3
3. Inflorescence compact, multi-flowered WA only .... *Clematicissus* (1)
3. Inflorescence loose, few flowered, Eastern States .... *Cissus opaca*
    4. Stigma large and 4-lobed .... *Tetrastigma* (5)
    4. Stigma small, not markedly lobed 5
5. Leaves simple or compound, if compound then palmate and ratio of central petiolule to adjacent lateral petiolule rarely greater than 1.5. Inflorescence an umbellate cyme .... *Cissus* (12)
5. Leaves compound, palmate or pedate, ratio of central petiolule to adjacent lateral petiolule rarely less than 2. Inflorescence a dichotomous cyme .... *Cayratia* (8)

Key using Vegetative Characters only

(*Clematicissus* which is endemic to the Geralton area of WA is omitted).

1. Leaves simple .... *Cissus* (9)
1. Leaves compound 2
    2. Leaves pedate 3
    2. Leaves 3-5(-7)-foliolate not pedate 5
3. Leaves glabrous, occur in rainforest .... *Tetrastigma* (2)
3. Leaves rarely glabrous, if so then not found in rainforest 4
    4. Leaves soft; margins serrated, hairs simple, erect 3-16 celled; usually found in moist areas .... *Cayratia* (3)
    4. Leaves firm; margins entire or shallowly serrated, sometimes tinged with red, hairs if unbranched, simple 1-3 celled, 2-armed hairs usually present; common in open forest or edge of vine thickets .... *Ampelocissus* (3)
5. Leaves glabrous, or if pubescent, then margins crenate and peg-like lenticels present on older stems, always trifoliolate, domatia absent, robust vines .... *Tetrastigma* (3)
5. Not the above combination of characters ... 6
    6. Leaves 3-foliolate, petiolules always present, ratio of central petiolule to adjacent lateral petiolule rarely less than 2. Lateral leaflets obviously oblique at base. Margins of mature leaves with numerous shallow to deep serrations on each side of the midrib ..... *Cayratia* (5)
    6. Leaves 3-5-foliolate, if petiolules present then ratio of central petiolule to adjacent lateral petiolule rarely greater than 1.5 and lateral leaflets not obviously oblique at base. Margins of mature leaves entire or with less than 5 indentations on each side of the mid-rib .... *Cissus* (4)

References

Gagnepain F. (1909) Classification des *Cissus* et *Cayratia*. *Notulae systematicae* **1**, 339-43.

Proc. Ecol. Soc. Aust. 15:273-279

Studies on the reproductive biology of some mosses from the Australian wet tropics

H. P. RAMSAY

School of Botany, University of New South Wales, Kensington, N.S.W. 2033

Abstract

The wet-tropical region of north Queensland is bryologically rich containing many unique moss and hepatic species. The variety of vegetation types, high rainfall and altitudinal range from sea level to above 1 500 m provide, in a relatively small area, habitats for many different bryophytes. The diversity of mosses is high, (ca 325 species) about one-third of the total for Australia, in the region from Cardwell to Cooktown including the Atherton and Windsor Tablelands. A high proportion of species are Australian endemics (35%), about half of these occurring only in the region. A significant component have relationships to Malesia and Oceania. Publications on bryophytes for the area are few but new, interesting and unique families, genera and species are being reported. Distribution patterns and research on Macromitrium *suggest that species fall into several categories and include relict species, authochthonous elements and recent arrivals. Studies of the family Sematophyllaceae show its centre of distribution (13 in 16 genera) is in north Queensland. Reproductive strategies show a high proportion of* dioicous *species including examples with dwarf males, a specialized adaptation to enhance spore production in some* dioicous *forms. In* Macromitrium *anisospory is a further adaptation for fertilization success in* dioicous *species with dwarf males. The capacity for species to spread into new niches is increased by asexual reproductive mechanisms of many kinds, from fragments to specialized structures. Many asexual strategies for survival are represented in mosses of the wet tropics. The uniqueness of north Queensland supports the view that the survival of bryophytes as well as vascular plants is endangered in the wet tropics.*

Introduction

The wet tropical region of north-east Queensland, a unique botanical region in Australia (Webb & Tracey 1981; Keto & Scott 1986) is bryologically rich, containing many unique moss and hepatic species. The variety of rainforest (closed forest) types in the region (Tracey 1982) and the altitudinal range from sea level to above 1 500 m provide, in a relatively small area, habitats for many different bryophytes. The paucity of knowledge of the bryophytes in Australian rainforests generally has been stressed by Ramsay *et al.* (1987). There have been no bryoecological studies in north-east Queensland, for example, there are no data on distribution of species in the various ecosystems present, no physiological studies on the contribution of bryophytes to the maintenance of humidity, no analysis of their importance for nutrient recycling in the ecosystems present, no detailed studies on the types of reproductive mechanisms which limit or enhance the spread and survival of bryophytes in the tropics here.

> "The Bryology of north Queensland is of particular interest to botanists, in part because of what has been found in part because of what may be found" (Dixon 1938).

Published works on north-east Queensland mosses is limited to some information in Watts & Whitelegge (1902, 1906), Bailey (1913), Brotherus & Watts (1918), Burges (1932, 1935) and Bartram (1952) with descriptions of new species in the last decade by Ilma Stone (e.g., 1976, 1985). Specimens from the Flecker Herbarium now housed in Canberra (CANB), contain collections examined by Dixon some of which have been described (Dixon 1938, 1942) others not. Brotherus & Watts (1918) and Dixon (1938) both comment on the interesting relationships of the mosses present in this region. In recent years some significant collections have been made by Stone, Weber & McVean, Norris, Streimann, Vitt, Ramsay, Schofield, Thiers & Holling, Kantack & Churchill and some others. Where these are located in overseas herbaria duplicates of many are being or have been sent here for study. These are providing new records and valuable new information on distribution. Recent taxonomic revisions are beginning to improve our knowledge of some taxa in the area, for example Vitt & Ramsay (1985a,b) — Orthotrichaceae; Nowak (1980), Reese & Mohamed (1984), Mohamed & Reese (1985), Reese *et al.* (1985) — Calymperaceae.

Distribution of Mosses in North Queensland

A preliminary estimate of the mosses present in the area from Cardwell north to Cooktown and west to the Atherton and Windsor Tablelands (<100 km inland) reveals a broad representation of families (42) genera (130) and species (*ca* 325). The species recorded so far represent about one-third of the total for Australia estimated at *ca* 1 000 (D. G. Catcheside pers. comm.). By comparison, the total diversity is higher than that for the subtropical North Coast of New South Wales, a larger geographical region (Ramsay 1984a,b). Many collections in Australian and overseas herbaria remain undetermined at present. Recent collections often add new data to earlier collections which have either not been identified or not published. The work of Ilma Stone, however, has shown that there are new and significant finds still to be made.

Among the species recorded are a high proportion of Australian endemics (118 — 35%) of which almost half (50-60) are known only from north-east Queensland: examples include *Macromitrium dielsii, M. funiforme* (Vitt & Ramsay 1985a,b), *Buxbaumia thorsbornei, Fissidens gymnocarpos, F. henryae, F. traversii* (Stone 1983a,b, 1984a). The distributional relationships of the moss flora are mainly to S.E. Asia, Malesia (22%) and Oceania (20%), with only 13% common to New Zealand, while 7.0% occur in Africa, and a similar proportion in the Americas. The majority of species in Oceania occur also in Southeast Asia but the pantropical component is small (*ca* 3-4%) (Wijk *et al.* 1959-69).

Particularly important study areas are the high altitude cloud forests on Mt Bartle Frere, Mt Bellenden Ker, Mt Lewis and Thornton Peak, which represent high altitude "islands" with certain differences in their vascular (Tracey & Webb 1975; Tracey 1982) as well as their non-vascular flora. These are illustrated in the uniqueness of species present on Bellenden Ker & Bartle Frere, close in distance but with distinct differences.

From records available at present, numbers of species in each of five arbitrary geographical areas within the region (Fig. 1) were calculated. The 2 500 mm isohyet was chosen as the boundary between the coastal and Atherton Tableland species (area 1). The two peaks Mt Bartle Frere and Mt Bellenden Ker (areas 2 & 3) were dealt with separately from each other and from the coastal lowlands below them including Russell River, Babinda south to Cardwell (area 4). The final area (5) ranges from Cairns north to Cooktown and west to include the Windsor Tableland. All areas have variations in vegetation types but the Atherton Tableland, which includes drier vegetation types as well as wet forests, had the greatest number of species (Fig. 1). The largest number of these were collected in the Ravenshoe, Millaa Millaa, Tully Falls section. Whilst there are similarities between the species present on all the high peaks, there are a few species confined to only one, for instance *Macromitrium dielsii* occurring only on Mt Bellenden Ker. The lower number of mosses recorded for the northern area (area 5), which includes the Daintree River, Cape Tribulation and Mossman Gorge, may be attributable to the type of vegetation being less suitable for mosses (i.e. lowland vine forests). The higher peaks, Thornton Peak, Mt Lewis and Mt Carbine within this same area, however, have some very interesting species not found elsewhere, for example *Groutiella tomentosa* (Vitt & Ramsay 1985a,b). Much of the mountain area has not yet been fully collected and the numbers of mosses recorded are likely to increase as more are identified.

Moss families highly represented in north-east Queensland but which are rare, less diverse, or

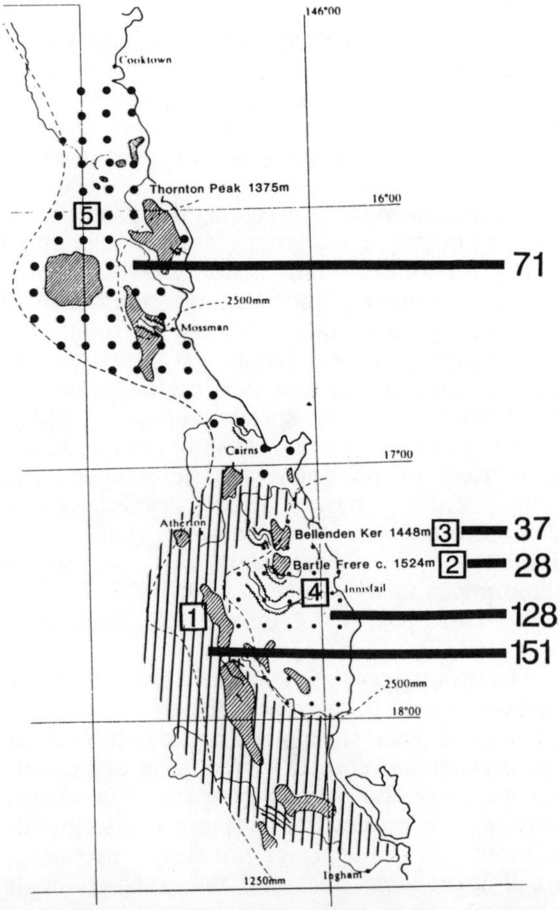

FIG. 1. Distribution of mosses in five arbitrary geographical areas of north Queensland. *Area 1:* Artherton Tableland between 2 500 mm and 1 250 mm isohyets south of Cairns. *Area 2:* Mt Bartle Frere. *Area 3:* Mt Bellenden Ker. *Area 4:* Areas above 2 500 mm rainfall between Cardwell and south of Cairns. *Area 5:* The northern section of Cairns and northwards including tableland coastal rainforests and mountain peaks, rainfall above 1 250 mm.

KEY:
Basic map based on Webb & Tracey (1981) showing refugia.
Area 1 Areas 2 & 3 as named (refugia)
Area 4 Area 5
Areas indicated thus: 1 Numbers of species: 37

do not occur in southern Australian states are the Calymperaceae, Pterobryaceae, Sematophyllaceae, Sorapilaceae, Neckeraceae, Meteoriaceae, Trachypodaceae, Myuriaceae.

The family Sematophyllaceae, at present under revision by Ramsay & Schofield, represents one with its centre of diversity in north Queensland, where all 16 genera occur and 70% species. Only three of the 16 genera *(Sematophyllum, Wijkia, Rhaphidorrhynchium)* occur south of Queensland. Most of the Australian endemics are found there while the distribution patterns of the non-endemic species extend to New Caledonia, south-east Asia, Malesia and Oceania. A number of new records for the Calymperaceae whose only distribution in Australia is in Queensland, have been reported for north Queensland by Reese *et al.* (1986). At the generic level studies on the *Macromitrium* complex (Orthotrichaceae) in Australasia have highlighted north Queensland as the major centre of diversity for this genus (Fig. 1 Vitt & Ramsay 1985b): 17 out of 21 species or 80% occurring in this region. Of these, *Macromitrium dielsii* and *M. funiforme* are endemic to the area. The other species are either distributed within Australia along the eastern coastal and tablelands, or extend to tropical regions e.g., Malesia or Oceania (*M. incurvifolium, M. involutifolium, M. leratii, M. microstomum* and *Groutiella tomentosa* (Vitt & Ramsay 1985a,b)).

Problems Raised by Distribution Patterns

The fact that Queensland contains a significant number of mosses not present in the southern states (Scott & Stone 1976) raises the question of their origins and of the mechanisms which have enabled them to successfully establish in the tropics. The north Queensland bryoflora include a small percentage of relictual components, showing residual taxonomic affinity with the floras of similar habitats elsewhere as suggested by Barlow (1981) for the Australian vegetation as a whole. The high degree of endemism including paleo-endemics and neoendemics suggests evolutionary diversification from ancestral Gondwanan stocks. Studies on *Macromitrium* and the Calymperaceae provide evidence of more recent intrusion of tropical taxa.

Groutiella tomentosa is known in Australia from only a few collections made by Dan Norris on Thornton Peak and Mt. Carbine (Vitt & Ramsay 1985a,b). Only female plants with archegonia have been found and no sporophytes or male plants collected (the males are dwarf and epiphytic on females). Present evidence suggests successful long distance dispersal, possibly from New Guinea its nearest centre of distribution rather than relictual isolation of female plants. As only females are known the species must have arrived as either a spore or vegetative propagule.

Establishment and further dispersal must have depended on asexual means. The endemic species of *Macromitrium (M. dielsii* and *M. funiforme)* represent advanced (apotypic) taxa within the genus and are considered recent examples of speciation (neoendemics) while other species represent ancient gondwanic stock (paleoendemics) (Vitt & Ramsay 1985b). The families Calymperaceae and Sematophyllaceae include Australian endemics and species distributed widely in the tropics. Do all endemics represent recent speciation or are they relicts isolated here by climatic changes in the Tertiary? Are there authochtonous components among the mosses present in the wet tropics?

There are many examples of endemism and disjunct or unusual distribution patterns within the region. Some can be explained by geographical isolation in closed forests on mountains after the Tertiary with increase in aridity (Barlow 1981). Recent arrivals would need to establish themselves in existing ecosystems in competition with mosses already present. Would species with a preadapted ability to withstand dryness have greater survival potential or be more likely to be successful invaders as they occupy uncolonized niches? *Macromitrium* is considered an example of such a preadapted xerophyte (Vitt & Ramsay 1985b). Can reproductive and dispersal strategies found in the tropical mosses present account for the various distribution patterns?

Reproductive Mechanisms and Dispersal Strategies

Sexual

Reproductive strategies can be classified into sexual and asexual mechanisms. Haploid spores dispersed by wind are produced as a result of successful sexual reproduction and sporophyte development. These carry the genetic potential for variation and adaptation. Mosses are either *dioicous* or *monoicous* although the variations in strategies used are many. In the mosses from north-east Queensland for which data are available, there are a high proportion of *dioicous* species. In mosses generally, dioicy is most common (Schofield 1985). As fertilization is dependent on a continuous film of water between the antheridium and archegonium, in *dioicous* species both sexes would need to grow in close proximity. Maximum distance calculated by Anderson & Lemmon (1974) is a few centimetres. Dispersal of male and female spores would be expected to be random with equal opportunity for germination, but in spite of large numbers being released, successful establishment of both sexes in close proximity would not be assured, and would be particularly difficult for epiphytic species.

A mechanism for overcoming this problem has evolved in many moss groups in the form of dwarf males, less than 1.0 mm in size, growing epiphytically on female plants. A large proportion of these are in epiphytic species and are found in a wide range of mosses (Ando 1977). At least 23 genera of the mosses present in north Queensland have many, if not all, species with dwarf males. Dwarf males have been shown to be of two kinds, genetically or physiologically determined, and both types have been experimentally demonstrated in *Macromitrium* (Une 1985; Ramsay 1979). For summaries on sexual dimorphism in mosses see (Ando 1977; Ramsay 1979, 1985; Ramsay & Berrie 1982; Ramsay & Vitt 1984).

The dwarfing of the male which grows from the spore wherever it falls on the female, for instance on the leaf surface, stem, even sometimes among the archegonia, ensures closeness to archegonia for the successful transfer of sperm. The energy expended in production of sperm is also greatly reduced, for the spore germinates to produce a small protonemal mat and the complete male plant consists of only a few leaves with one or two perigonia. Rarely do the plants produce innovations which develop new perigonia the next year (H. P. Ramsay unpubl.). The genetic disadvantage of this mechanism, females fertilized by "sons", results in inbreeding unless incompatability mechanisms are present. Random outcrossing, even occasionally, could provide some genetic diversity for selection while the disadvantage of any inbreeding must surely be greatly outweighed by the increase in reproductive success (Ramsay 1985).

Some genera within the family Orthotrichaceae have an additional specialization not present in other mosses — anisospory with the female spore larger and the male smaller, (see Ramsay 1979, 1985; Une 1985) differences being as great as 3:1 by volume (H. Ramsay, *et al.* unpublished). This system is likely to ensure that the male spore is dispersed further, that is away from the female, enabling outbreeding to occur perhaps more often than thought. Thus the mechanism of dwarfing of males has all the advantages of the *monoicous* condition, increase in successful fertilization without its disadvantages.

Asexual

Mosses are highly adapted for asexual reproduction which occurs more frequently than sexual reproduction (During 1979). Regeneration, to found new colonies by production of secondary protonemata, may occur from almost any type of fragment: whole or part of the leaf, stem apices, fragile leaf apices, rhizoids, flagellate branches or by whole plants from clumps being broken and dispersed by wind or rain, from soil, rocks or tree branches. In addition, many species have specialized propagules for this purpose. Examples include gemmae at leaf or stem apices, on costa or surface of leaves, from the stem as bulbils in leaf axils, or as tubers on rhizoids or protonemata (Correns 1899; Brotherus 1924; Watson 1971; Schofield 1985). Most mosses are perennial although in some the above ground portions die down seasonally being regenerated from underground or surface parts, by breaking up of old colonies or by "rhizomes".

Some examples of asexual reproductive mechanisms in north Queensland mosses include plantlets on leaf tips *(Octoblepharum albidum)*; fragile leaf apices *(Groutiella tomentosa)*; gemmae on leaf surfaces *(Macromitrium brevicaule, Syrrhopodon* sp., *Calymperes tenerum, Clastobryella)*, gemmae on leafless stem apices *(Trachyloma diversinerve)* or axillary towards stem apex *(Trachyloma wattsii)*; flagellate julaceous shoots *(Taxithelium wattsii, Wijkia extenuata)*; shedding of whole leaves *(Arthrocormus* sp.).

Growth Form and Sporophyte Production

There are two major methods of growth patterns in mosses, acrocarpous and pleurocarpous, which influence sporophyte production. In acrocarps where the sporophyte is terminal, gametophores are upright and determinate with continued growth by lateral innovations, or plants are annual. The pleurocarps have indeterminate growth of the vegetative gametophore since they produce sporophytes from short lateral branches. Buck & Vitt (1986) suggest that pleurocarpy is commonest in mesophytic, stable ecosystems whereas acrocarpy is favoured in habitats which are xerophytic and unstable (temporary). Broad niche overlap and long growing seasons also favour pleurocarpy. The ratio of pleurocarpous mosses in north Queensland is higher than average; almost equal numbers of families and genera and about 40% species. Acrocarpous forms far outnumber pleurocarpous types in mosses generally. The pleurocarpous growth form is considered derived, occurring in all members of certain families, or in families other than those generally classified as pleurocarps and include taxa occurring on forest floor, wet rocks as well as epiphytic forms. Since north Queensland has wet closed forests which are highly mesophytic the data agree well with the view expressed by Buck & Vitt (1986).

Discussion

Studies on tropical bryophytes are discussed in detail by Magdefrau (1982), Pocs (1982) and Richards (1984). The means of reproduction present in individual species depends on a variety of factors (During 1979). Asexual and sexual reproduction seem to occur concomitantly in some of the mosses studied, e.g., *Wijkia extenuata,* in some populations but not in others but detailed studies have only begun. We have no data on relative survival values of gemmae and spores in northern Australian species. *Monoicous* species are thought to have greater ability for long distance dispersal since only a single spore is needed to establish itself and be successful in sporophyte production for future dispersal within a new environment. The presence of both sexual and asexual mechanisms in the one species and at the same time would overcome disadvantages of dioicy as isolation in time and space is less severe (Zanten & Pocs 1981).

The ratio of *monoicous* to *dioicous* species in a moss flora appears to give no reliable indication of whether the flora originated by long distance dispersal or arose from a flora of more continuous range where land masses were in contact or close to each other (Schofield 1974). Zanten & Pocs (1981) put forward some possible explanations for this in their discussion of dispersal and relationships of the Australian, New Guinea and New Zealand bryofloras. Movement is more likely to have been from Australia to New Guinea. The similarities in the bryoflora of these two land masses is at the generic rather than species level. Smith (1980) proposed migration in late Tertiary times from the eroding mountains in Queensland to the newly uplifted peaks in New Guinea. *Groutiella* is one example of a migration in reverse into Queensland (Vitt & Ramsay 1985a,b).

There is much to be learnt about the mosses, indeed all bryophytes, of the Australian wet tropics. The need is urgent. Not only do we not yet fully realize what species are present, we have little or no data on their role in the closed forest ecosystems as a whole in Australia. Whilst they have many mechanisms for survival and regeneration, they are an integral part of whole ecosystems and are dependent on the vascular flora for the habitats necessary for their continuing existence. We must know what their role is before the ecosystems are lost.

It is clear that rainforests of north-east Queensland have been subject to massive disturbance on an evolutionary timescale (Kershaw 1978) as well as disturbance by man in recent times. Webb & Tracey (1981) consider that the elements of the modern rainforest were conserved in humid refugia which acted as nuclei to recolonize available habitats as climate became more favourable. Bryophytes would also occur in these refugia. If they are similar in behaviour, being associated with and dependent on the particular vegetation types, for example some occurring in temperate others in tropical rainforests, their greatest diversity would be in these refugia. Newly colonized rainforests would have a simpler bryoflora dependent on establishment by disseminules. Webb & Tracey (1981) see the rainforests of north-east Queensland as "a mosaic of relict communities containing species which survived from Gondwanic times surrounded by recolonized areas west of the main range and on the humid lowlands". Vitt & Ramsay (1985b) have suggested that studies on the moss *Macromitrium* support this view for mosses. Damage to the canopy may have the severest effect on epiphytic species all kinds including mosses. The great diversity of mosses and hepatics (M. L. Hicks unpubl. results) in these refuge areas, indeed in all of north-east Queensland, provide an argument for greater research effort on bryophytes.

At the International Association of Bryologists (IAB) Conference on Bryoecology in Budapest August 1985 concern was expressed about the lack of knowledge of tropical bryophytes on a world scale. A resolution was passed by the IAB to prepare lists of endangered bryophytes and of endangered bryophyte habitats or ecosystems. The wet tropics of north-east Queensland surely qualify for inclusion on such a list. Schuster (1986), in his evaluation of the report on Tropical Rainforests of North Queensland: their conservation significance, criticized the lack of fieldwork and study of bryophytes in the region. Some work is in progress but more is needed urgently.

Acknowledgements

Many have helped in making collections available for study: I. G. Stone, H. Streimann, D. H. Norris, W. B. Schofield, D. H. Vitt, B. Thiers & R. Holling, Kandak & Churchill and their associated Herbaria. The National Parks & Wildlife Service of Queensland has kindly provided permits to collect in 1981 and 1985. I am indebted to ABRS and to ARGS for travel grants at various times to enable me to carry out field work.

References

Anderson L. E. & Lemon B. E. (1974) Gene flow distances in the moss *Weissia controversa* Hedw. *J. Hattori Bot. Lab.* **38,** 67-90.

Ando H. (1977) Topics on the Sexuality of Bryophytes. *J. Jap. Bot. Soc.* **2,** 30-32.

Bailey F. M. (1913) *Comprehensive Catalogue of Queensland Plants.* Govt. Printer, Brisbane.

Barlow B. A. (1981) The Australian Flora: Its Origin and Evolution. In: *Flora of Australia,* Vol. 1 (ed. A. S. George) pp. 25-75. Aust. Govt. Publ. Service, Canberra.

Bartram E. B. (1952) North Queensland mosses collected by L. J. Brass. *Farlowia* **4,** 235-47.

Brotherus V. F. (1924) Musci (Laubmoose). In: Engler & Prantl *Die Naturlichen Pflanzenfamilien* 2nd Edition. Vol. 10. Engelmann Leipzig.

Brotherus V. F. & Watts W. W. (1918) The mosses of North Queensland. *Proc. Linn. Soc. N.S.W.* **43,** 544-67.

Buck W. R. & Vitt D. H. (1986) Suggestions for a new familial classification of pleurocarpous mosses. *Taxon* **35,** 21-60.

Burges A. (1932) Notes on the mosses of New South Wales I. Additional records and description of a new species of *Buxbaumia. Proc. Linn. Soc. N.S.W.* **57,** 239-44.

Burges A. (1935) Notes on the mosses of New South Wales II. Additional records. *Proc. Linn. Soc. N.S.W.* **60,** 83-93.

Correns C. (1899) *Untersuchungen über die Vermehrung der Laubmoose durch Brutorgane und Stecklinge.* Gustav Fischer, Jena.

Dixon H. N. (1938) Mosses in North Queensland. *N. Qd. Nat.* **6,** 2-4.

Dixon H. N. (1942) Additions to the mosses of North Queensland. *Proc. Roy. Soc. Qd.* **53,** 23-40.

During H. J. (1979) Life strategies of bryophytes: a preliminary review. *Lindbergia.* **5,** 2-18.

Kershaw (1978) A record of the last interglacial-glacial cycle from north-east Queensland, Australia. *Nature (Lond.)* **272,** 159-61.

Keto A. & Scott K. (1986) *Tropical Rainforests of North Queensland, their conservation significance.* Special Australian Heritage Publications No. 3, Australian Govt. Publishing Service, pp. 159-62.

Magdefrau K. (1982) Life forms of Bryophytes. In: *Bryophyte Ecology* (ed. A. J. E. Smith) pp. 45-58. Chapman & Hall, London.

Mohamed H. & Reese W. R. (1985) *Syrrhopodon* (Musci: Calymperaceae) in Malaysia and adjacent regions. *Bryologist* **88,** 223-54.

Nowak H. (1980) Revision der Laubmoosgattung *Mittythyridium* (Mitten) Robinson fur Ozeanien (Calymperaceae). *Bryophytum Bibliobotheca* **20,** 1-236.

Pocs T. (1982) Tropical Forest Bryophytes. In: *Bryophyte Ecology* (ed. A. J. E. Smith) pp. 45-104. Chapman & Hall, London.

Ramsay H. P. (1979) Anisospory and sexual dimorphism in Musci. In: *Bryophyte Systematics* (eds G. C. E. Clark & J. G. Duckett) *Systematics Association Special Volume 14* pp. 281-316. Academic Press, London.

Ramsay H. P. (1984a). Census of New South Wales mosses. *Telopea* **2,** 455-534.

Ramsay H. P. (1984b). Phytogeography of New South Wales mosses. *Telopea* **2,** 535-47.

Ramsay H. P. (1985) Cytological and sexual characteristics of the moss *Dicranoloma.* In: *Studies in Botany dedicated to Lewis E. Anderson* (ed. M. R. Crosby) pp. 93-110. Monographs in Systematic Botany, Missouri Botanic Garden. **11.**

Ramsay H. P. & Berrie G. K. (1982) Sex determination in Bryophytes. *J. Hattori Bot. Lab.* **52,** 255-74.

Ramsay H. P. & Vitt D. H. (1984) Distribution, cytotaxonomy and sexuality of *Macromitrium* in the Pacific. In: Proceedings of the World Conference on Bryology, Tokyo. *J. Hattori Bot. Lab.* **55,** 23-54.

Ramsay H. P., Streimann H. & Harden G. M. (1987) The Bryoflora of Australasian Rainforests. *Symp. Biol. Hungarica* **35,** 605-20.

Reese W. D. & Mohamed H. (1985) A synopsis of *Calymperes* (Musci: Calymperaceae) in Malaysia and adjacent regions. *Bryologist* **88,** 98-109.

Reese W. R., Mohamed H. & Mohamed A. D. (1986). A Synopsis of *Mittythyridium* (Musci: Calymperaceae) in Malaysia and adjacent regions. *Bryologist* **89,** 49-58.

Richards P. W. (1984) The Ecology of Tropical Forest Bryophytes. In: *New Manual of Bryology* (ed. R. M. Schuster) pp. 1233-70, Vol. 2. Hattori Botanical Laboratory, Nichinan.

Schofield W. B. (1974) Bipolar disjunctive mosses in the Southern Hemisphere, with particular reference to New Zealand. *J. Hattori Bot. Lab.* **38,** 13-32.

Schofield W. B. (1985) *Introduction to Bryology.* Macmillan, New York.

Schuster R. M. (1986) Referee's evaluation of Keto A. & Scott K. *Tropical Rainforests of North Queensland. Their Conservation Significance.* Special Australian Heritage Publication No. 3. Australian Government Printing Service, Canberra. pp 159-63.

Scott G. A. M. & Stone I. G. (1976) *The Mosses of Southern Australia.* Academic Press, London, New York, San Francisco.

Smith J. M. B. (1980) Origins, affinities and distribution of the high altitude flora. In: *The Alpine Flora of New Guinea* (ed. P. van Royen) pp. 111-32. Vol. 1. Cramer, Vaduz.

Stone I. G. (1976) A remarkable new moss from Queensland, Australia — *Viridivellus pulchellum* new genus and species (new family Viridivelleraceae). *J. Bryol.* **9,** 21-31.

Stone I. G. (1983a) *Buxbaumia* in Australia, including one new species, *B. thorsborneae. J. Bryol.* **12,** 541-42.

Stone I. G. (1983b) *Fissidens gymnocarpus,* a new species from Queensland, Australia. *J. Bryol.* **12,** 553-57.

Stone I. G. (1984a) *Uleobryum curtisii* sp. nov. (Pottiaceae) from Queensland, Australia. *J. Bryol.* **13,** 19-24.

Stone I. G. (1985). New records of mosses in Australia. *J. Bryol.* **13,** 475-78.

Tracey J. G. (1982) *The vegetation of the humid tropical region of North Queensland.* CSIRO, Melbourne.

Tracey J. G. & Webb L. J. (1975) *Vegetation of the humid tropical region of North Queensland.* CSIRO Aust., Long Pocket Labs. Indooroopilly, Qld. (maps).

Une K. (1985) Sexual dimorphism in the Japanese species of *Macromitrium* Brid. (Musci: Orthotrichaceae). *J. Hattori Bot. Lab.* **59,** 487-513.

Vitt D. H. & Ramsay H. P. (1985a) The *Macromitrium* complex in Australasia (Orthotrichaceae: Bryopsida). Part 1. Taxonomy and Phylogenetic relationships. *J. Hattori Bot. Lab.* **59,** 325-451.

Vitt D. H. & Ramsay H. P. (1985b) The *Macromitrium* complex in Australasia (Orthotrichaceae: (Bryopsida)) Part II. Distribution, Ecology and Paleogeography. *J. Hattori Bot. Lab.* **59,** 453-68.

Watson E. V. (1971) *Structure and Life of Bryophytes.* Hutchinson University Library, London.

Watts W. W. & Whitelegge T. (1902) Census Muscorum Australiensium. *Proc. Linn. Soc. N.S.W.* **27,** Supplement: 1-90.

Watts W. W. & Whitelegge T. (1906) Census Muscorum Australiensium. *Proc. Linn. Soc. N.S.W.* **30,** Supplement 91-163.

Webb L. J. & Tracey J. G. (1981) Australian Rainforests: Patterns and Change. In: *Ecological Biogeography of Australia* (ed. A. Keast) pp 605-94. W. Junk, The Hague.

Wijk R. van der, Margadant W. D. & Florschutz P. A. (1959-1969) *Index Muscorum* (5 Vols.), Int. Bureau Plant Taxonomy, Utrecht.

Zanten van B. O. & Pocs T. (1981) Distribution and Dispersal of Bryophytes. *Adv. Bryol.* **1,** 479-562.

Wood Anatomy of *Idiospermum australiense* (Diels) S. T. Blake

D. B. FOREMAN

Royal Botanic Gardens and National Herbarium of Victoria, South Yarra, Vic. 3141

Following the rediscovery in 1971 of *Calycanthus australiensis* Diels in the Daintree region of north-eastern Queensland, after it had been implicated in the poisoning of a number of cattle, there has been keen interest in the plant as it appeared to occupy an interesting position amongst the putatively primitive angiosperms.

Two features of *Calycanthus australiensis* which were said to be distinctive were firstly the peculiar embryo with its 3-4, quite distinct, massive, sub-sessile cotyledons and secondly the wood anatomy which amongst other features was said to have wood with vessel elements with multiple scalariform perforations.

Primarily on the basis of these two features S. T. Blake (1972) described the new genus *Idiospermum* and placed it in a new family the Idiospermaceae.

Lamont (1974) reported briefly on the genus calling it a living primitive angiosperm. According to this report the Idiospermaceae possessed 13 primitive characters including the feature "vessels with scalarifom end-plates and side walls".

Wilson (1979) described various aspects of the vegetative anatomy of *Idiospermum* and was unable to confirm the presence of multiple scalariform plates which had been reported previously. Wilson reported finding exclusively simple perforation plates in the material he examined. These are somewhat similar to those found in *Calycanthus*.

Cronquist (1981) places the Idiospermaceae in his Laurales but appears to have overlooked the earlier work of Wilson and states the vessel segments of *Idiospermum* have scalariform perforation plates.

In mature wood of *Idiospermum australiense* I confirmed that vessel elements with simple perforation plates having moderately oblique to almost transverse end walls are found exclusively. In younger stems, however, vessel elements with perforation plates with about five bars (i.e., scalariform) have been observed as well as simple perforation plates. The original report of multiple scalariform plates was probably due to the misidentification of the fields of scalariform pitting in the side walls of the vessel elements. Other members of the Laurales such as *Palmeria scandens* F. Muell. (Monimiaceae) also have the combination of simple perforation plates and scalariform pitting in the side walls of the vessels elements.

References

Blake S. T. (1972) *Idiospermum* (Idiospermaceae), a new genus and family for *Calycanthus australiensis*. *Contr. Qld. Herb.* **12,** 1-37.

Cronquist A. J. (1981) *An Integrated System of Classification of Flowering Plants*. Columbia University Press, New York.

Lamont B. (1974) *Idiospermum* a Living Primitive Angiosperm. *Search* **5,** 535.

Wilson C. L. (1979) *Idiospermum australiense* (Idiospermaceae) – Aspects of Vegetative Anatomy *Amer. J. Bot.* **66,** 280-89.

Proc. Ecol. Soc. Aust. 15:283-296

Initial effects of Tropical Cyclone 'Winifred' on forests in north Queensland

G. L. UNWIN[1], G. B. APPLEGATE[2], G. C. STOCKER[1] and D. I. NICHOLSON[2]

[1]*CSIRO Tropical Forest Research Centre, Atherton, Qld. 4883*

[2]*Queensland Department of Forestry, Atherton, Qld. 4883*

Abstract

Tropical cyclone 'Winifred' crossed the coast of north Queensland on the evening of 1st February, 1986 causing extensive damage to coastal forests, crops and property. The authors inspected damaged coastal and subcoastal forests during the first 10 days and again at 40 days following the disturbance. This paper describes observed initial effects of the cyclone and discusses some likely consequences within the diverse range of closed and open forest types of the affected region.

Forest damage was classified in four categories defined according to scale and intensity of tree injury. Patterns of damage in these classes are recorded by physiographic location and by forest type, with some site and species observations listed as examples. In rainforest, data on initial stem damage and tree deaths were obtained from permanent plots previously established by forest research agencies.

Patterns of forest damage were influenced by location in the cyclone's path, topographic exposure, characteristics of the forest type and site conditions such as soil drainage. The effects of previous disturbance e.g., past cyclone damage or logging were difficult to determine in the absence of replicated comparative data. Large gaps caused by previous cyclones were not extended.

Although vegetative regrowth was rapid and widespread, fires occurred later in some disturbed coastal forests, including fire sensitive rainforests and palm forests. Such fires in closed forests are an effective invitation to grass and weed invasion. In this and other humid tropical regions, the periodic disturbances of cyclone, drought, fire and man appear most potent to the distribution and succession of forest types when operating in close association. The initial impact of cyclone 'Winifred' is a timely reminder of irregular catastrophic events which influence the nature and distribution of Australia's tropical forests.

Introduction

Severe tropical cyclone 'Winifred' (central pressure 957 hPa), crossed the coast of north Queensland just to the south of Innisfail between 1800-1930 hours on the 1st of February 1986. Hurricane force winds which had an estimated maximum gust speed of 176 kmh^{-1} (Anon. 1986) affected large areas of forest from Cairns to Cardwell and Ravenshoe (Fig. 1). Cyclones of this severity have been recorded on several previous occasions and long term residents equated this catastrophe with the cyclones of 1956 and/or 1918. (The meteorological record of Lourensz 1981 indicates that in the Innisfail area, the 1956 occurrence was less severe than 'Winifred', and the 1918 cyclone more severe.)

Though relatively infrequent on human time scales such disturbances are likely to exert important influences on tropical forest dynamics (Whitmore 1974). There are therefore both immediate and longer term consequences for the related practices of forest management and forest conservation. 'Winifred' provides an opportunity to observe the immediate vegetation effects and subsequent recovery processes associated with this type of catastrophe. This information supplements existing knowledge about environmental mechanisms that control the structure, floristics and functioning of the region's forests.

The aim of this paper is to present an initial assessment of the nature and extent of cyclone damage to forests and to some individual tree

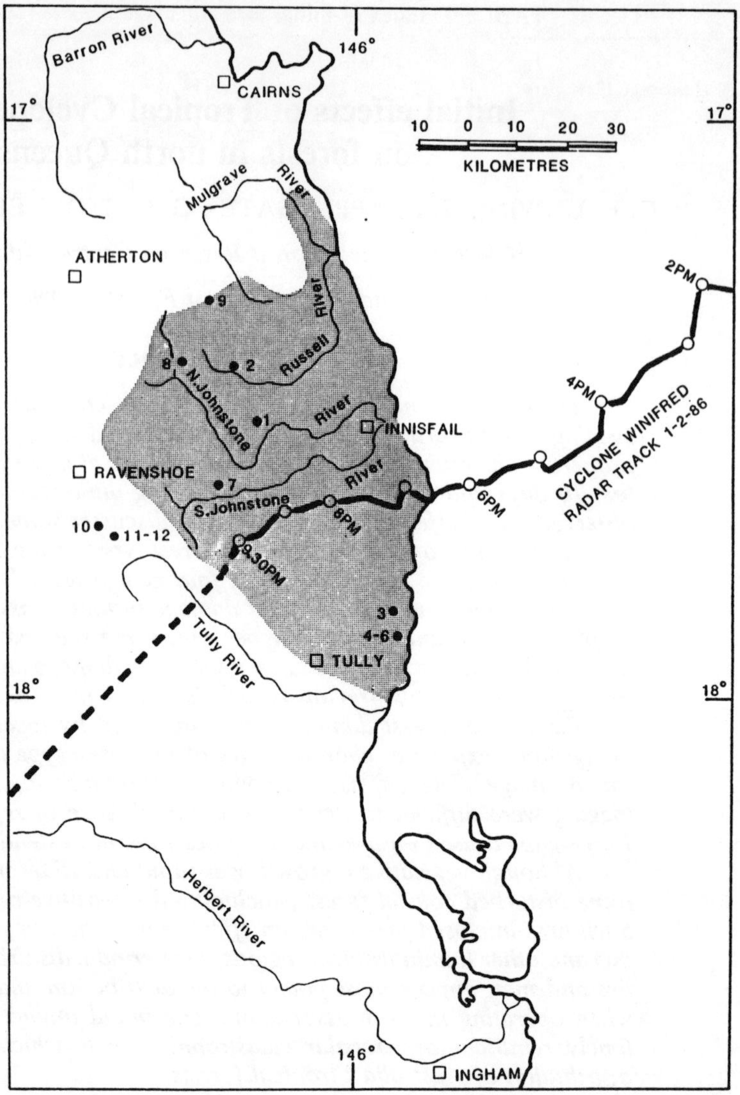

FIG. 1. Map showing the reported track of Cyclone 'Winifred' across the seasonally wet tropical coast of north Queensland on 1st February, 1986. Landfall occurred at about 1845 hours following late deviation to the west. Permanent sample plots in rainforest are indicated by closed circles numbered 1-12 and the post-cyclone distribution of moderate and severe forest damage is shaded.

species. Some longer term consequences are briefly discussed. Preliminary observations on such an extensive scale were, of necessity, qualitative. Where the cyclone impinged on an existing framework of permanent forest plots (for which pre-cyclone conditions were recorded), quantitative data were obtained.

Previous observation (e.g., Stocker 1976; Webb 1958), indicated a paramount need to document observed initial effects of the cyclone in the most effective manner immediately available. Prompt description of forest damage patterns across the whole of the area then accessible was required whilst indications of causal impact remained apparent. This once-only perspective can later be used to qualify data derived in longer term quantitative assessments, e.g., using established permanent plots affected by this and other disturbances (West *et al.* 1988 this volume), and where successful, using enhanced satellite imagery of cyclone damaged landscapes (Shea 1986).

Survey Methods

During the 10 days following the cyclone, accessible portions of the affected area were traversed using available roads and forest tracks. Notes and photographic records were made to illustrate important features of forest damage. Information was also obtained from observers who overflew inaccessible areas. Ground observations were repeated at 40 days to assess short-term

TABLE 1. Index of initial forest damage incurred by cyclone 'Winifred'.

Damage Category	Definition (by essential characteristics)	Occurrence in closed forests	Occurrence in open forests
1. Severe and extensive (Fig. 2)	Boles or crowns of most trees broken, smashed or windthrown. Impact multi-directional.	Extensive among coastal forests and remnants. Many emergent trees uprooted, canopy and understorey stems broken; Canopy (incl. vines epiphytes) reduced to heavy debris.	Rarely evidenced.
2. Severe and localized (Fig. 3)	Severe canopy disruption (as above) restricted to windward aspects. Direction of destructive winds clearly identifiable.	Defoliation, tree falls and stem breakage common on windward aspects only.	As above, (transitional understories defoliated as for closed forests.)
3. Moderate Canopy disturbance (Fig. 4)	Structural injury mostly branch and foliage loss. Some treefalls; most stems erect.	Extensive areas on exposed uplands, beyond most destructive winds. Wide-spread canopy thinning, exposure of epiphytes and understorey. Heavy litter and branch fall.	Most common injury in open forests of *Eucalyptus* and *Melaleuca*, on lowlands and foothills. Widespread defoliation and loss of twigs.
4. Slight canopy disturbance	Partial foliage loss on forest edges only. Subsequent leaf desiccation and heavy litter fall. Occasional stem or branch breakage.	Common on exposed margins of cyclone affected area. Canopy interior intact.	On main ridges and spurs peripheral to Cat. 3 areas above. Foliage desiccation short-lived and inconspicuous.

vegetative recovery. In rainforest, quantitative data on initial stem damage and tree deaths were obtained from permanent plots previously established by Department of Forestry and CSIRO personnel (Fig. 1).

Forest damage was broadly indexed according to four categories defined in Table 1. Within rainforest plots, individual tree damage was classified as trees smashed (CS) or fallen (CF), with additional records of stem scar and degree of foliage loss. Tree smash included boles or crowns broken or split either by direct wind load or indirectly through impact of other falling stems. Longer term effects such as disease increase, sunscorch, the opening of light to the understorey and the survival of tree coppice and seedling regeneration are being assessed at two-year intervals.

Results

The geographic limits of moderate and severe forest damage are indicated by the shaded area of Fig. 1. Typical scenes and some damage characteristics are illustrated in Figs. 2-5. Damage descriptions, site and species observations of particular physiographic locations are listed in Table 2. Table 3 provides a brief summary of rainforest plot data derived since the cyclone. Immediate structural effects observed in particular forest types are outlined below.

Category 1 Damage: Severe and Extensive (Fig. 2)

This category describes extensive areas of forest or forest remnants (in farmland), *where either the boles or crowns of most trees were severely damaged.* In rainforests (the most widespread of the forest assemblages in this region), many of the emergent and subcanopy stems were smashed and most of the canopy vines torn free and hurled with crown debris to the forest floor. Associated open forests and woodlands (dominated by either *Eucalyptus* or *Melaleuca* species) were not so severely affected for many of the trees in these assemblages tended to lose branches, twigs and leaves rather than suffer bole damage. In severely affected transitional forest, the young rainforest understorey was noticeably more susceptible to wind damage than the emergent eucalypts. The

FIG. 2. Severe and extensive damage (Cat. 1) to coastal rainforest in Kurrimine Beach environmental park, photographed on 10th February, 1986. The smashed tree at centre is *Acacia aulacocarpa,* a common and relatively short-lived canopy emergent in cyclone-prone rainforests of the region.

FIG. 3. Severe and localized Category 2 damage of regrowth *Acacia/Eucalyptus* woodland near Cowley Beach. In this scene, one week after disturbance, most stems or crowns have been severely damaged and direction of windthrow is clearly defined. Fallen trees, canopy debris and replacement foliage have since burned in late dry season fires.

closed subcanopy virtually disappeared from many of these ecotones on exposed coastal foothills.

In coastal and subcoastal parts of the region, the area of damaged forest extended about 80 km from near the mouth of Russell River in the north to Tully River in the south (Fig. 1). Although

FIG. 4. Large canopy gap created by windthrow of strangling fig and consequent smash of other adjacent stems. This view, four months following the cyclone, illustrates recovery from moderate canopy disturbance (Cat. 3) in the unlogged rainforest Plot 1. Most trees remain erect and lost foliage has been replaced by coppice shoots.

some similarly disturbed forests were found 50 km inland, the hilly to mountainous terrain of the region provided some shelter. Even adjacent to Mission Beach in the central path of the cyclone, aspect and topographic exposure were strong determinants of the degree of rainforest damage (Plots 4-6, Table 3).

Accounting for this topographic pattern on coastal ranges and foothills, and for the area already occupied by residential or farm land, the area of rainforest moderately and severely damaged (Cats. 1-3, Table 1) is estimated at 30 000 to 70 000 ha. By comparison, this area is similar to the estimate of 46 000 ha of severe damage caused by cyclone 'Tracy' in the Darwin area in 1974 (Stocker 1976). A more precise estimate requires an objective quantitative definition of damage, comprehensive aerial surveys and detailed ground checking throughout the region of damage. Landsat photo-imagery may assist as evidence suggests that damaged vegetation can be detected by comparing "false colour scenes" before and after the cyclone (Shea 1986; Preston 1987). However, for ground surveys, rapid vegetative regrowth now confounds precise damage recording even in the most severely affected forests.

Within severely damaged coastal rainforests, a heavy accumulation of tree debris occurred in the understorey (Fig. 2). Riverine rainforests and hillside forests, dominated by *Acacia,* were also extensively damaged (Table 2). On the lowland, rainfall ranged from 200 mm to 250 mm for the 24 hours to 9.00 a.m. on 2.2.86. Stream rises were rapid and short-lived. Flooding may have protected some inundated riverine patches from peak wind velocities, though many larger riverine species were uprooted.

In some coastal areas, especially to the south of the cyclone path, beachfront foliage sometimes remained intact on trees sheltered from the strongest gusts. Subsequent scorch of complete tree crowns suggested that salt spray may have been an important influence near the foreshore. The 1.5 m tidal surge south of Innisfail approximated highest astronomical tide (H.A.T.), hence low fringing mangroves were unaffected while the foliage of taller species was scorched and thinned from the south-east. Littoral shrubs on the most exposed foredunes were severely abraded by salt and sand.

Category 2 Damage: Severe and localized (Fig. 3)

This class of damage occurred in those areas close enough to the cyclone's path to experience

TABLE 2. Most common incidence of initial forest damage by cyclone 'Winifred', listed by physiographic location, forest type and where applicable, effects on conspicuous taxa. Bracketed values refer to sites or aspects more (or less) exposed than generally evidenced. Locality names appear on Australia 1:100 000 Topographic Survey Sheets 8062, 8063, 8162 and 8163.

Site/Relief	Forest type	Initial damage category (Table 1 refers)	Effects on dominant or conspicuous taxa
A. Littoral zone, foreshores, tidal swamps and old dune swales:Cowley Beach, Kurrimine, Mission Beach including Clump and Ninny Points	Mangrove forest	4 Slight	Wind abrasion, defoliation of taller emergents only.
	Strand forest	3 Moderate	Sand abrasion, foliage scorch by salt spray (*Calophyllum inophyllum* and *Casuarina equisetifolia* resisted structural injury).
	Paperbark swamp forest	3 (2) Moderate	Bark, foliage and small branches shed from *Melaleuca* spp. canopy. Shrubby understorey stripped of foliage.
	Remnant rainforest	2 Severe localized	Foreshore rainforest patches either windthrown or defoliated by wind or saline scorch.
	Mixed acacia/eucalypt woodland	2 (3) Severe localized	Eucalypts resistant to stem or crown breakage. Low *Acacia* spp. vulnerable to stem smash.
B. Nearshore lowlands, hillsides and low coastal ranges:- Brampston Beach, Cooper Point, Bingle Bay, Clump Mt, Double Mt and Tam O'Shanter Ra.	Coastal rainforest including acacia and palm forests	1 and 2 widespread (3 where protected)	Acacia canopies smashed and stripped of foliage. Palms (e.g. *Archontophoenix, alexandrae* and *Liculala ramsayi*) resistant. *Acacia mangium* soon producing vigorous coppice regrowth.
	Open eucalypt forests	3 (2 on waterlogged or shallow soils)	Eucalypts (e.g., *E. pellita* and *E. intermedia*) remained erect, shed foliage and small branches.
C. Lowlands, foothills and adjacent ranges of North and South Johnstone, Moresby and Russell Rivers including Babinda Ck. and Mena, Liverpool, Downey and upper Jarra Ck. catchments on Walter Hill Ra.; Graham, Moresby and Seymore Ra.	Coastal rainforest often including acacias	3 (2)	Cat. 2 severe where blowdown smashed other stems. Vines, especially *Calamus* spp., well adapted. *Alstonia scholaris* open grown relics shed foliage avoiding stem or branch failure.
	Riverine rainforests	2 (3)	Flood damage included large trees uprooted. Blowdown severe on waterlogged sites.
	Rainforest/eucalypt transitions	2 (understorey) 3 (eucalypt emergents)	Transition forests reverted to open forests on exposed ridges. Regression reinforced by any subsequent fires.
	Amenity pine plantations	2 Severe localized	Late-thinned *Pinus caribaea* plantations on poorly drained sites 90% flattened.

TABLE 2. — *continued*

Site/Relief	Forest type	Initial damage category (Table 1 refers)	Effects on dominant or conspicuous taxa
D. Upper tablelands, dissected scarps and upper catchments of Beatrice, North and South Johnstone Rivers: Mt Fisher, Malanda, Topaz to upper Palmerston Hwy.	Upland rainforest	4 (3-W aspect)	Moderate canopy disturbance on exposed W margins only.
	Windbreak plantations	3 (2-W aspect)	Windbreak eucalypts (e.g. *E. microcorys*) uprooted on exposed moist basalt sites. *Araucaria cunninghamii* vulnerable to stem break. *A. bidwillii* and *Agathis robusta* showed moderate to good resistance to injury and blowdown.
	Orchards	4 (3)	Orchard species uprooted or smashed on exposed W aspects. Lychees and bananas especially susceptible.
E. Exposed highland slopes of Bellenden Ker Ra., Mulgrave and Tully R. catchments excluding Jarra Ck. (see above).	Highland rainforests	3 (2)	Foliage removal and browning strongly topographic, usually on southern and western slopes.
	Eucalypt emergents on ridges	4 (3)	As above

FIG. 5. Initial stem coppice on windthrown *Syzygium alliiligneum* 40 days following cyclone impact. Mature fallen stems such as this example near East Palmerston will subsequently die due to severe damage of the bole and root system.

hurricane force winds, but far enough removed from the eye of the cyclone for the direction of destructive winds to be relatively constant. This direction was thus clearly reflected in the pattern of destruction (e.g., on steep highland slopes north-west of Innisfail). While patches of trees on ridges and windward slopes were smashed or severely defoliated (as for Category 1 damage), they were surrounded by sheltered forest which was only slightly affected.

Another cause of geographically restricted damage was the local concentration of turbulence and gustiness within the more general wind field (Oliver 1986). This recognized feature of 'Winifred' (and other tropical cyclones) contributed to a diversity of impact on vegetation, especially where topography was also variable.

Category 3 Damage: Moderate canopy disturbance (Fig. 4)

This damage intensity was observed in areas where either species' characteristics, topography or distance from the cyclone's centre reduced the impact of strong winds sufficiently to lessen structural effects. In rainforests this damage class was characterized by a thinning of canopy and understorey foliage, the loss of some branches and vines from emergents and some treefalls, mainly of previously weakened or heavily loaded stems (e.g., with epiphytes or strangling emergents). Many of the remaining leaves were damaged to the extent that they 'browned' during the subsequent dry two weeks. Although other forest types were affected, structural damage to branches and stems was less than in rainforests. Category 3 damage was evident in most open forest or transitional assemblages of *Eucalyptus* or *Melaleuca* spp. on exposed coastal lowlands.

The North and South Johnstone Rivers bisect the forested area most affected by cyclone 'Winifred', from west to east (Fig. 1). Where these streams descend from the broken south-eastern scarp of the Tablelands, their central catchment includes steep ridges and gorges largely covered by rainforest. Damage varied from moderate canopy disturbance (Cat. 3) on exposed parts of

TABLE 3. Summary of initial tree losses by cyclone 'Winifred', in virgin, logged and/or silviculturally treated rainforest plots within the affected area. L – Logged; T – Silvic. Treated.

Plot No.	Source	Plot area (ha)	Elev (m)	Aspect	History	No. Trees/ha (≥10 cm DBH)	Initial cyclone damage % stems fallen smashed, scarred or severely leaning			Total % stems lost
							10-30 cm	30-60 cm	60 + cm	
1	EP31	0.5	80	Riverine exposed SE	Virgin	490	23.7	18.6	19.0	22.8
2	EP34	0.5	380	Protected Riverine	Virgin	590	Nil	Nil	Nil	Nil
3	431	0.2	30		L.1950 T.1964	613	13.2	21.5	Nil	14.5
4	614/1	0.2	30	V. exposed SE	L.50	695	35.7	50.0	Nil	37.2
5	614/2	0.2	30	Exposed SE	L.1950 T.1952	547	15.6	31.5	Nil	20.0
6	614/3	0.2	30	Protected SW	L.1950 T.1955/66	677	0.9	5.0	Nil	1.4
7	640	5.0	720		L.1969/80	709	4.8	4.4	2.0	4.6
8	598	2.0	520		L.1963	555	5.9	2.0	12.8	5.3
9	608	0.2	760		L.1949	697	3.5	4.5	Nil	3.5
10	609	0.2	770		L.1950	975	0.8	Nil	Nil	0.6
11	618/1	0.2	740		L.1952 1967	624	1.2	Nil	5.0	1.6
12	618/2	0.2	740		L.1967 T.1956	795	2.3	3.5	Nil	2.5

the upper Tablelands and upper Palmerston, increasing to severe but topographically localized blowdown at lower elevations (Cat. 2).

An unusual feature of this cyclone's landfall was the persistence and strength of airflow from the west, drawn in on the northern (and usually weaker) side of the track. This resulted in strong wind funnelling from west to east and from north to south (the reverse of the normal prevailing airflow), between adjacent mountain blocks and down the valleys of most coastal streams. Also the storm weakened rapidly as it moved inland. Hence the most damaging winds in subcoastal vegetation were those experienced as the cyclone approached. Patterns of moderate canopy disturbance appeared dependent on the first incidence of destructive winds rather than when the storm centre passed over the location.

Category 4 Damage: Slight canopy disturbance

This was observed mainly on the edge of the affected area or on sheltered slopes near the cyclone track (Plots 2 and 6, Table 3). Again, aspect and exposure to the direction of maximum wind strength appeared to be the main determinants of damage. This category was characterized by slight or patchy thinning of canopy foliage, 'windburn' or rainforest edges (particularly from the west), very occasional stem or branch breakage and unusually heavy litter fall.

Western aspects of the upper Tablelands (east of Ravenshoe, Fig. 1) were most affected in this manner. Foliage on exposed western edges of rainforest patches was stripped from the canopy or windburnt, with subsequent desiccation resembling patterns of peripheral frost damage. Apart from a few (overmature) treefalls and heavy accretion of leaf and branch litter, fallen vines and epiphytes, damage to stand interiors was minimal. Radiation levels in the understorey were enhanced for several months.

Discussion

Tropical cyclones are an inherent feature of subequatorial climates. Wadsworth and Englerth (1959) reported the effects of a large cyclone on Puerto Rican closed forests during 1956. Maximum wind velocities were recorded at 150-185 kmh^{-1} and forest damage extended over an area of 150 000-200 000 ha. This disturbance was of similar total area and intensity to 'Winifred'. Although damage in north Queensland illustrated some of the conclusions of the Puerto Rican study, there were some apparent differences: in the latter, aspect was irrelevant. The absence of topographic influence may be attributed to the island geography of Puerto Rico in contrast to the substantial land mass and mountain terrain of coastal north Queensland. When a cyclone (or "hurricane") passes across an island area to continue unabated, most if not all aspects are exposed in turn and regional wind funnelling is less probable. Similarly, there is little or no decline in cyclone intensity with landfall, as frequently occurs in north Queensland.

The Puerto Rican study also concluded that all-aged mixed (closed) forests frequently suffered slight and selective damage according to tree species. In the present study some floristic differences in structural resistance and breakage habit were indicated (as discussed later). However, damage patterns were influenced more by exposure and forest type (Tables 1-3).

History of disturbance

About two tropical cyclones per season are recorded in the Coral Sea adjoining this part of north Queensland. (Average data are underestimates because of under-reporting prior to 1959; see Table 3, Lourensz 1981). Fewer than half of those which track near or cross the coast, exert any destructive influence on forest vegetation. Moderate or severe forest damage has been reported or observed as a consequence of at least eight tropical cyclones in the coastal region from Mackay to Cooktown since the turn of the century. Undoubtedly a few localized examples of forest damage will have passed unreported.

The extent and severity of vegetation damage by 'Winifred' appear similar to that reported between Babinda and Innisfail in March 1918 (Anon. 1981; T. D. McGeehan, pers. comm.) and between Gordonvale and Port Douglas following the cyclone of 2 February, 1920 (Anon. 1920). Webb (1958) estimated that none of the rainforests of the lowlands and foothills of north Queensland could escape severe or general cyclone damage for more than 40 years. The reported incidence of forest damage since the 1918 event supports this view. Indeed, prior to 'Winifred', the closed forests of the Innisfail area had experienced 30 years of relative calm since previous severe cyclone damage.

There were few, if any signs that large gaps caused by previous cyclones had been extended. Most of these 'cyclone scrub' patches were left structurally unscathed by 'Winifred', apart from foliage loss from existing 'climber tower' vines and any remaining emergent trees.

Interaction with logging

There is little evidence from which to determine the effects of previous artificial disturbance on current cyclone damage. Comparisons can be made within virgin stands (Plots 1-2, Table 3) and within logged and/or treated stands (Plots 4-6). Aspect and exposure appear strong determinants of initial stem loss among both virgin and disturbed forest plots. Recent logging did not appear to predispose forests to more damage than occurred in similarly exposed unlogged forest. However, adjacent plots of logged and unlogged forest subjected to equivalent cyclonic exposure were not available for direct comparison (Table 3).

The visual impression and structural impact of the two disturbances occurring in brief succession are likely to be additive rather than compounding. Likewise, the post-cyclone opening of damaged rainforest to salvage logging has altered the nature of understorey disturbance, especially on snig tracks and landings where the soil surface has been bared of the remaining understorey and heavy burden of tree debris.

If, as suggested by Keto (1986), previous logging increases susceptibility to cyclone damage, such risk would be maximized when logging was recent or where intensities of both logging and cyclonic exposure (rain and wind combined) were high. This concurrence of disturbance patterns was found in some areas dominated by *Backhousia bancroftii* (e.g. adjacent to Plot 1; see Figs. 1, 4 and Table 3). It was observed in the moderately damaged unlogged plot of that forest type that much of the current tree smash was caused by mature canopy emergents with heavy epiphytic crowns falling on other stems of the mid-canopy, producing substantial additional tree loss by way of a 'domino effect'. Where either logging intensity or cyclone damage is abnormally severe or frequent, the effect of one disturbance predisposing susceptibility to the other is probably of slight consequence, as little remains of the mature forest structure from either perturbation.

Structure and floristics

Individual species observations are included in Table 2. Coastal eucalypts, *Melaleuca* spp., several foreshore trees (especially *Calophyllum inophyllum* and *Casuarina equisetifolia*) and most mangrove species showed excellent tolerance to cyclone impact. Little is recorded of comparative stem or crown anatomy to explain floristic differences other than general morphology of crown shape, foliage density and leaf shedding habit.

Some life-forms appeared to have withstood cyclone impact better than others. In the lowland closed forest the two palm species (*Archontophoenix alexandrae* and *Licuala ramsayi*) withstood the high winds remarkably well. These palms are often a conspicuous feature of the subcanopy (occasionally even as emergents) within the region's lowland rainforests. At sites such as Pin Gin Hill, *Archontophoenix* appears now as a prominent emergent, whereas before the disturbance, its presence was barely apparent within the canopy. (Although the palm flowered conspicuously during the three months following the cyclone, similar behaviour was observed in unaffected areas of coastal rainforest.)

Heavy vine and epiphyte loadings on emergent trees appear to have increased the likelihood of trees being toppled. While many upper canopy vines were torn from the crowns, those which have survived (mainly *Calamus* spp.) should be favourably placed to continue their growth.

Open forest species were better adapted to damage avoidance than were closed forest species, apparently by shedding foliage prior to branch and stem failure. Accordingly, exposed open forests indicated moderate canopy disturbance (Cat. 3, Tables 1 and 2) or less.

Fire

Large areas of disturbed coastal forest were at risk from fire following the cyclone. Canopy disruption increased insolation near the forest floor, and associated lower humidities and higher temperatures desiccated understorey foliage and debris, predisposing susceptibility to burning. Seasonal behaviour of moist onshore winds introduced a strongly stochastic influence on fuel condition and flammability and the immediate fire threat passed. Canopy redevelopment and litter decay gradually reduced fire risk during the rainy season.

Wind-susceptible acacias often dominate disturbed rainforest on exposed coastal ridges. Occasionally during the subsequent dry season (July to November), fires escaped cleared foothills and sugar cane fields to penetrate severely damaged strips of this forest type. On steep hillsides, fires from open forests eroded disturbed rainforest margins by 20 m or more.

By December 1986, the protracted dry spell combined with high summer temperatures to increase fuel availability in cyclone disturbed forests. Hundreds of hectares of cyclone damaged coastal forests near Brampston Beach, Kurrimine and Mission Beaches were burned. Fires in disturbed rainforest proved difficult to control and were finally extinguished by rain showers from scattered thunderstorms. Although by this time, the impact of the cyclone was difficult to determine among other seasonal factors (e.g., the late and patchy occurrence of thunderstorms), only the most cyclone disturbed closed forests were observed to sustain fires for several days (as opposed to transient burning of the edges). In times past, repeated fires in cyclone damaged rainforests probably resulted in development of grassy slopes and eucalypts, the latter now characteristic of many ridgetop assemblages in the region (Stocker & Unwin 1986).

Recovery

During the first week following the cyclone, attached foliage damaged by wind began to wilt, turn brown and senesce prematurely. Weather conditions during this interval were mostly calm, dry and hot as the resulting rain depression moved far inland to the south-west. This circumstance imposed a severe diurnal microclimate on damaged leaf and stem tissues whose water conducting system was probably disrupted by physical injury. In damaged forests, much of the remaining foliage succumbed initially through water stress and associated problems of excessive radiation load, high leaf temperature and impaired physical support.

Post-cyclone defoliation was therefore rapid and widespread across most of the severely affected region. The damaged canopy was noticeably more brown, wilted and sparse on the third and fourth days following impact compared with day one. These conditions served to increase even further, effects due to exposure and added radiation load on the heavily disturbed understorey. By contrast, little if any sunlight reached the forest floor in severely damaged rainforest because of the heavy burden of debris.

By 14 days, some species defoliated or smashed had commenced new shoot growth. Subsequently, a small number of rainforest tree species, for example, *Pseudoweinmannia lachnocarpa, Xanthostemon chrysanthus* and *Brombya platynema* flowered conspicuously during one to three months following the cyclone. (They also flowered heavily but less conspicuously beyond the cyclone affected area). By 40 days, vigorous new leaf shoots and/or stem coppice were widespread in virtually all tree species (Figs. 4 and 5), providing an overwhelming mode of short term recovery. In the severely damaged Plot 1, large overmature stems which had been smashed were the only structural group not consistently displaying some indication of initial vegetative response.

Coppice ability among rainforest species has been observed previously in response to foliage removal through fire and/or mechanical injury (Stocker 1981; Unwin *et al.* 1985). Factors which seem important in successful vegetative regrowth include age, vigour, species and nature of disturbance rather than floristics alone. Many of the cyclone damaged stems displaying initial coppice or shoots may yet succumb to disease, injury or sun scorch of the bark cambium. However, even if coppiced stems later fail, their rapid initial recovery has for a period, replaced some of the foliage lost. The most severe disruption of forest microclimate has been ameliorated.

At time of writing, the vegetative mode is the major source of early regeneration. (One exception in disturbed coastal rainforests is emergent *Acacia aulacocarpa* which commonly has died following foliage removal.) Unless the rainforest floor is cleared of most accumulated debris (e.g. by fire or physical removal such as on logging tracks and landings), opportunities for seedling establishment will depend on breakdown of abnormally heavy leaf litter. Observations of seasonal behaviour of litterfall suggest this may await warm humid conditions of the following summer wet season. Such processes will be truncated dramatically where fire clears the forest margins and understorey of new shoot growth. This would reduce further the canopy cover and facilitate the early intrusion of grasses and introduced weeds.

Conclusions

1. Cyclone 'Winifred' caused extensive damage to coastal forests of humid tropical north Queensland across an area of approximately 150 000 ha. Although broad in extent, initial patterns of forest injury appeared no more severe than those of several previous disturbances affecting north Queensland vegetation this century.

2. Patterns of forest damage were influenced by the following factors:
   (a) location and proximity to the area of maximum cyclone intensity during landfall;
   (b) topographic exposure to the transient wind field as the cyclone approached, passed inland and disintegrated. The cyclone tracked across dissected terrain close to the wettest part of Australia. Accordingly there was pronounced topographic variation in patterns of forest damage.
   (c) Structural and floristic characteristic of the forest type. Closed canopy rainforests of high tree density were damaged more severely (Damage Categories 1-3), than were open *Eucalyptus* or *Melaleuca* forests on a range of adjacent sites of equivalent exposure. Some rainforest species and life forms were more susceptible to crown breakage and stem smash than were others.
   (d) Site drainage and soil condition, in so far as anchorage was impaired. Riverine rainforest and conifer plantations appeared more susceptible to windthrow whilst waterlogged than did paperbark or eucalypt forests.
3. The effects of previous disturbance history on current cyclone damage to forests were inconclusive, largely because of the absence of replicated comparative samples of previously disturbed and undisturbed forest of equivalent site conditions and cyclonic exposure.
4. The cyclone produced abnormally high litter loads in disturbed coastal rainforests and transitions. During the following dry season, the normally fire-proof understorey provided a heavy source of dry fuel available for combustion. Fires occurred on the edges of and later within the most disturbed coastal rainforests. Historically, such fires in rainforest are recognized as a probable means of conversion to grassy open forest, especially where the effects of cyclone and subsequent drought are combined.
5. Vegetative recovery of trees defoliated by the cyclone was rapid and widespread among all forest types. In addition to direct physiological benefits likely to arise, this immediate response ameliorated the harsh forest microclimate resulting from canopy disruption of closed forests. The possible floristic impact of coppice growth and survival versus the incidence of seedling regeneration will be expressed only in longer term studies.
6. Cyclone impact on forests of Australia's seasonally wet tropics is a timely reminder that such catastrophic influences, though irregular, are an intrinsic component of the forests' climatic history. This factor is not easily accommodated in experimental design for forest management research. Nonetheless such repeated irregular disturbances ought to be recognized for both the structural resilience and regenerative responses indicated and the climatic and biological limitations they impose on forest use.

Acknowledgements

The authors acknowledge local information provided by residents and landholders of the cyclone affected area. Observations, data and comments by Mr P. Fletcher of the Bureau of Meteorology, Brisbane and Prof. J. Oliver, also of Brisbane, were invaluable in defining the physical aspects of the cyclone. We are grateful to Mr E. Mannion, District Forester, Atherton for access to plot sites in affected forests and to Messrs A. N. Gillison and P. V. Woods for constructive comments on the manuscript.

References

Anonymous (1918) Annual Report of the Director of Forests for the year ended 30th June 1918. In: *Annual Report of the Department of Public Lands for the year 1917.* Government Printer, Brisbane.

Anonymous (1920) Annual Report of the Director of Forests for the year ended 30th June 1920. In: *Annual Report of the Department of Public Lands for the year 1919.* Government Printer, Brisbane.

Anonymous (1986) *Report on Cyclone 'Winifred', February 1986.* Australian Bureau of Meteorology, Canberra.

Keto A. (1986) The devastation of cyclone 'Winifred'. Rainforest Conservation Society of Queensland. *Liane* **20,** 5-34.

Lourensz R. S. (1981) Tropical Cyclones in the Australian Region July 1909 to June 1980, *Met. Summary, Bureau of Meteorology, Canberra.*

Oliver J. (1986) Evaluating tropical cyclone Winifred, 1986, as an environmental threat. *Proc. Workshop on the Offshore Effects of Cyclone Winifred.* Workshop Series No. 7, G.B.R.M.P.A., Townsville.

Preston R. A. (1987) Use of Landsat 5 MSS imagery to map cyclone damage to rainforest in north Queensland. Paper to IVth Australasian Remote Sensing Conference, Adelaide, September 1987.

Shea G. M. (1986) *Rainforest conservation and management in the aftermath of Cyclone Winifred.* Department of Forestry, Brisbane.

Stocker G. C. (1976) Report on damage to the natural vegetation after Cyclone 'Tracy', Darwin area December 25, 1974. *Forestry and Timber Bureau, Leaflet 127.*

Stocker G. C. (1981) The regeneration of a north Queensland rainforest following felling and burning. *Biotropica* **13,** 86-92.

Stocker G. C. & Unwin G. L. (1986) Fire and the functioning of tropical plant communities. In: *Tropical Plant Communities* (eds H. T. Clifford & R. L. Specht) pp. 91-103. Botany Department, University of Queensland, St. Lucia.

Unwin G. L., Stocker G. C. & Sanderson K. D. (1985) Fire and the forest ecotone in the Herberton Highland, north Queensland. *Proc. ecol. Soc. Aust.* **13,** 215-24.

Wadsworth F. H. & Englerth G. H. (1959) Effects of the 1956 hurricane on forest in Puerto Rico. *Carrib. For.* 38-51.

Webb L. J. (1958) Cyclones as an ecological factor in tropical lowland rainforest, north Queensland. *Aust. J. Bot.* **6,** 220-28.

West P. W., Stocker G. C. & Unwin G. L. (1988) Environmental relationships and floristic and structural change in some unlogged tropical rainforest plots of north Queensland. *Proc. Ecol. Soc. Aust.* **15,** 49-60.

Whitmore T. C. (1974) Change with time and the role of cyclones in tropical rain forest on Kolombangara, Solomon Islands. *Commonwealth Forestry Institute, Paper 46.*

Proc. Ecol. Soc. Aust. 15:297-301

Recovery of subtropical rainforest following storm damage

M. OLSEN and D. LAMB

Botany Department, University of Queensland, St. Lucia, Qld. 4067

Introduction

Storms and strong winds are widely recognized as important ecological influences in tropical forests but there are relatively few accounts of their long term effects. Such accounts that do exist suggest they may influence not only the structure but also the composition of forests for long time periods (Whitmore 1975; Dittus 1985; Wyatt-Smith 1954; Anderson 1964; Lugo *et al.* 1983). Apart from Webb's (1958) description of the "cyclone scrubs" of north Queensland there are few accounts of the effect of storms on Australian rainforests.

We report here the effect of a severe storm which damaged extensive areas of sub-tropical rainforest in Lamington National Park in the McPherson Ranges of southern Queensland and on some of the early patterns of recovery.

The Storm

A severe thunderstorm passed through the region on 8 September 1983. It was accompanied by strong winds and hail (up to 10 cm diameter). There were no direct measurements of wind velocity in the Lamington National Park area itself, but wind gusts of up to 150 km hr^{-1} were recorded at Amberley, 80 km to the north. A comparison of damage to vegetation at the Park and at Amberley suggests local wind gusts at the former must have greatly exceeded 150 km/hr. The few metereological data describing the storm are outlined in Olsen & Lamb (Unpublished Report).

The storm caused severe damage to forest along a strip up to 2 km wide and stretching from Lost World on the western boundary of the Park to Mt Hobwee and the south-eastern boundary, a distance of some 15 km. The damage along this strip was variable. In places the entire canopy was destroyed, and all trees were snapped off or uprooted. The uprooting caused extensive soil disturbance. At other sites much of the canopy was destroyed but some trees remained standing with all or part of their canopy intact. In still other areas along the path of the storm the canopy remained unbroken except for isolated tree falls. There was no strong relationship between any of these classes of damage and topography although there was a tendency for the most severe damage to be found on the windward side of ridges. Olsen & Lamb (Unpublished Report) have mapped the damaged forest and estimated that 640 ha of forest were completely destroyed (i.e., total canopy removal) and a further 2 600 ha were severely damaged (i.e., considerable canopy damage but some intact residual trees).

Forest and Environment

The forests of the McPherson Ranges (latitude 28°14′S longitude 153°7′E) have been described by McDonald and Whiteman (1979). Most of the area affected by the storm they have classified as tall closed forest of the *Argyrodendron trifoliolatum-A. actinophyllum-Caldcluvia paniculosa* alliance. This corresponds with the complex notophyll vine forest class of Webb (1959). The areas affected by the storm are mostly above 850 m elevation and are found on kraznozems developed from tertiary basalts (Beckman & Thompson 1977). Rainfall in the area is about 1 880 mm and mostly falls in the summer months between November and April. Other details of the general environment in the area are summarized by McDonald and Whiteman (1979).

Study Site

Forest regrowth was monitored in a large canopy gap in an area of severely disturbed forest. The gap was about 50 m wide and 100 m long, and was about 2 km south-east of O'Reilly's Guest house near the western border of Lamington National Park. The nearest vegetation other than rainforest lies outside the Park border near the Guest house.

All of the forest canopy at the site was destroyed although some trees remained standing. Many of these residual trees had lost all of their crowns or had been severely damaged. A 50 m × 25 m plot containing 50 subplots each 5 m × 5 m was established within the gap. The long axis of the plot stretched from one side of the gap to the other such that the boundaries of the plot were within a few metres of intact forest. All plant species within the plot (except *Dianella caerula, Lomandra spicata, Hydrocotyle pedicellosa* and *Oplismenus imbecilus*) were identified and their heights measured. Seedlings were recorded as growing on undisturbed soil or bared mineral soil. Damaged plants showing signs of vegetative regrowth were recorded separately. All vine species in the plot were identified but the numbers were not counted. Instead the vine cover in each subplot was estimated on a five point scale (1= absent, 5=dense cover over whole plot).

An estimate of the original forest composition was made by identifying all logs rooted within the plot and counting the number of mature shrub and vine species present when the plot was first assessed.

The plot was first measured in January/February in 1984 (five months old) and September 1985 (24 months old).

Results

The storm occurred during the dry season and little regeneration occurred before the wet season later in the year. With the onset of the wet season there was a vigorous flush of vegetative and seedling growth which averaged 2 m height in 1985 when 24 months old and reached 4.6 m in 1986 when it was 34 months old.

Regrowth composition

No species were immediately lost from the site as a result of the storm as all the original species were present as residual trees, coppicing stumps, young advanced growth or seedlings. In fact species richness increased from 87 species prior to the storm to 100 species in 1984 and 119 species in 1985 (Table 1). These additional species included native trees, shrubs and vines as well as a number of exotic 'weed' species. One new tree species was recorded in 1984. This was *Acacia melanoxylon,* a common early colonizer in the area and seedlings were found scattered throughout the plot. In 1985 a further 10 tree species were recruited to the site as seedlings. (*Acronychia oblongifolia, Alphitonia excelsa, Euodia micrococca, Euodia* sp. nov., *Lomatia arborescens, Pentaceras australis, Pittosporum undulatum, Polyscias elegans, Polyscias murrayi,* and *Premna lignum-vitae.* Of this total of 10 new species, four are secondary species and six are primary forest species. Two of the tree species originally present appeared to have disappeared from the site in 1985 *(Clerodendrum floribundum, Endiandra muelleri)* although these may have been hidden by other regrowth.

TABLE 1. Number of species in forest and forest regrowth in 0.124 ha plot.

Number of species	Original forest (estimated)	1984	1985	Change (1984-85)
Total	87	100	119	23-4=19
Trees	44	45	53	10-2= 8
Shrubs	17	21	22	2-1= 1
Vines	26	27	34	7
Exotics	0	7	10	4-1= 3
Vegetative tree regrowth	—	35	no change	
Number of species present as seedlings				
Total	?	30	43	13
Trees		12	25	13
Shrubs		4	8	4
Vines		7	4	-3
Exotics		7	6	-1

Shrub species also increased over the period. Four new shrubs appeared as seedlings in 1984 (*Omalanthus populifolius, Trema aspera, Solanum aviculare* and *Solanum inaequilaterum*) and another two *(Cuttsia viburnea, Kunzea phylicoides)* were recorded in 1985. One original species *(Harpullia alata)* could not be found in 1985 but, like the two tree species referred to earlier, it too may have been overlooked in the regrowth.

One new vine species *(Cissus antartica)* was present as a seedling at the site in 1984 and another seven species appeared as seedlings in 1985 *(Celastrus subspicatus, Cissus hypoglauca, Rubus rosifolius, Parsonsia ventricosa, Rubus moluccanus, Ripogonum elseyanus, Pandorea pandorana).*

Seven exotic species appeared at the site in 1984 *(Cirsium vulgare, Conyza canadensis, Crassocephalum crepidioides, Gomphocarpus physcocarpa, Hypochoeris radicans, Solanum mauritianum, Sonchus oleraceus).* A further four appeared in 1985 *(Ageratum houstonianum, Eupatorium adenophorum, Phytolacca octandrum, Sigsbeckia orientalis)* but by then one of the original species *(Gomphocarpus physcocarpa)* had disappeared.

The pattern of seedling regeneration reflected these changes in total species numbers. Thirty species were present as seedlings in 1984 and this increased to 43 in 1985 (Table 1). While there was an initial input of secondary species and short lived exotic herbs in 1984, the species regenerating in 1985 were mostly primary forest trees. The numbers of tree and shrub species present as seedlings increased with time after the storm but the number of vines and exotic species present as seedlings decreased.

A striking feature of the forest regrowth was the ability of most tree species to sucker or sprout when damaged. Thirty-five tree species of the total of 44 originally present were observed to have this property. The other nine species may also have been able to regenerate in this way but were not observed to do so within the plot.

Regrowth density

Although much of the area was covered by fallen logs and a heavy layer of leaf litter, dense regrowth quickly developed. Five months after the storm there were 3.54 trees or shrubs per m^2 and this increased to 3.76 per m^2 within 24 months (Table 2). Initially new seedlings formed the largest proportion of this total density. These seedlings were predominantly *Omalanthus populifolius* but other common species were *Duboisia myoporoides, Orites excelsa, Randia benthamiana* and *Solanum inaequilaterum.* Only *O. populifolius* was found as a seedling in the disturbed soil associated with uprooted trees. In 1985 the numbers of new seedlings declined sharply.

TABLE 2. Density of trees and shrubs in forest regrowth.

Density (m^2)	1984	1985
All plants	3.540	3.757
Residual trees	0.111	0.106
Vegetative regrowth	0.424	0.424
Shrubs and advanced growth	1.175	2.952
Seedlings	1.830	0.272

The majority of the residual trees were still present after two years. (< 5% death by 1985). This was despite a brief snow storm in July 1985 which affected some tree foliage. Vegetative regrowth accounted for 0.424 plants per m^2 or 11% of the plants present.

Vine growth

Vine growth was initially very rapid. Five months after the storm only 23 of the 50 subplots were free of vines and 11 of the 50 had a dense vine cover (a rating of 4 or 5) (Table 3). Subsequent changes were more gradual. By 1985 30 plots remained unchanged, 16 plots had increased their vine cover by one density unit and four plots had increased their vine cover by two density units. In most areas covered by vines there were seedlings, coppice or residual shrubs under and emergent through this vine cover.

TABLE 3. Number of 5 m × 5 m subplots in various vine density classes. (Density class 1 = vines absent, density class 5 = vines cover whole plot).

Year	Density Rating					Total plots
	1	2	3	4	5	
1984	23	12	4	7	4	50
1985	14	14	8	5	9	50
Change	-9	2	4	-2	5	

Discussion

These early observations reveal the recovery process after the storm has been quite rapid. Interestingly, it seems few if any species have been lost from the site. In fact a significant number of new species have apparently been able to become established. Assuming a total species richness before the storm of 87 species, a further 36 species appear to have reached the site and been present at some time during the two years following the disturbance. Of the native species most of the newcomers are commonly dispersed by birds and could have been brought in from the surrounding intact forest. These new recruits may therefore contribute to the probabilistic stage of successional development in these forests referred to by Webb *et al.* (1972). But at the same time, nearly half of these new arrivals are also present in the soil seed bank in undisturbed forest adjoining the plot and were presumably also present at the study site prior to the storm (Abdulhadi, pers. comm.). Thus the soil seed bank contained six of the 11 new tree species, three of the six new shrubs species and three of the eight new vine species. Some of these soil stored species such as *Omalanthus populifolius* are probably permanent members of the soil seed bank and thus contribute to the determinate phase of post-disturbance succession described by Webb *et al.* (1972).

Most of the 11 exotic species that became established at the site are wind dispersed and are present in the agricultural land surrounding the park. However, six of these are also present in the soil seed store in nearby undisturbed forest

(Abdulhadi, pers. comm.) and are almost certain temporary colonizers of any new gaps in these forests.

It is well known that soil seed stores are dominated by early secondary species (e.g., Whitmore 1983). Notwithstanding the importance of the soil seed store however, an increasing number of the tree seedlings regenerated at the site were primary forest species. In 1984, 83% of the 12 tree species were primary forest species and, in 1985, 88% of the 25 tree species were primary forest species. These proportions are high in a gap which might be expected to favour secondary species because of its size. Several explanations seem possible. One is that some of these primary forest species were shed from residual trees within the gap. Not much is likely to have originated from these trees to date because most sustained substantial crown damage. However, virtually all the residual trees have survived and may have a disproportionate effect on seedling recruitment in the future as their crowns recover. Their survival contrasts with a report of 35% death amongst residual trees suffering minor damage after logging in Malaysia (Marn & Jonkers 1982). A second explanation is that these residual trees may have acted as important focal points for frugivorous birds and thereby hastened the arrival of new tree seed. Finally, in microclimatic terms, the "large" gap may really be the equivalent of a "small" gap in a tropical forest because of the lower solar angle. In a gap with residual trees the overall effect of these factors may be to discriminate against some secondary species.

One of the striking features of the recovery process has been the extensive vegetative regrowth on many tree species. Vegetative regrowth has been reported in other tropical and subtropical rainforests in Australia. Stocker (1981), for example, noted that many tropical rainforest species could coppice. He also reported that the early growth of these plants was slower than that of trees regenerating from seed. On the other hand, Webb *et al.* (1972) found that plants with coppice growth were taller than plants regenerating from seed. They concluded that the ability to regenerate from coppice conferred a considerable competative advantage to a plant. Whitmore (1983) has argued that where extensive vegetative regrowth occurs, secondary succession will be suppressed. At this stage it is still too early to draw many conclusions about the proportion of coppicing plants at the site that will survive, or whether their growth will be such as to dominate and exclude other species. The number of new primary forest seedlings being added to the site suggest, however, that the coppicing plants may not have the suppressing effect suggested by Whitmore.

No species were found to prefer especially disturbed sites or the bared soils associated with uprooted trees and the only species found as seedlings on these sites was *Omalanthus populifolius*. In this case a mere 16 individuals of a total population in 1984 of 1 814 seedlings occurred on mineral soil. This contrasts with the observation by Putz (1983) that such microsites were more commonly colonized in gaps in Panamanian rainforest than elsewhere. On the other hand, new microsites created by the storm that did apparently favour a particular group of species were the organic mounds formed by epiphytes brought down from the canopy. Tree species such as *Caldcluvia paniculosa, Geissois benthamii* and *Quintinia sieberi* were observed on such elevated sites. (In undisturbed forest these species are often found on tree fern trunks).

The vine species were the least affected by the storm. In fact, they may have been advantaged by this disturbance. They suffered very little defoliation and would have benefitted from the improved radiation levels and moisture regime. This early advantage appears to have diminished with time and already tree seedlings are emerging through vine tangles. It will be interesting to compare this situation with that in tropical Queensland following the passage of cyclone Winifred in January 1986 which devastated large areas of tropical rainforest. In that case vines such as *Faradaya splendida* and *Merremia peltata* now dominate the vegetation in many regrowth areas. These have much larger leaves than the subtropical vines and had an opportunity to cover large areas of growth before subsequent rain allowed seedling regeneration to occur. Observations by one of us (M.O.) suggest they may well reduce the diversity of forest regrowth for some time.

Acknowledgements

We would like to acknowledge the considerable help of Paul Grimshaw and the assistance of John Jones, David Doley and David Yates. We are also thankful for the material and financial support from the Queensland National Parks and Wildlife Service.

References

Anderson J. A. R. (1964) Observations on climatic damage in peats swamp forest in Sarawak. *Common. For. Rev.* **43**, 145-58.

Beckman G. G. & Thompson C. H. (1977) The Soils. In: *The Border Ranges: A land use conflict in regional perspective* (eds R. Monroe & N. C. Stevens) pp. 7-10. Royal Society of Queensland, Brisbane.

Dittus W. P. J. (1985) The influence of cyclones on the dry evergreen forest of Sri Lanka. *Biotropica* **17**, 1-14.

Lugo A., Applefield M., Pool D. & McDonald R. (1983) The impact of hurricane David on the forests of Dominica. *Can. J. For. Res.* **132**, 201-11.

Marn H. M. & Jonkers W. (1982) Logging damage in tropical high forest. In: *Tropical Forests: Source of Energy through Optimisation & Diversification* (eds P. B. L. Srivastava *et al.*) pp. 27-38. Penerbit Universiti Pertanian Malaysia, Serdang, Malaysia.

McDonald W. J. F. & Whiteman W. G. (1979) Moreton Region Vegetation Map Series: explanatory booklet for Murwillumbah Sheet. Botany Branch, Department Primary Industries, Brisbane.

Putz F. E. (1983) Treefall pits and mounds, buried seed and the importance of soil disturbance to pioneer trees on Barro Colorado Island, Panama. *Ecology* **64**, 1064-74.

Stocker G. C. (1981) Regeneration of a north Queensland Rain Forest following felling and burning. *Biotropica* **13**, 86-92.

Webb L. J. (1958) Cyclones as an ecological factor in tropical lowland rainforest, north Queensland. *Aust. J. Bot.* **6**, 220-28.

Webb L. J. (1959) A physiognomic classification of Australian rainforests. *J. Ecol.* **47**, 551-70.

Webb L. C., Tracey J. G. & Williams W. T. (1972) Regeneration and pattern in the sub tropical rainforest. *J. Ecol.* **60**, 675-95.

Whitmore T. C. (1975) *Tropical Rainforests of the Far East.* Clarendon Press, Oxford.

Whitmore T. C. (1983) Secondary Succession from seed in tropical rainforest. *For. Abstr.* **44**, 767-79.

Wyatt-Smith J. (1954) Storm forest in Kelantan. *Malay. Forest.* **17**, 5-11.

Proc. Ecol. Soc. Aust. 15:303-305

Forest successions following European settlement and mining in the Rossville area, north Queensland

G. L. UNWIN, G. C. STOCKER and K. D. SANDERSON

CSIRO Tropical Forest Research Centre, Atherton, Qld. 4883

Abstract

The Rossville area south of Cooktown provided an exceptional opportunity to examine the short- and long-term effects of mining and other disturbances on the interaction between closed and open forests. The present study was conducted within a compact forested mosaic on the fringe of the humid tropical region's main belt of rainforest.

The floristics of vegetation developing after disturbance depended on the composition of available seed sources and the scale and intensity of mining disturbance over the past 100 years.

All previously mined sites, other than those subjected to permanent flooding, supported vigorous regeneration of tree species. Under these particular circumstances, forest regeneration has been accelerated rather than impeded by mining disturbance.

Mining disturbance has added to the perceived shift in fire regime associated with European settlement and displacement of tribal Aborigines. The combined effect has accelerated the advance of closed forest into open forest margins. This process continues to gather momentum, further disrupting previous fire patterns. The observed succession is such that much previously open eucalypt forest of the Rossville area, particularly on upland slopes, may remain closed rainforest.

The matrix of natural and man-made disturbances and changes in human settlement and land use combines to make the Rossville area one of the most complex examples of the rainforest-eucalypt forest ecotone observed in north-eastern Australia.

Introduction

In the Rossville area, 30-35 km south of Cooktown, alluvial tin mining has been the dominant land use for almost 100 years. Mining began with European settlement, displacing an Aboriginal society of hunter-gatherers. In many parts the extent, severity and duration of the vegetation disturbances created by this dramatic change in land use are documented (e.g., Saint-Smith 1916) or can be deduced from available records and some knowledge of the various forms of mining activity. Disturbance due to mining has occurred in both rainforest and eucalypt forest, mostly along thin strips of riverine or "gallery" forest.

An area of 8×8 km surrounding Rossville and east of the upper Annan River (Lat. 15°46′S, Long. 145°16′E) was selected for a detailed vegetation survey (Stocker, Unwin & Sanderson, Unpublished Report). Heavily disturbed sites and several intact blocks of forest were included. The floristics and structure of forests on both undisturbed and successional sites were of particular interest for the area is located on the highly dynamic and little studied north-western fringe of the extensive belt of rainforest between Cooktown and Townsville. In addition to the vegetation survey, transects were used to indicate characteristics of different forest successions on sites of particular disturbance histories.

The aim of this paper is to examine the ecological and successional status of forests within the Rossville tin mining area. Transect observations are discussed briefly and summary conclusions are listed.

Methods

Survey methods included a traverse of all accessible sites using a 1 000×1 000 m grid, augmented by stereo-interpretation of full colour air photographs at a nominal scale of 1:25 000. Initial observation of floristics and forest structure were stratified intuitively for subsequent detailed study of representative locations. Physical site characteristics, canopy structural parameters (e.g., canopy height, height of emergent trees, canopy texture, species size classes) and detailed floristic composition were recorded together with evidence of type and extent of disturbance and subsequent forest regeneration.

Floristic, vegetation structural and geomorphological data were used to define 12 forest types, including closed forests, open forests, riverine galleries and transitions. Structural forest patterns were delineated on the air photo mosaic and a vegetation map of 1:25 000 scale was plotted on the existing grid. Where necessary, the precise location of closed forest-open forest boundaries was verified on the ground. In addition to 12 forest types described, two disturbed conditions (forest clearance and another indicating a windswept rainforest canopy) were also identified.

Forest successions

Two localities, one intact and one disturbed, were selected from within the mapped area to identify the influence of a defined disturbance on forest structure and floristics. The first was a 100 ha patch of rainforest on a steep western catchment which remained virtually undisturbed, except for localized sluicing of minor stream banks. The perimeter forest was mapped at 1:2 500 and vertical profiles were recorded of canopy structure and composition across the intact rainforest-eucalypt forest boundary.

Within the same area, additional transects were used to compare present day vegetation along mined and unmined tributaries of Leswell Creek. Mining disturbance (*ca.* 1915) was locally severe but spatially restricted to manual removal and stacking of stream-bed materials. Plant species lists and size class distributions of the mined stream-bed indicate a substantial recovery of floristics and structure when compared with the undisturbed riverine forest.

By contrast, the second, intensively surveyed area had suffered repeated disturbance by different types of mining and associated activity (e.g., timber-getting for fuelwood and fluming). At Rossville itself, disturbance has varied from localized turning of streambanks to extensive dredging of alluvial deposits and in places, removal of the whole riverine complex.

Results and Discussion

The present distribution of forest types was found to be complex. The survey area contained many disturbed examples of the closed forest-open forest ecotone, similar in structural variation to forest boundary conditions studied elsewhere in north Queensland (Unwin *et al.* 1985; Stocker & Unwin 1986). Even in locations not directly affected by mining, local topography, forest fuel conditions and fire patterns have strongly influenced the structure and location of ecotonal forest types.

We interpret the observed advance of the intact rainforest edge into open eucalypt forest as evidence of a dynamic release of five sensitive rainforest species following demise of traditional Aboriginal land use. Although this trend may have begun before European settlement and mining, it appears to have been maintained by the subsequent change in fire regime. For many years, mining disturbance has deflected recurrent fires of low to moderate intensity away from the rainforest-eucalypt forest transition. Domestic grazing has served to reinforce protection of the expanding forest ecotone by reducing the amount of grass fuels and favouring the development of woody shrubs and tree regeneration.

The overall evidence and reconstruction of mining history from available records indicates that natural regeneration foliowed removal of the source of physical disturbance, on almost all previously mined sites. The availability of propagules appeared to be the most important factor in controlling the floristics of initial regeneration following severe disturbance. Rainforest patches generally reverted to rainforest; ecualypt to eucalypt. At sites where propagules from both assemblages were available a hybrid assemblage developed. Examples of this mixture now aged about 70 years show signs of changing to rainforest as the eucalypts die out and are not replaced. The portal road constructed next to Leswell Creek in 1984/85 passed through this mixed species regeneration. Floristic history has been repeated as the same list of shade intolerant tree genera (e.g., *Eucalyptus, Aleurites, Alphitonia, Blepharocarya, Acacia* and many others) has colonized the new gravel embankment vigorously to sapling heights of 4.0 m, by July 1986.

Elsewhere before 1915, some of the open forest was heavily cut over for firewood. Although regeneration was typical of open forest floristics, canopy structure was unusual as the short-lived *Casuarina* component was almost as tall as the dominant eucalypts, rather than forming a distinct "layered" subcanopy. Under present conditions however, this succession appears destined to restore the more common structure of the open forest and woodland, as many of the larger regrowth *Casuarina* are dying.

In contrast to the localized primary effects of various forms of mining disturbance, secondary effects are more widespread and often less obvious. They are related to an altered fire regime, a centralized approach to human settlement and land use, and the introduction of domestic grazing animals. Immediate successions following mining have done little to thwart the general advance of the closed rainforest, reflecting a shift in the fire regime from that established by the Aborigines. Many forms of mining have aided the rainforest advance by creating fire-proof shelters for regeneration. There are also longer term, probably cyclic, changes which may be linked with cyclone/fire interactions (West *et al.* 1988; Unwin *et al.* 1988).

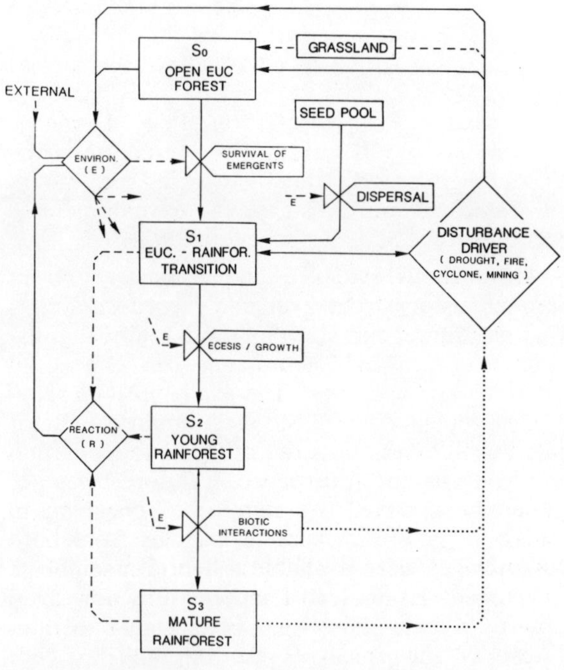

FIG. 1. A model of the change in status of the eucalypt forest margin during succession to rainforest. Boxes S_0 to S_3 represent state of the margin at particular points in time. Diamonds are components of the environment including disturbance events. Control gates (X) indicate groups of processes each of which is influenced by the environment of the margin. At present the reversion of S_2 and S_3 to S_0 is extremely infrequent except in very small scale changes to the margin. The figure has been adapted from the general successional model of MacMahon 1979.

McMahon's (1979) successional model provides a useful analogue for the trend observed in this study (Fig. 1). The model has been adapted to accommodate apparently cyclic changes within each forest type (S_0 to S_3), as well as the major succession of present interest i.e., eucalypt forest to rainforest via an expanding transition. The model allows a possible but presently infrequent return to open forest in marginal exposed areas. Intense environmental disturbance on the rainforest edge would be required to destroy the closed forest canopy and encourage repeated fire and invasion by grasses. Only this concentration of recurrent disturbance is likely to reverse the present widespread advance of closed forest.

Acknowledgements

The authors are grateful for the assistance provided by the Shell Company of Australia, especially Messers Trevor Jones and Barry Middleton.

References

MacMahon J. A. (1979) Ecosystems over time: succession and other types of change. In: *Forests: Fresh Perspecitves from Ecosystems Analysis.* (ed. R. H. Waring). Oregon State University Press, Corvallis, Oregon.

Saint-Smith E. C. (1916) Geology and mineral resources of the Cooktown District Tinfields, North Queensland, 1914. *Queensland Geological Survey, Publ. No. 250,* Department of Mines, Brisbane.

Stocker G. C. & Unwin G. L. (1986) Fire and the functioning of tropical plant communities. In: *Tropical Plant Communities* (eds H. T. Clifford & R. L. Specht) pp. 91-103. University of Queensland, Brisbane.

Unwin G. L., Applegate G. B., Stocker G. C. & Nicholson D. I. (1988) Initial effects of tropical cyclone 'Winifred' on forests of north Queensland. *Proc. Ecol. Soc. Aust.* **15,** 283-96.

Unwin G. L., Stocker G. C. & Sanderson K. D. (1985) Fire and the forest ecotone in the Herberton Highland, north Queensland. *Proc. ecol. Soc. Aust.* **13,** 215-24.

West P. W., Stocker G. C. & Unwin G. L. (1988) Environmental relationships and floristic and structural change in some unlogged tropical rainforest plots of North Queensland. *Proc. Ecol. Soc. Aust.* **15,** 49-60.

Proc. Ecol. Soc. Aust. 15:307-311

Plant succession under repeated disturbance: establishment and growth of strandline forbs and creepers on unstable beaches in the wet tropics

R. BUCKLEY

Centre for Resource and Environmental Studies, Australian National University, Canberra, A.C.T. 2601

Abstract

This study examined the alternate growth and disturbance of wet-tropical strandline vegetation from two perspectives: disturbance ecology *and* coastal processes.

This strandline vegetation undergoes a seasonal sawtooth succession of disturbance and recovery. It has two main components. Runners from mature creepers change considerably in spatial extent but little in assemblage composition. Small herbaceous species and free seedlings establishing directly from seed, however, differ greatly in abundance and assemblage composition between sites and seasons, depending largely on propagule availability. Post-disturbance recovery paths thus contain a conservative component and a random component.

The seasonal patterns of plant colonization and destruction, coupled with the seasonal changes in wind and wave regime, produce seasonal pulses in sand movement: physical coastal processes are modified by biological processes as well as vice versa.

Introduction

Scope, site and environment

This study examines the dynamics of terrestrial herbaceous plants on sandy strandlines of islands in the northern sector of the Great Barrier Reef Province, Australia: north of 15°S. It is also generally applicable to calcareous sandy beaches on the adjacent mainland of eastern Cape York. Most of the experimental work described here was carried out on Lizard Island and Eagle Cay between March 1979 and August 1981, supplemented by observations during short visits to over 50 other islands in June and December 1979.

Mean annual rainfall in this area ranges from 1 250 to over 2 000 mm a^{-1} and mean maximum January temperatures from 30-33°C. Tidal range varies from 1.8-2.5 m and median wave height is 0.5 m. Storm surges and cyclone generated waves can reach 3 m in height. There is strong seasonal variation in wind and wave energy, with the southeasterly winter trade winds reaching 30-40 km hr^{-1} from March to November, and calms and occasional cyclones from December to February.

The strandline zone is a straightforward concept for practical field use, but hard to define with precision. For the present purpose it may be taken as that sector of the shoreface zone which has been reached by swash at least once during, say, the past three years. This definition includes areas of accreting sediment which were deposited or reworked by swash recently, but which are beyond its present reach.

Disturbance ecology: Recovery after repeated disturbance

During recent decades considerable attention has been focused on the role of disturbance in shaping biotic assemblages. Earlier classifications of communities as seral or climax have been replaced by the field of disturbance ecology, which recognizes the wide range of different disturbance types, regimes, and effects, and which is attempting to establish a general framework of disturbance-response patterns from analysis of numerous individual systems.

One general problem within this framework is as follows. If a biological community is subjected to a regime of repeated similar disturbances of a particular type, does the community follow the same successional recovery path after each disturbance, or is the effect of each successive disturbance modified by the previous ones? And if so, are such cumulative effects greater for severe or minor disturbances, frequent or infrequent, and so on? And by what mechanisms do they operate?

Coastal processes: Sand movement and plant growth

From a coastal process viewpoint, the strandline vegetation forms a trap for sand in its movement in the shoreface zone; either parallel to the shore or perpendicular to it, between nearshore zone, beach and aeolian foredune systems. The physical processes of sand movement by swash and wind affect the biological process of plant establishment and growth by providing new substrate for colonization; the plant cover affects the physical processes by hindering aeolian transport. But which processes are dominant and how do they interact? This has major implications for the stability of small sand cays such as those in the Great Barrier Reef, for example: does the initial establishment of vegetation increase the chance that the island — and hence the plants — will persist, or not?

Methods

The principal technique employed was the repeated monitoring of fixed transects across the shore zone and fixed plots on the strandline, plotting the precise position of each individual plant (over 1 000 in all) at intervals of days, months and years.

I also measured seedling growth and creeper elongation rates over various periods, for self-sown and planted individuals of various sizes; and examined plant root systems in relation to moisture/salinity profiles and soil nutrient accumulation.

Results

The full results of these investigations are voluminous and are therefore not given here in detail. Some examples are shown in Figures 1 to 3: changes in seedling populations in three of the fixed plots in Figure 1, and changes in vegetation patterns in two sectors of one transect in Figures 2 and 3. Patterns were recorded at intervals from April 1979 to August 1981: these figures show results for April 1979, December 1979 and April 1980. Overall, results may be summarized as follows.

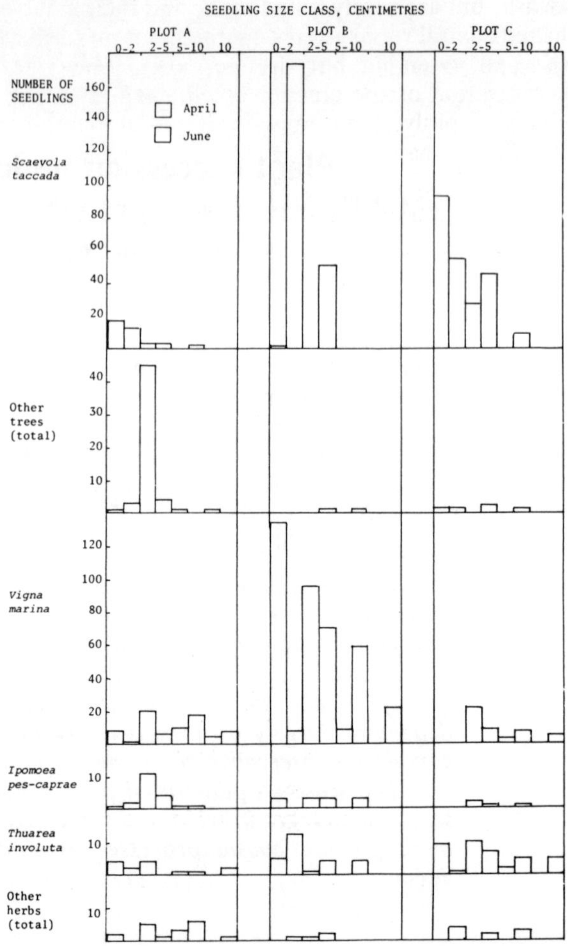

FIG. 1. Example of changes in strandline seedling populations with time: Data from three adjacent 2 m^2 fixed plots, April-June 1979.

Colonizing species of sandy strandline areas may usefully be considered in two main classes. Mature plants of species such as *Canavalia maritima, Vigna marina, Ipomoea pes-caprae* and *Spinifex longifolius,* established at the upper margin of the strandline zone, can extend new runners over bare sand very rapidly. These runners may be dislodged and swept back by swash, but are rarely killed by it; and their established central root systems enable them to survive through extended hot dry periods. Hence the composition of the creeper assemblage does not change rapidly, even though the spatial extent of its runners may.

Other species, which establish directly from seed, are much more vulnerable to wind, water and desiccation, and have a high seedling attrition rate. This group includes seedlings of characteristic strandline-fringing trees such as *Scaevola taccada, Tournefortia argentia, Guettarda speciosa, Casuarina equisetifolia* and *Terminalia* species. It also includes shortlived forbs such as *Salsola kali* and *Euphorbia* species, and new seedlings of the creeping species listed above. Tree seedlings generally do not survive to maturity unless they establish above the uppermost swash limit. For seaborne seeds, this will generally occur only on strandline sectors that are prograding or accreting, at least to some degree. Short-lived herbaceous species, on the other hand, need only survive long enough to flower and fruit in order to contribute to future generations. These species often form the main component of the strandline plant cover during periods when tidal range and wave energy are low, but are easily removed by storm swash at high tides.

In addition to these two main groups there are a few strandline species with different life history characteristics: for example, sedges such as *Remirea maritima,* and succulents such as *Sesuvium portulacastrum*. These may be abundant locally, but are not major components of the strandline vegetation overall.

Elongation rates for the strandline creepers vary widely. Most of these species have three broadly defined growth modes: an extension or

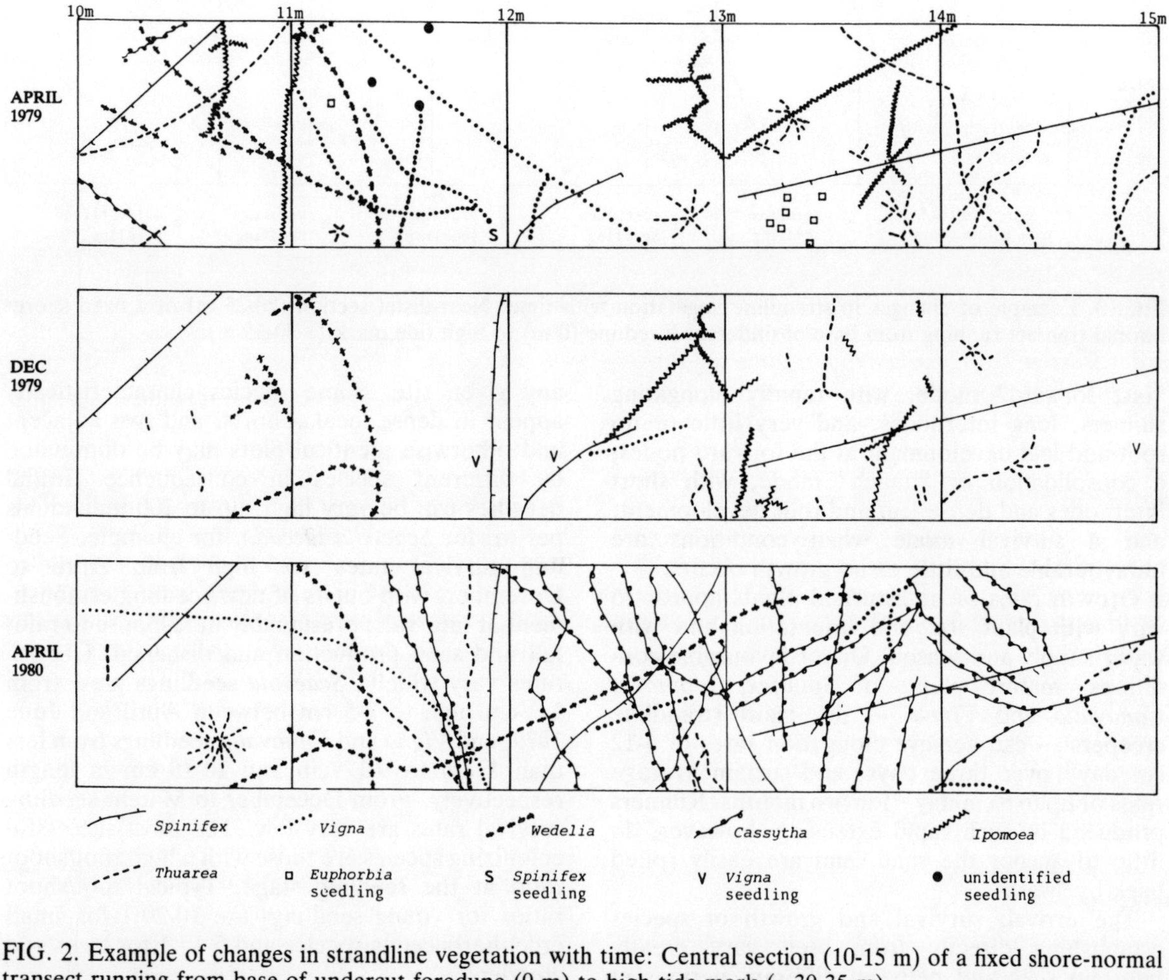

FIG. 2. Example of changes in strandline vegetation with time: Central section (10-15 m) of a fixed shore-normal transect running from base of undercut foredune (0 m) to high tide mark (~30-35 m).

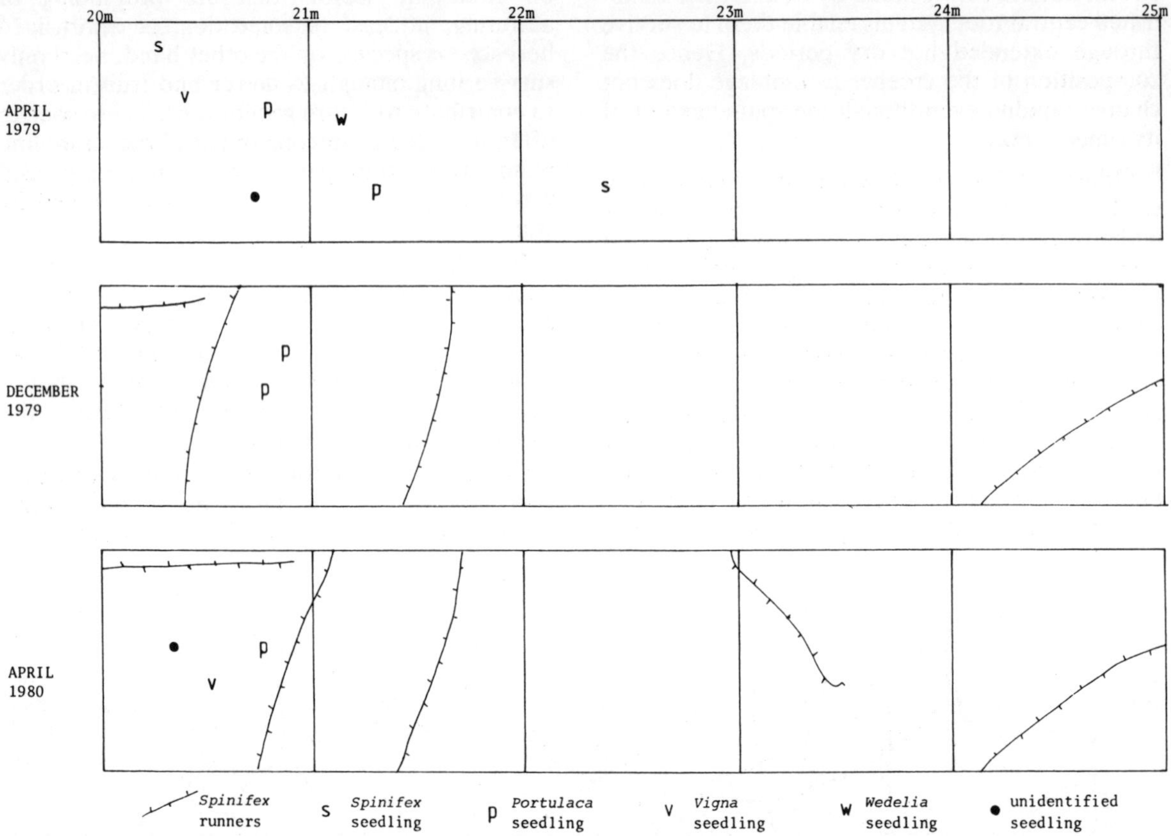

FIG. 3. Example of changes in strandline vegetation with time: Near-distal section (20-25 m) of a fixed shore-normal transect running from base of undercut foredune (0 m) to high tide mark (~30-35 m).

"fast forward" mode, with rapidly elongating runners, long internodes, and very little or no root and leaf development at the forward nodes; a consolidation or "bunch" mode, with short internodes and dense leaf and root development; and a survival mode when conditions are unfavourable and little or no growth occurs.

Growth rates of undisturbed plants appear to vary with plant size and orientation and with water supply and season. Under favourable conditions, mature plants of *Spinifex, Ipomoea, Canavalia* and *Vigna* — the main strandline creepers — can achieve elongation rates of 5-12 cm day^{-1} over three days, and sustain average rates of up to 5 cm day^{-1} for two months. Runners produced by such rapid extension, however, do little to anchor the sand, and are easily rolled back by swash.

The arrival, survival and growth of species establishing directly from seed vary greatly between sites and between different periods at any given site. Some species characteristically appear in dense local cohorts, and two adjacent and otherwise identical plots may be dominated by different species in consequence. Initial densities can be very high: up to 100 individuals per m^2 for *Scaevola taccada,* for example. Seedling survival rates are high from April to December, with bursts of new seedling establishment at intervals; presumably in response to rainfall and seed production and dispersal. Growth rates vary widely: *Scaevola* seedlings grew from 1-2 cm high to 2-5 cm between April and June 1979, and *Vigna* and *Canavalia* seedlings from less than 1 cm to 9-12 cm and 10-15 cm in length respectively. From December to March, seedling survival rates are very low. The most successful colonizing species are those with a high root/shoot ratio at the seedling stage. Typical root/shoot ratios for young seedlings are 10-20:1 for small erect herbaceous species and 5-10:1 for trees and creepers.

A number of relatively fine differences in habitat preference were noted. Most of the strandline creepers, for example, seem to grow equally well on pure calcareous sand, as on the cays, or on mixed calcareous-siliceous sand, as on Lizard Island or mainland beaches. The stoloniferous grass *Thuarea involuta,* however, is distinctly more common on mixed-sand beaches. Differences in propagule dispersal also seem to influence the distribution of species between island beaches: *Spinifex longifolius,* for example, is generally either abundant or entirely absent, perhaps because its large fruit heads cannot survive prolonged immersion and are too large to be carried between islands by wind or birds.

Conclusions

1. The study contrasted the very different strategies of different strandline plant species, and in some cases revealed fine differences in habitat preferences.
2. It showed that vegetation on the tropical beaches studied undergoes a seasonal "saw-tooth" succession — increasing in cover and complexity during the lower-energy summer season when winds are gentle (unless cyclones occur), and then being cut back heavily by the stronger waves associated with the south-easterly trade winds. The degree to which it is cut back varies from year to year.
3. The successional recovery path after each disturbance may differ substantially between different sites and disturbance events, but this is due principally to random differences in the assemblage of seaborne plant propagules available for new colonization which modify the assemblage carried over from the pre-disturbance vegetation.
4. The seasonal pattern of plant colonization and destruction, coupled with the seasonal changes in wind and wave regime, produce seasonal pulses in sand movement: physical coastal processes are modified by biological processes as well as *vice versa.*

Acknowledgements

Field work on Lizard Island and nearby reef islands was carried out from Lizard Island Research Station. Field work on the more remote islands was carried out from the M.V. *Forth*, chartered by the Australian Institute of Marine Science. The author was supported by a CSIRO Postdoctoral Fellowship at AIMS in 1979 and by a Rothmans Research Fellowship at the Australian National University in 1980 and 1981.

Proc. Ecol. Soc. Aust. 15:313-326

Conservation and management of Australia's tropical rainforests: local realities and global responsibilities

D. S. CASSELLS[1], M. BONELL[2], D. A. GILMOUR[3] and P. S. VALENTINE[2]

[1]*Department of Ecosystem Management, University of New England, Armidale, N.S.W. 2351*

[2]*Department of Geography, James Cook University of North Queensland, Townsville, Qld. 4811*

[3]*Nepal-Australia Forestry Project, Australian National University, Canberra, A.C.T. 2600*

Abstract

The wet tropics of north-eastern Australia have experienced and indeed to some extent continue to experience many of the development pressures currently being felt elsewhere in the developing tropical world. In particular, the rainforests of the wet tropical coast and hinterland between Townsville and Cooktown have historically been viewed largely as a barrier to development and closer settlement, and until quite recently, the community in general gave little thought to their conservation and management. However, in the past 10 to 15 years, there have been dramatic changes in community attitudes and the rainforests are now highly valued for their intrinsic qualities.

This paper outlines the historical development of land-use policies and practices that have influenced patterns of exploitation and conservation management in the north Queensland rainforest. A number of contemporary resource management issues are discussed and some future policy directions are suggested. The need for Australia to place the conservation and management of its tropical rainforests in the context of the wider global changes to the tropical moist forest biome is stressed. A proposal for a Rainforest Research Institute formally linked to a Tropical Forest Management Centre in north Queensland is also discussed.

Introduction

The wet tropics of north-eastern Australia have experienced and indeed to some extent continue to experience many of the development pressures currently being felt elsewhere in the tropical world. As such, the experience with the development of land management policies and practices in this region provides an appropriate focus for considering some of the future problems and possibilities for the conservation and management of the tropical rainforest biome as a whole. In this paper we will consider the conservation and management of the north Queensland rainforest in the context of the global changes to tropical rainforests. In our discussions we will seek to clarify what we see as some of the important implications of the terms "conservation" and "management". We will also seek to outline our perceptions of the global problem of tropical deforestation. We will then review the historical factors that have influenced past and contemporary rainforest management in north-eastern Australia as a precursor to discussing future options and directions for rainforest conservation and management in this region.

The focus of our discussions is the rainforests of the wet tropical coast and hinterland between Townsville and Cooktown. This particular region represents the largest contiguous area of rainforest in the Australian tropics. It is also the area that has experienced the most intense pressure for economic development and land use change since European settlement in northern Australia (Courtenay 1982). The authors have also had the bulk of their research and land management experience in this region. We therefore feel more confident of constructively contributing to the current policy debate on future directions for the management of this region's resources than we would elsewhere. However, this focus of attention on the wet tropical coast does not imply any judgement that other areas containing tropical rainforest communities, such as the Mackay/Proserpine region or the extensive closed forests in the more seasonal tropical environments north of Cooktown, are in any way less worthy of conservation and management. Indeed, we feel that the general thrust of our arguments about the wet tropical coast could readily be extended to these areas.

Conservation and Management: A Perspective

Considerable confusion exists in both the academic literature and wider community with regard to the meaning of the term "conservation" in the context of land management and land use planning. Austin & Miller (1978) have suggested that much of the confusion about the term results from the apparently antagonistic "utilization" and "preservation" aspects of the term. Similarly, Usher (1973) has also noted these different strands of conservation philosophy and has sought to integrate them together by equating "conservation" with "biological management" which he has suggested can take many forms ranging from "direct human interference in an ecosystem" to the "controlled neglect of a wilderness or any degree of human interference between these extremes". O'Riordon (1971) has similarly equated "resource management" with "the new conservation" which he has suggested tends to "emphasize rationality over emotionalism, ethics over economics and ecology over engineering".

Elsewhere, O'Riordon (1971) has noted that resource management is ultimately "a decision-making process where optimal solutions regarding the manner, timing and allocation of resources are sought within the economic, political, social and institutional framework afforded by any given culture at any particular time". Thus, all resource management problems can be viewed fundamentally as institutional problems that demand institutional responses in the form of participatory planning systems which the various competing interests can accept as legitimate, irrespective of their outcomes (Cassells & Valentine 1987). Gilbert (1984) has further noted that any "decision making process of management" should consist of a continual process of problem identification; goal definition; strategy design; implementation design; and evaluation of outcome results.

In the context of the north Queensland rainforest, we would suggest that the objectives of the maintenance of ecological processes and life support systems, the preservation of genetic diversity and the sustainable utilization of species and ecosystems defined in the World Conservation Strategy (Allen 1980) provide an appropriate starting point for the development of a strategy for conservation management. We would also suggest that the strategies to realize these objectives should be developed in the context of the wider changes and challenges facing the tropical rainforest biome on a global scale. These strategies should also be developed through participatory planning procedures that can legitimately resolve the conflicting resource use interests at the local, state and national levels.

Global Prospects for Tropical Moist Forests: An Overview

It is now widely known that the world's tropical rainforests are being rapidly exploited and that the area they occupy is contracting markedly (Whitmore 1984). However, there is considerable uncertainty with regard to the actual rate of tropical forest deforestation (Spears 1979; Grainger 1984). At the most optimistic end of the spectrum, some writers have until quite recently viewed the tropical forest resource as vast and effectively inexhaustible, with estimates that current levels of tropical hardwood exports can be maintained for periods of up to 400 years (Richardson 1970). By way of contrast however, the World Bank suggested in 1978 that at current rates of deforestation, the remaining tropical rainforests could disappear in some 60-80 years (Spears 1979). The most pessimistic scenario suggests that tropical rainforest ecosystems will have virtually disappeared from the face of the earth by the end of the current century (Richards 1973). However, despite debate over the precise rate of tropical deforestation (Myers 1980; Lanly 1981; Shepherd & Richter 1985), it is clear that substantial areas of forest are being destroyed or degraded, and that both national and international action is needed to ensure appropriate conservation and development of tropical forest resources and the enhancement of their contribution to human welfare (FAO 1985).

The consequences of this potential loss of tropical forest is seen by some as a "biological debacle" (Myers 1983a) which may mean the "demise of one million of the 5-10 million species of organism" (Myers & Ayensu 1983). Wider environmental impacts beyond the question of species loss have also been documented (e.g., Spears 1979).

Myers (1983a) has also described some of the casual agents that are leading to the current high rates of tropical deforestation. He lists the commercial exploitation of timber, the cattle rancher, the forest farmer or subsistence shifting cultivator, and the fuelwood gatherer. Many of the commercial timber operations in rainforest ecosystems in the Third World are aimed at export markets in the affluent communities of the First World, as are the products of the cattle ranchers, particularly in Latin America. Thus, many authors (e.g., de Condolle 1981; Rubinoff 1983) have drawn attention to the need for First World countries to contribute to their solution if for no other reason than their role as partial contributors to these problems.

However, while the validity of these arguments can be readily accepted, the ready identification of First World villains and Third World victims should not be allowed to divert attention from other causal agents of tropical deforestation. FAO (1985) has noted that while they account for more than half of the world's forests, tropical forests provide less than half of the world's total output of wood and only about one-fifth of the world output of wood for industrial purposes. Indeed, of the 1 400 million m^3 of wood taken from tropical forests each year, only some 200 million m^3 are used for purposes other than fuel. Thus, we note Westoby's timely reminder that on a global scale tropical deforestation is as much a problem about rural landlessness, poverty and powerlessness within Third World societies as it is about First World exploitation, and that the truly important action to save the tropical forests is political action (Westoby 1983).

The authors of this paper recognize the complexity of the social and political problems behind the current patterns of tropical deforestation and support the multi-pronged approach outlined in the FAO Tropical Forestry Action Plan. This plan aims to secure the conservation of key rainforest areas as part of a wider attack on rural poverty through sustainable patterns of forest and rural land use (FAO 1985), and we feel that governments and agencies should use the action plan as an initial common reference for both the formulation of their own tropical forest programmes, and the harmonization of action between the various national plans at an international level. We hope that this paper will help to stimulate debate about a policy response through which Australia can make a significant contribution to the solution of the wider global problem of tropical deforestation. We feel that as one of the few relatively affluent countries that possess tropical rainforests, Australia should be playing a leadership role in rainforest conservation, rainforest management and rainforest research.

Pressures on the North Queensland Rainforest: A Historical Review

A number of authors have outlined both the general history of north Queensland (Bolton 1963) and the changing community attitudes towards its rainforest environments (Cassells & Gilmour 1978; Frawley 1983a). Earnest European exploration of the region commenced with the 1873 expedition of G. E. Dalrymple. Among the important discoveries of this expedition were a series of river valleys between Cardwell and Cooktown that were covered with rainforest rich in the prized timber species red cedar (*Toona australis* (F. Muell.) Harms.). Within a year of Dalrymple's expedition, the cedar cutters had moved into the region, and land settlement quickly followed.

By 1879, sugar cane growing had commenced in the Innisfail district and with the depletion of the readily accessable coastal cedar supplies, timber cutting began in the rainforests of the Atherton Tablelands. Settlement was encouraged by government and large tracts of coastal and tableland rainforests were cleared for agriculture, principally in the form of sugar, maize and dairying. These trends have continued to the present day though large scale alienation of rainforest land has largely ceased in the last 10-15 years, due to both changes in community attitudes and the physical location of the bulk of the remaining rainforests in the more elevated and broken areas of the region. By 1983, some 779 600 ha of rainforest remained in the region, representing some 80% of the estimated cover of some 965 314 ha at the time of the first European settlements (Winter *et al.* 1984).

In a manner not too dissimilar to what one could easily imagine as the attitude of many rural dwellers in today's developing tropical economies, the first European residents of north Queensland perceived the rainforests largely as an obstacle to economic development and closer settlement. The monetary value of a few prime cabinet timbers was recognized, but even this recognition was strongly tempered by the desire for land for settlement (Payne *et al.* 1931).

Thus, the early land settlement patterns in north Queensland brought with them considerable forest destruction which sowed the seeds of a long running conflict between utilitarian conservation interests and the prevailing land settlement ethos. In the 1930's, this conflict between forestry authorities and land settlement interests had heightened to such an extent that a Royal Commission into Land Settlement and Forestry was necessary. The findings of the Commissioners reflected both the low esteem with which the forestry authorities were held and the low value the community attached to the rainforest resource at that time (Payne *et al.* 1931). However, by the early 1960's, the findings of the North Queensland Land Classification Committee (Sloan *et al.* 1962) reflected both a greater community recognition of the value of the rainforest resource, and their increasingly residual nature.

Over the last 10-15 years, there appears to have been a further re-evaluation of the value of the rainforest resource to the community, with a much greater emphasis on the rainforest's intrinsic rather than its utilitarian values. Indeed, since the mid 1970's, there has been growing public pressure for the complete preservation of the remaining rainforest areas in north Queensland (Frawley 1983b), and it is clear that a significant section of society now views the rainforest as being more important for values such as environmental protection, recreation and tourism than for more utilitarian values such as timber production (Cassells & Gilmour 1978). However, it is equally clear that there is still a wide range of community attitudes towards the rainforest resource (Anon. 1985), and that these community attitudes still include both the utilitarian and the antagonistic perspectives, despite the more recent expansion of an alternative romantic perspective (Valentine, in press).

These conflicting community values and perceptions of the relative benefits of different rainforest resource use possibilities are translated into conflicting community resource use demands which present policy-makers and resource managers with a classical multiple use dilemma. Zivnuska (1961) has noted that managing forests and wildlands for competing demands only becomes a significant problem when forest resources become sufficiently scarce that conflicting demands occur for the same areas. In north Queensland, it is clear that the conflicting demands on the rainforest resource are such that the era of the wildland frontier has passed. Conflict resolution is now a major resource management problem.

Zivnuska (1961) has also noted that for effective multiple land use to occur, "the administering agency must be empowered to respond to all the values and costs involved". Cassells and Valentine (1987) have developed a participatory land use planning framework which addresses the problem of legitimately resolving competing community demands on forest resources. In the context of the north Queensland rainforest, such a framework should be used to help resolve the critical resource use conflicts that exist at present, so as to enhance their overall contribution to community welfare. We would suggest that such a planning exercise is needed to help clarify important resource management issues such as the requirements for biological conservation, the problem of sustained yield in rainforest timber production, the need for landscape rehabilitation, the possibilities for rainforest recreation and tourism and the potential role the region could play in relation to the global problem of rainforest conservation and management through an expanded research, education and training infrastructure.

Requirements for Biological Conservation

The biological significance of the north Queensland rainforest is now well established (Anon. 1985; Australian Heritage Commission 1986), and from the perspective of biological conservation, a strong case can be made for the total preservation of the remaining rainforests (Adamson & Fox 1982). However, biological conservation is only one of many community demands being placed on the rainforests in north Queensland, and total preservation is unlikely to receive sufficient social support to become an immediate reality. Thus, some rigorous assessment of the rainforests' conservation status is required both to assess the adequacy of existing reserves, and to help define priorities for future key area reservations. A rigorous contemporary assessment of the conservation status of the north Queensland rainforest would seem to be particularly important given its established biological significance, and the fact that most of the region's National Parks were established prior to the mid-1970's in an era when aesthetics and emotion often had a stronger influence on park selection than ecology (Specht

1975). National Park authorities have also indicated that at present they consider that the existing system of National Parks and equivalent reserves falls far short of achieving the aim of an adequate representation of rainforest ecosystems (Anon. 1985), though the number of broad rainforest formations "reasonably well conserved" in existing reserves has risen from just 50% in 1971 to some 87% in 1979 (Specht 1981).

The complexity and diversity of tropical rainforest ecosystems makes any ultimate assessment of conservation value or status a difficult and imprecise task (Myers 1981, 1983a). However, carefully planned and executed surveys, combined with further research into the propagation and size of breeding populations of tropical plants (and their dependent wildlife), can produce a sound biological basis for an enhanced conservation programme (Ashton 1981). Clearly, such an expanded programme of biological survey and research activities has inevitable time lags, though broad scale habitat classification can be used to indicate immediate priorities (Tracey 1981).

The problem of a limited ecological knowledge base for undertaking evaluations of conservation status or value is not restricted to tropical rainforests. Elsewhere in Australia, methodologies have been developed both to assess the existing conservation status of ecological communities, and to assign priorities for conservation reservation (e.g., Specht *et al.* 1974; Stanton & Morgan 1977; Austin & Miller 1978; Bolton & Specht 1983). While there are some differences in approach, all these methods provide some sort of explicit framework for considering the important conservation criteria of representativeness; diversity; naturalness; effectiveness or reserve viability; and the immediacy of any threat that would modify conditions of naturalness.

The work by Stanton & Morgan (Unpublished report) has already indicated some reservation priorities for the wet tropical coast and hinterland, which could usefully provide a starting point for a more detailed and specific evaluation of the conservation priorities in the region. In addition, despite some obvious limitations of scale and detail, the regional vegetation maps of Tracey & Webb (1975) provide an extremely useful tool for helping to determine interim conservation priorities. Any experience in developing conservation evaluation methodologies in Australia's wet tropics should provide valuable lessons for assessing conservation priorities elsewhere in the tropics, where the problems of an inadequate information base and rapidly changing land use are much more immediate. However, it must also be recognized that application of these techniques can only provide provisional assessments of the conservation status of these communities, and long-term biological monitoring should be seen as an important management priority on National Parks and other public rainforest lands.

The Problem of Sustained Yield

The possibility of a sustained yield of forest products from tropical rainforests is a question of considerable importance at both the local and the international levels. Locally, the absence of the possibility of a sustained timber yield would significantly reduce the need to retain extensive areas of rainforests in State Forests at the expense of alternative land uses. Similarly, in the absence of such a possibility, the retention of large areas of natural or semi-natural rainforest on a global scale becomes much more problematical (Lamb 1977; Leslie 1977), and in tropical enviornments one of the key concepts behind the World Conservation Strategy, the conservation of living resources for sustainable development, becomes little more than an illusionary dream.

Since the publication of the landmark paper by Gomez-Pompa *et al.* in 1972, the perception of tropical rainforests as "non-renewable resources" has become quite widespread. Indeed, this perception has been one factor which has led a number of influential tropical foresters to argue for the extension of more productive plantation monocultures in lieu of timber production from natural forests (e.g., Spears 1979). In north Queensland, similar perceptions have also developed with the Queensland Department of Forestry's long serving senior ranger/inspector Sam Dansie being widely quoted as saying "rainforests the world over are in general one cut relatively low production forests and ours are no exception to the rule" (Rainforest Information and Research Group 1985). Clearly there is a need to clarify the meaning of the sustained yield concept and the concept of renewable resources in the context of tropical forestry.

The sustained yield concept together with the ideal of natural regeneration after timber harvesting to reproduce a forest of similar structure and composition to the undisturbed forest has formed part of the "classical forestry" paradigm that has influenced the direction of forest management in many tropical countries including Australia (Leslie 1977). However, the relevance of these concepts to tropical forestry and indeed to industrial forestry in much of the developed temperate world has frequently been questioned by forest managers and scientists (e.g., Smith 1962, 1969; Leslie 1977; Spears 1979).

Clawson & Sedjo (1984) have outlined the history of the sustained yield concept and discussed its applicability to forest management in Europe, the United States and the countries of the developing tropics. They noted that the concept essentially evolved in Germany during the mid 19th Century. By this time, the German forests had experienced a long history of purposeful management. When trees had reached a desired degree of maturity they were cut, the forest was regenerated and the trees grown for a new cycle, leading to another forest ready to harvest at some future date. By having equal volumes of timber in each forest age class — the so-called normal or regulated forest — the harvest each year or at each interval could be approximately equal.

It is important to recognize that some centuries of previous forest utilization had produced the conditions that had allowed the development of the relatively even age class distributions which made this particular form of regulated forestry practical. Significant volumes of old growth trees were largely absent. It is also important to recognize that at the time, Germany was a wood deficit country where local demand always exceeded supply such that there was a constant market for the outputs from German forests. The sustained yield concept also developed in a social setting characterized by small processing units located in small villages where the fairly constant workforce generally lived within a walking distance from both the forest and the processing mill. The world view in which the concept was developed also envisaged fairly static harvesting and processing technologies. Clawson and Sedjo (1984) have noted that "one need only list the assumptions, basic to sustained yield as it developed, to realize that the concept might not be fully applicable in countries that lacked a long history of managed forests and where economic, social and technological conditions were dynamic and changing".

The sustained yield concept has been less successfully transferred to forestry practice in the new world forests of north America and Australia, despite its intellectual appeal to foresters and its emphasis in formal forestry education. In the United States, Clawson & Sedjo (1984) have noted that historically the U.S. Forest Service had to modify the original concept in practice because of the dominant presence of old growth forest stands. Thus, the initial forest management problem became to oversee the orderly liquidation of the accumulated forest capital in old growth stands during an extended transition period that would eventually lead to a yield allocation that could be sustained by the growth increment of the managed forest. In this transitional period, allowable cuts often exceeded the annual growth of the forest, though this "excess" forest capital was usually disposed of in an orderly fashion through the application of various "allowable cut" formulas such as the Hanzlick Formula (CHEC 1982). Early forestry writings in this country suggest similar priorities for dealing with the accumulated old growth in Australian forests (e.g., Hutchins 1916). Clawson & Sedjo (1984) have suggested that similar situations now apply to many tropical forest areas and that for many developing tropical countries the sustained yield concept is "unlikely to be practically applicable for at least one, perhaps two, human generations".

Since the 1940's, the sustained yield concept in the United States has been increasingly formalized around the idea of "nondeclining yields", though this has not occurred without controversy (Clawson & Sedjo 1984; CHEC 1982). In addition, with the passage of the U.S. Multiple Use-Sustained Yield Act in 1960, the sustained yield concept was broadened to include non-wood forest values as well as simple timber volumes, and it made specific reference to the ecological concept of maintaining productivity. This legislation defined sustained yield as "the achievement and maintenance in perpetuity of high-level annual or regular periodic output of the various renewable resources without impairment of the productivity of the land" (Clawson & Sedjo 1984).

The concern with maintaining output without impairing land productivity is particularly relevant to any consideration of sustained yield in tropical rainforests, because of the widespread perception of the tropical rainforest as a non-renewable resource which was mentioned earlier. However, in this context, it is important to read beyond the title of the Gomez-Pompa paper to consider the specific land use context in which he and his co-workers were discussing the non-renewability of rainforest ecosystems. Gomez-Pompa *et al.* (1972) were writing about the large scale clearing of rainforests for intensive and extensive agriculture. They felt that the regeneration characteristics of many primary rainforest species were such that they would not be able to recolonize any abandoned farmlands and that because of this, there was a great danger of a mass extinction of thousands of species. However, Gomez-Pompa *et al.* (1972) did not suggest that tropical rainforest ecosystems were non-renewable in any absolute sense. Indeed, they looked favourably on traditional shifting cultivation systems as a natural way of using the regenerative properties of the rainforest for human benefit. Furthermore, Gomez-Pompa *et al.* (1972) maintained that under "the shifting cultivation system, the genetic pool of primary trees is retained, and from this pool comes the raw material for the successional process". However, Gomez-Pompa *et al.* did make the important distinction between traditional shifting cultivation systems and the more intensive shifting cultivation patterns that are now commonly practised in many parts of the tropical world due to demographic pressures. They also noted the fundamental importance of scale in terms of determining the impact of human disturbance of tropical rainforest ecosystems — "We cannot overstress the importance of the space factor in these considerations, because it makes an enormous difference if we use or destroy thousands of square kilometres or if we destroy one or two".

In the context of timber production from the north Queensland rainforest, the scale of the relatively low intensity selective logging operations would seem to preclude any immediate *a priori* concern about exceeding any absolute limits of the renewability of the resource. In addition, a body of regional research information is available to suggest sufficient increment levels in logged forests to sustain some level of timber production in purely volume terms (e.g., Nicholson, Henry, Rudder & Anderson unpublished), and this empirical increment data has been used in modelling stand development for yield prediction purposes under the mixed age class selection system practised in north Queensland (Bragg & Henry 1985). However, the levels of increment are low by world standards, and the potential for increasing the effective productivity of the natural forest by silvicultural modification is limited (Stocker 1985). There is also considerable uncertainty as to whether or not this potentially available long term timber volume increment can be translated into an economically relevant available supply for industrial utilization.

Much of this uncertainty can be attributed to the difficulties associated with forest mensuration in tropical rainforest environments (Stocker *et al.* 1977). This is reflected in the widely varying estimates given for the long term sustainable yield from the north Queensland rainforest over the last 30-40 years. With only limited inventory data to guide it, the 1949 McLean Commission of Inquiry into the timber industry estimated that the annual growth of the north Queensland rainforest was in the order of 75 000 m^3 and noted that the standing sawlog resource at the time was thought to be some 8.7 million m^3 (QFD 1981), a volume that could maintain the then current levels of exploitation for some 30 years without relying on any increment from the forest. After that date, a period of silvicultural optimism seemed to set in with Volck in 1968 (unpublished) suggesting that silvicultural treatment of the rainforest could lift yields to the projected year 2010 annual regional requirements of 600 000 m^3, though he did note the need for more financial investment to achieve this result.

By 1975, Volck (unpublished) had tempered this optimism somewhat and noted that current estimates of the long term sustainable yield then ranged between 160 000 and 200 000 m^3, and suggested that a regional plantation programme would be necessary to meet projected regional timber demands. Stocker *et al.* (1977) warned that many field staff felt that even the then current minimum estimate was a gross overestimate of the rainforests' effective productive capacity, and pointed to the problem of uncertainty with regards to the area of accessible, productive rainforest that could effectively contribute to any sustainable yield. Subsequent inventory work has drastically revised estimates of both the effective area of productive rainforest and the potentially available sustainable yield, with estimates of only some 143 000 ha of the public forest estate of some 440 000 ha south of the Daintree River, contributing to an estimated sustainable yield of some 80 000 m^3 per annum (QFD 1981).

In addition to this uncertainty about the level of available yield in purely volume terms, there is also considerable uncertainty as to whether or not this potentially available increment can be translated into the spatial, species and wood quality distributions that will be economically relevant to the region's timber industry. All natural forests face increasing cost disadvantages relative to plantation forests because of their location in physically difficult enviornments that are generally remote from processing centres and key markets (Groome unpublished). In addition, Stocker *et al.* (1977) have questioned whether the existing north Queensland industry would be able to adapt to the likely changes in log sizes and species composition following the cessation of logging in "virgin" stands, and many observers have noted the importance of high quality logs with high value added end products to the continued viability of the north Queensland timber industry (QFD 1981; Anon 1985; Rainforest Information Group 1985).

Thus, consideration of the question of sustained yield in the north Queensland rainforest must extend beyond a simple discussion of potential timber volumes to include discussions of timber quality distributions, logging costs and logging impacts through a broader assessment of land capability and land suitability for long term cabinet timber production. Such an expanded concept of sustainable yield would also imply a considerable research commitment to monitor and understand the impact of re-cut logging on non-wood forest values. Indeed, re-cut logging could not be considered consistent with modern sustained yield concepts unless it is backed by sound research programmes on the impact of logging on flora and fauna and important ecological processes such as the forests' hydrological dynamics and nutrient cycling. Experience from these programmes of scientific research, and particularly those programmes that emphasize investigation of basic ecological processes, should again provide valuable lessons for other tropical countries who have yet to pass the wildland frontier so that they have to rely solely on the rainforests' renewability to provide them with their basic needs.

The Need for Landscape Rehabilitation

On a global scale, there is considerable uncertainty about the ability of rainforest lands to support permanent agriculture and it is now widely recognized that the luxuriant growth of tropical rainforests does not always coincide with rich agricultural soils. Indeed some authors have emphasized the fragility and non-renewability of rainforest soils (Gomez-Pompa *et al.* 1972; Caufield 1985), and this is certainly a major issue on the wet tropical coast of north-east Queensland. In a detailed review of the region's hydrology, Bonell, Gilmour and Cassells (1986) concluded that this area occupies part of the extreme wet spectrum in storm runoff hydrology because of its extreme meteorological conditions. The extremely high rainfalls of the summer wet season causes the frequent occurrence of widespread surface runoff on even the undisturbed rainforest slopes, and as a result, a very high proportion of gross rainfall, often in excess of 45%, is discharged out of rainforest catchments during storm events. This is an unusual response for undisturbed forests in view of the high surface infiltration capacities and it ensures high rates of soil loss on removal of the forest. Caplin & Prove (1983) have reported that former rainforest land now covered by sugar cane monocultures experience annual soil losses in the order of 100 tonnes ha^{-1}.

In contrast, other writers (e.g., Sanchez 1976) have emphasized the potential of even poor tropical rainforest soils for productive management, and in this regard, perhaps not enough emphasis has been put on the variability of tropical soils (Jordon & Herrera 1981) and hence, the variability of consequences following forest clearing for agriculture. Padoch & Vayda (1983) have reported that permanent field agriculture based on intensive management of soil moisture, slope conditions and soil fertility once supported relatively high population densities in many neotropical areas, though today this once sustainable type of traditional field agriculture survives in only a few limited sites. Similarly, in S.E. Asia, Whitmore (1984) has noted that many areas cleared by western colonial interests for plantation agriculture have subsequently been abandoned and over-run by secondary forest regeneration.

An important key to maintaining the productive stability of agro-ecosystems in the tropics appears to be the maintenance of appropriate inputs ranging from the application of natural or artificial fertilizers to erosion control and the use of appropriate weed control measures. In north Queensland, both the physical and economic difficulties of maintaining appropriate levels of inputs on steep slopes and marginal soils have resulted in considerable areas of degraded farmlands on private agricultural holdings. On the Atherton Tablelands, Gilmour and Reilly (1970) estimated that there were some 57 000 ha of these degraded lands, and two decades later this broad pattern remains essentially unchanged (e.g., see Kent & Tanzer 1983).

Such areas could well be appropriate targets for land rehabilitation or farm forestry. It is unfortunate that Gilmour and Reilly's proposals (1970) for expanding forestry extension programmes or public purchase of degraded lands for forest plantations were not adopted at the time, because land prices then were dramatically cheaper. Both options could have helped to reduce the job losses associated with the now reduced availability of rainforest logs from public forests (Anon 1985).

While the cost of large scale public land purchases may now be prohibitive, the option of increased extension services to foster private farm forestry still warrants immediate government attention. If co-operation between private land holders and government agencies could be fostered in the region, a sustainable pattern of productive forest land use could well be developed. As well as diversifying farm incomes, this land use pattern would have the advantage of numerous small plantation units within a relatively compact production zone, thus offering greater protection from the effects of tropical cyclones than larger contiguous areas of long rotation forest plantation crops (Stocker & Gilmour 1976).

The Potential for Tourism and Recreation

Many writers have suggested that tropical rainforests have considerable potential to boost regional economies by becoming popular recreational and tourist destinations. Caufield (1985) has noted that "educational tourism" based on rainforest areas has become a highly profitable industry in Costa Rica and has suggested that this provides a strong economic argument for the protection of rainforest resources from other forms of exploitation. Similarly, Allen (1975) has claimed that globally "the tourist potential of rainforest fully equals that of the East African game parks". However, other authors have been rather more reserved in their assessment of the tourist potential of rainforest ecosystems (e.g., Budowski 1976; Valentine 1982; Myers 1983b).

In northern Queensland, very few recreational preference studies have been undertaken, though the few that have do suggest that rainforests could potentially play an important role in the development of a regional tourism infrastructure. Valentine (1982) has reported that rainforests are perceived as important secondary attractions for visitors to the Mission Beach area, whereas some 45% of respondants to a survey of day visitors on Dunk Island specified rainforests as a prime reason for their visit. Elsewhere, the specific attraction of rainforests *per se* has not been investigated, though there are some indirect indications of the potential recreational importance of rainforest. In a survey of visitors to the largely rainforest clad Danbulla State Forest near Tinaroo Dam, Cassells (1984) found that some 67% of respondants participated in sightseeing or pleasure driving while some 5% of respondents specifically nominated nature study as one of the recreational pursuits they participated in while visiting the reserve.

Valentine (1983) suggests that in excess of 300 000 visitors per year currently visit rainforest park areas in north Queensland. However, he also reports personal observations that suggest that much of this current visitation is dominated by short-term activity targets that are only loosely connected to the presence of rainforest ecosystems. While these activities are no doubt enjoyable for the visitors and contribute to both the region's social and its economic well being, they are almost incidental to their location in rainforest National Parks. They certainly do little to promote either an appreciation of the rainforest environment or a unique rainforest experience that might help to strengthen any distinctive attractiveness of the region as a tourist destination in a competitive domestic and international tourist market.

With the exception of one or two interpretive walks and forest drives, current recreational use patterns and current recreational management emphases do little to foster distinctive regional recreational experiences. There is only a rudimentary support system for visitors and little or no active environmental interpretation for either local or tourist visitors to rainforest areas. With appropriate management, north Queensland's rainforest environments clearly have the potential to make a greater contribution to the regions social and economic well being through outdoor recreation and tourism. However, this will not occur by accident. This potential can only be realized if considerable attention is given to recreation research, management and planning. Indeed, given the growing public and in some instances entrepreneurial interest in the tourist and recreational potential of the rainforest, there would appear to be an urgent need for an expanded programme of rainforest recreation research to help ensure appropriate management responses in this important area.

The Need for an Expanded Research, Education and Training Infrastructure

In terms of formulating policies for the management of Australia's tropical rainforests, this paper has stressed the need for policy makers to consider the wider changes being made to the tropical moist forest biome as a whole. As outlined earlier, there is no doubt that globally tropical forest biomes are experiencing the most intense pressures for land use change and that the tropical moist forest is the biome about which there is the least information, either in terms of understanding the implications of these changes or in terms of being able to respond to perceived problems with appropriate land management practices (see FAO 1985; Farnworth & Golley 1974). The need for international co-operation in research, education and training was also stressed by FAO (1985) and Farnworth and Golley (1974), as was the need to translate existing and developing ecological knowledge into improved land management practices.

In this context, it seems that there are particular opportunities for Australia to play a regional leadership role in tropical rainforest conservation, management and research through an expanded research and training infrastructure in north Queensland. As noted earlier, Australia's wet tropics have historically experienced many of the development pressures currently being experienced to such a heightened extent elsewhere in the tropical world. However, the region has now passed the wildland frontier and entered a period of relative land use stability where the implications of different forms of tropical land use can be scientifically assessed.

The north Queensland region has also had a long history of rainforest research (QFD 1983) and in the past two decades or so this traditional silvicultural focus has been considerably expanded with work by CSIRO, James Cook University and a limited number of state government scientists in fields such as rainforest taxonomy and ecosystem classification, catchment hydrology, regeneration dynamics and disturbance ecology, wildlife management and ecophysiology.

In his unpublished 1978 UNESCO review of the suitability of the north Queensland region to support an integrated research and training programme in the context of the Man and the Biohere Programme, Hamilton (*pers. comm.*) noted both the potential of the scientists working in the region to contribute to such a programme and the then fading government support for rainforest research in north Queensland. He also noted a number of bureaucratic and institutional barriers that could prevent the realization of the region's research and training potential. Bonell (unpublished report) made similar observations about both the potential of the region to support integrated land management research, and the barriers preventing the region realizing this potential. He stressed the region's unique advantages in terms of accessibility, political stability and the ready availability of western technology and noted how this offered opportunities for long-term, integrated tropical rainforest research that were not available in the many short term research programmes that were operated out of the United States and western Europe.

Bonell also critized the lack of commitment to rainforest research by both the then Federal and State Governments, and suggested the establishment of a co-ordinating body or institution which had a mandate for intergrating the widest range of disciplines in environmental research in tropical rainforests. He suggested that such an institution should aim both to establish an integrated, multi-discipline research programme in north Queensland, and to offer facilities for scientists from developing tropical countries to participate in or train in such work. Given the budgetary problems of the time, he also suggested that such an institution could be funded by a small redirection of funds from Australia's overseas aid programme.

The recent consolidation and expansion of CSIRO activity at the Tropical Rainforest Research Centre in Atherton seems to represent a welcomed strengthening of the present Federal Government's commitment to forest ecology research in the region, though there is also a clear need for the Queensland Government to revitalize and expand its regional forest research efforts. Given the public's growing interest in rainforest conservation and management, it is difficult to understand why the Queensland Government allowed its graduate forest research staff in north Queensland to drop from seven in the early 1970s to just two in the early 1980s. Given the particular growth in public concern about the environmental values of the rainforest, it is even more difficult to understand why in 1981, the government allowed the closure of both the internationally recognized catchment hydrology programme it ran co-operatively with the Department of Geography at the James Cook University (see Hamilton 1983), and the programme of regional recreation research and planning which was initiated by the Queensland Department of Forestry in 1978.

With the establishment of the Tropical Rainforest Research Centre in Atherton, some of the barriers preventing integrated rainforest research and management may have potentially been overcome, though many inter-governmental and institutional problems that prevent Australia from realizing its potential regional leadership role still remain. An important consideration in terms of overcoming these problems is the regional presence of the James Cook University with its speciality interest in tropical studies. This university could play a crucial integrative role in the region, particularly if it became the focus for interaction between research, education and resource management personnel as suggested by Daniel Janzen in his recent review of CSIRO research in tropical Australia (Janzen unpublished report).

In this context, it is relevant to note the recommendations of a group of middle to upper level forest managers from a wide range of Asia-Pacific countries who attended the Tropical Forest Management Workshop which was held by the Australian Development Assistance Bureau in Queensland in 1983 (Shepherd & Richter 1985). These participants recommended that the Australian Government should give consideration to the establishment of a regional Tropical Forest Management Centre which would be responsible for collating and summarizing existing and developing research information; developing information clearing and dissemination systems; and formulating operational guidelines for forest management.

Conclusions and Recommendations

This paper has sought to summarize the historical development of land-use policies and practices that have influenced the patterns of exploitation and conservation management in the north Queensland rainforest. A number of contemporary resource management issues have been discussed and the need for Australia to place the conservation and management of its tropical rainforests in the context of the wider global changes to the tropical moist forest biome has been stressed.

The different community value systems that govern both the community's perception of the rainforest and the pattern of resource use demands it places on rainforest ecosystems have also been outlined. To some, the rainforest is just a place to obtain timber or mineral resources. To others, it is a place to scrape away for agricultural, urban or tourist infrastructure development. To an increasing number of people, the rainforest is something more. To many, it is a sacred place to be kept as free as possible from the impacts of human disturbance in its various forms.

In this complex perceptual environment, developing acceptable patterns of rainforest land use is a difficult task, and a degree of community conflict is perhaps inevitable. However, one model that could create a more united community vision about the future of rainforest management is the concept of the Biosphere Reserve which arose from the UNESCO Man and the Biosphere Programme (McMichael 1979).

Biosphere Reserves are conceived of as protected areas of land which constitute a world-wide network, representative of the major biomes of the earth. They will include areas containing ecosystems in an essentially natural state together with areas of harmonious managed landscape resulting from traditional patterns of land use and examples of modified or degraded ecosystems capable of being restored to natural conditions. The prime uses envisaged for Biosphere Reserves include ecological research, education and training. The fundamental aim of these reserves is to provide objective information on which sound, sustainable land management can be based.

The authors believe that the Biosphere Reserve provides an appropriate model for developing a viable pattern of land use which will protect the various values of the wet tropics and its rainforests. Essentially, we would envisage an expanded pattern of protected areas managed in harmony with productive managed forests and farm lands, together with areas of rehabilitated waste lands. The benefits of achieving this pattern of land use would not be sought just for ourselves. A major aim would be to develop a regional research and management training infrastructure that would allow Australia to make a significant contribution to the solution of the problems associated with tropical deforestation in the Asia-Pacific Region.

The most immediate problem preventing the establishment of a pattern of land use consistent with this Biosphere model is the continuing conflict between timber production and biological conservation interests. However, the existing timber industry is being rapidly "rationalized" in response to changing resource availabilities and significant job losses are occurring (Anon 1985). The authors would suggest that this process of rationalization should not be the result of a "laissez-faire" accident. Rather, we feel it should be the result of deliberate planning undertaken co-operatively by the State and Federal Governments in close consultation with the north Queensland community to achieve the best possible pattern of regional land use. Such a planning exercise should aim to:

- Expand the area of rainforest in National Parks or equivalent reserves to ensure that viable samples of all rainforest types are preserved and that key refugia areas are identified and conserved.

- Allow for rainforest timber production only on lands which are both capable and suitable for this purpose. Where the biophysical characteristics of rainforest lands are such that they are capable of supporting timber production without land degradation, their suitability for this purpose should be determined in relation to factors such as community attitudes and the economics of production. A major constraint in determining the suitability of rainforest land for timber production will be the need to maintain viable protected reserves for biological conservation purposes.
- Prevent any further rainforest clearing or further alienation of publicly owned rainforest lands.
- Rationalize the timber industry based on the rainforest resource both to facilitate rainforest conservation and to maximize long-term employment opportunities in the communities most dependant on this resource.
- Create new employment opportunities in the communities disadvantaged by the timber industry rationalization programme. Emphasis here could, for example, be given to reafforestation of degraded farmlands and the establishment of rainforest-based tourist enterprises in both the public and the private sectors.
- Ensure that Australia can make a significant contribution to the global conservation of rainforests by promoting rainforest research within north Queensland and extending the results of this research to rainforest management both in Australia and the developing Asia-Pacific region.

Such a planning programme would in the view of the authors produce more acceptable land use outcomes for both local communities and wider national interests. However, it would require some injection of human and financial resources and with the diversity of local, state, national and even international considerations involved, there is a clear case for co-operation between State and Federal Governments.

In the past, there has tended to be considerable antipathy and mistrust between the State and Federal Governments, just as there has been between the various protagonists in the resource utilization-biological conservation debate. To help overcome these problems, the authors would in conclusion suggest two immediate steps that could be taken to overcome these problems and establish the environment of trust necessary to help facilitate this considerable exercise in community planning.

The first would be for the Federal Government to make the resources of the former CSIRO Division of Land Use Research available to the Queensland Government. The experience of this unit would help to accelerate the technical aspects of the planning process and would aid State Government agencies to improve their own technical land use planning skills. The second would be for the Queensland Government to place a moratorium on any logging of "virgin" rainforest areas until the completion of the planning process. As well as preventing the immediate closure of reservation options for biological conservation purposes, the moratorium period would help to resolve some of the uncertainties associated with re-cut logging in rainforests, to the considerable benefit of forest industry planners and the wider community.

References

Adamson D. A. & Fox M. D. (1982) Change in Australasian vegetation since European settlement. In: *A History of Australasian Vegetation.* (ed. J. M. B. Smith) pp. 109-46. McGraw-Hill Book Company, Sydney.

Allen R. (1975) The year of the rainforest. *New Scientist* **66**, 178-80.

Allen R. (1980) *World Conservation Strategy: Living Resource Conservation for Sustainable Development.* International Union for the Conservation of Nature and Natural Resources, Geneva.

Anonymous (1985) *Rainforest Conservation in Australia.* A Report to the Minister for Arts, Heritage and Environment by a Working Group on Rainforest Conservation.

Ashton P. S. (1981) Techniques for the Identification and Conservation of Threatened Species in Tropical Forests. In: *The Biological Aspects of Rare Plant Conservation* (ed. H. Synge) pp. 155-64. John Wiley and Sons, Chichester.

Austin M. P. & Miller J. P. (1978) Conservation. In: *Land Use on the South Coast of New South Wales.* (eds M. P. Austin & D. J. Cocks) pp. 161-95. CSIRO, Australia.

Australian Heritage Commission (1986) *Tropical Rainforests of North Queensland: Their Conservation Significance.* Special Australian Heritage Publication, Series No. 3.

Bolton G. G. (1963) *A Thousand Miles Away: A History of North Queensland to 1926.* Australian National University Press, Canberra.

Bolton M. P. & Specht R. L. (1983) *A Method for Selecting Nature Conservation Reserves.* Australian

National Parks and Wildlife Service, Occasional Paper Number 8.

Bonell M., Gilmour D. A. & Cassells D. S. (1986). *The Storm Runoff Response to Various Rainfall Systems on the Wet Tropical Coast of North-East Queensland, Australia.* East-West Environment and Policy Institute Working Paper, East-West Center, Honolulu, Hawaii.

Bragg C. T. & Henry N. B. (1985) Modelling stand development for prediction and control in tropical forest management. In: *Managing the Tropical Forest.* (eds K. R. Shepherd & H. V. Richter) pp. 281-97. Development Studies Centre, Australian National University, Canberra.

Budowski G. (1976) Why save tropical rainforests? Some arguments for campaigning conservationists. *Amazonia* **5,** 529-38.

Caplin M. A. & Prove B. G. (1983) Soil conservation problems of the humid coastal tropics of north Queensland. *Proceedings of the Australian Society of Sugar Cane Technologists Conference* pp. 87-93. Mackay.

Cassells D. S. (1984) *Recreational Use of Tinaroo Dam: Some Resource Management Implications for the Danbulla State Forest.* M.Sc. Thesis, James Cook University of North Queensland.

Cassells D. S. & Gilmour D. A. (1978) The Changing Role of the North Queensland Rainforest. *Proceedings of the Eighth World Forestry Congress, Jakarta.* 8 pp.

Cassells D. S. & Valentine P. S. (1987) From conflict to consensus: towards a framework for community control of public forests and wildlands. In: *Proceedings of the Ecopolotics Conference* (eds R. Toplin and P. Tighe) pp. 201-09. School of Australian Environmental Studies, Griffith University, Nathan, Brisbane, August 1986.

Caufield C. (1985) *In The Rainforest.* Picador Press, London.

CHEC (1982) *The Citizens' Guide to Forest Planning.* Cascade Holistic Economic Consultants, Eugene, Oregon.

Clawson M. & Sedjo R. (1984) History of Sustained Yield Concept and Its Application to Developing Countries. In: *History of Sustained Yield Forestry.* (ed. H. K. Steen) pp. 3-15. Proceedings of a IUFRO Symposium, Portland, Oregon.

Courtenay P. P. (1982) *Tropical Australia: Patterns and Problems of Tropical Development in an Advanced Country.* Longman Chesire, Melbourne.

de Condolle R. (1981) A plea for the world's tropical forests. *Environ. Cons.* **8,** 2-3.

Farnworth E. G. & Golley F. B. (eds) (1974) *Fragile Ecosystems: Evaluation of Research and Applications in the Neotropics.* Springer-Verlag, Berlin.

FAO (1975) *Tropical Forestry Action Plan.* A Report of the Committee on Forest Development in the Tropics, FAO Rome.

Frawley K. J. (1983a) *A History of Forest and Land Management in Queensland with particular Reference to the North Queensland Rainforest.* A Report to the Rainforest Conservation Society of Queensland.

Frawley K. J. (1983b) The North Queensland rainforest: retrospect and prospect. *Taminga* **20,** 107-20.

Gilbert G. R. (1984) The study of policy formation and the conduct of policy analysis and evaluation. In: *Making and Managing Policy: Formulation, Analysis and Evaluation* (ed. G. R. Gilbert) pp. 1-9. Marcel Dekker Inc. New York.

Gilmour D. A. & Reilly J. J. (1970) Productivity survey of the Atherton Tableland and suggested land use changes. *J. Aust. Inst. Agri. Sci.* **36,** 259-72.

Gomez-Pompa A., Vazquez-Yanes C. & Guevara S. (1972) The tropical rainforest: a nonrenewable resource. *Science (N.Y.)* **177,** 762-765.

Grainger A. (1984) Quantifying changes in forest cover in the humid tropics: overcoming current limitations. *J. World For. Res. Manag.* **1,** 3-63.

Hamilton L. S. (1983) *Tropical Forested Watersheds: Hydrologic and Soils Responses to Major Uses and Conversions.* Westview Press, Boulder, Colorado.

Hutchins D. E. (1916) *A Discussion of Australian Forestry with Special References to Forestry in Western Australia.* Government Printer, Perth.

Jordon C. F. & Herrera R. (1981) Tropical rainforests: are nutrients really critical? *Amer. Nat.* **117,** 167-80.

Kent D. J. & Tanser J. M. (1983) *Evaluation of Agricultural Land Eacham Shire North Queensland.* Department of Primary Industries Division of Land Utilization Report, Brisbane.

Lamb D. (1977) Conservation and management of tropical rainforest: a dilemma of development in Papua New Guinea. *Bio. Cons.* **4,** 121-29.

Lanly J. P. (ed.) (1981) *Tropical Forest Resources Assessment Project (GEMS): Tropical Africa, Tropical Asia, Tropical America,* FAO/UNEP, Rome, (4 volumes).

Leslie A. (1977) Where contradictory theory and practice co-exist. *Unasylva* **29,** 2-17.

McMichael D. (1979) An international perspective. In: *The Value of National Parks to the Community* (eds J. Messer & G. Mosley) pp. 34-44. Australian Conservation Foundation, Malvern, Victoria.

Myers N. (1980) *Conversion of Tropical Moist Forests.* U.S. National Research Council, Washington D.C.

Myers N. (1981) Conservation needs and opportunities in tropical moist forests. In: *The Biological Aspects of Rare Plant Conservation* (ed. H. Synge) pp. 141-54. John Wiley and Sons, Chichester.

Myers N. (1983a) Conversion rates in tropical moist forests. In: *Tropical Rainforest Ecosystems: Structure and Function* (ed. F. B. Golley) pp. 289-300. Elsevier Scientific Publishing Company, Amsterdam.

Myers N. (1983b) Conservation of rainforests for scientific research, for wildlife conservation and for recreation and tourism. In: *Tropical Rainforest Ecosystems: Structure and function* (ed. F. B. Golley) pp. 325-34. Elsevier Scientific Publishing Company, Amsterdam.

Myers N. & Ayensu E. S. (1983) Reduction of biological diversity and species loss. *Ambio* **12,** 72-74.

O'Riordon T. (1971) *Perspectives on Resource Management.* Pion Ltd., London.

Padock C. & Vayda A. P. (1983) Patterns of resource use and human settlement in tropical forests. In: *Tropical Rainforest Ecosystems: Structure and Function* (ed. F. B. Golley) pp. 301-13. Elsevier Scientific Publishing Company, Amsterdam.

Payne W. L., Campbell W. W. & Atheton J. G. (1931) *Report of the Royal Commission on the Development of North Queensland (Land Settlement and Forestry).* Government Printer, Brisbane.

QFD (1981) *Timber Production from North Queensland's Rainforests.* A Position Paper by the Queensland Department of Forestry, Brisbane.

QFD (1983) *Rainforest Research in North Queensland. A Position Paper.* Released by the Queensland Department of Forestry, Government Printer, Brisbane.

Rainforest Information and Research Group (1985) *A Review of Rainforest Conservation in Australia: A Report to the Minister of Arts, Heritage and Environment.* CRES Working Paper 1985/24, Centre for Resource and Environmental Studies, Australian National University, Canberra.

Richards P. (1973) The tropical rainforest. *Scient. Amer.* **229,** 58-68.

Richardson S. D. (1970) The end of forestry in Great Britain. *Common. For. Rev.* **49,** 324-35.

Rubinoff I. (1983) A strategy for preserving tropical rainforests. *Ambio.* **12,** 255-58.

Sanchez P. A. (1976) *Properties and Management of Soils in the Tropics.* John Wiley and Sons, New York.

Shepherd K. R. & Richter H. V. (1986) *Managing the Tropical Forests.* Development Studies Centre, Australian National University, Canberra.

Sloan W. J. S., Owens A. J. & Johnston M. A. (1962) *Report of the North Queensland Land Classification Committee, 1960-1961.* Government Printer, Brisbane.

Smith J. H. G. (1962) Sustained yield is not ideal. *Forest. Chron.* **38,** 167-72.

Smith J. H. G. (1969) An economic view suggests 'sustained yield' should have gone out with the cross-cut saw. *Forest. Chron.* **45,** 167-72.

Spears J. S. (1979) Can the wet tropical forest survive? *Common. Forest. Rev.* **58,** 165-80.

Specht R. L. (1975) The report and its recommendations. In: *A National System of Ecological Reserves in Australia* (ed. F. Fenner) pp. 11-21. Australian Academy of Science Report No. 19, Canberra.

Specht R. L. (1981) Conservation of vegetation types. In: *Australian Vegetation* (ed. R. H. Groves) pp. 393-410. Cambridge University Press, London.

Specht R. L., Roe E. M. & Boughton V. H. (1974) Conservation of major plant communities in Australia. *Aust. J. Bot.* Supplementary series. **7,** 667 pp.

Stocker G. C. (1985) Aspects of gap regeneration theory and the management of tropical rainforests. In: *Managing the Tropical Forest* (eds K. R. Shepherd & H. V. Richter) pp. 225-28. Development Studies Centre, Australian National University, Canberra.

Stocker G. C. & Gilmour D. A. (1975) Environmental and social influences affecting land use planning in the forested areas of Far North Queensland. *Aust. For.* **37,** 225-32.

Stocker G. C., Gilmour D. A. & Cassells D. S. (1977) The future of our northern rainforests in the face of economic and political reality. In: *Focus on the Forester.* Proceedings of the 8th Institute of Foresters of Australia Conference, Adelaide. 10 pp.

Tracey J. G. (1981) Australia's rainforests: where are the rare plants and how do we keep them? In: *The Biological Aspects of Rare Plant Conservation* (ed. H. Synge) pp. 165-78. John Wiley and Sons, Chichester.

Tracey J. G. & Webb L. J. (1975) *A Key to the Vegetation of the Humid Tropical Region of North Queensland-1:100 000 Sheets.* CSIRO Rainforest Ecology Unit, Division of Plant Industry, Indooroopilly, Brisbane.

Usher M. P. (1973) *Biological Management and Conservation: Ecological Theory, Application and Planning.* Chapman and Hill, London.

Valentine P. S. (1982) Tropical rainforests and the wilderness experience. In: *Wilderness* (ed. V. Martin) pp. 123-32. The Findhorn Press, Scotland.

Valentine P. S. (In press) Rainforest recreation: a review of experience in Northern Australia. In: *The Rainforest Legacy: Australian National Rainforest Study,* Special Australian Heritage Publication Series Number 7, Australian Government Printing Service, Canberra.

Volck H. E. (1968) *Silvicultural Research and Management in North Queensland Rain Forests.* Department of Forestry, Brisbane.

Volck H. E. (1975) *Problems in the Silvicultural Treatment of the Tropical Rainforests of North Queensland.* Department of Forestry, Brisbane.

Westoby J. C. (1983) Saving the tropical forests: some facts which have to be faced. *Aust. For.* **46,** 241-46.

Whitmore T. C. (1984) *Tropical Rainforests of the Far East.* 2nd Edition, Clarendon Press, London.

Winter J. W., Bell F. C., Pahl L. T. & Atherton R. G. (1984) The Distribution of Rainforest in North-eastern Queensland. In: *Australian National Rainforest Study Report* (eds G. L. Werren & A. P. Kershaw) pp. 145-54. Vol. 1, World Wildlife Fund/Australian Conservation Foundation, Melbourne.

Zivnuska J. A. (1961) The multiple problems of multiple use. *J. Forest.* **59,** 555-60.